Land Use Management

Land Use Management

Edited by Kyrie Hamilton

SYRAWOOD
PUBLISHING HOUSE

New York

Published by Syrawood Publishing House,
750 Third Avenue, 9th Floor,
New York, NY 10017, USA
www.syrawoodpublishinghouse.com

Land Use Management
Edited by Kyrie Hamilton

International Standard Book Number: 978-1-68286-507-1 (Hardback)

Cataloging-in-Publication Data

Land use management / edited by Kyrie Hamilton.
    p. cm.
Includes bibliographical references and index.
ISBN 978-1-68286-507-1
1. Land use--Management. 2. Land use--Environmental aspects.
3. Land use--Planning. I. Hamilton, Kyrie.
HD108.3 .L36 2018
333.73--dc23

TABLE OF CONTENTS

PREFACE

Land use management can be referred to as the planning and implementation of processes with a view to develop an area of land. Modern land use management seeks to integrate urban development trends along with sustainable measures to protect the environment. Resource management and geospatial characteristics of a region are taken into consideration in this field. Smart growth policies that allocate green corridors and sustainable transport systems are other aspects of land use management. The various advancements in this field are glanced at and their applications as well as ramifications are looked at in detail. This book is meant for students who are looking for an elaborate reference text on land use management.

All of the data presented henceforth, was collaborated in the wake of recent advancements in the field. The aim of this book is to present the diversified developments from across the globe in a comprehensible manner. The opinions expressed in each chapter belong solely to the contributing authors. Their interpretations of the topics are the integral part of this book, which I have carefully compiled for a better understanding of the readers.

At the end, I would like to thank all those who dedicated their time and efforts for the successful completion of this book. I also wish to convey my gratitude towards my friends and family who supported me at every step.

Editor

Are Species Coexistence Areas a Good Option for Conservation Management? Applications from Fine Scale Modelling in Two Steppe Birds

Rocío Tarjuelo*, Manuel B. Morales, Juan Traba, M. Paula Delgado

Terrestrial Ecology Group (TEG), Department of Ecology, Universidad Autónoma de Madrid, Madrid, Spain

Abstract

Biotic interactions and land uses have been proposed as factors that determine the distribution of the species at local scale. The presence of heterospecifics may modify the habitat selection pattern of the individuals and this may have important implications for the design of effective conservation strategies. However, conservation proposals are often focused on a single flagship or umbrella species taken as representative of an entire assemblage requirements. Our aim is to identify and evaluate the role of coexistence areas at local scale as conservation tools, by using distribution data of two endangered birds, the Little Bustard and the Great Bustard. Presence-only based suitability models for each species were built with MaxEnt using variables of substrate type and topography. Probability maps of habitat suitability for each species were combined to generate a map in which coexistence and exclusive use areas were delimited. Probabilities of suitable habitat for each species inside coexistence and exclusive areas were compared. As expected, habitat requirements of Little and Great Bustards differed. Coexistence areas presented lower probabilities of habitat suitability than exclusive use ones. We conclude that differences in species' habitat preferences can hinder the efficiency of protected areas with multi-species conservation purposes. Our results highlight the importance of taking into account the role of biotic interactions when designing conservation measurements.

Editor: Antoni Margalida, University of Lleida, Spain

Funding: R. Tarjuelo was supported by a PhD grant from the Spanish Ministry of Education (FPU). This paper is a contribution to project CGL2009-13029/BOS of the Spanish Ministry of Science, as well as to the REMEDINAL2 network of the CAM (S-2009/AMB/1783). Field work in Ciudad Real was partly supported by Aeropuerto de Ciudad Real, S. A. The funders had no role in study design, data collection and analysis, decision to publish, or preparation of the manuscript.

Competing Interests: The authors have declared that no competing interests exist.

* E-mail: rocio.tarjuelo@gmail.com

Introduction

The distribution of species is the result of evolutionary, ecological or anthropogenic processes that operate at different spatial and temporal scales [1–4]. Climate has been described to play a major role in shaping the distribution of the species at continental and regional scales, while biotic interactions are generally considered secondary [5,6] but see [7,8]. Land use and biotic interactions become relevant at local scale, at which they exert a major effect in the configuration of population and community dynamics [9,10].

The presence of heterospecifics has been proposed as a factor influencing the habitat use of organisms at local scale [11]. Coexistence of sympatric species may be mediated by the segregation of shared resources [12], for example, the differentiation of habitat preferences at landscape or at microhabitat scale [13], or changes in a species' behavioural and food resource-use patterns [14]. Thus, direct or indirect interactions may condition the occurrence of heterospecifics in space and further, the fitness of the individuals [14]. This may be especially relevant for species subject to conservation efforts, since potential changes in habitat use patterns due to biotic interactions may affect their distribution at local scale [9,15,16].

In recent years, conservation issues from both theoretical and applied approaches have been increasingly addressed by the use of species distribution models (SDMs) [4,17–20]. SDMs use species occurrence records to infer the environmental conditions under which a species exists in a particular context and further, they allow to predict potential geographic distribution areas. Despite the potential importance of biotic interactions in determining the spatial distribution patterns of species at fine scale, SDM studies usually focus on single, often keystone or umbrella species [4,21]. However, the efficacy of umbrella and flagship species as conservation tools for protecting other species in the community has been questioned [22,23], and several studies have highlighted the importance of considering more than one species in designing successful conservation measures [24,25]. Conservation efforts should be directed towards groups of interacting species, focusing on areas that encompass species assemblages despite the lack of information about interaction networks [26].

In this context, the present study focuses in two steppe bird species which coexist in many areas of their distribution range: The Little Bustard (*Tetrax tetrax*) and the Great Bustard (*Otis tarda*). The two species are of high conservation concern since both are globally endangered and classified as near threatened and vulnerable respectively [27]. Nowadays, Spain holds more than half of their global population [28,29], being agricultural intensification and the increase of infrastructure development two major causes of population decline and distribution shrink [28,30]. The Little Bustard is a medium sized steppe bird, which prefers heterogeneous agricultural landscapes that maintain a high proportion of fallows and short natural vegetation [31,32]. The Great Bustard is one of the heaviest flying birds and shows preference for stubbles, leguminous crops and fallows, although its habitat selection pattern changes seasonally and differs greatly between regions [33,34]. Both species avoid man-made structures, such as buildings, roads and tracks [33,35,36]. To the best of our knowledge, no studies have been conducted at local scale on the Little and the Great Bustards together in order to integrate their habitat preferences for the management of areas in which both species coexist.

Therefore, the aim of this study is to provide useful guidelines for the conservation of these two sympatric species with different habitat preferences through the identification and environmental characterization of coexistence areas at landscape scale. The delimitation of areas in which species are more likely to coexist might help focusing management efforts on the benefit of both species. We discuss the implications of using coexistence areas to conserve species that differ in their habitat preferences.

Methods

Ethics Statement

The present study did not required the capture or handling of protected or endangered animals. All data about species' locations were collected by observation at distance using binoculars. The described field studies were carried out on privately-owned farms with the permission of farmers.

Study sites

The study was carried out in two localities of central Spain, Campo Real sited in Madrid province (40°19′N, 3°18′W. 1 145 ha) and Calatrava, in Ciudad Real province (38°54′N, 3°53′W. 9 016 ha). Both regions are under a Mediterranean climate with annual mean precipitations around 550 mm. These sites are flat to slightly undulated, encompassing mosaics of different agrarian substrates. Extensive dry cereal croplands and ploughed lands make up more than 50% of the surface, with a varying cover of fallows of different ages, leguminous crops and interspersed patches of olive groves, vineyards and fruit tree orchards. Pasturelands and scrublands are also present but in a low percentage.

Little and Great Bustard data

Little and Great Bustard data were collected between March–April 2008 and 2009 in Calatrava and April–May 2011 and 2012 in Campo Real, during the period of reproductive activity of both species [37]. Surveys were made by car routes throughout the available roads and tracks that cover the entire study site, stopping at every 500 m to ensure the record of all individuals, which were geo-referenced. Each study site was surveyed simultaneously by two car-teams, each composed by two experienced observers and covering a half of the study area, in order to fully cover the study site in a single bustard daily activity period. Surveys were made within the first three hours after daybreak and the three hours before sunset since this is the moment of highest activity, and thus individuals are easier to detect [37]. Only Little Bustard males were considered in this study since females are very difficult to observe due to their secretive behaviour. The detectability of Little Bustard males and Great Bustard males and females were almost complete since the vegetation height is relatively low at this time of the year. In addition, Little Bustard males were also detected acoustically. The Great Bustard presents a lek mating system in which individuals tend to aggregate around conspecifics [30,38].

Thus, Great Bustard individuals observed in the same flock were considered as a single occurrence record in subsequent analyses in order to avoid the potential effects that conspecific clustering could have in the modelling process.

Environmental predictors

We used as environmental predictors variables related to substrate types and topography according to existing ecological knowledge on the species [32–35]. All the environmental variables were rasterized for model calibration, considering a cell size of 50×50 m. Land-use variables were extracted from land-use maps elaborated from field surveys in each study site and year. Fields on land-use maps were classified regarding their potential to affect the presence of Little and Great Bustards. Thus, agricultural habitat types were: 1) arable lands, including cereal crops and ploughed lands, 2) leguminous crops, which are important for both Little and Great Bustards [33,39,40], 3) young fallows (hereafter referred to as fallows), 4) fallows of more than two years and low height scrublands (hereafter called natural vegetation), 5) dry woody cultures which include olive groves and vineyards, 6) others, which comprises urban substrates, fruit tree orchards and forest (Fig. 1). Land-use rasters reflect the proportion of the corresponding land use inside each cell. Land-use proportion was calculated taking into account all land-use categories, so that the sum of all of them was 1 for each cell. As it is highly recommended to reduce the number of variables for model calibration [41], the variable Others was not considered for the analysis since both species avoid the agricultural substrates enclosed in this category [37,40].

A topography position index (TPI) was also calculated from the digital elevation model, constructed from maps of five meter elevation contour lines. This index was calculated as the elevation value of each cell minus the mean elevation of the neighbouring cells within a particular radius. In this study, a radius of 250 m was selected according to the biological characteristics of the species, given their size and their lek mating system [30,42]. Therefore, it classifies each cell regarding the elevation of the neighbour cells,

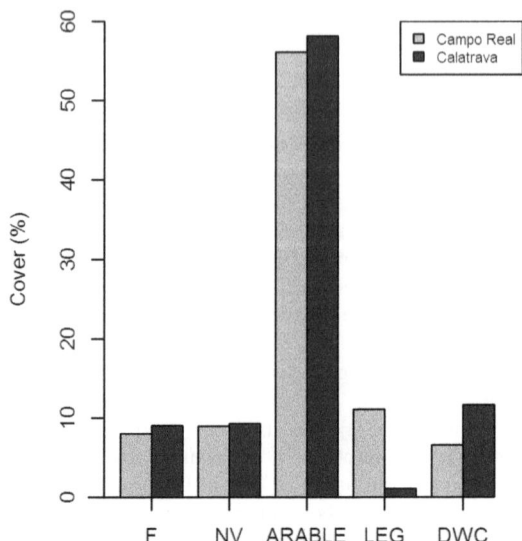

Figure 1. Land use cover in the study sites. Cover percentage of the land uses considered for Maxent modelling in 2011 in Campo Real and 2008 in Calatrava (F: short term fallows, NV: natural vegetation encompassing long term fallows and low height scrubs; Arable: cereal fields and ploughed lands; LEG: leguminous crops; DWC: dry woody cultures).

reflecting how visible a particular location is. From a behavioural point of view, the selection of areas according to their visibility could result from a trade-off between being detected by conspecifics and concealment from potential predators [43].

Habitat suitability models of Little and Great Bustards

MaxEnt was selected for modelling the spatial distribution of each study species since it is a presence-only approach. This is a machine-learning method based on the principle of maximum entropy [44] that has been employed widely in many ecological studies (for further details see [41,45]). MaxEnt models have been proved to yield one of the highest quality predictions among several modelling methods and the best performance at low sample sizes [46–48].

The species distribution modelling required two independent set of observations, one for calibrating the model and the other for evaluating model predictions [26]. Models were built separately for each species and study site with datasets from years 2008 and 2011 for Calatrava and Campo Real respectively. The regularization parameters to reduce model over-fitting were selected automatically by the program [41]. Predictive maps of probability of habitat suitability for each species and study site were built from calibration datasets and subsequently transformed to Boolean maps of presence/absence by selecting a threshold. We decided to use the average suitability approach [49], which fixes the threshold at the mean of all predicted cell values from the calibration dataset. This approach was chosen because it does not require true absence data and because of its effectiveness and simplicity [50].

Models were evaluated using 2009 and 2012 datasets respectively for Calatrava and Campo Real. Model evaluation should deal with two aspects, the performance and the significance of the model [26]. First, model performance shows how well or poorly the model classifies presence and absence of the species. Omission error rate (the proportion of presence occurrence records of the evaluation dataset that fall in an area predicted as unsuitable for the species) was used as a measure of model performance, expecting low omission rates for good models [26]. This measure of model performance was selected because it does not need true absence records for its calculation [26]. Second, it is also necessary to assess model significance, ie. whether the model predicts presence occurrence records from the evaluation dataset better than expected under random prediction [26]. Thus, one-tailed binomial tests (one per model) were performed to evaluate whether the proportion of correctly classified occurrences differs from the proportion of area predicted as presence by the model.

Coexistence and exclusive use areas of Little and Great Bustards

Since we were mainly interested in the delimitation of areas in which both species might coexist, a coexistence map was built in each study site. Coexistence maps were generated by superimposing both the Little and the Great Bustard Boolean maps, generating a new one with four cell types: 1) cells predicting presence of both species, 2) cells predicting only Little Bustard presence, 3) cells predicting only Great Bustard presence and 4) cells predicting absence of both species. The surface and density of each species for coexistence, exclusive use and absence areas were calculated in each study site. In addition, means of each land use cover in coexistence and exclusive use areas were calculated in order to describe the land use composition of each area type. Finally, we evaluated habitat suitability differences between coexistence and each species exclusive use areas. In order to eliminate the spatial trends of the data we used a third order polynomial regression with the spatial coordinates [51]. Residuals of the regression were analysed by a Student t test to determine whether probabilities of habitat suitability differ between these area types for both species.

Environmental predictors were generated using ArcGis v9.3 program [52]. TPI was built by the extension "Topographic Position Index (TPI) v 1.2" [53] and MaxEnt modelling was performed by the package "dismo" [54] for the R software v2.14 [55].

Results

Campo Real presented densities of 4.02 Little Bustards and 5.6 Great Bustards/km^2 in 2011, higher than the 2.48 Little Bustards and 1.9 Great Bustards/km^2 found in Calatrava 2008.

Habitat suitability models of Little and Great Bustards predicted the distribution of evaluation points accurately and better than random for the two study sites (Table 1). Little Bustard models predicted a greater extension of presence area than Great Bustard models for both study sites. Little Bustard model in Campo Real showed the highest predicted presence area as well as the lowest omission error rate, predicting correctly almost all the evaluation data set (Table 1).

Models for Little Bustard were influenced mainly by the presence of dry woody cultures and fallows as shown by their contribution percentages (ie. the relative contribution of each variable to the model. Table 2). The response was positively related to fallow cover while the cover of dry woody cultures was negatively related with the predicted probabilities of habitat suitability in both study sites (Fig. 2). The cover of leguminous crops was also an important variable, positively related with the

Table 1. Percentage of predicted presence area of Little and Great Bustards in Campo Real 2012 and Calatrava 2009 (corresponding with the evaluation datasets).

	Campo Real		Calatrava	
	Little Bustard	**Great Bustard**	**Little Bustard**	**Great Bustard**
Predicted area (%)	72.07	49.73	58.50	54.55
Omission error rate	0.09	0.33	0.21	0.11
P	0.003	0.0375	<0.001	<0.001

Omission error rates (proportion of presence occurrence records of the evaluation dataset that fall in an area predicted as unsuitable for the species) and p-values of one-tailed binomial test for evaluating model performance and significance respectively are provided.

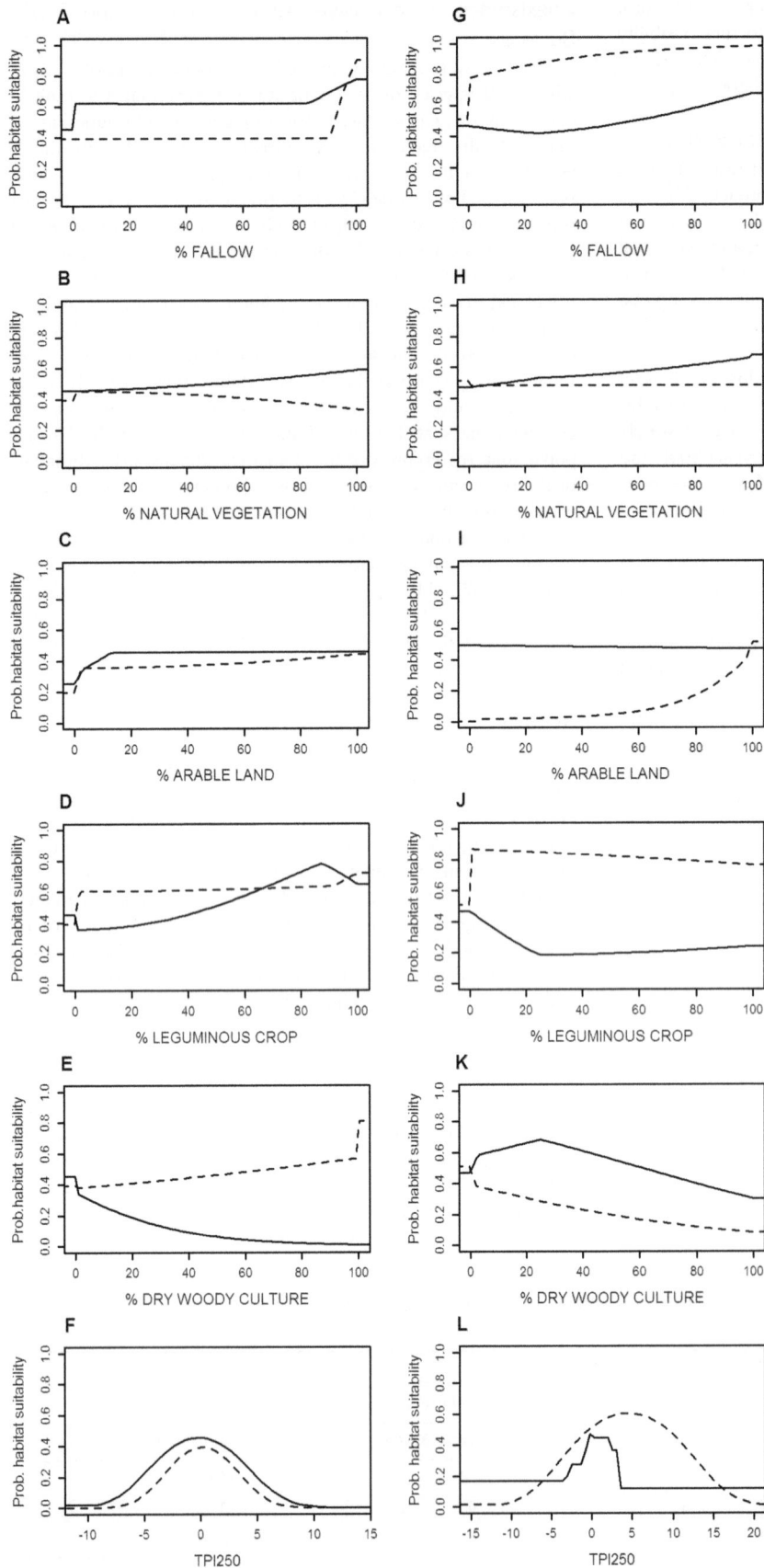

Figure 2. Probabilities of habitat suitability for the environmental predictors. Maxent response curves representing the probability of habitat suitability for each environmental predictor (percentage of land uses and Topographic position index at 250 m resolution, TPI250) for the study species in Campo Real (A–F) and Calatrava (G–L). Solid lines correspond to Little Bustard response curves while broken lines correspond to Great Bustard response curves.

Little Bustard predicted distribution in Campo Real (Fig. 2). TPI was one of the most relevant environmental predictors in Calatrava, with highest predictive power at values around 0, indicating a preference for flat zones (Fig. 2).

Differences between study sites were greater in Great Bustard models. Arable land appeared as one of most relevant predictors, especially in Calatrava's model (Table 2). Campo Real's model was highly influenced also by the presence of fallows and leguminous crops, both showing a positive relationship with the predicted probability of habitat suitability (Fig. 2).

In Campo Real, 45.33% of the surface corresponded to the coexistence area (Fig. 3). The Little Bustard exclusive use area presented a cover value of 20.78%, whereas the Great Bustard exclusive area reached a lower cover of 12.62%. In Calatrava, the predicted coexistence area accounted for the 36.15% of the surface (Fig. 3), lower than the value found in Campo Real. The area predicted as exclusively used by the Little Bustard in Calatrava reached 22.38% cover, while the predicted Great Bustard exclusive area was 20.80%.

In Campo Real, the density of Little Bustards in the predicted coexistence area was slightly higher than in the exclusive use area (Table 3). The same pattern was found for Great Bustards in Calatrava site. However, densities in coexistence area were lower than in exclusive use area in the case of Little Bustard in Calatrava and Great Bustard in Campo Real (Table 3). Regarding land use composition, Little Bustard exclusive use areas showed a higher cover of fallows and natural vegetation than Great Bustard exclusive use and coexistence areas in both study sites (Fig. 4). Little Bustard exclusive use area showed a lower value of arable surface in Calatrava than in Campo Real. In addition, this value was also lower than the cover of Great Bustard exclusive use and coexistence areas in both study sites (Fig. 4).

The residuals of the polynomial regression were significantly different between coexistence and exclusive use areas for both species in both study sites. The Little Bustard showed higher probabilities of habitat suitability in areas where only this species was predicted as present than in areas in which it might coexist with the Great Bustard (Campo Real: $t_{0.05;1868.391} = 12.047$, $p<0.001$; Calatrava: $t_{0.05; 9703.717} = 98.200$, $p<0.001$, Fig. 5). The same pattern was found for Great Bustards in Campo Real ($t_{0.05; 1150.884} = 21.817$, $p<0.001$), although this species showed

higher probabilities of habitat suitability in coexistence areas than in areas of exclusive use in Calatrava ($t_{0.05; 13177.676} = -27.053$, $p<0.001$, Fig. 5).

Discussion

The models yielded by MaxEnt for two endangered bird species linked to pseudo-steppe landscapes, the Little and the Great Bustards, were able to predict suitable areas accurately. It is important to note that Little Bustard results correspond only to males and conclusions may not apply to females which might show a different habitat selection pattern. Our results showed that models are not only species-specific but also context-dependent. Little Bustard presence areas seem to be the result of a more complex combination of different substrate types while the Great Bustard shows a higher dependence on arable fields. Coexistence areas are also context-dependent at local scale and tend to harbour less suitable habitat than areas of exclusive use. The results found in this study have implications for conservation and management strategies.

The Little and the Great Bustards have been the object of many habitat selection studies due to their interest as lekking species and their worrying conservation status caused by changes in agricultural practices during the last decades. Our models showed that both species benefit from the presence of short term fallows in accordance with previous studies [32,34,56]. Little Bustard males' preference for short term fallows as habitats that ensure conspicuousness for sexual displaying and food supply [32,56], is reflected in our models by their high contribution percentages. In the case of Great Bustard, the importance of fallow cover in explaining the distribution pattern seems particularly context-dependent. In Campo Real, fallows appear as a relevant substrate type for Great Bustard while the effect on its distribution is minimal in Calatrava. Leguminous crops play also an important role for both species when they are present in the landscape. In the case of Little Bustard, leguminous crops reach a similar importance in the model as fallow lands in Campo Real, but remain as a minor variable in Calatrava, where the presence of this substrate is clearly marginal.

However, these species differ in their responses to other landscape variables, indicating some level of niche segregation at

Table 2. Contribution percentages of each environmental predictor (percentage of each land use type, and Topographic position index at 250 m resolution) to each species and study site models yielded by MaxEnt.

	Campo Real		Calatrava	
	Little Bustard	**Great Bustard**	**Little Bustard**	**Great Bustard**
Fallows	20.60	37.41	43.16	5.59
Natural Vegetation	0.56	5.98	4.49	1.33
Arable	11.39	16.32	2.10	77.82
Dry woody cultures	44.94	5.45	20.24	10.36
Leguminous crops	19.74	28.08	0.50	1.10
TPI250	2.77	6.78	29.49	3.81

Models were built using Little and Great Bustard observations from 2011 for Campo Real and 2008 for Calatrava.

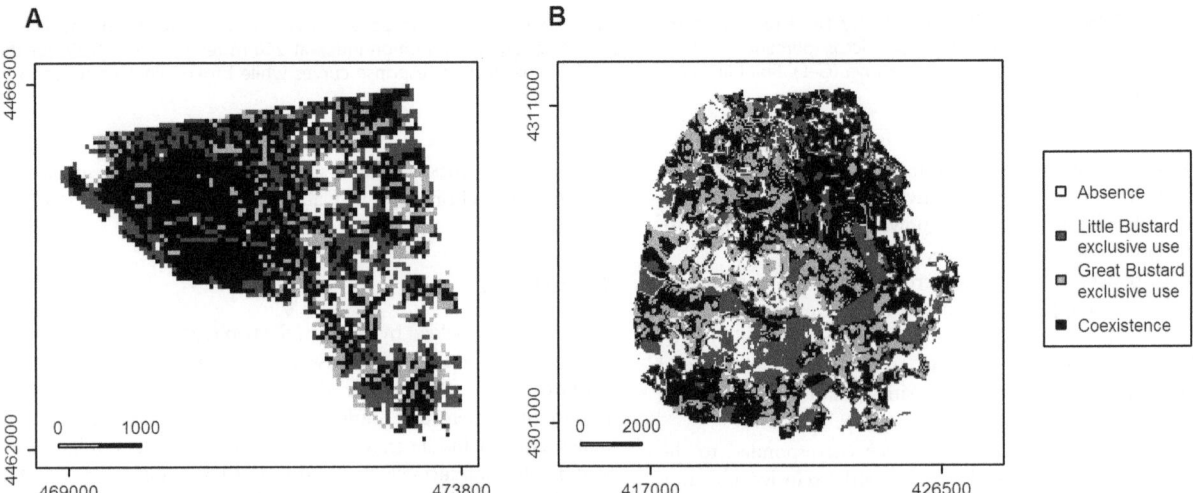

Figure 3. Coexistence maps of Little and Great Bustards. Maps of Little Bustard and Great Bustard coexistence for 2011 in Campo Real (A) and 2008 in Calatrava (B), showing also areas of exclusive use and areas in which both species were predicted to be absent. The scale bar is given in meters.

local scale. For instance, the relevance of arable lands is clearly different between species, being the cover of this land use more important for the Great Bustard. The presence of dry woody cultures plays a minor role in the distribution pattern of the Great Bustard but not for the Little Bustard, which avoids vineyards and olive groves in accordance with previous studies [32,57]. Finally, the importance of topography varies between species and study sites. The Little Bustard shows in both study sites the same preference for flat zones where they are visible to other individuals during the sexual display season. However, the relevance of flat zones changes with the study site, being especially high in Calatrava, which might be due to its higher variability in topography. In the case of the Great Bustard, its distribution pattern is hardly affected by topography, while land use variables acquire a major role in determining the species' distribution in both study sites. The differences found between study sites might

be indicating that habitat selection depends on the particular landscape composition. This is especially noteworthy for the Great Bustard, which may be explained by its greater niche width [58]. Nevertheless, results might also be influenced by the SDMs' dependency on the environmental context, since the model calibration process depends on the particular combination of variables that occurs in each study site [26]. Although the spatial scale selected may influence observed response patterns, this seems to occur only at high cover values of some land uses (Fig. 2). In any case, results are consistent with the existing habitat selection knowledge for the species, as pointed out previously.

Our results show that concentrating conservation efforts on preserving the habitats most preferred by one species at local scale may be detrimental for the other given their different requirements, leaving habitats relevant to that species without protection. Therefore, a multi-species approach may help prioritize conser-

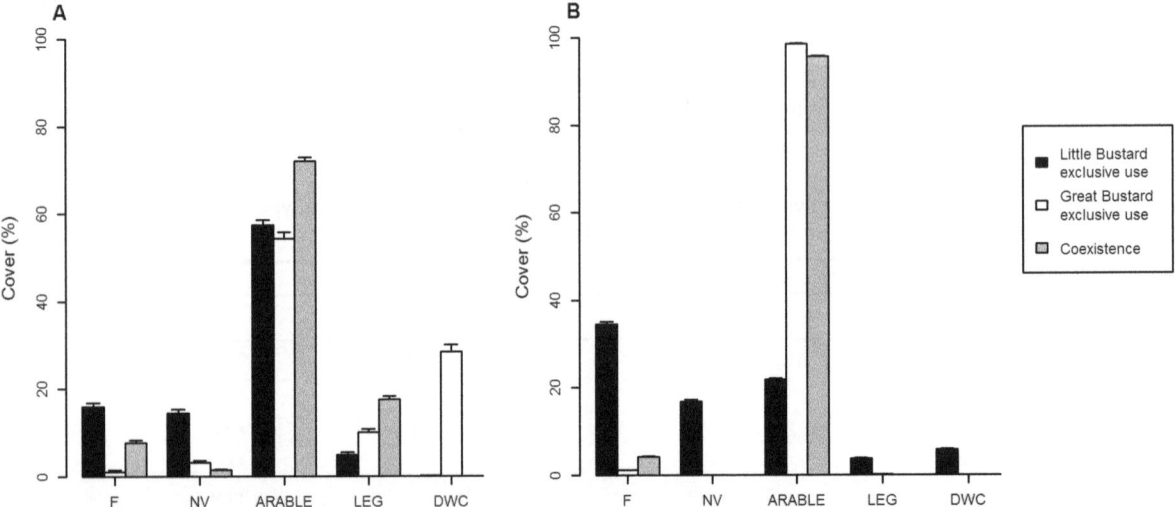

Figure 4. Land use cover in each area type. Mean and standard error of land use cover in the predicted Little and Great Bustard exclusive use and coexistence areas for 2011 in Campo Real (A) and 2008 in Calatrava (B) (F: short term fallows, NV: natural vegetation encompassing long term fallows and low height scrubs; Arable: cereal fields and ploughed lands; LEG: leguminous crops; DWC: dry woody cultures).

Table 3. Densities of Little (males/km²) and Great Bustards (individuals/km²) in the different area types generated by superimposing the predicted presence maps of Little and Great Bustards for 2011 in Campo Real and 2008 in Calatrava.

	Campo Real		Calatrava	
	Little Bustard	**Great Bustard**	**Little Bustard**	**Great Bustard**
Absence area	2.46	0.41	0.81	0.05
Little Bustard exclusive area	5.04	0.84	4.71	0
Great Bustard exclusive area	0.69	5.54	1.23	1.01
Coexistence area	5.20	3.66	2.52	1.35

vation efforts on coexistence areas. Our study shows that coexistence and exclusive use areas of Little and Great Bustards differ in their habitat features, which may also vary in relation to the local environmental context. The area predicted as suitable for the coexistence of these species is greater than the surface of each species exclusive use in both study sites. However, different situations emerge when looking at probabilities of habitat suitability and actual densities. In two cases, Little Bustard in Calatrava and Great Bustard in Campo Real, the corresponding exclusive use area harbours better habitat conditions for the target species and also higher density. Thus, the coexistence area might correspond to suboptimal zones for the species. However, we cannot disentangle whether the low probabilities of habitat suitability predicted for coexistence areas are due to poor habitat quality or to the avoidance of heterospecifics since both factors can affect distribution patterns [11]. The other two cases (Little Bustard in Campo Real and Great Bustard in Calatrava) present similar densities but different habitat suitability for each area type. The exclusive Little Bustard area in Campo Real shows higher habitat suitability than the coexistence area. It seems that Little Bustards might occupy less suitable areas in the absence of enough space or good quality habitats. However, the pattern for Great Bustards in Calatrava is the opposite, with higher probabilities of habitat suitability in coexistence areas. Therefore, the coexistence area in Calatrava seems to reflect Great Bustard habitat preferences whereas Little Bustards concentrate mainly in their

exclusive use area. Low density might allow Little Bustards to occupy their most preferred habitat features without using areas suitable for the Great Bustard. It is noteworthy that each species presents lower densities in the absence and exclusive use areas of the other species, a fact that might support the hypothesis of segregation between these two steppe-birds. Consequently, by prioritizing the preservation of coexistence areas, we may be protecting low quality habitats that are being used by the two (or more) species because higher quality exclusive areas are scarce, thus preventing natural between-species avoidance.

Some interesting conservation consequences arise from this study. Both species seem to benefit from high percentage of short-term fallows and leguminous crops at landscape scale, so that promoting the application of agri-environmental schemes that favour the concentration of these habitats in small areas in the landscape is desirable. In this context, Concepción and Díaz [59] emphasized that the effects of agri-environmental schemes are limited by their application at field level, and plans designed at landscape level are needed to maintain the mosaic structure of this extensive cereal croplands. For instance, the traditional two-year rotation system known as Iberian dry-farming would benefit species linked to extensive cereal croplands since it maintains a complex and dynamic structure of different and complementary land uses [60]. However, their different habitat preferences constrain the potential delimitation of coexistence areas encompassing high quality habitats at local scale. In order to meet

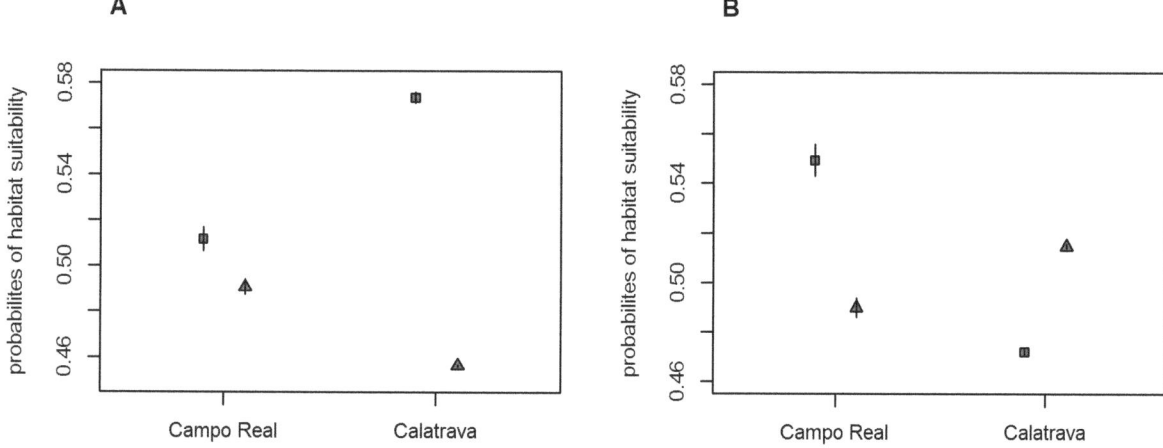

Figure 5. Probabilities of habitat suitability of coexistence and exclusive use areas. Mean and 95% confidence interval of probabilities of habitat suitability in coexistence and exclusive use areas for the Little (A) and Great Bustards (B) in Campo Real 2011 and Calatrava 2008. Student t tests were performed with the residuals of the polynomial regression although original probabilities are shown for the sake of interpretation. Probability means of coexistence areas are represented as triangles and probability means of exclusive use areas are represented as squares.

species' spatial requirements, protected areas for these (and probably other) steppe birds should cover territories large enough to allow their coexistence by the selection of their preferred areas, or their tendency to segregate in space. Therefore, the role played by biotic interactions in a community should be considered when designing conservation strategies at least at local scale. Finally, the context-dependence of habitat selection in these species advices designing conservation measures for particular landscape situations.

Spatial distribution modelling is a useful tool for species conservation since it can integrate behavioural traits and landscape measurements and helps identifying general responses to environmental variables. In addition, it allows the extrapolation of results to other regions in order to preserve non-occupied areas of suitable habitat that could be potentially colonized at long term [61]. This is important even in the case of the Great Bustard whose strong breeding philopatry constrains the colonization of unoccupied areas [62].

Conclusions

The identification of coexistence areas of two farmland birds at local scale described in this study provides insightful results that might apply in other cases. Concentrating efforts on one umbrella species may be hazardous if that species does not adequately reflect the ecological requirements of sympatric heterospecifics.

Hence, a multi-species approach may be more adequate, and the identification of coexistence areas may provide an idea of the spatial requirements of a particular assemblage. However, when coexistence areas correspond to suboptimal habitats for species that would be otherwise segregated due to their different ecological requirements, focusing efforts on these areas may be misleading at local scale. Moreover, the influence of the local environmental context in determining coexistence areas is not detected at broader scales, at which species sharing requirements overlap in their distribution ranges. Finally, integrating information of species distribution models built at local scale might lead to a better understanding of general patterns at broader scales [7].

Acknowledgments

We wish to thank J. Viñuela, S. Herrero, I. Hervás, F. Casas, E. García and J. Caro who collaborated in field work. We are grateful to C.P. Carmona and S. Suárez-Seoane for their helpful advice on the modelling process, as well as the three anonymous reviewers for their comments.

Author Contributions

Conceived and designed the experiments: RT MM JT MPD. Performed the experiments: RT MM JT MPD. Analyzed the data: RT. Contributed reagents/materials/analysis tools: RT MM. Wrote the paper: RT MM JT MPD.

References

1. Gaston KJ (2003) The Structure and Dynamics of Geographic Ranges. Oxford: Oxford University Press. 266 p.
2. Wiens JJ, Donoghue MJ (2004) Historical biogeography, ecology and species richness. Trends Ecol Evol 19: 639–644.
3. Ricklefs RE (2007) History and Diversity: Explorations at the Intersection of Ecology and Evolution. Am Nat 170: S56–S70.
4. Braunisch V, Patthey P, Arlettaz R (2011) Spatially explicit modeling of conflict zones between wildlife and snow sports: prioritizing areas for winter refuges. Ecol Appl 21: 955–967.
5. Pearson RG, Dawson TP (2003) Predicting the impacts of climate change on the distribution of species: are bioclimate envelope models useful? Glob Ecol Biogeogr 12: 361–371.
6. Hampe A (2004) Bioclimate envelope models: what they detect and what they hide. Glob Ecol Biogeogr 13: 469–471.
7. Araújo MB, Luoto M (2007) The importance of biotic interactions for modelling species distributions under climate change. Glob Ecol Biogeogr 16: 743–753.
8. Heikkinen RK, Luoto M, Virkkala R, Pearson RG, Körber J-H (2007) Biotic interactions improve prediction of boreal bird distributions at macro-scales. Glob Ecol Biogeogr 16: 754–763.
9. Martin TE (2001) Abiotic vs. Biotic Influences on Habitat Selection of Coexisting Species: Climate Change Impacts? Ecology 82: 175–188.
10. Pearson RG, Dawson TP, Liu C (2004) Modelling species distributions in Britain: a hierarchical integration of climate and land-cover data. Ecography 27: 285–298.
11. Morris DW (2003) Toward an ecological synthesis: a case for habitat selection. Oecologia 136: 1–13.
12. Chesson P (2000) Mechanisms of Maintenance of Species Diversity. Annu Rev Ecol Syst 31: 343–366.
13. Kotler BP, Brown JS (1988) Environmental Heterogeneity and the Coexistence of Desert Rodents. Annu Rev Ecol Syst 19: 281–307.
14. Martin PR, Martin TE (2001) Ecological and fitness consequences of species coexistence: a removal experiment with wood warblers. Ecology 82: 189–206.
15. Morris DW (1989) Density-dependent habitat selection: Testing the theory with fitness data. Evol Ecol 3: 80–94.
16. Delgado MP, Sanza MA, Morales MB, Traba J, Rivera D (2013) Habitat selection and coexistence in wintering passerine steppe birds. J Ornithol 154: 469–479.
17. Araújo MB, Cabeza M, Thuiller W, Hannah L, Williams PH (2004) Would climate change drive species out of reserves? An assessment of existing reserve-selection methods. Glob Chang Biol 10: 1618–1626.
18. Martínez-Meyer E, Peterson AT, Servín JI, Kiff LF (2006) Ecological niche modelling and prioritizing areas for species reintroductions. Oryx 40: 411–418.
19. Titeux N, Dufrene M, Radoux J, Hirzel AH, Defourny P (2007) Fitness-related parameters improve presence-only distribution modelling for conservation practice: The case of the red-backed shrike. Biol Conserv 138: 207–223.

20. Kremen C, Cameron A, Moilanen A, Phillips SJ, Thomas CD, et al. (2008) Aligning Conservation Priorities Across Taxa in Madagascar with High-Resolution Planning Tools. Science 320: 222–226.
21. Wilson CD, Roberts D (2011) Modelling distributional trends to inform conservation strategies for an endangered species. Divers Distrib 17: 182–189.
22. Andelman SJ, Fagan WF (2000) Umbrellas and flagships: Efficient conservation surrogates or expensive mistakes? Proc Natl Acad Sci U S A 97: 5954–5959.
23. Caro T, Engilis AJ, Fitzherbert E, Gardner T (2004) Preliminary assessment of the flagship species concept at a small scale. Anim Conserv 7: 63–70.
24. Carroll C, Noss RF, Paquet PC (2001) Carnivores as focal species for conservation planning in the Rocky Mountain Region. Ecol Appl 11: 961–980.
25. Zipkin EF, Royle JA, Dawson DK, Bates S (2010) Multi-species occurrence models to evaluate the effects of conservation and management actions Biol Conserv 143: 479–484.
26. Peterson AT, Soberón J, Pearson RG, Anderson RP, Martínez-Meyer E, et al. (2011) Ecological Niches and Geographic Distributions. Oxford: Princeton University Press. 314 p.
27. IUCN (2012) Red List of Threatened Species. www.iucnredlist.org. 02-10-2013.
28. García de la Morena EL, Bota G, Ponjoan A, Morales MB (2006) El Sisón Común en España. I Censo Nacional (2005). Madrid: SEO/BirdLife. 155 p.
29. Palacín C, Alonso JC (2008) An updated estimate of the world status and population trends of the Great Bustard Otis Tarda. Ardeola 55: 13–25.
30. Morales MB, Martín CA (2002) The Great Bustard Otis tarda. The Journal of Birds of the Western Palearctic 4: 217–232.
31. Wolff A, Paul J-P, Martin J-L, Bretagnolle V (2001) The benefits of extensive agriculture to birds: the case of the little bustard. J Appl Ecol 38: 963–975.
32. Morales MB, García JT, Arroyo B (2005) Can landscape composition changes predict spatial and annual variation of little bustard male abundance? Anim Conserv 8: 167–174.
33. Lane SJ, Alonso JC, Martín CA (2001) Habitat preferences of great bustard Otis tarda flocks in the arable steppes of central Spain: are potentially suitable areas unoccupied? J Appl Ecol 38: 193–203.
34. López-Jamar J, Casas F, Díaz M, Morales MB (2011) Local differences in habitat selection by Great Bustard Otis tarda in changing agricultural landscapes: implications for farmland bird conservation. Bird Conserv Int 21: 328–341.
35. Suárez-Seoane S, Osborne PE, Alonso JC (2002) Large-scale habitat selection by agricultural steppe birds in Spain: identifying species–habitat responses using generalized additive models. J Appl Ecol 39: 755–771.
36. Silva JP, Pinto M, Palmeirim JM (2004) Managing landscapes for the little bustard Tetrax tetrax: lessons from the study of winter habitat selection. Biol Conserv 117: 521–528.
37. Cramp S, Simmons KEL (1980) The Birds of the Western Palearctic. Vol II. Oxford: Oxford University Press. 696 p.
38. Alonso J, Martín C, Alonso J, Palacín C, Magaña M, et al. (2004) Distribution dynamics of a great bustard metapopulation throughout a decade: influence of conspecific attraction and recruitment. Biodivers Conserv 13: 1659–1674.

39. Martínez C (1994) Habitat selection by the Little Bustard *Tetrax tetrax* in cultivated areas of central Spain. Biol Conserv 67: 125–128.

40. Salamolard M, Moreau C (1999) Habitat selection by Little Bustard *Tetrax tetrax* in a cultivated area of France. Bird Study 46: 25–33.

41. Elith J, Phillips SJ, Hastie T, Dudík M, Chee YE, et al. (2011) A statistical explanation of MaxEnt for ecologists. Divers Distrib 17: 43–57.

42. Jiguet F, Arroyo B, Bretagnolle V (2000) Lek mating systems: a case study in the Little Bustard *Tetrax tetrax*. Behav Processes 51: 63–82.

43. Aspbury AS, Gibson RM (2004) Long-range visibility of greater sage grouse leks: a GIS-based analysis. Anim Behav 67: 1127–1132.

44. Jaynes ET (1957) Information Theory and Statistical Mechanics. Physical Review 106: 620–630.

45. Phillips SJ, Dudík M, Schapire RE (2004) A maximum entropy approach to species distribution modeling. Proceedings of the twenty-first international conference on Machine learning. Banff, Alberta, , Canada: ACM.

46. Hernandez PA, Graham CH, Master LL, Albert DL (2006) The effect of sample size and species characteristics on performance of different species distribution modeling methods. Ecography 29: 773–785.

47. Pearson RG, Raxworthy CJ, Nakamura M, Peterson AT (2007) Predicting species distributions from small numbers of occurrence records: a test case using cryptic geckos in Madagascar. J Biogeogr 34: 102–117.

48. Wisz MS, Hijmans RJ, Li J, Peterson AT, Graham CH, et al. (2008) Effects of sample size on the performance of species distribution models. Divers Distrib 14: 763–773.

49. Cramer JS (2003) Logit models: from economics and other fields. Cambridge: Cambridge University Press. 173 p.

50. Liu C, Berry PM, Dawson TP, Pearson RG (2005) Selecting thresholds of occurrence in the prediction of species distributions. Ecography 28: 385–393.

51. Legendre P, Legendre L (1998) Numerical Ecology. Second English Edition. Amsterdam: Elsevier Publishers. 852 p.

52. ESRI (2007) GIS and mapping software. www.esri.com. 02-10-2013.

53. Jenness J (2006) Topographic Position Index (tpi_jen.avx) extension for ArcView 3.x, v. 1.2. Jenness Enterprises. Available at: http://www.jennessent.com/arcview/tpi.htm.

54. Hijmans RJ, Phillips SJ, Leathwick J, Elith J (2012) dismo: Species distribution modeling. R package version 0.7-17.

55. R Development Core Team (2013) R: A Language and Environment for Statistical Computing. Vienna, Austria.

56. Delgado MP, Traba J, García de la Morena E, Morales MB (2010) Habitat Selection and Density-Dependent Relationships in Spatial Occupancy by Male Little Bustards *Tetrax tetrax*. Ardea 98: 185–194.

57. Lapiedra O, Ponjoan A, Gamero A, Bota G, Mañosa S (2011) Consequences of agricultural intensification on the ranging behavior and breeding success of threatened steppe-land birds: the case of little bustard. Biol Conserv: 2882–2890.

58. Morales MB, Suárez F, García de la Morena EL (2006) Responses des oiseaux de steppe aux differents niveaux de mise en culture et d'intensification du paysage agricole: une analyse comparative de leurs effets sur la densite de population et la selection de l'habitat chez l'Outarde Canepetière *Tetrax tetrax* et l'Outarde Barbue *Otis tarda*. Rev Écol (Terre et Vie) 61: 261–269.

59. Concepción ED, Díaz M (2010) Relative effects of field- and landscape-scale intensification on farmland bird diversity in Mediterranean dry cereal croplands. Asp Appl Biol 100: 245–252.

60. Suárez F, Naveso MA, de Juana E (1997) Farming in the drylands of Spain: birds in the pseudosteppes. In: Pain DJ, Pienkowski MW, editors. Farming and Birds in Europe. Cambridge: Academic Press. pp. 297–330.

61. Hanski I (1999) Metapopulation Ecology. Oxford: Oxford University Press. 313 p.

62. Martín B (2009) Dinámica de población y viabilidad de la avutarda común en la Comunidad de Madrid. Ph.D. Thesis. Madrid: Universidad Complutense de Madrid.

2

Spatial Heterogeneity in Human Activities Favors the Persistence of Wolves in Agroecosystems

Mohsen Ahmadi[1], José Vicente López-Bao[2,3], Mohammad Kaboli[1]*

1 Department of Environmental Sciences, Faculty of Natural Resources, University of Tehran, Karaj, Iran, **2** Research Unit of Biodiversity (UO/CSIC/PA), Oviedo University, Mieres, Spain, **3** Grimsö Wildlife Research Station, Dep. of Ecology, Swedish University of Agricultural Sciences (SLU), Riddarhyttan, Sweden

Abstract

As human populations expand, there is increasing demand and pressure for land. Under this scenario, behavioural flexibility and adaptation become important processes leading to the persistence of large carnivores in human-dominated landscapes such as agroecosystems. A growing interest has recently emerged on the outcome of the coexistence between wolves and humans in these systems. It has been suggested that spatial heterogeneity in human activities would be a major environmental factor modulating vulnerability and persistence of this contentious species in agroecosystems. Here, we combined information from 35 den sites detected between 2011 and 2012 in agroecosystems of western Iran (Hamedan province), a set of environmental variables measured at landscape and fine spatial scales, and generalized linear models to identify patterns of den site selection by wolves in a highly-modified agroecosystem. On a landscape level, wolves selected a mixture of rangelands with scattered dry-farms on hillsides (showing a low human use) to locate their dens, avoiding areas with high densities of settlements and primary roads. On a fine spatial scale, wolves primarily excavated dens into the sides of elevated steep-slope hills with availability of water bodies in the vicinity of den sites, and wolves were relegated to dig in places with coarse-soil particles. Our results suggest that vulnerability of wolves in human-dominated landscapes could be compensated by the existence of spatial heterogeneity in human activities. Such heterogeneity would favor wolf persistence in agroecosystems favoring a land sharing model of coexistence between wolves and people.

Editor: Clinton N. Jenkins, Instituto de Pesquisas Ecológicas, Brazil

Funding: The study was supported by Department of Environmental Sciences, Faculty of Natural Resources, University of Tehran, and provincial bureau of Department of Environment, Hamedan province. JVLB was supported by a "Juan de la Cierva" research contract from the Spanish Ministry of Economy and Competitiveness. The funders had no role in study design, data collection and analysis, decision to publish, or preparation of the manuscript.

Competing Interests: The authors have declared that no competing interests exist.

* Email: mkaboli@ut.ac.ir

Introduction

As human populations expand, there is increasing demand and pressure for land (characterized by an increment and expansion in settlements, habitat transformation and extension of agricultural lands, and industrial development) and, consequently, different impacts on wildlife are expected. Under this scenario, behavioural flexibility and adaptation are important processes leading to the persistence of viable animal populations in human-dominated landscapes, including urban environments (e.g. mammalian carnivores [1,2]). For species like large carnivores, with remarkable large spatial requirements, low reproductive rates or low densities [3], as well as a high potential for conflict (e.g. livestock attacks [4,5]), such behavioural processes are key elements determining their persistence in human-dominated landscapes. In fact, the capability of these species to persist in this scenario, and its behavioural, demographic and ecological consequences, have attracted a great attention in recent times [2,6,7,8].

Existing evidence shows how wolves (*Canis lupus*) are able to persist in contrasting human-dominated landscapes [7,9,10,11,12] as soon as legislation is favourable and human pressure is low [13], and minimum food and refuge requirements are fulfilled [3]. Several mechanisms are behind this ability such as the spatio-temporal segregation between wolves and human activities [9,14], their capacity to use different human-related sources of food [15,16] or other behavioural adaptations such as den shifting [17]. All this information suggest that wolves are highly capable to persist in humanized landscapes by perceiving mortality risk associated with humans, adjusting, for instance, the use of the space at different scales over time accordingly [7,17,18] (see [2] for an example with the red wolf). Thus, the spatial and temporal heterogeneity in human activities would emerge as a major environmental factor modulating vulnerability and persistence of wolves in human-dominated landscapes, resulting in wolf persistence even in areas completely transformed by humans [2,7,18,19].

In agroecosystems, ecological systems modified by human beings to produce food, fibre or other agricultural products [20], such heterogeneity in human activities may provide wolves with places of low human use where they can go unnoticed and, more importantly, can reproduce. Although the impact of humans on wolf persistence has been inferred using different surrogates such as human population density, infrastructures, level of transformation of the landscape or the spatial distribution of activities [7,21,22], how these human-related factors interact with the persistence of wolves in agroecosystems remains poorly understood. However, this knowledge becoming particularly important owing to the recent expansion of wolf populations and human activities, particularly agriculture [8,23], being crucial to adopt a balanced landscape planning ensuring both, human needs and wolf persistence [22]. Moreover, understanding the abilities of wolves to persist in each particular local context is a pressing need to reach a context-dependent conservation and management approach in agroecosystems, since heterogeneity is the norm across human-dominated landscapes [24].

Reproductive success is a cornerstone for the persistence of any species. For large carnivores, reproductive success is highly influenced by humans [3]. Because the highest mortality rate of wolves occur in the first months of their life [25,26], selection of the place where to locate the den site is crucial for wolves, being particularly important in human-dominated landscapes [17,27]. Available information suggests that, in agroecosystems, exposure risk to humans will exert the strongest effect on den site selection, with wolves aiming to minimize such risks. As a result, even in completely transformed landscapes wolves may place their den sites in areas where human activities are low [2,18,19]. In addition, the strength of human activities driving the selection of den sites by wolves in these systems may force other natural components of this selection process to the background. For example, in many areas wolves select for sites where they can dig easily [9,28], but in agroecosystems, where intensive cultivation practices are preferable on good soil conditions, wolves may be forced to dig in low-quality sites in terms of soil conditions.

In this study, we aimed to identify patterns of den site selection by wolves in agroecosystems of western Iran (Hamedan province), and provide insights into the behavioural response of wolves to the spatial heterogeneity in human activities. Since large-scale approaches may disregard fine-scale patterns affecting different components of the selection processes we were interested, we evaluated the requirements of denning wolves at large (den area) and fine (den site) spatial scales. In particular, we hypothesized that wolves are able to assess the type and intensity of human activities over a wide geographic range selecting den areas with low human use, minimizing the risk of mortality. Thus, on a landscape level, we first expect that wolves will avoid areas with high densities of infrastructures and humans and, second, we also predict that, in absence of natural dense vegetated areas in this agroecosystem acting as refuge and where to locate the den sites, wolves will select farmlands with the lowest intensity of human activity. On a fine scale, we expect that although wolves will select for den sites fulfilling previous known environmental requirements for the species (e.g. water availability, refuge, human inaccessibility, [9,28,29,30,31]), the strength of humans activities influencing den site selection in agroecosystems may push some components of the selection process into the background as a response to minimize the risk of exposure to humans.

Materials and Methods

Study area

Despite extensive studies on wolf distribution, biology, ecology and behaviour (see review in [7,11,32,33,34]) and conflict with humans (e.g. [4,5]) in Europe, North America or India, wolves are less studied in the Middle East. However, conflicts between wolves and humans are considerable in anthropogenic landscapes of Iran, affecting the attitudes of rural communities and the conservation status of the species [35,36,37].

This study was carried out in Hamedan province, a human-dominated landscape located in western Iran (88 inhabitants/km^2; Fig. 1) [38] and covering an area of 19,546 km^2 ($47°34' - 49°36'$

E and $35°25' - 35°15'$ N; Fig. 1). The region has a cold semi-arid climate with an average annual precipitation of 325 mm and a mean annual temperature of 11°C. The landscape in Hamedan province is severely transformed because traditionally rural community has been mostly engaged in agriculture and livestock rearing and husbandry. Consequently, agricultural lands dominate this semi-arid landscape ([39], Fig. 2; Figure S1). The very few (2% of the whole province), and small in size, patches of natural vegetation - composed by shrub species such as *Astragalus* spp. and *Bromus* spp. and with scattered trees such as Persian oak (*Quercus brantii*), Dogwood (*Cornus australis*) or Cherry plum (*Prunus divaricata*) [40] - are distributed within a heterogeneous agricultural matrix composed by intensive irrigated potato and corn farms, dry-farms (cereals) and rangelands – which are used for extensive grazing - with scattered dry-farms (Fig. 2, Figure S2). Landscape transformation has been dramatic in this area in recent times resulting in an increase of agriculture lands from 20,468 ha to 550,264 ha during the past 30 years [39]. Consequently, rangelands covered by perennial bushes and grasses decreased from 539,697 ha to 164,679 ha [39]. The expansion of agriculture lands have significantly reduced the amount of natural refuge for wolves in this open landscape and, at the same time, have also reduced wild prey populations [39], thus increasing human-wolf encounters and associated conflicts [37].

Small variations in topographic attributes - altitude and slope - in this plateau (most of the area ranges between 1,500 and 2,000 m.a.s.l and slope changes between 0 to 41 degrees) strongly determine the use of the landscape by local people. Thus, while flat areas (slope <10 degrees) are the most preferred landscape for settlements, development and human activities (84.5% of the study area), rugged landscapes (slope> 10 degrees) only encompass 15.5% of the whole landscape and is mainly used as rangelands and, sometimes, dry-farms. As a result, human activities are heterogeneously distributed across different types of farmlands. Based on cultivation and livestock practices and land use, intensity of human activities differ across farmlands as follow: irrigated farms> dry farms> rangeland with scattered farms> rangelands. For example, in irrigated farmlands (e.g. potatoes, corn), the use of heavy equipment and mechanized cultivation is quite common and these type of crops requires a continuous human presence during many months of the year, including the peak of reproductive activity of denning wolves. On the other hand, cultivation strategies of other types of farms such as dry-farms require human presence only in two specific periods, plant and harvest, resulting in low human presence especially during denning activities and rearing of immature pups.

Data collection

We used information from 35 den sites detected between 2011 and 2012 (5 den sites in 2011 and 30 in 2012; all den sites were different). Wolf dens were located using information from local sources in the rural areas, especially observations from sheep-herders and game guards of the Department of Environment of Hamedan province, as well as field patrols conducted by motorcycle in those areas where we expected to find wolf dens according to previous local knowledge in the area. Since all issues subject to wildlife care and animal welfare regulations is handled by Department of Environment (DOE) In Iran, as well as the study was in collaborated with Hamedan Provincial Bureau of Department of Environment (43106/140), all our fieldwork procedures was adhered to the animal welfare regulations. Our data sampling was carried on after confirming that wolf packs left their dens. Our field survey did not involve chasing the wolves to locate their dens. We also did not destroy or damage wolf dens.

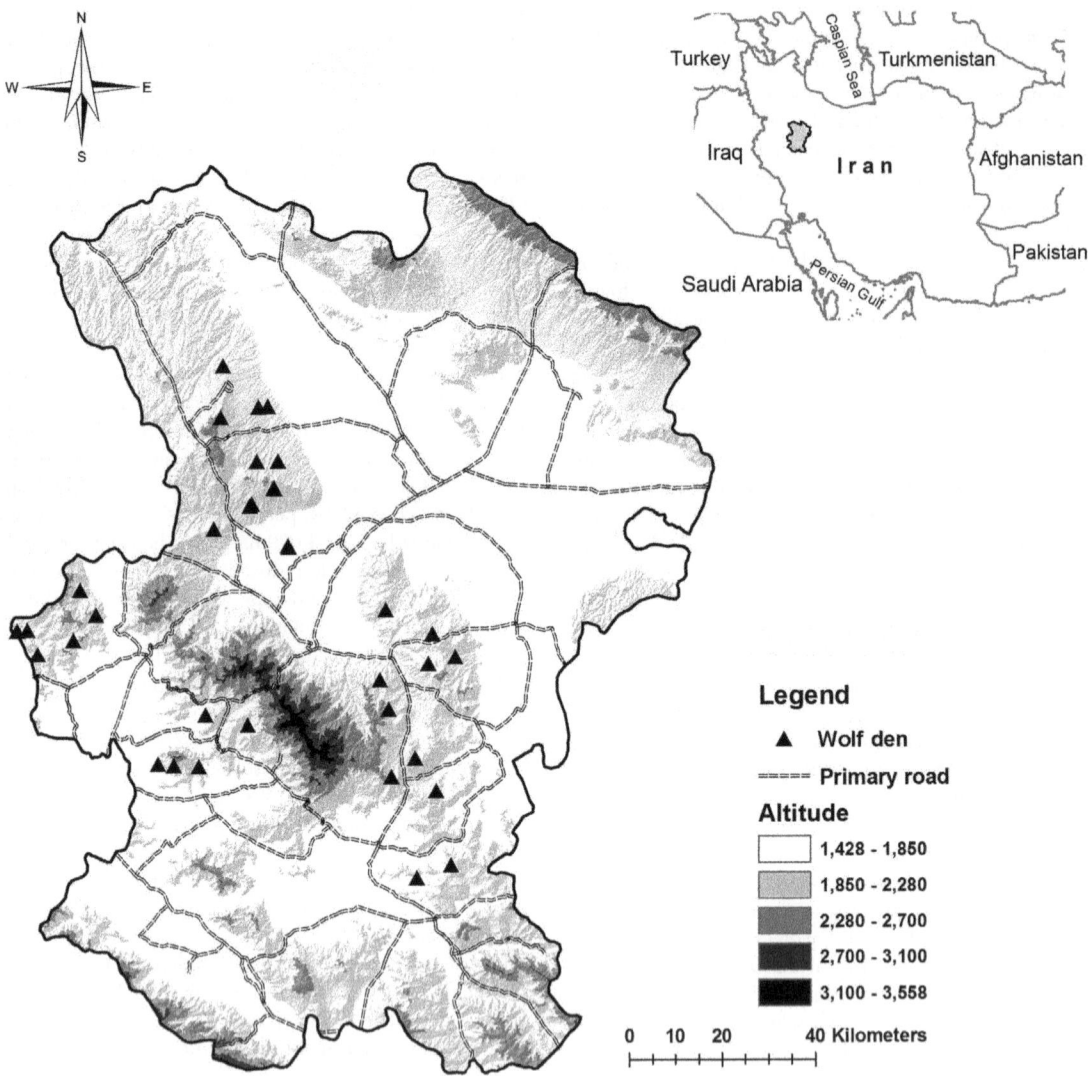

Figure 1. Distribution of gray wolf dens detected between 2011 and 2012 in Hamedan province, Iran. Wolf dens were overviewed in a context of topography and main roads in Hamedan province, Iran.

Since the breeding season is the most sensitive period for wolves [32], once a potential den site was found, we approached to the site when the pups were out of the den (between May and June) to confirm wolf reproduction. After dens were located and absence of wolves and pups was ensured, we took the location of the den sites with a GPS unit and measured the fine-scale variables we were interested (see below).

Data sampling and measurement of environmental variables were performed in two different spatial scales and using different protocols: i) den area (12.5 km²; landscape scale), where environmental variables were measured by using GIS; and ii) den site (0.01 km²; fine scale), where variables were measured *in situ*. On a landscape level, we estimated the spatial heterogeneity in human activities around den areas using a 2 km circular buffer centered on the den sites. The lack of information on wolf territory size in the study area confined us to consider a 2 km buffer size based on literature review [28,41], which well-describes landscape characterization of den areas [42]. For non-den areas we randomly selected 100 non-overlapping circular plots with the same radius excluding the largest cities and areas with an altitude

of higher than 3,000 m.a.s.l. Because of the extensive movements of wolves, the distance between random and observed (den sites) points was controlled not to fall below 15 km. This conservative distance was selected based on published empirical values of the nearest neighbor distance for active breeding dens of wolves [33,42].

The spatial heterogeneity in human activities was inferred using three different surrogates (Table 1). First, we calculated the proportion of each land use type on a landscape-level (2 km circular buffer) using the Iranian Forests, Range and Watershed Management Organization National land use/land cover map [43]. We focused on four categories of land use representing the above-mentioned gradient in the intensity of human activities (irrigated farms> dry-farms> rangeland with scattered farms> rangelands). We excluded bare lands and rocks areas due to its anecdotic representation in the area (Fig. 2). Second, we used density of settlements and length of roads as a surrogate of human intrusion and risk of mortality in the landscape. These factors are well-known affecting wolf habitat selection in general [7,11,34], and den site selection in particular [9,31]. Density of settlements and length of roads were calculated from topographic military maps of Iran with a 1:25,000 scale. Because of the different response of wolves to road networks with varied level of human activity [11,33], we classified road networks into two categories: primary roads, including national primary roads and highways with bound> 45 m, and secondary roads, including regional and district roads with bound <30 m.

Third, using the Shuttle Radar Topography Mission elevation model with 100 m resolution, we compiled mean altitude and roughness as the main factors describing the topographic context of each area which is expected to be correlated with human activities as mentioned above (human activities decrease with the increment in altitude and roughness; [7,44]). For each den area, we then calculated the mean altitude (m) by averaging altitudes of all raster cells included in this area, and roughness (m) was estimated as the standard deviation of the altitudes of all the 100 m raster cells included in each den area. Both measures reflect different types of human use; i.e. flat areas are preferred for intensive agriculture whereas rough surfaces are more inappropriate to use farm machinery being used for extensive livestock practices and dry-farms. Vegetation types providing structural protection to wolves, such as scrubs or forests, are often selected as refuge [7,34]. But semi-arid agroecosystems of Iran, as well as other open semi-arid landscapes within the wolf's range [18], lacks such suitable cover types to provide concealment for wolves. Hence roughness of terrain that is taken into account in this study could be a representative of concealment for wolf movements [7,35].

On a fine-scale (100 m radius), we measured thirteen variables related to the vulnerability of wolves (vegetation types and slope as surrogates of refuge, human activity – existence of farmlands -), ease to dig (soil/petrology; soil type and rock density can affect den site selection by wolves [17,47]), water availability, which may be a determinant factor to locate the den [28,30], particularly in arid environments, along with solar insulation. These variables were

chosen based on their suggested importance for wolf den site selection in other temperate study areas [9,28,29,30,31]. Excepting for solar insulation, all fine-scale variables were measured in five 20 × 20 m plots, one centred at the den opening and the other four plots 50 m far from the den opening in the cardinal directions [29]. We averaged all variables measured in the five plots, excluding water availability and existence of farmlands that were categorized as a binary factor, to get a general overview of the surroundings of the den and to provide a realistic distribution of the selected variables in den sites. We used hillshade as a surrogate of solar insulation [45]. Hillshade was calculated by combining slope and aspect in the den site and using ArcGIS 9.3 [46]. Hillshade values represent the average amount of shade per year received at any point. Thus, warmer slopes (facing southwest) will receive the greatest hillshade values, whereas cooler northeastern slopes will correspond to the lowest hillshade values. Due to the lack of information on accurate home range size of wolves in the study area, we conservatively selected absence plots to measure the same variables for the fine-scale analysis 1 km away from the den in a random direction (i.e. random points; equal number of points per known den sites), where we were ensured of the absence of wolf dens [29,30]. Out of the 35 den sites detected, fine-scale data sampling was carried out in 32 dens (3 den sites were destroyed before we could measure fine scale variables).

Statistical Analyses

In a first step, we carried out univariate analyses (Mann–Whitney U-tests) testing for significant differences between wolf den areas/sites and non-wolf den areas/sites for all the explanatory variables, excepting for the proportion of den sites with water bodies and farmlands within 100 m radius, where Z-proportions tests were used (Table 1, Table 2). At fine scale, we also used principal component analysis (PCA) to extract orthogonal multivariate axes on fine-scale soil-petrologic variables (Table 2). PCs obtained were used to identify the combination

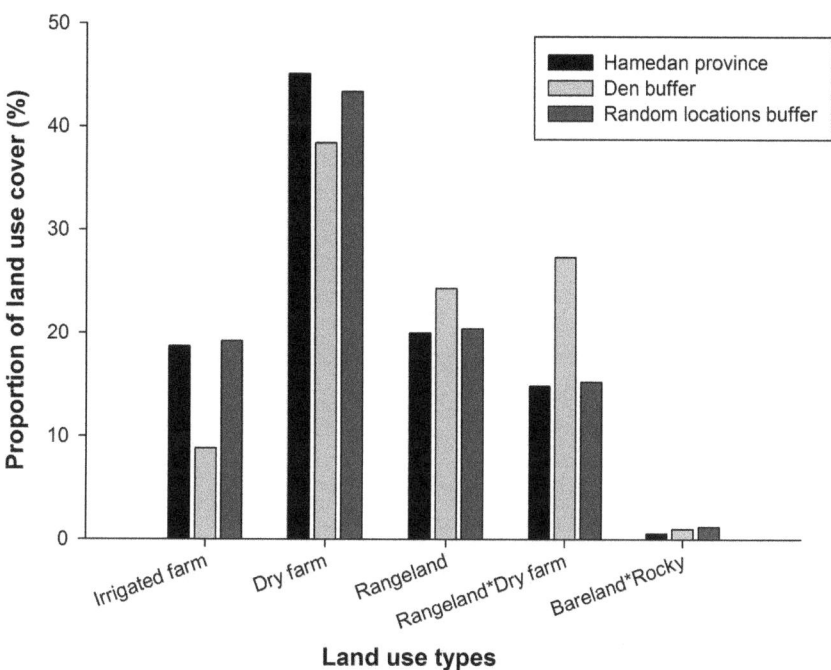

Figure 2. Proportion of land use/land cover categories used in this study. Proportion of each cover type was calculated within 2 km circular buffers around wolf den sites (den areas), random locations (random areas) and proportion of the whole study area (Hamedan province).

Table 1. Mean (SE) values of variables measured at the level of the den area, in 2 km circular buffers with and without wolf dens in Hamedan province, Iran.

Variables (unit)	Abbreviation	Den areas	Random areas	P-value
Dry farms (%)	Dry	39.8 (3.5)	40.3 (3.3)	0.770
Irrigated farms (%)	Irgt	8.5 (2.4)	22.3 (3.2)	0.050
Rangeland (%)	Rng	22.1 (4.4)	19.5 (2.9)	0.259
Rangeland with scattered farms (%)	Rng_Dry	28.3 (4.5)	13.6 (2.3)	0.001
Bareland and Rocks (%)	Bare	0.01 (0.01)	1.3 (0.8)	0.912
Altitude (m)	Alt	2116.0 (20.4)	1999.4 (25.8)	0.000
Roughness (m)	Rough	55.9 (5.1)	49.7 (5.4)	0.022
Length of primary roads (km)	Prim	0.4 (0.2)	1.6 (0.3)	0.010
Length of secondary roads (km)	Scond	2.5 (0.4)	2.2 (0.3)	0.190
Density of settlements (%)	Setl	0.0027 (0.0005)	0.0130 (0.0026)	0.034

Comparisons between den areas and random areas were done by Mann–Whitney U-tests.

of inter-correlated petrologic measurements into organized components that best separate used/unused wolf den sites. We extracted the first two components (PC1soil and PC2soil, Table 2) which explained 73% of soil characteristic variance in measured plots and used them as den site descriptive variables for soil conditions. PC1soil was related to coarse particles of soil and rocks and PC2soil indicated fine soil particles (i.e. optimum areas for cultivation; Table 2).

For both spatial scales, we built separate Generalized Linear Models (GLMs) with binomial error distribution and logit link to assess the influence of human activities on den site selection patterns by wolves in this semi-arid agroecosystem. For each spatial scale, Pearson correlation coefficients were used to test for multicollinearity among predictors, but no significant correlation

between any pair of explanatory variables was detected. At the landscape scale, because of the inherent relationship between topographic contexts with land use, we first examined the possible interactions between elevation and roughness against land use types and length of primary and secondary roads (Table S1), and significant interactions were included in the full model. To do this, we generated a set of additional GLMs containing the pairwise interaction of each land use and type of roads with elevation and roughness (Table S1). We then used the "*anova*" function of the "car" package for R [48] to calculate Likelihood-Ratio χ^2 and Wald χ^2 in order to evaluate the significance level of each interaction. Akaike's Information Criterion corrected for small sample sizes (AICc) [49] was used for model selection and multi-model inference. For each spatial scale, we selected models with

Table 2. Mean (SE) values of fine-scale variables measured in sample plots with and without wolf dens in Hamedan province, Iran.

Variables	Description	Den sites	Random points	P-value
Slope	Measured by a clinometers	15.4 (6)	9 (3.3)	0.000
Hillshade	Measured by a combination of slope and aspect	175.5 (2.11)	177.8 (3.24)	0.234
Herbaceous	Vegetation height less than 25 cm (percentage)	53.3 (20.6)	70.9 (21.4)	0.008
Shrub	Vegetation height between 25 to 200 cm, (percentage)	43.6 (18.3)	25.1 (20.4)	0.003
Tree	Vegetation height above than 200 cm, (percentage)	3.1 (5.3)	3.9 (5.6)	0.406
Soil/Petrology (proportion)	Sable: Particles of clay and sand	35.8 (11.5)	60.6 (7.5)	
	Mm: Soil particle ≤ 1cm	33.8 (7.8)	28.7 (5.3)	
	Cm: Pebbles with size of ≤ 10 cm	15.9 (6.3)	8.4 (3.2)	
	Dc.m: Pebbles with size of ≤ 1 m	9.5 (6.4)	3.2 (2.3)	
	M: Rock with size of ≤ 10 m	5 (6.5)	1.4 (1.5)	
	Dca.m: Rocky materials with size of> 10 m	1.7 (3.5)	0.1 (0.7)	
	PC1 soil: first component of PCA analysis preformed on Soil/Petrology - coarse soil particles -			0.006
	PC2 soil: second component of PCA analysis preformed on Soil/ Petrology -fine soil particles -			0.004
Water availability	Proportion of sites with water bodies within 100 m radius	0.75	0.31	0.001
Farm	Proportion of sites with farmlands within 100 m radius	0.56	0.72	0.283

Comparisons between den sites and random sites were done by Mann–Whitney U-tests excepting for the proportion of presence of water bodies and farmlands within a 100m radius, which were evaluated using Z-proportions tests.

ΔAICc <2, and we calculated Akaike weights (AICc wi) [49]. Moreover, for each predictor selected in the set of models with ΔAICc <2, we calculated its estimated importance (or relative evidence weight), computed as the sum of the relative evidence weights of all models in which the variable appears, as well as model-averaged estimates and their unconditional standard errors (SE). Using this approach we reduced model selection bias effects on regression coefficient estimates in all selected subsets [49]. Finally, to verify how well the selected models described our dataset, we performed a Goodness-of-fit test using Hosmer-Lemeshow (HL) procedure [50]. The Area Under the Curve (AUC) of ROC was also calculated as a measure of discrimination capacity of selected candidate models. All analyses were carried out in R version 3.0.1 [51].

Results

Breeding in agroecosystems

Den areas were located in agricultural matrix with a significantly less proportion of irrigated farms (Mann–Whitney U-test, P <0.05; Table 1) and a higher proportion of mosaics of rangelands with scattered dry-farms than random areas (Mann–Whitney U-test, P <0.001; Table 1). We did not find significant differences between den and random areas for the rest of land uses (Table 1). Wolves tended to select elevated and rough areas (where intensive agricultural practices, such as irrigated farms, are less probable; altitude: Mann–Whitney U-test, P <0.0001; roughness: Mann–Whitney U-test, $P=0.022$; Table 1, Table S1). Finally, as predicted, wolves also avoided areas with abundant primary roads and density of settlements (primary roads: Mann–Whitney U-test, $P=0.017$; settlements: Mann–Whitney U-test, $P=0.016$; Table 1). However, location of den sites was not influenced by the development of the network of secondary roads in the den area (Table 1).

We found a significant interaction between irrigated farms and roughness ($\chi^2 = 6.147$, $P=0.013$; Table S1), and between altitude and secondary roads ($\chi^2 = 3.967$, $P=0.043$; Table S1). Hence these two interactions were included in the set of predictors for the landscape scale models. Seven candidate models showed ΔAICc <2 (Table 3), with the best model including rangelands with scattered dry-farms, altitude, roughness, human settlements, primary roads and the interaction between irrigated farms and roughness (Table 3). The probability of a given area being selected as a den area by wolves in this semi-arid agroecosystem raised with an increase in the proportion of rangelands with scattered dry-farms, located at high altitudes and with low human presence (negative estimation for length of primary roads and density of human settlements; Table 4). Model-averaged coefficient estimates indicated that rangelands with scattered dry farms, altitude, roughness, primary roads and human settlements were the most important predictors determining the probability of a given area being selected as a den area by wolves (Table 4). AUC of ROC curve showed good discrimination capacity of selected candidate models and, we did not find evidence of lack of fit in the different models (HL tests, Table S2).

Fine-scale den site selection patterns in agroecosystems

Wolves were prone to excavate dens in rough hillsides with moderate shrub cover (Mann–Whitney U-test, P <0.05; Table 2). At fine-scale, the strongest significant difference between occupied and unoccupied sites was slope (15.4 ± 6.0 vs. 9.0 ± 3.3; Mann–Whitney U-test, P <0.0001; Table 2). In addition, den sites were characterized by significantly lower percentage of open areas (dominated by herbaceous) as well as higher shrub cover (43.6 ± 18.3 vs. 25.1 ± 20.4; Mann–Whitney U-test, $P=0.003$; Table 2). Water availability was significantly higher in den sites ($Z=3.276$; P <0.001; Table 2) and wolves tended to locate them in areas with a high proportion of coarse soil particles (Mann–Whitney U-test, $P=0.0004$; Table 2). As expected, because the study area was dominated by humans, the presence of farm-lands did not differ between occupied/unoccupied sites at fine scale ($Z=1.073$; $P=0.283$; Table 2). Also, the difference of the amount of shade received at wolf den and random points was not significant (Table 2).

For den sites, eight candidate models showed ΔAICc <2 (Table 5) and the best model included slope, soil/petrologic terms (PC2soil; fine soil particles) and water availability. These three variables were the most important fine-scale predictors of den site selection by wolves based on their relative importance (Table 6). Averaging the coefficient estimates of the selected candidate models revealed that wolves selected for sites with availability of water bodies, placed in stepper hills and with coarse soil particles (Table 6). Based on AUC, we found a very good discrimination capacity for the selected candidate models ranging from 0.915 to 0.933, and no evidence of lack of fit was detected (HL tests, Table S2).

Discussion

Humans are the main source of disturbance for large carnivores affecting, for example, the composition and security of their habitats [52]. Wolf distribution and habitat suitability is mainly influenced by human-associated factors [32]. Such human influence can be both direct (i.e. mortality; legal hunting, poaching, road kills) [32,53] and indirect (behaviour), for example, wild prey depletion or availability of human-related sources of food [15,16]. However, wolves, as many other large carnivores [2,6,8], do not strictly required areas devoid of humans, showing a high ability adapting to multiple used landscapes. This phenomenon is particularly interesting in agroecosystems where virtually all habitats are agricultural and transformed and wild prey can be rare, with wolves usually feeding on livestock, waste or animal carcasses [7,15,18,54].

In agroecosystems, simply avoiding transformed land cover types is impossible, such as the case of western Iran with the almost complete loss of natural habitats (2%) [39]. As a consequence, wolves are relegated to utilize non-natural land cover types while avoiding negative interactions with humans [2,7,18,19]. So, understanding how wolves adjust the use of space in agricultural lands (one of the most widespread habitats worldwide), adapting to human activities, is therefore a critical step to ensure the persistence and conservation of this species in agroecosystems minimizing human-wolf conflicts. This is particularly important since the occurrence of this contentious species in agroecosystems is beyond anecdotic, with several packs occurring, for example, in our study area, as reflected by the number of wolf dens [35] used here [17,18] (see also [55] for a similar scenario in Spain).

Based on the comparison of human land use between den areas, random areas and the whole study area (Hamedan province) we found that the mixture of rangelands with scattered dry-farms (accounting around 15% of the whole study area; Fig. 2) was preferred by denning wolves, whereas irrigated farms were actively avoided and no patterns were found for extensive and homogeneous dry-farms or rangelands (Fig. 2; Figure S2). The proportion of mixed rangelands with dry-farms was the most predictive variable identifying wolf den areas along with a combined preference for hillsides. Two non-exclusive explanations may be behind of this result. By one hand, dry farming practices requires

Table 3. Selected candidate Generalized Linear Models explaining gray wolf den area selection patterns in Hamedan province, Iran, at the landscape level.

Model	AICc	ΔAICc	AIC w_i
Rng_Dry + Alt + Rough + Setl + Prim + (Irgt × Roug)	125.12	0.02	0.18
Rng_Dry + Alt + Rough + Setl + Prim	125.28	0.18	0.16
Rng_Dry + Alt + Rough + Setl + Prim + (Irgt × Roug) + (Alt × Scond)	126	0.90	0.11
Rng_Dry + Alt + Rough + Setl + Prim + (Alt × Scond) + Dry	126.21	1.11	0.10
Rng_Dry + Alt + Rough + Prim + (Irgt × Roug) + (Alt × Scond)	126.45	1.35	0.09
Rng_Dry + Alt + Rough + Setl + Prim + (Irgt × Roug) + Rng	126.81	1.71	0.07
Rng_Dry + Alt + Rough + Setl + Prim + (Irgt × Roug) + Irgt	127.06	1.96	0.06

Models were ranked according to AICc, and only models with ΔAICc <2 are shown for simplicity. For variables description see Table 1.

low levels of human activity, with human presence not overlapping with the most sensitive period for wolves (denning period) because human activity is limited to only the planting and harvesting seasons. On the other hand, rangelands, which can also show a low intensity of human use depending on livestock practices, can also provide wolves with human-related sources of food (e.g. livestock, carrion, waste). Because of the low abundance of wild prey in the area [56] and the use of human-related food sources by wolves in such ecosystems [54,57], traditional herd roaming in rangelands adjacent to dry-farms by local community may favor food availability (higher density of livestock close to farms), affecting den site selection. On the other hand, this scenario (i.e. the presence of scattered dry-farms) may also increase food availability for scarce wild prey. Further analyses are needed to test these hypotheses.

As we expected, wolf den areas were characterized by lesser density of settlements and primary roads compared with random areas [9,58]. The lack of difference between den and random areas in the length of secondary roads suggests that having lesser disturbance from main surrogates of human activity (primary roads and settlements; areas with an intense human land use), secondary roads may be a less important limiting factor for den site selection by wolves. In fact, because secondary roads generally show a lower human use, wolves may use these linear infrastructures for ease of travel within their territories [2,33].

The lack of refuge - considering the well-established link between the concept of refuge and certain vegetation structures providing safe places to wolves such as forests or scrublands [7,34] - in our study area highlights the importance of rouged terrains with low human use providing good concealment for denning wolves in open areas [7,58,59]. Therefore, although wolves selected for den sites located in places with a higher proportion of shrubs compared to random sites in this agroecosystem (Table 2), on a landscape level, vegetation/habitat types becomes a secondary factor for den selection processes, being strongly modulated by the level of human activities.

On a fine spatial scale, our results indicated that wolves primarily excavated dens into the sides of elevated steep-slope hills (Figure S2), selecting sites with steeper slopes, which is consistent with the selection patterns found in other studies (e.g. similar average values for slope, ca. 15 degrees) [29,60]. The slope in these places will also cause more drainage – in case of torrential rain - than surrounding regions that have gentle slope [29,31,47]. Apart from slope, fine soil particles –PC2 soil- (negative selection) and existence of water bodies (positive selection) were the most important variables affecting den site selection patterns. In an

Table 4. Relative importance (W+), model-averaged coefficient estimates (Estimate), and unconditional standard errors (SE) for the predictors included in the selected candidate models determining the probability of a given area being selected as a den area by wolves in Hamedan province, Iran (models with ΔAICc <2).

Variable	W+	Estimate	SE
Intercept		−4.71	3.71
Rng_Dry	1	0.02	0.01
Alt	1	0.002	0.001
Roug	1	0.02	0.01
Setl	1	−0.01	0.05
Prim	1	−0.003	0.002
Irgt × Roug	0.95	0.001	0.001
Alt × Scond	0.83	0.0004	0.002
Irgt	0.35	−0.001	0.002
Rng	0.27	0.02	0.01
Dry	0.10	−0.02	0.01

For variables description see Table 1.

Table 5. Selected candidate Generalized Linear Models explaining gray wolf den site selection patterns in Hamedan province, Iran, at the fine spatial scale.

Model	AICc	ΔAICc	AICc w_i
Slope + PC2 soil + Water	55.31	0	0.17
Slope + PC2 soil + Water + Hillshade	55.88	0.56	0.13
Slope + PC2 soil + Water + Herbaceous + Shrub	56.33	1.02	0.10
Slope + PC2 soil + Water + Hillshade + PC1 soil	56.36	1.05	0.10
Slope + PC2 soil + Water + PC1 soil	56.45	1.14	0.10
Slope + PC2 soil + Water + Hillshade + Shrub	56.58	1.27	0.09
Slope + PC2 soil + Water + Shrub	56.66	1.35	0.09
Slope + Water + Herbaceous + Shrub	57.25	1. 94	0.06

Models were ranked according to AICc, and only models with ΔAICc <2 are shown for simplicity. For variables description see Table 2.

unusual pattern, we found that the existence of farmlands did not affect selection patterns by denning wolves [18]. High tendency of local communities to place dry-farms in areas with unsuitable topographic conditions for other cultivation practices may also explains why many dens were located in the vicinity of farmlands. We found a significant difference between den and random sites (that were often located within agricultural lands) in terms of soil variables. Most of the areas with a gentle slope and rich soil (PC2 soil) are used for farming by local people. Accordingly, rangelands adjacent to farms are less usable for agriculture and wolves were forced to den in places with coarse soil particles. Finally, we found that the availability of water bodies in the vicinity of den sites is an important factor for denning wolves. As expected, due to high water requirement of lactating females, den sites were selected relatively close to water sources [28,29,30,60]. In semi-arid landscapes, we predict that the dependency of both, denning wolves and humans, to scarce water bodies may have increased human-wolf conflict locally, being an important limiting factor for the persistence of the species.

Our findings at different spatial scales show how wolves can be tolerant to placing their dens in agricultural lands, which demonstrates their resilience to persist in agroecosystems. As agricultural lands dominated this landscape, wolves selected for den areas with low human use irrespective whether such areas were profoundly transformed or not. In our case, this is possible because small dry-farms adjacent to rangelands require minimum human intervention, consequently having a low impact on habitat security and decreasing the risk of mortality for wolves during the breeding period. Thus, spatial and seasonal heterogeneity in human activities become an important factor explaining the persistence of wolves in agroecosystems [61].

As in other regions of the Middle East, agricultural activities in Hamedan province started more than 5000 years ago [62]. Moreover, contrary to European and North American wolf ranges [63] where wolves were exterminated from huge areas during the 19th and 20th centuries [24,32,65], and only began to recolonize some of their former range in recent times [32], such pattern of eradication/re-colonization did not occur in Iran, with wolves persisting in this area continuously over time. Thus, here wolves and human activities have been interacting for a much longer period of time than in other parts of the current and historical wolf range leading to a unique scenario of wolf adaptations to humans.

Effective large carnivore conservation in human-dominated landscape matrix and outside of formally protected areas is of paramount importance in the Anthropocene [64,65]. Successful conservation strategies requires minimizing conflicts between large carnivores and humans, understanding where and when to establish limits of sharing the landscape with these contentious species. Alternatives range from a focus on fencing large carnivores to allowing them to share the landscape with humans (e.g. [66]). However, this debate also requires determining to what extent large carnivores can tolerate living in human-dominated landscapes

Table 6. Relative importance (W+), model-averaged coefficient estimates (Estimate), and unconditional standard errors (SE) for the predictors included in the selected candidate models determining the probability of a given site being selected as a den site by wolves in Hamedan province, Iran (models with ΔAICc<2).

Variable	W+	Estimate	SE
Intercept		−6.68	8.09
Slope	1	0.31	0.09
Water	1	3.10	1.04
PC2 soil	0.94	−0.87	0.45
Hillshade	0.48	−0.03	0.02
Shrub	0.39	0.11	0.10
Herbaceous	0.29	0.14	0.09
PC1 soil	0.20	0.29	0.25

For variables description see Table 2.

considering different spatial and ecological constraints and levels of conflict. Along these lines, our results show how the heterogeneity in human activities emerges as a key factor favoring the persistence of wolves in agroecosystems. Thus, vulnerability of wolves, and other large carnivore species, in human-dominated landscapes could be compensated by the existence of spatial heterogeneity in human activities, favoring a land sharing model of coexistence between large carnivores and people.

However, despite the ability of wolves to persist in agroecosystems, with much of the landscape being devoted to agricultural and livestock activities, human-wolf encounters and conflicts can also increase. As a consequence, because of the high accessibility to wolf dens by people in agroecosystems, lactating wolves and their pups can be very vulnerable to active illegal human persecution [35]. Since wolf core use areas, including den areas, are used by wolf packs more intensively throughout the year and wolves are even prone to use the same den in subsequent years [29,47], there is a pressing need to adopt efficient measures to mitigate human-wolf conflicts in agroecosystems (e.g. discouraging people from destroying wolf dens, changing human behaviors and livestock practices) in order to keep acceptable levels of tolerance and favoring wolf persistence.

Supporting Information

Figure S1 General views of the agroecosystems of Hamedan province, Iran.

Figure S2 Fine-scale pictures showing the environment around den sites in rangelands of Hamedan province, Iran.

Table S1 Results from Generalized Linear Models testing for significant effects of the pairwise interactions between land use and type of roads with elevation and roughness.

Table S2 Results of the assessment of goodness-of-fit and discrimination capacity of selected candidate models explaining the selection of den areas/sites by wolves in Hamedan, Iran.

Acknowledgments

We would like to thank the many local people and the staff of Hamedan DoE who helped us with the field work. Our special thanks go to the wardens who were a great help in finding and surveying dens as well as Elham Nourani for her helps. JVLB was supported by a "Juan de la Cierva" research contract from the Spanish Ministry of Economy and Competitiveness.

Author Contributions

Conceived and designed the experiments: MA JVLB MK. Performed the experiments: MA MK. Analyzed the data: MA JVLB MK. Contributed reagents/materials/analysis tools: MA MK. Wrote the paper: MA JVLB MK.

References

1. Bateman PW, Fleming PA (2012) Big city life: Carnivores in urban environments. J Zool 287: 1–23.
2. Dellinger JA, Proctor C, Steury TD, Kelly MJ, Vaughan MR (2013) Habitat selection of a large carnivore, the red wolf, in a human-altered landscape. Biol Conserv 157: 324–330.
3. Gittleman JL, Funk SM, Macdonald D, Wayne RK (2001) Carnivore conservation. Cambridge University Press, Cambridge, UK.
4. Woodroffe R, Thirgood S, Rabinowitz A (2005) People and wildlife, conflict or co-existence? Cambridge University Press, UK.
5. Treves A, Karanth KU (2003) Human-carnivore conflict and perspectives on carnivore management worldwide. Conserv Biol 17: 1491–1499.
6. Woodroffe R (2011) Ranging behaviour of African wild dog packs in a human-dominated landscape. J Zool 283: 88–97.
7. Llaneza L, Lopez-Bao JV, Sazatornil V (2012) Insights into wolf presence in human dominated landscapes: the relative role of food availability, humans and landscape attributes. Divers Distrib 18: 459–469.
8. Athreya V, Odden M, Linnell JDC, Krishnaswamy J, Karanth U (2013) Big cats in our backyards: persistence of large carnivores in a human dominated landscape in India. PLoS One 8: e57872. doi:10.1371/journal.pone.0057872.
9. Theuerkauf J, Rouys S, Jędrzejewski W (2003) Selection of den, rendezvous, and resting sites by wolves in the Białowieza Forest, Poland. Can J Zool 81: 163–167.
10. Blanco JC, Cortés Y (2007) Dispersal patterns, social structure and mortality of wolves living in agricultural habitats in Spain. J Zool 273: 114–124.
11. Eggermann J, da Costa GF, Guerra AM, Kirchner WH, Petrucci-Fonseca F (2011) Presence of Iberian wolf (Canis lupus signatus) in relation to land cover, livestock and human influence in Portugal. Mamm Biol 76: 217–221.
12. Chavez AS, Gese EM (2006) Landscape use and movements of wolves in relation to livestock in a wildland-agriculture matrix. J Wildl Manage 70: 1079–1086.
13. Boitani L (2003) Wolf conservation and recovery. In: Mech LD, Boitani L, editors. Wolves, behavior, ecology and conservation. The University of Chicago Press, Chicago and London. 317–340.
14. Latham ADM, Latham MC, Boyce MS, Boutin S (2011) Movement responses by wolves to industrial linear features and their effect on woodland caribou in northeastern Alberta. Ecol Appl 21: 2854–2865.
15. Meriggi A, Lovari S (1996) A review of wolf predation in southern Europe: does the wolf prefer wild prey to livestock? J Appl Ecol 33: 1561–1571.
16. López-Bao JV, Sazatornil V, Llaneza L, Rodríguez A (2013) Indirect effects on heathland conservation and wolf persistence of contradictory policies that threaten traditional free-ranging horse husbandry. Conserv Lett 6: 448–455.
17. Habib B, Kumar S (2007) Den shifting by wolves in semi-wild landscapes in the Deccan Plateau, Maharashtra, India. J Zool 272: 259–265.

18. Agarwala M, Kumar S (2009) Wolves in agricultural landscapes in Western India. Tropical Resources: Bulletin of the Yale Tropical Resources Institute 28: 48–53.
19. Mech LD (2006) Prediction failure of a wolf landscape model. Wildl Soc Bull 34: 874–877.
20. Conway GR (1987) The properties of agroecosystems. Agricult Sys 24: 95–117.
21. Blanco JC, Cortés Y, Virgós E (2005) Wolf response to two kinds of barriers in an agricultural habitat in Spain. Can J Zool 83: 312–323.
22. Falcucci A, Maiorano L, Tempio G, Boitani L, Ciucci P (2013) Modeling the potential distribution for a range-expanding species: Wolf recolonization of the Alpine range. Biol Conserv 158: 63–72.
23. Tilman D, Balzer C, Hill J, Befort BL (2011) Global food demand and the sustainable intensification of agriculture. PNAS 108: 20260–20264.
24. Boitani L (2000) Action plan for the conservation of wolves (Canis lupus) in Europe. Council of Europe Publishing, Strasbourg, France.
25. Harrington FH, Mech LD (1982) Patterns of home-site attendance in two Minnesota wolf packs. In: Harrington FH, Paquet PC, editors. Wolves of the world: perspectives of behavior, ecology, and conservation. Noyes Publications, New Jersey. 81–104.
26. Frame PF, Cluff HD, Hik DS (2007) Response of Wolves to Experimental Disturbance at Homesites. J Wildl Manage 71: 316–320.
27. Frame PF, Cluff HD, Hik DS (2008) Wolf reproduction in response to caribou migration and industrial development on the Central Barrens of mainland Canada. Arctic 61: 134–142.
28. Norris DF, Theberge MT, Theberge JB (2002) Forest composition around wolf (Canis lupus) dens in eastern Algonquin Provincial Park, Ontario. Can J Zool 80: 866–872.
29. Trapp JR, Beier P, Mack C, Parsons DR, Paquet PC (2008) Wolf, Canis lupus, den site selection in the Rocky Mountains. Can Field Nat 122: 49–56.
30. Person DK, Russell AL (2009) Reproduction and den site selection by wolves in a disturbed landscape. Northwest Sci 83: 211–224.
31. Unger DE, Keenlance PW, Kohn BE, Anderson EM (2009) Factors Influencing Home site Selection by gray wolves in Northwestern Wisconsin and East-Central Minnesota. In: Wydeven AP et al., editors. Recovery of gray wolves in the Great Lakes Region of the United States. Springer Science + Business Media. 175–189, doi: 10.1007/978-0-387-85952-1-11.
32. Mech LD, Boitani L (2003) Wolves: Behavior, Ecology and Conservation. Chicago, University of Chicago Press.
33. Jędrzejewski W, Niedziałkowska M, Nowak S, Jędrzejewska B (2004) Habitat variables associated with wolf (Canis lupus) distribution and abundance in northern Poland. Divers Distrib 10: 225–233.

34. Jędrzejewski W, Jędrzejewska B, Zawadzka B, Borowik T, Nowak S, et al. (2008) Habitat suitability model for Polish wolves based on long-term national census. Anim Conserv 11: 377–390.

35. Ahmadi M, Kaboli M, Nourani E, Alizadeh Shabani A, Ashrafi S (2013) A predictive spatial model for gray wolf (*Canis lupus*) denning sites in a human-dominated landscape in western Iran. Ecol Res 28: 513–521.

36. Ziaie H (2008) A field guide to mammals of Iran. 2nd ed. Iranian Wildlife Center, Tehran (in Persian).

37. Behdarvand N, Kaboli M, Ahmadi M, Nourani E, Salman Mahini A, et al. (2014) Spatial risk model and mitigation implications for wolf–human conflict in a highly modified agroecosystem in western Iran. Biol Conserv 177: 156–164.

38. Reyahi-Khoram M, Fotros MH (2011) Land use planning of Hamadan province by means of GIS. International conference on chemical, biological and environment sciences (ICCEBS, 2011) Bangkok.

39. Imani Harsini J (2012) Study on change detection of land use/cover in Hamedan province considering wolves potential habitats during the past 30 years. M.Sc. dissertation. Department of Environmental Sciences, University of Tehran.

40. Safikhani K, Rahiminejhad MR, Kalvandi R (2007) Presentation of flora and life forms of plant species in Kian region (Hamadan province). Watershed Management Research Journal 74: 138–154 (in Persian).

41. McLoughlin PD, Walton LR, Cluff HD, Paquet PC, Ramsay MA (2004) Hierarchical habitat selection by tundra wolves. J Mammal 85: 576–580.

42. Iliopoulos Y, Youlatos D, Sgardelis S (2014) Wolf pack rendezvous site selection in Greece is mainly affected by anthropogenic landscape features. Eur J Wildl Res 60: 23–34.

43. Forest, Range and Watershed Management Organization I.R. of Iran, FRWMO 2010. Iranian Forests, Range and Watershed Management Organization National Land use/Land cover map.

44. Glenz C, Massolo D, Kuonen D, Schlaepfer R (2001) A wolf habitat suitability prediction study in Valais (Switzerland). Landsc Urban Plan 55: 55–65.

45. Ciarniello LM, Boyce MS, Heard DC, Seip DR (2005) Denning behavior and den site selection of grizzly bears along the Parsnip River, British Columbia, Canada. Ursus 16: 47–58.

46. ESRI (2010) ArcGis 9.3. Environmental Systems Research Institute. Redlands, CA.

47. Ballard WB, Dau JR (1983) Characteristics of gray wolf, *Canis lupus*, den and rendezvous sites in Southcentral Alaska. Can Field Nat 97: 299–302.

48. Fox J, Weisberg S (2011) An R Companion to Applied Regression, Second Edition. Thousand Oaks CA: Sage. Available: http://socserv.socsci.mcmaster.ca/jfox/Books/Companion.

49. Burnham KP, Anderson DR (2002) Model selection and inference: a practical information theoretic approach. Springer-Verlag, New York, New York, USA.

50. Hosmer DW, Lemeshow S (2000) Applied logistic regression. Wiley Series in Probability and Statistics. John Wiley and Sons, New York, USA.

51. R Core Team (2013) R: A Language and Environment for Statistical Computing. R Foundation for Statistical Computing, Vienna, Austria. http://www.R-project.org/

52. Weaver JL, Paquet PC, Ruggiero LF (1996) Resilience and conservation of large carnivores in the Rocky Mountains. Conserv Biol 10: 964–976.

53. Liberg O, Chapron G, Wabakken P, Pedersen HC, Hobbs NT, et al. (2012) Shoot, shovel and shut up: cryptic poaching slows restoration of a large carnivore in Europe. Proc Roy Soc Lond B Biol 279: 910–915.

54. Tourani M, Moqanaki EM, Boitani L, Ciucci P (2014) Anthropogenic effects on the feeding habits of wolves in an altered arid landscape of central Iran. Mammalia 78: 117–121.

55. Blanco JC, Cortés Y (2002) Ecología, censos, percepción y evolución del lobo en España: análisis de un conflicto. SECEM, Málaga.176 pp.

56. Darvishsefat AA (2006) Atlas of Protected Areas of Iran (English-Persian), University of Tehran Press, Tehran.

57. Hosseini-Zavarei F, Farhadinia MS, Beheshti-Zavareh M, Abdoli A (2013) Predation by grey wolf on wild ungulates and livestock in central Iran. J Zool 290: 127–134.

58. Capitani C, Mattioli L, Avanzinelli E, Gazzola A, Lamberti P, et al. (2006) Selection of rendezvous sites and reuse of pup raising areas among wolves *Canis lupus* of northeastern Apennines, Italy. Acta Theriol 51: 395–404.

59. Corsi F, Dupre E, Boitani L (1999) A large-scale model of wolf distribution in Italy for conservation planning. Conserv Biol 13: 150–159.

60. Unger DE (1999) A multi-scale analysis of timber wolf den and rendezvous site selection in northwestern Wisconsin and east-central Minnesota. M.Sc. dissertation, University of Wisconsin.

61. Schuette P, Wagner AP, Wagner ME, Creel S (2013) Occupancy patterns and niche partitioning within a diverse carnivore community exposed to anthropogenic pressures. Biol Conserv 158: 301–312.

62. Farshad A, Barrera-Bassols N (2003) Historical anthropogenic land degradation related to agricultural systems: case studies from Iran and Mexico. Geogr Ann A 85: 277–286.

63. Young SP, Goldman EA (1944) The wolves of North America. Dover, New York, USA.

64. Wikramanayake E, McKnight M, Dinerstein E, Joshi A, Gurung B, et al. (2004) Designing a conservation landscape for tigers in human-dominated environments. Conserv Biol 18: 839–844.

65. Muntifering JR, Dickman AJ, Perlow LM, Hruska T, Ryan PG, et al. (2006) Managing the matrix for large carnivores: a novel approach and perspective from cheetah (*Acinonyx jubatus*) habitat suitability modelling. Anim Conserv 9: 103–112.

66. Packer C, Loveridge A, Canney S, Caro T, Garnett ST, et al. (2013) Conserving large carnivores: dollars and fence. Ecol Lett 16: 635–641.

3

A Process-Based Approach to Predicting the Effect of Climate Change on the Distribution of an Invasive Allergenic Plant in Europe

Jonathan Storkey[1]*, Pierre Stratonovitch[2], Daniel S. Chapman[3], Francesco Vidotto[4], Mikhail A. Semenov[2]

1 AgroEcology Department, Rothamsted Research, Harpenden, Hertfordshire, United Kingdom, 2 Computational and Systems Biology Department, Rothamsted Research, Harpenden, Hertfordshire, United Kingdom, 3 Centre for Ecology and Hydrology, Edinburgh, United Kingdom, 4 University of Turin, Grugliasco, Italy

Abstract

Ambrosia artemisiifolia is an invasive weed in Europe with highly allergenic pollen. Populations are currently well established and cause significant health problems in the French Rhône valley, Austria, Hungary and Croatia but transient or casual introduced populations are also found in more Northern and Eastern European countries. A process-based model of weed growth, competition and population dynamics was used to predict the future potential for range expansion of *A.artemisiifolia* under climate change scenarios. The model predicted a northward shift in the available climatic niche for populations to establish and persist, creating a risk of increased health problems in countries including the UK and Denmark. This was accompanied by an increase in relative pollen production at the northern edge of its range. The southern European limit for *A.artemisiifolia* was not expected to change; populations continued to be limited by drought stress in Spain and Southern Italy. The process-based approach to modelling the impact of climate change on plant populations has the advantage over correlative species distribution models of being able to capture interactions of climate, land use and plant competition at the local scale. However, for this potential to be fully realised, additional empirical data are required on competitive dynamics of *A.artemisiifolia* in different crops and ruderal plant communities and its capacity to adapt to local conditions.

Editor: Bruno Hérault, Cirad, France

Funding: The research leading to these results has received funding from the European Union's Seventh Framework Programme (FP7/2007–2013) under grant agreements n°282687 – Atopica. http://cordis.europa.eu/fp7. The funders had no role in study design, data collection and analysis, decision to publish, or preparation of the manuscript.

Competing Interests: The authors have declared that no competing interests exist.

* E-mail: jonathan.storkey@rothamsted.ac.uk

Introduction

Climate change may impact the severity of pollen induced atopic disease by affecting the large scale distribution and local prevalence of allergenic species, the timing and amount of pollen produced and the allergenicity of individual pollen grains. A species of particular concern in Europe is *Ambrosia artemisiifolia* L. (common ragweed), an alien plant in Europe that has expanded its range over recent decades. It is now responsible for significant health and economic impacts in the most infected areas namely a) in the Pannonian Plain in Central Europe including Hungary and neighbouring countries especially Serbia, Croatia, Slovenia, Slovakia and Romania [1,2,3,4], b) in the Rhône Valley in France [5] and c) in Western Lombardy, Italy [6]. *A.artemisiifolia* is an annual plant with origins in North America. Although it was first observed in Europe in the mid 19th Century, it began to spread rapidly in Europe after 1940 via transportation networks and contaminated crop seed [7]. It is highly invasive with allergenic pollen that causes hay fever, asthma and atopic dermatitis [8]. Once established in a country, control measures are labour intensive and expensive [9] and there are benefits to anticipating the potential future distribution and impact of the species under climate change to inform surveillance of regions that are vulnerable to populations establishing.

The conventional approach to predicting changes in plant distribution under climate change has been to develop species distribution models (SDMs) based on the current range of a species characterised by their habitat requirements. Climate change scenarios can then be applied to these models to predict the change in the bioclimatic envelope and the resulting shift in community composition [10]. A number of habitat models have been developed for *A.artemisiifolia* that also incorporate dispersion dynamics [11,12], these models are valuable for predicting future distribution based on physiological thresholds and rate of spread. A more mechanistic phenological approach was also recently developed by Chapman *et al.* [13] who predicted the future distribution of *A.artemisiifolia* based on the likelihood of different stages of the life cycle being completed. However, as well as being constrained by physiological tolerance thresholds, the probability of a species establishing in a locality will be determined by other factors such local soil properties and management and biotic interactions including competition with crops and the native flora [14,15]. These additional drivers of community assembly effectively mean that the realised niche for a species may be considerably smaller than its fundamental niche determined by physiological tolerances. Local environment and management factors are particularly relevant for modelling the fitness of populations of a ruderal species such as *A.artemisiifolia* that is adapted to managed habitats such as cropped fields and can be rapidly out competed by native species [16]. Therefore, while SDMs are a useful first approximation, ideally, process-based models that can operate on a smaller scale are required to predict the effects of local resource heterogeneity, land management and biotic interactions on plant community assembly [17]. In the context of predicting the future impact of allergenic plant species, process-based models also have the advantage of quantifying the response of plant growth and, therefore, pollen production to changes in climate or management.

A process-based model of plant growth, competition and population dynamics, Sirius 2010, has previously been developed to predict the impact of climate change on the damage niche of an agricultural weed, *Alopecurus myosuroides* (Huds.) [18]. The model predictions highlighted the importance of considering local environmental conditions as the response of the weed to climate change was confounded by variation in soil properties. Sirius 2010 has a modular structure and the algorithms which describe plant growth and development are generic. Assuming sufficient empirical data are available to parameterise and calibrate the model, therefore, it can be used to simulate the response of any annual plant species to climatic and environmental variation. In this paper, we present an adaptation of Sirius 2010 for *A. artemisiifolia*, growing in monoculture stands, and predict the potential for range expansion under climate change based on the population growth rate (λ) calculated across Europe. Although, in reality, *A. artemisiifolia* will be found growing in plant communities (either alongside transport routes or in crops), our aim in this first iteration of the model was to quantify the potential niche given the most favourable growing conditions. The model is also used to predict the relative change in pollen production under climate change scenarios.

Materials and Methods

Modelling *A. artemisiifolia* growth, development and population dynamics

Sirius 2010 is a plant growth simulation model capable of modelling inter-plant competition for light, water and nutrients based on a wheat growth model [19,20] and using functions for competition for resources from the INTERCOM model of crop-weed competition [21,22]. A mechanistic approach is taken to modelling the transition between life stages which captures the interactions of weather and management on biological processes. Plant growth and development is simulated on a daily time step and plants described as a collection of organs (roots, leaves, stems, flowers and seeds). The model has the capability to simulate competition for resources between multiple species in the canopy; for the initial model runs presented here, however, *A. artemisiifolia* is assumed to be growing in a single species stand. The growth of individual organs is predicted from resource demand, calculated as a function of developmental stage and potential growth rate under unlimiting conditions, and resource supply. The equations describing resource capture and the impact of limiting factors on resource use efficiency (CO_2, temperature and water) are described in detail in Stratonovitch *et al.* [18]. Here, we focus on the modifications made to the model to simulate specific aspects of the biology of *A. artemisiifolia* and sources of data for parameterisation.

A. artemisiifolia is a frost sensitive, summer annual that is adapted to avoid emergence in the autumn. This is achieved by seeds having primary dormancy at the time of shedding and requiring a period of chilling to break the dormant state. The optimum temperature for breaking dormancy has been deter-

mined experimentally as 4°C; slower dormancy release was observed at temperatures above 5°C or below 0°C [23]. In the model, a period of 12 weeks is required at this temperature before seeds are able to germinate. Sub-optimal temperatures are also effective but take longer. A function describing the rate of dormancy release (d^{-1}) at different temperatures based on the results of Willemsen [23] was included in Sirius 2010. Once the threshold for the breaking of dormancy has been reached, the rate of germination was modelled as a Weibull function using hydrothermal time that integrates the effect of temperature and soil moisture (Equation 1):

$$Y = M(1\text{-}exp[\text{-}k(\theta HT \text{ - } a)^c]) \tag{1}$$

Where Y is cumulative germination (%) at a hydrothermal time (θ_{HT}), M is maximum germination, k is rate of increase, a is lag phase and c is shape parameter.

The calculation of θ_{HT} requires parameters for the base temperature and moisture content below which θ_{HT} does not accumulate. These values have been determined experimentally as 3.6°C and −0.8 mPa respectively [24]. The proportion of seeds emerging from the seed-bank was modelled as a stochastic function with a left skewed distribution and a median of 26%. This was calculated as a generic distribution using data from multiple weed species across several sites and years.

Phenological development of *A. artemisiifolia* is primarily driven by temperature [25] with estimated base and optimum temperatures of 0.9°C and 31.7°C [26]. Previous authors have also found a relationship with photoperiod; *A. artemisiifolia* is characterised as a short daylength plant and flowering is predicted to be delayed when daylength exceeds 14.5 hours [13]. These functions predict a positive correlation between the onset of flowering, measured in Julian days, and latitude. However, data from the US on *A. artemisiifolia* flowering times [27] and from Europe on pollen production do not support this prediction and imply that regional biotypes are adapted to local conditions, synchronising flowering time. Our model, therefore, used a standard calendar date for the onset of flowering (1st of August). *A. artemisiifolia* is frost sensitive and the end of flowering has been found to be correlated spatially and temporally with the onset of the first frost [27]. We imposed a number of rules that simulated the end of flowering and plant senescence based on physiological thresholds and resource availability: 1) threshold of accumulated thermal time reached; 2) low temperature thresholds, i.e. daily minimum temperature is below 0°C, or 5-day mean minimum temperature is below 7°C; 3) severe drought, i.e. 10-day mean of the ratio between actual and potential transpiration is below 0.1. These rules terminate the flowering season and therefore stop pollen and seed production. They have been parameterized to reproduce observed end of pollen seasons for a range of latitudes in North America and Europe. From germination, the plant is killed by frost (daily min air temperature ≤0°C) and drought when it transpires less than 10% of the potential transpiration over 10 days. The relationship between thermal time and transition between developmental stages, with the associated shift in allocation of resources between plant organs, was based on empirical data from field experiments in Michigan, US [28]. *A. artemisiifolia* is a monoecious plant (having the female and male reproductive organs separated in different floral structures on the same plant). The relationship between plant mature biomass and pollen and seed production is allometric and has been quantified from empirical data [29]. Losses of fresh seed from predation were modelled as a stochastic function with a normal distribution (mean 0.9 and standard deviation 0.1) and decline of the old seedbank as an exponential function with a half-life of 17 years [30]. The model was used to calculate the fitness of a local population using a measure of the population growth rate, λ, calculated as seed number at time t_{+1}/seed number at time t_0.

The data for initial parameterisation of the model were largely derived from studies in the US. It was important, therefore, to calibrate the model for European populations of *A. artemisiifolia*. Growth analysis data from two field experiments in 2006 and 2007 at Grugliasco (Turin, Italy: 45°03′53″N, 7°35′38″E) sampling monoculture stands of *A. artemisiifolia* were, therefore, used in the final model development. The model was fitted to observed data for emergence date, flowering time, height and biomass allocation to different plant organs.

Local scale climate scenarios and modelling European distribution

To model the European distribution of *A. artemisiifolia*, Sirius 2010 was used to evaluate the population growth rate (λ) as a function of seed germination rates, production of new seed and seed losses driven by local climate variables. From the ELPIS dataset [31], 479 locations were randomly selected with a minimum neighbouring distance of 100 km. For each location, 100 years of daily synthetic weather were generated using LARS-WG [32]. The asymptotic population growth rate (Λ) was calculated as the mean of $\log(\lambda)$ over 1000 yearly simulations, each using a random combination of weather data generated for one growing season and one germination and seed loss rate. Two additional statistics were also computed: end of pollen season and seasonal pollen production. From the 479 random locations, a subset of 25 sites across Europe was used to categorise Λ values into a suitability index: U.0 – highly unsuitable, U.1 – unsuitable, C.0 – casual (less likely), C.1 – casual, E.0 – established, E.1 – well established. Ranges for these categories were derived by comparing generated Λ in baseline climatic conditions at these 25 sites to the level of abundance of *A. artemisiifolia* from the map of current distribution.

Results

The observed data of the start of the *A. artemisiifolia* pollen season only covered a relatively small latitudinal gradient and the correlation between latitude and the onset of flowering expected from a previous phenological model was not observed (Figure 1). Although it would be instructive to perform a similar analysis with data on flowering times across a wider latitudinal gradient in Europe, our results agree with data on *Ambrosia spp.* flowering times from the US where the onset of flowering could largely be predicted on the basis of calendar day suggesting local populations may have adapted to synchronise development [27]. The observed data from European pollen monitoring stations had high year to year variability but predicted values for the end of pollen production, using the three rules, generally fell within the range of observed values (Figure 1).

The predicted distribution defined using the six categories of suitability from 'unsuitable' to 'well established' using baseline climatic data corresponded reasonably well with the latest observed distribution maps (Figure 2). Centres of high likelihood of occurrence were identified as the Rhône valley in France, an area in and around Hungary and Switzerland and the Netherlands. Although giving an indication of the validity of the model output, Figures 1a and 1b cannot be directly compared statistically, because they represent different quantities. Figure 1a shows observed grids or sites where ambrosia has been detected and, for the majority of grids/sites, there is no information

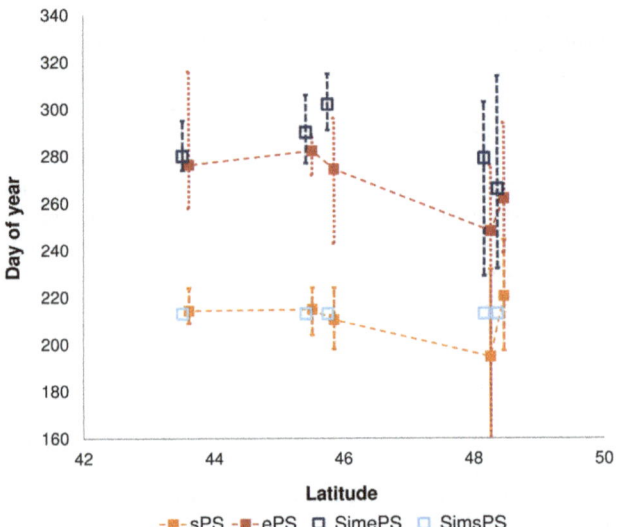

Figure 1. Mean values for observed start (sPS) and end (ePS) of pollen season of *A.artemisiifolia*, from up to 20 years of data, along a latitudinal gradient and start and end simulated by the process-based model (SimsPS, SimePS). Error bars represent maximum and minimum observed and simulated values.

provided on whether the population is established or casual. In addition, because the invasion event is on-going, the observed distribution is expected to change over coming decades whereas Figure 2b predicts the potential distribution once the invasion is complete. It is also likely that the distribution map in Figure 1a is biased by sampling effort; for example, it is likely that populations in the South-eastern corner of the maps in Ukraine may have been underrepresented [6].

Here we assume that there are no barriers to dispersal and all areas have suitable land use (ruderal habitats, especially spring sown annual crops) for *A.artemisiifolia* to persist. For some regions lacking suitable habitat the map of potential distribution predicts established populations of *A.artemisiifolia* where it has not been observed; for example, the south west of England and Southern Ireland are dominated by permanent grassland. Incorporating land use factors at a finer scale would further refine the predictions but will not affect the predicted northern, southern and eastern limits of its range. These correspond well with reported data on the extent of the range of *A.artemisiifolia* with isolated casual populations extending as far as south as northern Spain and Italy and north up to the UK and Sweden [33,34]. The model predicted the southern limit to be driven by moisture stress and the northern limit by insufficient thermal time for seed to mature. Importantly, there were no areas where *A.artemisiifolia* is currently an established problem and where the model predicts unsustainable populations with the possible exception of Russia and Ukraine where data quality on observed populations used to generate the distribution map is poor [35].

Under the HadCM3(A1B) scenarios for 2010–2030 and 2050–2070, the southern limit of the range of *A.artemisiifolia* was predicted to change very little, as future rainfall patterns meant most of Spain, southern Italy and Greece remained too dry for *A.artemisiifolia* to maintain viable populations (Figure 3). However, the available suitable habitat for *A.artemisiifolia* was predicted to extend further north and east under climate change as increasing summer temperatures resulted in a faster accumulation of thermal time and a higher probability of the plant

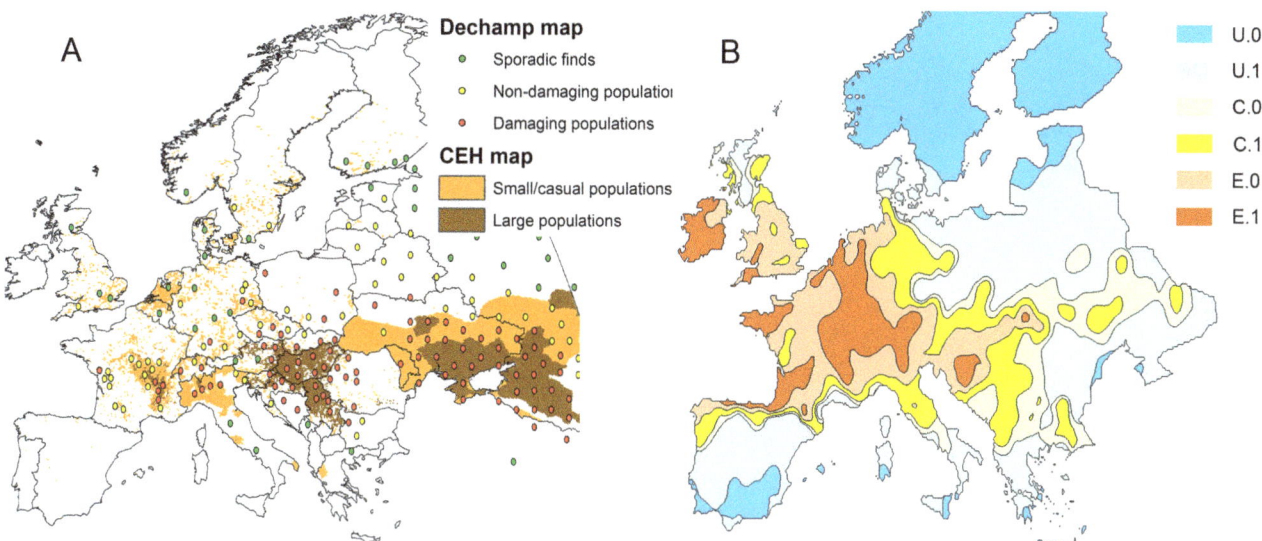

Figure 2. Potential climatic niche of *Ambrosia artemisiifolia* **under baseline conditions compared to observed distribution.** A. Combination of two recent maps showing the approximate distribution of *A.artemisiifolia* in Europe compiled by Déchamp [44] and Bullock et al at the Centre for Ecology and Hydrology (CEH) [35]. The quality of observed data used to generate the maps varies with good quality data obtained from northern, central and western Europe and poorer quality data from Russia and Ukraine. B. Predicted potential range of *A.artemisiifolia* from output of process based model assuming that appropriate habitat in terms of land use is available at all sites. Actual distribution will be modified by cropping patterns and level of control. The categories are: U.0 - highly unsuitable, U.1 - unsuitable, C.0 - casual (less likely), C.1 - casual, E.0 - established, E.1 - well established.

completing its life cycle and producing mature seed. Under these scenarios, it is likely that populations in Scandinavian countries and Britain that are currently casual may become established (Figure 3). Where *A.artemisiifolia* is currently well established and found at high densities, in Hungary, Croatia and the French Rhône Valley, the model predicted only a small increase in the capacity for increased pollen production under climate change (Figure 4). However, at the Northern extent of the range relatively large increases in pollen production were predicted potentially

creating a health problem where *Ambrosia* pollen concentrations are currently below the threshold for inducing clinical symptoms.

Discussion

The model output predicted a wider European distribution of *A.artemisiifolia* than is currently observed. Although, the actual distribution will be further modified by land use factors (see below), it is likely that the species has not yet filled the potential climatic space as the spread of *A.artemisiifolia* through Europe represents

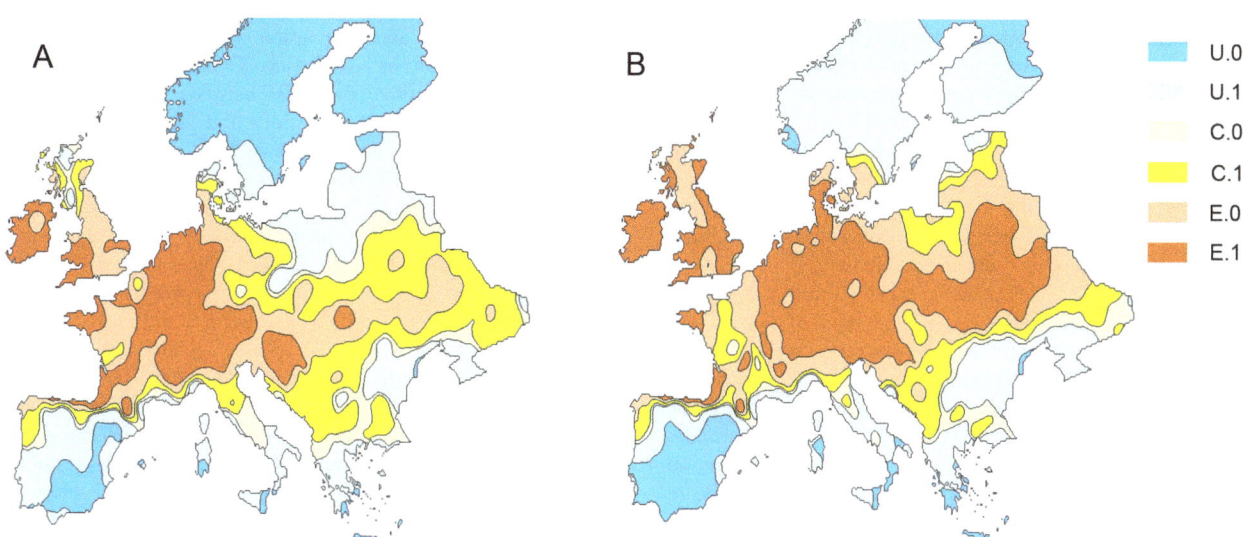

Figure 3. Distribution of *Ambrosia artemisiifolia* **(common ragweed) in Europe under climate change as predicted by the process based model.** A. Using HadCM3(A1B) scenarios for near future 2010–2030 and B. long-term future 2050–2070. The categories are: U.0 - highly unsuitable, U.1 - unsuitable, C.0 - casual (less likely), C.1 - casual, E.0 - established, E.1 - well established.

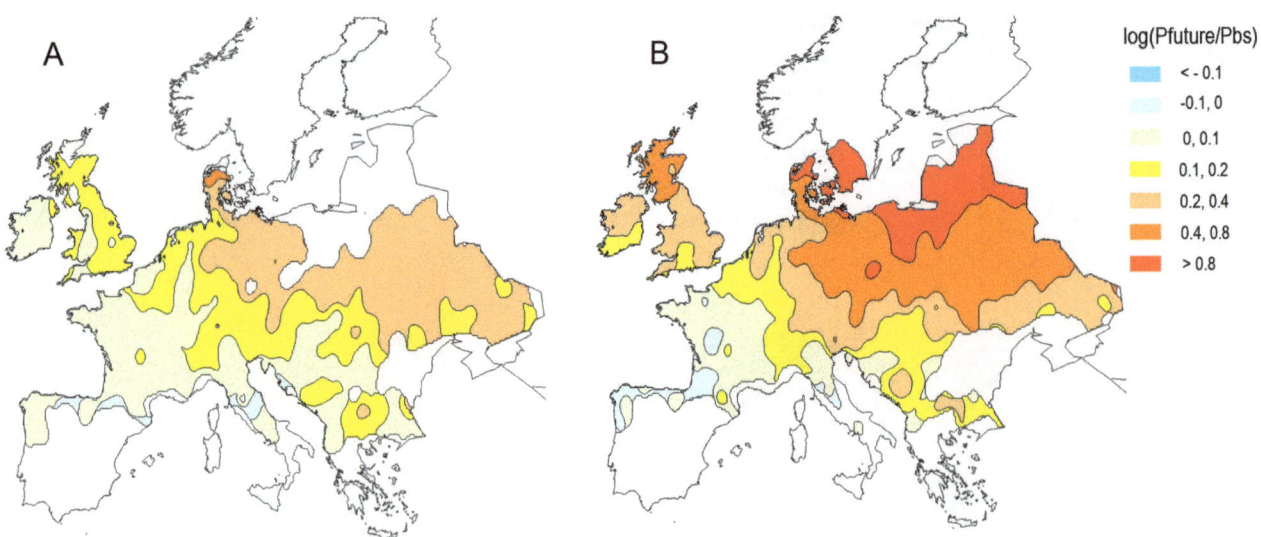

Figure 4. Logarithm of relative increase in *Ambrosia artemisiifolia* pollen production, Pfuture/Pbs, for the HadCM3 (A1B) scenarios. A. Near future (2010–2030) and B. long-term future (2050–2070), where Pbs and Pfuture are pollen production for the baseline and future climate scenarios. The model output is pollen production by the plant, there is no prediction of pollen emission or transportation included in the model. Only areas predicted to be suitable for *A.artemisiifolia* (casual or established) are shown. Where populations are predicted to colonise new areas, increased pollen production will therefore be indicated as >0.8 although absolute numbers of pollen grains may be relatively low when compared to well established populations.

an on-going, dynamic invasion event [11]. Developing predictive models for the potential stable state of such a system under current conditions and future climate poses fundamental challenges for traditional correlative approaches. Specifically, models based on correlations between species occurrence and habitat parameters depend on the assumption that its current distribution is in equilibrium with its environment [36]. The alternative of using habitat data from the species' native range may also be problematic because of evidence of niche shifts between native and naturalised environments due to processes such as enemy release [37]. In contrast, forward process models, of the type developed here, based on first principles and parameterised from independent eco-physiological experiments have the potential to predict potential realised niches under novel environments with greater confidence as it can be assumed that the biological principles driving the model output are constant across environments [38]. However, these models fall at the extreme end of the tractability/complexity trade-off, with a high demand for data and parameterisation [39]. As a consequence, their development and application to species distribution modelling lags behind correlative models and they also often fall back on correlating missing parameters with the observed distribution [38].

Independently of the development of species distribution models in invasion ecology, detailed process based models of crop growth and development have been progressed over recent years to better understand the environmental and physiological constraints on crop yield [20]. These models have the power to capture the interaction of multiple management and environmental variables on a small temporal and spatial scale and they represent an important resource for predicting realised niches under future scenarios of invasive species. This potential has recently been demonstrated for an annual weed at a national scale [18]; here we have used the same approach to model the potential niche for a species currently invading a new continent. In so doing, we have addressed a number of challenges facing the application of forward models to invasion ecology discussed by Chapman *et al.* [13].

Specifically, the processes in the model are all parameterised using independent data sets and the use of local scale climate scenarios generated on a daily time step allows the impact of short term environmental stress on plant growth to be captured.

The simulation of the future distribution of *A.artemisiifolia* indicated a risk of established populations spreading to Northern European states where the species is currently only recorded as a casual alien, including the UK and Denmark, due to more favourable growing conditions and delayed frost. Whether this threat is realised will be a function of future land use and levels of control. *A.artemisiifolia* is a 'ruderal' species that relies on regular disturbance to persist [16]; land use is, therefore, an important driver of occurrence and it is commonly found in high densities along transport routes and as an arable weed in crops of sunflower (in the same family as *Ambrosia*, Compositae), maize and soybean. The competitive dynamics between plant species will be specific to these different situations and potentially affect the population growth rate and fitness of local populations of the weed. In addition, the level of control (either through physical removal or herbicides) will vary between region and country and have a large impact on the capacity of the weed to establish in a new location. For example, the herbicide load calculated as the total weight of active herbicide ingredients applied to cereals for Hungary (where *A.artemisiifolia* has become the most economically damaging arable weed) is 0.11 kg ha^{-1} compared to 0.78 for Germany and 1.34 for the United Kingdom [40]. Differences in the types of crops grown and level of weed control will not only mitigate the spread of *A.artemisiifolia* but also determine the regional health impacts through the amount of pollen produced. As such, the effects of climate change on the realised niche for *A.artemisiifolia* will be manifested as much by indirect effects on farmer behaviour as direct climatic effects on plant biology.

The process based approach to modelling the response of plant populations to climate change developed here has sufficient mechanistic detail to capture these interactions between local weather, management conditions and biotic interactions with neighbouring species. However, at present insufficient empirical data are available to calibrate the model for multispecies plant communities, including crop competition under future climate scenarios [41]. Future development of the model will allow the predictions of spread and impact (in terms of pollen emission and crop yield loss) to be further refined by incorporation of data on land use and crop management at the field scale. An additional advantage of our process based growth over conventional SDMs is the capacity to simulate biomass production and allocation, including the amount of pollen produced, based on published allometric relationships. This has enabled us to generate initial predictions of relative changes in the regional potential pollen production of *A.artemisiifolia* under climate change. However, the situation is complicated by the capacity of *A.artemisiifolia* to shift the allocation of resources between seed and pollen in response to the environment; *A.artemisiifolia* is a monoecious species (having the male and female reproductive organs separated in different floral structures on the same plant). As a consequence, the allometric relationships between biomass and seed and pollen numbers are not constant but plastic. Incorporation of these processes into the mechanistic model would not only improve its practical application but potentially contribute to ecological theory seeking to explain the optimal strategy under contrasting environments [42].

In addition to the problem of morphological plasticity, a further challenge to a process based approach to modelling the impact of change on plant populations dynamics is the capacity of biotypes at the limit of the current range to adapt and expand the available niche for colonisation [43]. It is likely that this is the explanation for the poor predictive power of the photothermal time model of *A.artemisiifolia* developed from a single population at a single site, to predict flowering times at the continental scale. Although some progress has been made in terms of predicting the temporal and spatial variability in *A.artemisiifolia* phenology [13], further work is required to understand the importance of adaptation to local conditions of spatially discrete biotypes. It appears that previous assumptions that a relatively simple function of thermal time and daylength is inadequate and a more mechanistic understanding of the interactions of local environment and genetic adaptation is required. This highlights the dangers of basing generic models on data derived from a single population measured at a single site. Incorporating variance in model parameters based on empirical data from populations sampled across an environmental gradient in combination with a sensitivity analysis of the relative importance of processes operating on different stages of the weed life cycle would, therefore, constitute a major advance and improve confidence in model predictions.

Author Contributions

Conceived and designed the experiments: JS PS FV MS. Performed the experiments: PS MS FV. Analyzed the data: JS PS MS. Contributed reagents/materials/analysis tools: FV DC. Wrote the paper: JS PS MS.

References

1. Kazinczi G, Beres I, Pathy Z, Novak R (2008) Common ragweed (*Ambrosia artemisiifolia* L.): a review with special regards to the results in Hungary: II. Importance and harmful effect, allergy, habitat, allelopathy and beneficial characteristics Herbologia 9.
2. Pinke G, Karacsony P, Czucz B, Botta-Dukat Z (2011) Environmental and land-use variables determining the abundance of Ambrosia artemisiifolia in arable fields in Hungary. Preslia 83: 219–235.
3. Gallinza N, Baric K, Scepanovic MG, Ostojic Z (2010) Distribution of invasive weed Ambrosia artemisiifolia L. in Croatia. Agriculturae Conspectus Scientificus 75: 75–81.
4. Makra L, Juhasz M, Beczi R, Borsos E (2005) The history and impacts of airborne Ambrosia (Asteraceae) pollen in Hungary. Grana 44: 57–64.
5. Chauvel B, Cadet E (2011) Introduction and spread of an invasive species: Ambrosia artemisiifolia in France. Acta Botanica Gallica 158: 309–327.
6. Rodinkova V, Palamarchuk O, Kremenska L (2012) The most abundant Ambrosia pollen count is associated with the southern, eastern and the northern-eastern Ukraine. Alergologia et Immunologia 9: 181.
7. Chauvel B, Dessaint F, Cardinal-Legrand C, Bretagnolle F (2006) The historical spread of Ambrosia artemisiifolia L. in France from herbarium records. Journal of Biogeography 33: 665–673.
8. Mitich LW (1996) Ragweeds (Ambrosia spp) - The hay fever weeds. Weed Technology 10: 236–240.
9. Bohren C, Mermillod G, Delabays N (2006) Common ragweed (Ambrosia artemisiifolia L.) in Switzerland: development of a nationwide concerted action. Journal of Plant Diseases and Protection: 497–503.
10. Berry PM, Dawson TP, Harrison PA, Pearson RG (2002) Modelling potential impacts of climate change on the bioclimatic envelope of species in Britain and Ireland. Global Ecology and Biogeography 11: 453–462.
11. Smolik MG, Dullinger S, Essl F, Kleinbauer I, Leitner M, et al. (2010) Integrating species distribution models and interacting particle systems to predict the spread of an invasive alien plant. Journal of Biogeography 37: 411–422.
12. Vogl G, Smolik M, Stadler LM, Leitner M, Essl F, et al. (2008) Modelling the spread of ragweed: Effects of habitat, climate change and diffusion. European Physical Journal-Special Topics 161: 167–173.
13. Chapman DS, Haynes T, Beal S, Essl F, Bullock JM (2014) Phenology predicts the native and invasive range limits of common ragweed. Global Change Biology 20: 192–202.
14. Dunnett NP, Grime JP (1999) Competition as an amplifier of short-term vegetation responses to climate: an experimental test. Functional Ecology 13: 388–395.
15. Brooker RW (2006) Plant-plant interactions and environmental change. New Phytologist 171: 271–284.
16. Ziska LH, George K, Frenz DA (2007) Establishment and persistence of common ragweed (Ambrosia artemisiifolia L.) in disturbed soil as a function of an urban-rural macro-environment. Global Change Biology 13: 266–274.
17. Pearson RG, Dawson TP (2003) Predicting the impacts of climate change on the distribution of species: are bioclimate envelope models useful? Global Ecology and Biogeography 12: 361–371.
18. Stratonovitch P, Storkey J, Semenov MA (2012) A process-based approach to modelling impacts of climate change on the damage niche of an agricultural weed. Global Change Biology 18: 2071–2080.
19. Lawless C, Semenov MA, Jamieson PD (2005) A wheat canopy model linking leaf area and phenology. European Journal of Agronomy 22: 19–32.
20. Jamieson PD, Semenov MA, Brooking IR, Francis GS (1998) Sirius: a mechanistic model of wheat response to environmental variation. European Journal of Agronomy 8: 161–179.
21. Kropff MJ, Spitters CJT (1992) An eco-physiological model for interspecific competition, applied to the influence of *Chenopodium album* L. on sugar beet. I. Model description and parameterization. Weed Research 32: 437–450.
22. Storkey J, Cussans JW (2007) Reconciling the conservation of in-field biodiversity with crop production using a simulation model of weed growth and competition. Agriculture Ecosystems & Environment 122: 173–182.
23. Willemsen RW (1975) Effect of stratification temperature and germination temperature on germination and induction of secondary dormancy in common ragweed seeds. American Journal of Botany 62: 1–5.
24. Shrestha A, Roman ES, Thomas AG, Swanton CJ (1999) Modeling germination and shoot-radicle elongation of Ambrosia artemisiifolia. Weed Science 47: 557–562.
25. Bassett IJ, Crompton CW (1975) Biology of canadian weeds 11. Ambrosia artemisiifolia (L.) and A. psilostachya (DC). Canadian Journal of Plant Science 55: 463–476.
26. Deen W, Hunt T, Swanton CJ (1998) Influence of temperature, photoperiod, and irradiance on the phenological development of common ragweed (Ambrosia artemisiifolia). Weed Science 46: 555–560.
27. Ziska L, Knowlton K, Rogers C, Dalan D, Tierney N, et al. (2011) Recent warming by latitude associated with increased length of ragweed pollen season in central North America. Proceedings of the National Academy of Sciences of the United States of America 108: 4248–4251.
28. Gleeson SK (1987) Biomass allocation in Ambrosia artemisiifolia L. [PhD]: Michigan State University.
29. Fumanal B, Chauvel B, Bretagnolle F (2007) Estimation of pollen and seed production of common ragweed in France. Annals of Agricultural and Environmental Medicine 14: 233–236.
30. Toole EH, Brown E (1946) Final results of the Duvel buried seed experiment. Journal of Agricultural Research 72: 201–210.
31. Semenov MA, Donatelli M, Stratonovitch P, Chatzidaki E, Baruth B (2010) ELPIS: a dataset of local-scale daily climate scenarios for Europe. Climate Research 44: 3–15.

32. Semenov MA, Stratonovitch P (2010) Use of multi-model ensembles from global climate models for assessment of climate change impacts. Climate Research 41: 1–14.

33. Dahl A, Strandhede SO, Wihl JA (1999) Ragweed - An allergy risk in Sweden? Aerobiologia 15: 293–297.

34. Rich TCG (1994) Ragweeds (Ambrosia L) in Britain. Grana 33: 38–43.

35. Bullock JM, Chapman DS, Schafer S, Roy DB, Girardello M, et al. (2012) Mapping the distribution of Ambrosia artemisiifolia across the EU27. NERC, UK Government. 71–121 p.

36. Guisan A, Thuiller W (2005) Predicting species distribution: offering more than simple habitat models. Ecology Letters 8: 993–1009.

37. Genton BJ, Kotanen PM, Cheptou PO, Adolphe C, Shykoff JA (2005) Enemy release but no evolutionary loss of defence in a plant invasion: an inter-continental reciprocal transplant experiment. Oecologia 146: 404–414.

38. Dormann CF, Schymanski SJ, Cabral J, Chuine I, Graham C, et al. (2012) Correlation and process in species distribution models: bridging a dichotomy. Journal of Biogeography 39: 2119–2131.

39. Thuiller W, Albert C, Araujo MB, Berry PM, Cabeza M, et al. (2008) Predicting global change impacts on plant species' distributions: Future challenges. Perspectives in Plant Ecology Evolution and Systematics 9: 137–152.

40. Storkey J, Meyer S, Still KS, Leuschner C (2012) The impact of agricultural intensification and land-use change on the European arable flora. Proceedings of the Royal Society B-Biological Sciences 279: 1421–1429.

41. Harrison PA, Butterfield RE (1996) Effects of climate change on Europe-wide winter wheat and sunflower productivity. Climate Research 7: 225–241.

42. Paquin V, Aarssen LW (2004) Allometric gender allocation in Ambrosia artemisiifolia (Asteraceae) has adaptive plasticity. American Journal of Botany 91: 430–438.

43. Clements DR, DiTommaso A (2011) Climate change and weed adaptation: can evolution of invasive plants lead to greater range expansion than forecasted? Weed Research 51: 227–240.

44. Déchamp C, Méon H, Reznik S (2009) Ambrosia artemisiifolia L. an invasive weed in Europe and adjacent countries: the geographical distribution (except France) before 2009. In: Déchamp C, Méon H, editors. Ambroisie: The first international ragweed review. Saint-Priest, France, AFEDA.

4

Re-Meandering of Lowland Streams: Will Disobeying the Laws of Geomorphology Have Ecological Consequences?

Morten Lauge Pedersen[1]*, **Klaus Kevin Kristensen**[2], **Nikolai Friberg**[3]

1 Department of Civil Engineering, Aalborg University, Aalborg, Denmark, **2** Ringkøbing-Skjern Municipality, Ringkøbing, Denmark, **3** Section of Freshwater Biology, Norwegian Institute for Water Research, Oslo, Norway

Abstract

We evaluated the restoration of physical habitats and its influence on macroinvertebrate community structure in 18 Danish lowland streams comprising six restored streams, six streams with little physical alteration and six channelized streams. We hypothesized that physical habitats and macroinvertebrate communities of restored streams would resemble those of natural streams, while those of the channelized streams would differ from both restored and near-natural streams. Physical habitats were surveyed for substrate composition, depth, width and current velocity. Macroinvertebrates were sampled along 100 m reaches in each stream, in edge habitats and in riffle/run habitats located in the center of the stream. Restoration significantly altered the physical conditions and affected the interactions between stream habitat heterogeneity and macroinvertebrate diversity. The substrate in the restored streams was dominated by pebble, whereas the substrate in the channelized and natural streams was dominated by sand. In the natural streams a relationship was identified between slope and pebble/gravel coverage, indicating a coupling of energy and substrate characteristics. Such a relationship did not occur in the channelized or in the restored streams where placement of large amounts of pebble/gravel distorted the natural relationship. The analyses revealed, a direct link between substrate heterogeneity and macroinvertebrate diversity in the natural streams. A similar relationship was not found in either the channelized or the restored streams, which we attribute to a de-coupling of the natural relationship between benthic community diversity and physical habitat diversity. Our study results suggest that restoration schemes should aim at restoring the natural physical structural complexity in the streams and at the same time enhance the possibility of re-generating the natural geomorphological processes sustaining the habitats in streams and rivers. Documentation of restoration efforts should be intensified with continuous monitoring of geomorphological and ecological changes including surveys of reference river systems.

Editor: Tomoya Iwata, University of Yamanashi, Japan

Funding: This work has been funded by the Aalborg University. The funder had no role in study design, data collection and analysis, decision to publish, or preparation of the manuscript.

Competing Interests: The authors have declared that no competing interests exist.

* Email: mlp@civil.aau.dk

Introduction

The vast majority of European streams and rivers have been altered by human activities; thus, 95% of all riverine floodplains have been lost to agriculture and urbanization and river systems have been fragmented by thousands of major and minor dams influencing flow conditions and the longitudinal migration of organisms [1]. Many European rivers and streams are also characterized by high levels of organic pollution and nutrient enrichment from agriculture, and most river basins suffer from the combined impacts of pollution and elevated nutrient concentrations as well as physical habitat degradation [2], [3], [4]. The consequence of the past and contemporary degradation of European stream ecosystems is that the majority of streams fail to reach the "good ecological status" stipulated within the legislative context of the Water Framework Directive (WFD). Therefore, improvement of ecological status poses a key challenge to water managers, and there is an urgent need to implement cost effective mitigation measures and restoration projects to improve the ecological status of water bodies in Europe.

Since the 1970s policies adopted at local, regional, national and international levels have improved the water quality of streams, primarily through improved waste water treatment [5]. This improvement has however, only to a certain degree, been matched, by enhanced diversity of stream biota, the most likely explanation being poor physical conditions and improper management of rivers and their flows and insufficient time to re-colonize polluted reaches [6], [7], [8]. Since the 1990s focus has also been dedicated to improving in-stream habitat conditions through river rehabilitation/restoration across Europe and North America [9], [10], [11]. The dominant paradigm in river restoration has been rehabilitation of the physical system with primary focus on habitat structure and water flow to enhance habitat heterogeneity and biodiversity. Physical habitat restoration schemes typically work on a local scale and the measures implemented are usually introduction of gravel bars and patches of large woody debris (LWD) in small sections. At the intermediate scale restoration schemes aim to restore degraded river sections to their natural condition through re-meandering of entire sections of the river. The main objective of hydrological restoration is to obtain near natural hydrological conditions in entire catchments [12], [13]. However, by emphasizing only in-stream habitat heterogeneity the success of restoration efforts may be compromised if the geomorphological processes (e.g. interaction between flow regime and morphological units at different scales) behind the heterogeneity are not well understood [10], [14]. Thus, water managers may risk restoring the habitats to conditions that cannot be sustained on a longer temporal scale because fundamental physical laws governing the dynamic interaction between flow regime and geomorphology in a particular stream/river are inadequately considered. To sustain a heterogeneous environment capable of supporting diverse ecological communities, a key issue is to determine how natural streams and rivers are structured in terms of physical habitats and flow and sediment regime and how these three factors interact [10].

Knowledge of the dynamic linkages between forms and processes across different scales in natural streams and rivers is the key to understanding in-stream heterogeneity, and such understanding is essential to restore streams/rivers to natural conditions. Seen within the perspective of river ecology, or restoration ecology, the all-important issue is how spatial and temporal physical heterogeneity creates a range of niches and micro habitats for the biota in natural streams and how this heterogeneity can be recreated within a rehabilitation context [14], [15], [16]. The morphology of a river depends on catchment-scale controls (hydrology, geology), differences in channel patterns at reach scale (i.e. local slope, geology) and micro-scale variations in the structure and composition (flow and turbulence structure, bank material) of the river, factors that all vary over different time scales [17]. Hence, rivers experience a predictable morphological pattern at both reach and habitat scale (dominant bed type, entrenchment ratio, sinuosity, width to depth ratio and water surface slope), with topography and catchment geology playing at multiple scales a major role in structuring the habitats [18], [19].

Even though the number of river restorations has increased over the last several decades in both Europe and North America [11], [20], [21], studies providing conclusive empirical evidence of its effects are lacking [13]. A comprehensive review by Feld et al. [11] provided almost no evidence of a long term (+5 years) positive effect of river restoration on biotic communities. A very recent paper by Lorenz et al. [21] described, though, a longer term positive response of macrophytes to restoration measures. A similar conclusion to that of Feld et al. [11] was reached by Miller et al. [22] in a review of 24 case studies. Roni et al. [9] conclude that the lack of firm evidence is primarily a consequence of limited spatial and temporal resolution of data on physical habitats and biota. Long term monitoring and comparisons with reference stream systems, serving as restoration targets, are clearly needed by water policy managers and stakeholders in order to assess the socio-economic and ecological success of stream restoration schemes. The very limited evidence of links between restoration activities and improvement in ecological status constitutes a substantial problem for water managers when having to select appropriate measures as the costs involved can be very high [23].

The overall objective of the present study is to highlight important drivers of restoration success in lowland streams with Danish sites serving as examples. Denmark provides a unique opportunity for more conclusive restoration studies to be carried out, as several restoration projects have been implemented since the late 1980s in the Danish lowland landscape exhibiting limited spatial variability compared to the rest of the world [1], [24]. The Danish landscape consists primarily of soft sediments of glacial origin, ranging from sandy soils to loamy soils with up to 30% clay in a sandy matrix. Hence, on a global scale Denmark is geologically relatively homogeneous, allowing comparison of spatial variation in the physical environment among many sites. The Danish landscape is thus well suited for undertaking long term evaluation of restoration projects linking physical processes over a

temporal scale ranging from years to decades with ecological recovery processes of invertebrate communities within a catchment context. We believe that our results may provide insight of general interest to both scientists and managers in both Europe and North America. To evaluate the effects of restoration on a longer time scale than previously, we evaluate restorations conducted in small Danish lowland streams involving both re-meandering and re-sectioning of the profile and in-stream habitat enhancement by gravel addition. We examine if physical habitats and macroinvertebrate communities in restored streams resemble those of channelized reaches and naturally meandering streams or whether a new ecological state has developed. We hypothesize that the physical conditions of restored streams will resemble those of natural streams and that this resemblance will be reflected in the macroinvertebrate communities.

Methods

Ethics statement

All reaches were located on public watercourses, hence no permission were required to access the sites. All field sites are identified by UTM coordinates in Fig. 1. Protected and endangered species were carefully sorted from the samples in the field. Given the problems of identifying macroinvertebrates in the field, only the largest specimens were sorted alive before preservation. Smaller specimens could only be identified in the laboratory.

Site selection

The study streams were all located in Jutland, Denmark (Fig. 1). The western part of Jutland remained ice-free during the last glaciation (ending 10,000 years ago) and the landscape is dominated by sandy soils developed on glacio-fluvial outwash plains and loamy sand soils on moraine hills from previous glaciations. The eastern part of Jutland was located close to the glacier margin and is mainly characterized by sub-glacial loamy moraine from the Weichsel glaciation. The dominant land use (app. 70%) in all catchments is agriculture with smaller areas of forest, heath land and wetlands (Table S1). Mean annual precipitation varies between 900 and 1000 mm and the hydrological regime is dominated by groundwater during summer, whereas increased precipitation and lower evaporation result in higher discharge during winter. The hydrological regime is also affected by drainage of agricultural areas in the catchments, and even naturally meandering streams are thereby also impacted by land use changes, leading to changes in hydrological regime and sediment dynamics.

Eighteen reaches were selected, each from a different stream to avoid interdependence (stream width 2–5 m, depth app. 0.50–0.70 m; Table 1) and irrespective of catchment geology. Streams were only included in the surveys, which were conducted in April and May 2002, if no point sources of nutrients and pollutants (fish farms, waste water treatment plants, lakes, reservoirs etc.) occurred upstream the surveyed reach. The reaches were 100 m long and covered approximately 5 riffle-pool sequences depending on stream width. Six streams were in a near natural meandering state with little physical alterations; hereafter referred to as "Natural" streams. Six were channelized and 6 were former channelized reaches that had been re-meandered minimum 3 years prior to the investigation (Fig. 1). The re-meandered stream reaches were selected to include a buffer of at least 200 m restored reach upstream of the study reach.

Field work was conducted over a 40 day period in April and May 2002. In order to minimize the effects of high flow event field work were only carried if no precipitation had occurred in the

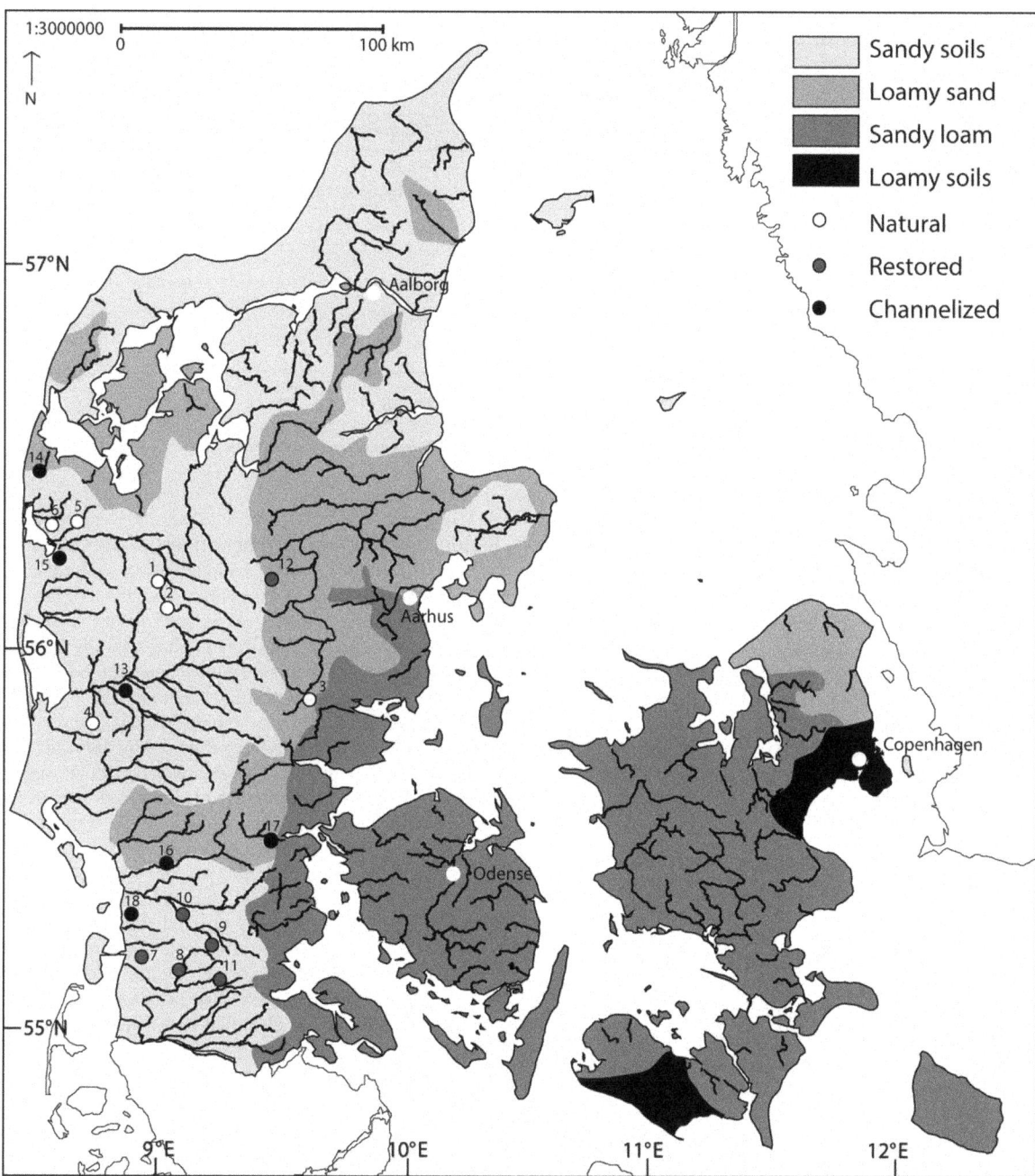

Figure 1. Location of the 18 stream reaches in Denmark. Natural streams (1–6); Restored streams (7–12); Channelized streams (13–18). UTM coordinates of the sites (UTM Zone 32, datum ED50). 1: Sunds Nørreå (N6231730; E496890), 2: Fjederholt (N6214415; E500939), 3: Linding (N6171439; E473283), 4: Gesager (N6190369; E543271), 5: Grydeå (N6243183; E471912), 6: Idom (N6243861; E468179), 7: Brøns (N6116409; E484998), 8: Lobæk (N6108125; E499423), 9: Surbæk (N6102701; E510372), 10: Jels (N6127631; E509606), 11: Gels (N6117435; E512790), 12: Lemming (N6233250; E532931), 13: Simmebæk (N6188424; E488854), 14: Fåre Mølleå (N6257940; E454624), 15: Madum (N6233919; E463860), 16: Hjortvad (N6137346; E494356), 17: Kongeå (N6141296; E519069), 18: Rejsby (N6121446; E483188).

previous 7 days prior to the sampling. One week was omitted from sampling in order to reduce the risk of influence from heavy rain showers.

Catchment and river corridor data

Data on catchment geology and land use were extracted from the national GIS data base using ArcGIS (ArcGIS Desktop 10, ESRI). River corridor land use was also extracted from a buffer covering a width of 50 meters on each side of the stream and

extending 1 kilometer up- and downstream from the field site. This analysis was also conducted in the ArcGIS environment.

In the re-meandered streams, pebble and gravel and to a lesser degree stone substrate had been added to the stream bed to increase habitat diversity. The banks had been re-profiled and, in some cases, the bed level had been raised to increase hydrological interactions with the floodplain. All restoration measures were aimed at creating a more heterogeneous and hence natural stream reach.

Table 1. Physical characteristics of the 18 streams.

	Stream type		
	Natural	**Channelized**	**Restored**
Catchment area (km^2)	61±32	44±22	81±28
Slope (‰)	1.6±0.6(ab)	0.6±0.3(b)	1.9±1.1(a)
Substrate heterogeneity	0.32±0.07	0.36±0.20	0.41±0.10
Current velocity (m/s)	0.34±0.04(a)	0.26±0.05(b)	0.30±0.04(ab)
Current velocity$_{CV}$ (%)	27±4(a)	14±5.38(b)	18±5(b)
Width (cm)	392±56	360±137	493±121
Width$_{CV}$ (%)	18±7(a)	7±1(b)	12±5(ab)
Depth (cm)	52±18	46±15	47±15
Depth$_{CV}$ (%)	44±3(ab)	34±7(b)	50±9(a)

Mean values are presented along with standard deviations (SD). Letters indicate significant differences among stream types using one-way ANOVA and pair-wise Bonferroni corrected *post hoc* tests.

Water chemistry

In order to characterize the water chemistry of the sites and to ensure that the water chemistry did not affect biological communities, nine chemical variables were analyzed in the laboratory. Biological oxygen demand over five day (BOD$_5$) was measured according to Danish Standard 1899:1 [25]. pH was measured on a PHM240 pH-meter and alkalinity was determined by Gran titration on 100 mL subsamples of stream water [26]. Ferro-iron (Fe^{2+}), total-N, NH$_4$$^{2+}$, NO$_3$$^-$, total-P and PO$_4$$^{2-}$ (all mg/L) were measured according to Danish Standards, DS 219 [27], DS 221 [28], DS 223 [29], DS 292 [30] and DS 291 [31], respectively.

Physical habitats

The physical habitats were measured in 120 plots (25×25 cm) placed side-by-side covering the entire width of the stream in 10 to 12 equally spaced cross sectional transects along the 100 m reach. Water depth was measured to nearest cm in the middle of each plot and mean depth was subsequently calculated. Stream width was measured from bank to bank at each transect and the mean width of the stream reach was calculated. In order to quantify the variation in stream reach dimensions, the coefficient of variation of depth and width was calculated [32].

The dominant substrate type in each plot was categorized according to a modified Wentworth scale [33] as: cobble (> 64 mm diameter), pebble/gravel (2–64 mm), sand (0.1–2 mm), silt/clay (<0.1 mm, inorganic particles, usually with a compact structure) and mud (<0.1 mm, a mixture of inorganic particles and organic debris (FPOM), typically brown or black, loosely structured). The relative frequency of the various substrate types was calculated from these recordings. Substrate heterogeneity (SH) was quantified from the spatial distribution of substrate types according to Pedersen et al. [24].

The average current velocity and the velocity heterogeneity were characterized by measuring the current velocity in four different depths in five vertical profiles equally spaced across the stream at the downstream end of the reach). Velocities were measured with a propeller current meter (Kleinflügel, OTT Instruments).

Biological sampling of macroinvertebrates

Macroinvertebrates were sampled using a stratified random sampling methodology. Two main meso-habitats were identified: the edge habitat, located close to the bank having current velocities below 0.1 m/s, and a riffle/run habitat typically located in the center of the stream with current velocities exceeding 0.1 m/s.

A total of six macroinvertebrate samples were collected at each reach. Within each of the two meso-habitats 3 surber (500 cm^2; 500 μm mesh size) samples were collected by disturbing the upper 5 cm of the stream bed. The sampling locations were selected randomly among the 120 surveyed habitat plots at each reach. All samples were preserved in 70% ethanol and transported to the laboratory for sorting and identification. Macroinvertebrates were identified to species level except for dipterans, which were identified to sub-family level, and oligochaetes, which were identified to sub-class level. Protected and endangered species were carefully sorted from the samples in the field. Given the problems of identifying macroinvertebrates in the field, only the largest specimens were sorted alive before preservation. Smaller specimens could only be identified in the laboratory.

Statistical analyses

For each sample macroinvertebrate community structure and diversity were expressed using several metrics. Species richness and total invertebrate abundance and total abundance of Ephemeroptera, Plecoptera, Trichoptera and Coleoptera (EPTC), Shannon-Wiener diversity (H'), were calculated for each sample [34]. All metrics were log-transformed prior to any further analyses to satisfy assumptions of normality. To test for differences among stream types (natural, restored, channelized) a nested analysis of variance (ANOVA) was used, where samples and streams were nested in type. This allowed us to test for differences in macroinvertebrate metrics among types and at the same time correcting for repeated sampling within types [35].

Macroinvertebrate community composition at the 18 sites was analyzed by means of Detrended Correspondence Analyses (DCA) using PC-ORD version 6 (MjM Software) and then related to environmental variables by means of Spearman rank correlation analysis. All the above mentioned macroinvertebrate metrics were also calculated for each reach by pooling the 6 samples. Using type (natural, restored or channelized) as a co-variate, a Spearman rank correlation analysis between reach-scale physical parameters and macroinvertebrate community variables was performed in order to elucidate the effects of channelization and restoration on community structure [35]. Significant relationships between physical parameters and macroinvertebrate metrics were further analyzed and quantified using ANOVA analysis and subsequently linear regression analysis. Residuals of all developed relationships were tested for normality to satisfy the assumptions of regression analysis [36]. Additional ANOVA analyses were carried in order to test for possible confounding factors. Factors included: water chemistry, catchment geology, river system location and years since restoration.

Results

Physical habitats

The composition of the stream bed substrate varied significantly among stream types, cobble and pebble being significantly dominant in restored streams and sand in natural and channelized streams (Table S2). A significant empirical relationship existed between stream bed slope and coverage of pebble and gravel in natural streams ($R^2 = 0.76$; $p = 0.025$; Fig. 2). In contrast the relationship in both channelized and restored streams were not significant; $R^2 = 0.01$ ($p = 0.87$) and $R^2 = 0.52$ ($p = 0.11$), respectively. In natural streams the heterogeneity of the stream bed substrate decreased with a linear increase in the coverage of sand as expected ($R^2 = 0.76$; $p = 0.025$; Fig. 3). Also in the channelized streams substrate heterogeneity was inversely related to the coverage of sand ($R^2 = 0.73$; $p = 0.031$; Fig. 3), while the regression line parameters differed from those of the natural streams. This relationship between substrate heterogeneity and sand cover was not detected in the restored streams (Fig. 3), indicating a de-coupling of natural physical processes structuring the stream bed composition; the de-coupling arise from the addition of gravel and stones to the restored stream beds.

Marked differences were also found in the heterogeneity of stream dimensions and current velocities. In natural streams width variation was significantly higher (18%) than in channelized streams (7%). In the restored streams some of the natural variation in width had been recreated through re-meandering, variation being intermediate (Table 1). Depth varied most markedly in the restored streams and was significantly higher than in the channelized streams. Natural streams exhibited the significantly highest variation in the flow environment, but velocity variation in restored reaches was more similar to that of channelized reaches, indicating a failure of restoration to restore a natural flow environment (Table 1).

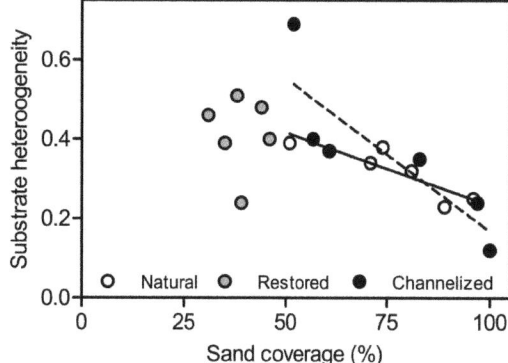

Figure 3. Relationship between stream bed sand coverage and substrate heterogeneity in the 3 stream types. The solid regression line describes the relationship in natural streams and the dotted line is the fitted line for channelized streams.

Macroinvertebrate communities

The DCA analysis of the macroinvertebrate communities revealed a distinction between stream types. The natural streams were located along the entire DCA axis 1 gradient and showed pronounced variation in width and large within-group variations in species composition (Fig. 4). Natural streams showed little variation in DCA axis 2 values. Coverage of pebble/gravel and slope were correlated with DCA axis 2, which is reflected in the species distributions of both channelized and restored streams. This reflects differences in species distribution among the stream types. A total of 129 taxa were encountered across the 18 sites (Table S3). When contemplating the 10 most dominant taxa in the three stream types little variation appeared (Table S4). The bulk of the community does not differ among stream groups; however, there are indications of differences in species composition, which probably reflects differences in the physical environment. Taxa groups normally associated with coarse grained substrates (Ephemeroptera, Plecoptera, Trichoptera and Coleoptera) are located and low and intermediate DCA axis 2 scores in the species plot (Fig. S1).

Macroinvertebrate community metrics varied negligibly among stream types, i.e. taxonomic richness, abundance, evenness and diversity measures showed no significant differences (Table 2). When combining the DCA analysis and results from Table 2, it is

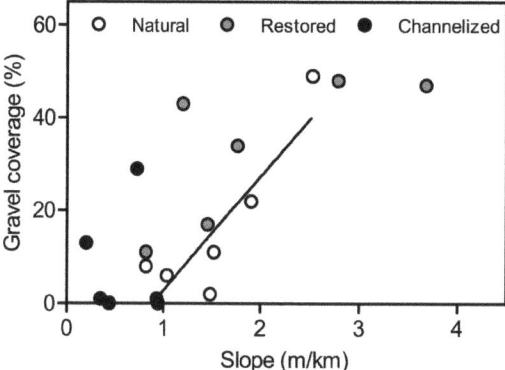

Figure 2. Relationship between stream bed slope and gravel coverage in the 3 stream types. The solid regression line describes the relationship in natural streams.

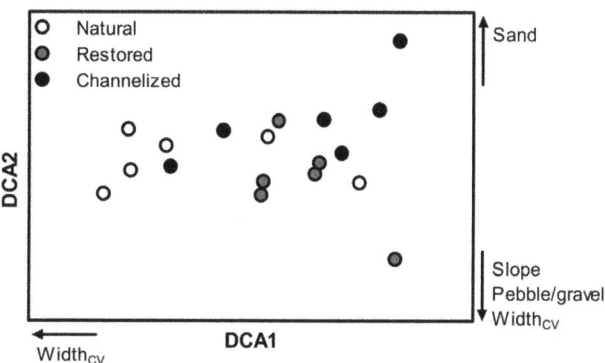

Figure 4. DCA ordination of macroinvertebrate communities in natural, channelized and restored stream reaches. Environmental parameters significantly (Spearman rank correlation, $p < 0.05$) correlated with in-stream physical characteristics are also shown.

evident that the community metrics used reveal no effects of channelization and restoration; however, endangered species occurred more frequently in the natural streams. Moreover, endangered species occurred in 5 out of 6 natural streams but only in 1 restored and 2 channelized streams (Fig. 5), reflecting the sensitivity of these species to channelization and thus the limited possibility of their re-colonization after restoration due to habitat modifications and hence low availability of adequate habitats along the stream.

It is generally assumed that high habitat heterogeneity is matched by high species diversity. Our study provided no conclusive evidence for the existence of such a relationship across all study sites combined. Instead, a relationship was established between habitat heterogeneity measured as substrate heterogeneity and species richness only in natural streams ($R^2 = 0.93$; $p = 0.0021$; Fig. 6). The same relationship did not emerge along either the channelized or the restored reaches despite the fact that variation in substrate heterogeneity was lower in natural streams compared to channelized and restored streams.

Possible confounding factors

We examined several possible confounding influences on our analyses: time since restoration, water chemistry, catchment geology, river system location, river corridor land use and variation in stream size. The analysis revealed no significant influence of river system location despite a tendency towards location of restored stream in the southern part of Jutland (ANOVA, p>0.05). The study streams were restored over a 10-year period and differences in recovery time may potentially have influenced the results. Using time since restoration as a co-variate, all the developed relationships were analyzed for this confounding factor, but no effect was demonstrated (ANOVA, p>0.05). No significant effects of catchment geology and land use on the results was found (ANOVA, p>0.05; Table S1). River corridor land use varied among the groups. The natural sites were dominated by wetlands and natural riparian areas, whereas channelized and restored streams were dominated by agricultural land use (Table S5). The water chemistry variables showed no significant variation among the groups (one way ANOVA, p>0.05; Table S6) and water chemical stress was therefore assumed to be uniformly distributed among the stream types. The natural, channelized and restored streams used in the present study were of similar size. No significant differences occurred in the catchment area or regarding stream width and depth among the stream types (Table 1). The stream bed slope varied significantly between the stream types; the

highest slopes appearing in the restored streams (mean = 1.9 m/km) and the lowest in the channelized streams (mean = 0.6 m/km). The natural streams were characterized by intermediate slopes.

Correlation coefficients among the physico-chemical parameters are given in the supplementary material (Table S7). There are few correlations among the parameters when analyzed across the entire dataset, probably reflecting the effects of channelization and restoration. Coarse substrate coverage increased with increasing catchment area and was inversely correlated with coverage of sand, as indicated by the results in Fig. 3. The variation in velocity increased with increasing variation in width, probably due to enhanced physical variation in heterogeneous streams.

Discussion

Restoration effect studies

The effects of restoring, re-meandering or re-habilitating river and streams have been documented in numerous studies conducted primarily in Western Europe and North America over the past 30–40 years [11], [20]. The results of most studies are, however, inconclusive. In a recent review of 345 projects Roni et al. [9] concluded that: "… firm conclusions …were difficult to make because of the limited information provided on physical habitat, water quality, and biota …". A similar conclusion was reached by Palmer et al. [37]. The lack of clear results is partly attributed to inadequate pre- and post-project monitoring, which is often neglected by water managers [13], [22], [38], [39], and partly to the focus of most re-habilitation schemes on reach scale and lacking consideration of catchment processes [40]. With this in mind we evaluated the success of restoration in 6 streams by comparing physical habitats and the response of macroinvertebrates to restoration in channelized and natural streams. We found significant effects of restoration on some physical habitat parameters and on the interactions between stream habitat heterogeneity and macroinvertebrate diversity.

Physical habitats

We found significant changes of the physical habitats in restored streams compared to natural (reference) and channelized streams. Pebble and gravel dominated the substrate composition in the restored streams, whereas sand was the most prominent substrate in the channelized and natural streams. This clearly shows that too much emphasis is given to gravel bed restoration in this setting, and the pebble/gravel coverage is significantly higher than in natural streams. Little effort is devoted to balancing substrate

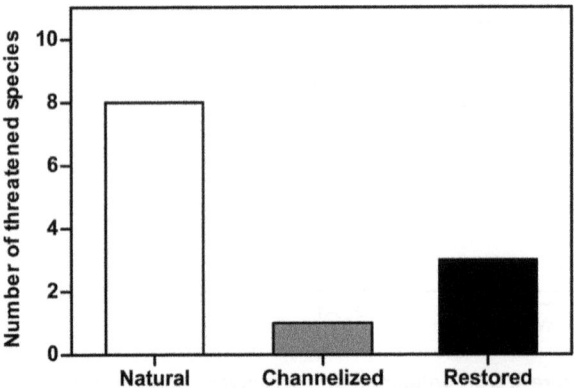
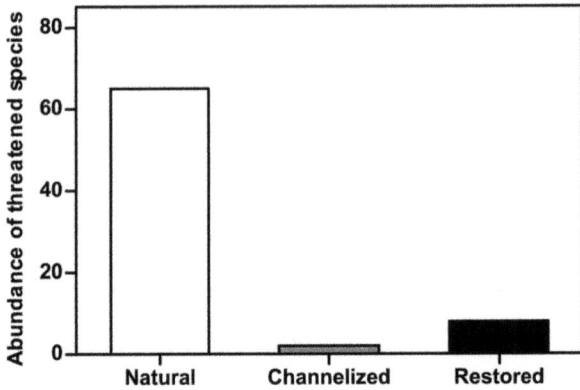

Figure 5. Number of threatened species and abundance according to stream type. In total, 75 individuals of 9 threatened species were found in the 18 streams.

Table 2. Macroinvertebrate community metrics in the three stream types.

	Stream type		
	Natural	**Channelized**	**Restored**
Taxa richness	47.2±7.2	36.5±12.1	43.8±6.5
Abundance	2828±1083	1873±1460	2683±707
Shannon diversity (H')	2.28±0.26	2.25±0.27	2.19±0.24
Evenness	0.59±0.06	0.64±0.08	0.58±0.06
EPTC taxa	24.5±2.2	24.3±2.8	24.5±4.7
ETPC abundance	845±268	423±145	787±540

Mean values are presented along with standard deviations (SD).

composition and stream slope. The stream bed slope varied significantly among the three stream types; the highest slopes occurring in the restored streams and the lowest in the channelized streams. This may seem contradictory but is a consequence of differences in catchment slope rather than of past effects of river regulation. The key to natural morphology and biodiversity in restored rivers is thus to ensure a balance between stream slope (or rather stream power) and substrate [41]. Channelization and dredging remove coarse substrate from the stream bed. The physical and biological impacts of this are well documented [42], [43]. Despite this it appears as if water managers install too much coarse substrate, affecting colonization and reestablishment of the macroinvertebrate community and, in consequence, community composition and long term ecosystem processes in the restored streams [11]. The time scale and set-up of our study did not allow us to quantify ecosystem processes (e.g. decomposition of organic matter) which have the potential of becoming a useful indicator of ecosystem change [11], [44], [45], [46].

The restoration work carried out in Denmark as well as in the rest of Europe and North America primarily rests on the principle of natural channel design (NCD) [18], [47]. This principle, and hence the restoration efforts, is primarily used at the reach scale [48]. NCD is basically a static design principle and critics argue that the form-based system ignores the processes in alluvial streams where form and substrate continuously adjust to varying water and material inputs [49], [50] – in other words, the principle oversimplifies the physical/geomorphological processes in the

streams [49], [51]. In-stream heterogeneity measured as the coefficient of variation in width, depth and velocity was only partly re-established by the stream restoration measures. At reach scale no significant differences appeared in substrate heterogeneity. Our results indicate that the natural functioning of the geomorphological processes is affected both by channelization and by restoration and, moreover, that restoration does not re-establish the natural functioning of the processes at the time-scale (3+ years) studied here. Studies over longer periods of time are required to document the recovery of natural physical processes in restoration schemes where in-stream substrate composition has been altered beyond natural conditions. In natural streams the recorded relationship between slope and pebble/gravel coverage indicated coupling between energy and substrate characteristics; the more energy the more coarse substrate. This relationship was probably not present in the channelized streams due to dredging and in restored streams due to the placement of large amounts of pebble/gravel, which distorts the natural relationship. This may on a longer time scale affect other geomorphological (sediment transport, shear stress interactions with stream bed) and biological processes (e.g. plant recolonization) in the streams and hence possibly also the recovery of biotic communities to a natural state. Moreover, the absence of a relationship between sand coverage and substrate heterogeneity in restored streams is a clear indicator of the disruption of natural dynamic processes in restored streams.

Macroinvertebrates

One indisputable result emerged from our study – namely the direct link between substrate heterogeneity and macroinvertebrate diversity in natural streams, a similar relationship being non-existing in channelized and restored streams. This result supports the general ecological hypothesis as well as the specific stream ecology hypothesis that biodiversity is closely linked with habitat heterogeneity [52], [53], [54], [55], [56], [57]. Habitat heterogeneity is, however, loosely defined [10], rendering comparisons of results difficult. In our study substrate heterogeneity is used as a surrogate for habitat heterogeneity. Mixed results have been reported for correlating in-stream heterogeneity and diversity in stream ecosystems [10]. Pedersen et al. [15] found an increase in invertebrate community diversity and evenness 3 years after restoring a large lowland river. Similarly, O'Connor [58] recorded an increase in habitat diversity and species richness from large woody debris in a study in Australia. The meta-analysis by Miller et al. [22] indicated that in some cases increased habitat complexity is matched by increases in macroinvertebrate community metric scores. Jähnig et al. [59] also reported the existence of

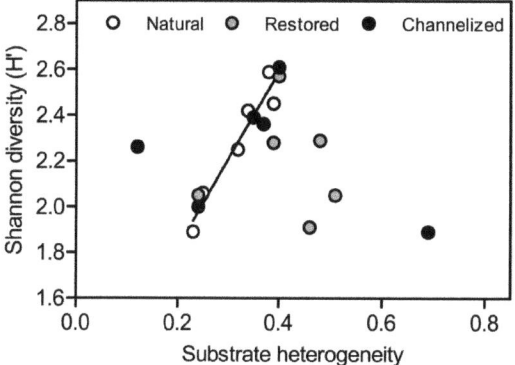

Figure 6. Relationship between substrate heterogeneity and macroinvertebrate community diversity for the 3 stream types. The solid regression line shows the significant relationship in natural streams.

a relationship between macroinvertebrate community and habitat diversity.

Despite significant differences in physical habitat conditions, macroinvertebrate taxonomic richness, abundance and diversity showed a similar lack of response channelized and restored reaches. A similar absence of response was reported from at meta-analysis study of 24 projects by Miller et al. [22]. Ernst et al. [40] found that only one macroinvertebrate metric responded to restoration in small forested headwater streams in the Catskill Mountains in New York State. Such a lack of response is consistent with the results of numerous other studies recording little or no response of macroinvertebrates to restoration. Lepori et al. [60] concluded that local scale restoration had little effect on macroinvertebrate communities compared to watershed scale factors. In a meta-analysis of stream restoration projects from 1975 to 2008, Palmer et al. [10] found that only 2 of 78 restoration projects generated increases in macroinvertebrate diversity.

Corroborating the conclusions reached by Lepori et al. [60], other studies have revealed that the positive effects of restoration can be short-lived because of catchment-scale impacts. Thus, Moerke and Lamberti [61] found that restoration of a channelized stream in the Midwest led to immediate improvement of habitat quality, but the improvement became less noticeable three years later because of continued high rates of erosion in the watershed. Similarly, Ernst et al. [40] concluded that catchment-scale factors were more important than restoration efforts in structuring the macroinvertebrate community. The reach by reach approach to restoration taken in our study did not address upstream stressors or catchment scale issues that may continuously affect in-stream biodiversity. We did, though, select our stream reaches in a way to ensure that chemical stress was at comparable levels for all reaches irrespective of stream type (natural, channelized or restored).

Effect studies of river restoration in agricultural landscapes are always subject to influence from confounding factors, such as the higher intensity of agriculture in the riparian zones of the channelized and restored streams compared with natural streams. This is an inherent problem in this type of studies as streams in lowland areas have been channelized to improve draining of riparian areas to create suitable conditions for farming. However, our substrate data from the restored streams suggest no major impacts of siltation by fine sediments or changed hydrology (erosion) that can be related back to riparian land use. Regarding the biota it is not possible to separate any additive effects of riparian land-use on community structure. Although studies have been able to link arable land-use with negative ecological status of rivers [4], in our case "agriculture" will include less intensive forms of farming (e.g. pastures) making it unlikely that riparian land-use should be the major driver of the patterns we observe.

Conclusions

Two main conclusions can be drawn from our work. Firstly, river restoration, as practiced in Denmark today, does not restore streams to natural conditions *per se*. Habitat diversity is somewhat enhanced compared to channelized reaches in terms of width and depth variations, but the addition of large quantities of gravel and pebble to the restored streams skews the substrate composition in a non-natural direction. Large quantities of coarse substrate will likely influence macroinvertebrate colonization and hence community composition. Secondly, relationships between slope, substrate composition, substrate heterogeneity and macroinvertebrate diversity are affected by the excessive use of gravel/pebble in these restored streams, potentially influencing geomorphologic and biological processes.

Recommendations

Our results clearly suggest that restoration schemes should aim at restoring natural structures and enhancing the possibility of regenerating the natural geomorphological processes sustaining the habitats in streams and rivers. The excessive use of pebble and gravel should be abandoned and replaced by generating a more natural substrate distribution, mimicking those of reference streams. More investigations should be carried out with focus on developing biological indicators of habitat improvements [62], [63]. Macroinvertebrates are an important organism/functional group in streams, but their mixed response to restoration and habitat improvement suggests than other organism groups should be included. Moreover, more emphasis should perhaps be given to developing functional and process based metrics. Even though documentation of restoration efforts is plentiful, the quality of the data is somewhat questionable as suggested by Palmer et al. [10] and Miller et al. [22]. Scientists must keep monitoring the effects and work together with water managers in an effort to increase monitoring activities both before and after restoration. Water managers and scientists need to collaborate on putting restoration schemes into the right perspective. Hence, catchment-scale restoration plans and schemes acting beyond the reach are important to obtain scientific documentation of river restoration on a larger scale.

Supporting Information

Figure S1   DCA Species plot from all 18 stream reaches. Abbreviations: Hydr.ind : Hydracarina indet.; Oreo.san : *Oreodytes sanmarkii*; Elmi.aen : *Elmis aenea*; Limn.vol : *Limnius volckmari*; Ouli.sp : *Oulimnius* sp.; Orec.vil : *Orectochilus villosus*; Hali.sp : *Haliplus* sp.; Elod.m.g : *Elodes minuta* gr.; Athe.ibi : *Atherix ibis*; Cera.ind : Ceratopogoninae indet; Chir.ind : Chironominae indet; Orth.ind : Orthocladinae indet; Prod.ind : Prodiamesinae indet; Tany.ind : Tanypodinae indet; Empi.ind : Empididae indet; Hexa.ind : Hexatominae indet; Pedi.ind : Pediciinae indet; Ptyc.sp : *Ptychoptera* sp.; Simu.ind : Simuliidae indet; Ostr.ind : Ostracoda indet.; Baet.nig : *Baetis niger*; Baet.rho : *Baetis rhodani*; Baet.sp : *Baetis* sp.; Baet.ver : *Baetis vernus*; Cent.lut : *Centroptilum luteolum*; Caen.riv : *Caenis rivulorum*; Ephe.ign : *Ephemerella* ignita; Ephe.sp : *Ephemerella* sp.; Ephe.dan : *Ephemera* danica; Hept.fus : *Heptagenia fuscogrisea*; Hept.sul : *Heptagenia sulphurea*; Lept.mar : *Leptophlebia marginata*; Para.sp : *Paraleptophlebia* sp.; Para.sub : *Paraleptophlebia submarginata*; Acro.lac : *Acroloxus lacustris*; Ancy.flu : *Ancylus fluviatilis*; Lymn.per : *Lymnaea peregra*; Phys.fon : *Physa fontinalis*; Velia.sp : *Velia* sp.; Erpo.oct : *Erpobdella octoculata*; Glos.com : *Glossiphonia complanata*; Helo.sta : *Helobdella stagnalis*; Hydra.sp : *Hydra* sp.; Pisi.sp : *Pisidium* sp.; Asel.aqu : *Asellus aquaticus*; Gamm.pul : *Gammarus pulex*; Sial.ful : *Sialis fuliginosa*; Sial.lut : *Sialis lutaria*; Olig.ind : Oligochaeta indet.; L.fu.di : *Leuctra fusca/digitata*; Amph.sp : *Amphinemura* sp.; Nemo.cin : *Nemoura cinerea*; Nemo.sp : *Nemoura* sp.; Isop.dif : *Isoperla difformis*; Brac.mac : *Brachycentrus maculatus*; Hydr.pel : *Hydropsyche pellucidula*; Hydr.sil : *Hydropsyche siltalai*; Lepi.hir : *Lepidostoma hirtum*; Athr.sp : *Athripsodes* sp.; Anab.ner : *Anabolia nervosa*; Eccl.dal : *Ecclisopteryx dalecarlica*; Hale.rad : *Halesus radiates*; Hale.sp : *Halesus* sp.; Limn.lun : *Limnephilus lunatus*; Limn.rho : *Limnephilus rhombicus*; Pota.cin : *Potamophylax cingulatus*; Pota.lat : *Potamophylax latipennis*; Pota.sp : *Potamophylax* sp.; Plec.con : *Plectrocnemia conspersa*; Poly.fla : *Polycentropus flavomaculatus*; Rhya.nub : *Rhyacophila nubile*; Rhya.sp : *Rhyacophila* sp.; Noti.cil : *Notidobia ciliaris*; Seri.per : *Sericostoma personatum*; Duge.gon : *Dugesia gonocephala*.

Table S1 Catchment geology and land use characteristics of the natural, channelized and restored streams. Mean values are presented along with standard deviations (SD). P-values for the one-way ANOVA analyses on arc sine transformed data are also shown.

Table S2 In-stream substrate composition. Mean values are presented along with standard deviations (SD). Upper case letters indicate significant differences among stream types using one-way ANOVA and pair-wise Bonferroni corrected *post hoc* tests.

Table S3 Benthic macroinvertebrate taxa encountered across the 18 sites included in the survey.

Table S4 Benthic macroinvertebrates –10 common taxa in the different stream types. Mean abundance (per m^2) is presented along with taxonomic names.

Table S5 River corridor land use characteristics of natural, channelized and restored streams. Mean values are presented along with standard deviations (SD).

Table S6 Water chemistry characteristics of the natural, channelized and restored streams. Mean values are presented along with standard deviations (SD).

Table S7 Spearman rank correlation coefficients among the physico-chemical parameters from the stream reaches. P values are also presented in brackets (N = 18). Significance levels: *: 0.05; **: 0.01; ***:0.001

Acknowledgments

The authors wish to thank Tina Pedersen for help with the field work and Søren Erik Larsen for valuable comments on the statistical analyses. We also thank the two anonymous reviewers for their astute comments on an earlier version of the manuscript.

Author Contributions

Conceived and designed the experiments: MLP NF KKK. Performed the experiments: KKK NF MLP. Analyzed the data: MLP NF KKK. Contributed reagents/materials/analysis tools: MLP NF KKK. Wrote the paper: MLP NF KKK.

References

1. Tockner K, Uehlinger U, Robinson CT, Tonolla D, Siber R, et al. (2009) Introduction to European Rivers. In: Tockner K, Robinson CT, Uehlinger U, editors. Rivers of Europe. Amsterdam: Academic Press. 1–21.

2. Johnson RK, Hering D (2009) Response of river inhabiting organism groups to gradients in nutrient enrichment and habitat physiography. J Appl Ecol 46: 175–186.

3. Hildrew AG, Statzner B (2009) European Rivers – A Personal Perspective. In: Tockner K, Robinson CT, Uehlinger U, editors. Rivers of Europe. Amsterdam: Academic Press. 685–698.

4. Friberg N (2010) Pressure-response relationships in stream ecology: introduction and synthesis. Freshwater Biol 55: 1367–1381.

5. Kristensen P, Hansen HO (1994) European Rivers and Lakes. Assessment of their environmental State. EEA Environmental Monographs 1. Copenhagen: European Environment Agency. 120 p.

6. Haury J (1996) Assessing functional typology involving water quality physical features and macrophytes in a Normandy river. Hydrobiologia 340: 43–49.

7. Pinder LCV, Marker AFH, Mann RHK, Bass JAB, Copp GH (1997) The river Great Ouse a highly eutrophic slow-flowing regulated lowland river in Eastern England. Regul River 13: 203–219.

8. Langford TEL, Shaw PJ, Ferguson AJD, Howard SR (2009) Long-term recovery of macroinvertebrate biota in grossly polluted streams: re-colonisation as a constraint ecological quality. Ecol Indic 9: 1064–1077.

9. Roni P, Hanson K, Beechie T (2008) Global review of the physical and biological effectiveness of stream habitat rehabilitation techniques. N Am J Fish Manage 28: 856–890.

10. Palmer MA, Menninger H, Bernhardt ES (2010) River Restoration, Habitat Heterogeneity and Biodiversity: A Failure of Theory or Practice? Freshwater Biol 55: 205–222.

11. Feld CK, Birk S, Bradley DC, Hering D, Kail J, et al. (2011) From natural to degraded rivers and back again: a test of restoration ecology theory and practice. Adv Ecol Res 44: 119–209.

12. Palmer MA, Allan JD, Meyer JL, Bernhardt ES (2007) River Restoration in the Twenty-First Century: Data and Experiential Knowledge to Inform Future Efforts. Restor Ecol 15: 472–481.

13. Bernhardt ES, Palmer MA (2011) River restoration: the fuzzy logic of repairing reaches to reverse catchment scale degradation. Ecol Appl 21: 1926–1931.

14. Vaughan IP, Diamond M, Gurnell AM, Hall KA, Jenkins A, et al. (2009) Integrating ecology with hydromorphology: a priority for river science and management. Aquat Conserv 19: 113–125.

15. Pedersen ML, Friberg N, Skriver J, Baattrup-Pedersen A, Larsen SE (2007) Restoration of Skjern River and its valley–short-term effects on river habitats, macrophytes, and macroinvertebrates. Ecol Eng 30: 145–156.

16. Beechie TJ, Sear DA, Olden JD, Pess GR, Buffington JM, et al (2010) Process-based principles for restoring river ecosystems. BioScience 60: 209–222.

17. Frissell CA, Liss WJ, Warren CE, Hurley MD (1986) A hierarchical framework for stream habitat classification: viewing streams in a watershed context. Environ Manage 12: 199–214.

18. Rosgen DL (1994) A classification of natural rivers. Catena 22: 169–199.

19. Downs PW (1995) River channel classification for channel management purposes. In: Gurnell A, Petts G, editors. Changing river channels. Chichester: John Wiley and Sons. 347–365.

20. Bernhardt ES, Palmer MA, Allan JD, Alexander G, Barnas K, et al (2005) Ecology – Synthesizing US river restoration efforts. Science 308: 636–637.

21. Lorenz AW, Korte T, Sundermann A, Januschke K, Haase P (2012) Macrophytes respond to reach-scale river restorations. J Appl Ecol 49: 202–212.

22. Miller SW, Budy P, Schmidt JC (2010) Quantifying macroinvertebrate responses to in-stream habitat restoration: application of meta-analysis to river restoration. Rest Ecol 18: 8–19.

23. Kristensen EA, Baattrup-Pedersen A, Jensen PN, Wiberg-Larsen P, Friberg N (2012) Selection, implementation and cost of restorations in lowland streams: A basis for identifying restoration priorities. Environ Sci Policy 23: 1–11.

24. Pedersen ML, Friberg N, Larsen SE (2004) Physical habitat structure in Danish lowland streams. River Res Appl 20: 653–669.

25. Danish Standard 1899:1 (1999) Water quality - Determination of biochemical oxygen demand after n days (BODn) - Part 1: Dilution and seeding method with allylthiourea addition. Copenhagen: Danish Standard.

26. Stumm W, Morgan JJ (1981) Aquatic Chemistry. An Introduction Emphasizing Chemical Equilibria in Natural Waters, 2nd ed. New York: John Wiley and Sons. 1040 p.

27. Danish Standard 219 (1975) Water analysis - Determination of iron. Copenhagen: Danish Standard.

28. Danish Standard 221 (1975) Determination of Nitrogen Content after Oxidation by Peroxodisulphate. Copenhagen: Danish Standard.

29. Danish Standard 223 (1975) Water Analysis – Determination of the Sum of Nitrite- and Nitrate-nitrogen. Copenhagen: Danish Standard.

30. Danish Standard 291 (1985) Water Analysis – Phosphate – Photometric Method. Copenhagen: Danish Standard.

31. Danish Standard 292 (1985) Water Analysis – Total phosphorous – Photometric Method. Copenhagen: Danish Standard.

32. Sokal RR, Rohlf FJ (1995) Biometry. The principles and practice of statstics in biological research. 3rd. ed. New York: W. H. Freeman. 880 p.

33. Wentworth CK (1922) A scale of grade and class terms for clastic sediments. J Geol 30: 377–392.

34. Washington HG (1984) Diversity, biotic and similarity indices. Water Res 18: 653–694.

35. Conover WJ (1980) Practical Nonparametric Statistics 2nd edition. New York: John Wiley and Sons. 584 p.

36. Snedecor GW, Cochran WG (1989) Statistical Methods. Ames: Iowa State College Press. 503 p.

37. Palmer MA, Bernhardt ES, Allan JD, Lake PS, Alexander G, et al. (2005) Standards for ecologically successful river restoration. J Appl Ecol 42: 208–217.

38. Muotka T, Paavola R, Haapala A, Novikmec M, Laasonen P (2002) Long-term recovery of stream habitat structure and benthic invertebrate communities from in-stream restoration. Biol Conserv 105: 243–253.

39. Jähnig SC, Lorenz AW, Hering D, Antons C, Sundermann A, et al. (2011) River restoration success: a question of perception. Ecol Appl 21: 2007–2015.

40. Ernst AG, Warren DR, Baldigo BP (2012) Natural channel design restorations that changed geomorphology have little effect on macroinvertebrate communities in headwater streams. Restor Ecol 20: 532–540.

41. Newson MD, Large ARG (2006) "Natural" rivers, "hydromorphological quality" and river restoration: a challenging new agenda for applied fluvial geomorphology. Earth Surf Proc Land 31: 1606–1624.

42. Brookes A (1988) Channelized Rivers: Perspectives for Environmental Management. Chichester: John Wiley and Sons. 326 p.

43. Petts GE (1988) Regulated rivers in the United Kingdom. Regul River 2: 201–220.

44. Feld CK, da Silva PM, Sousa JP, de Bello JP, Bugter R, et al. (2009) Indicators of biodiversity and ecosystem services: A synthesis across ecosystems and spatial scales. Oikos 118: 1862–1871.

45. Feld CK, Sousa JP, da Silva PM, Dawson TP (2010) Indicators for biodiversity and ecosystem services: Towards an improved framework for ecosystems assessment. Biodivers Conserv 19: 2895–2919.

46. Friberg N, Bonada N, Bradley DC, Dunbar MJ, Edwards FK, et al. (2011) Biomonitoring of human impacts in natural ecosystems: The good, the bad and the ugly. Adv Ecol Res 44: 1–68.

47. Rosgen DL (1996) Applied river morphology. Pagosa Springs: Wildland Hydrology. 390 p.

48. Niezgoda SL, Johnson PA (2005) Improving the urban stream restoration effort: identifying critical form and processes relationships. Environ Manage 35: 579–592.

49. Doyle MW, Miller DE, Harbor JM (1999) Should river restoration be based on classification schemes or process models? Insights from the history of geomorphology. Proceeding from ASCE International Conference on Water Resources Engineering, Seattle, Washington. Available: http://www.globalrestorationnetwork.org/uploads/files/LiteratureAttachments/466_should-river-restoration-be-based-on-classification-schemes-or-process-models.pdf. Accessed 2014 Aug 26.

50. Simon A, Doyle M, Kondolf GM, Shields FD Jr, Rhoads B, et al. (2007) Critical evaluation of how the Rosgen classification and associated "natural channel design" methods fail to integrate quantify fluvial processes and channel response. J Am Water Resour As 43: 1–15.

51. Kondolf GM (2006) River restoration and meanders. Ecol Soc 11: 42.

52. Hynes HBN (1970) The Ecology of Running Waters. Toronto: University of Toronto Press. 555 p.

53. Allan JD (1975) Distributional ecology and diversity of benthic insects in Cement Creek, Colorado. Ecology 56: 1040–1053.

54. Ricklefs RE, Schluter D (1993) Species diversity in ecological communities. Chicago: University of Chicago Press. 414 p.

55. Muotka T, Syrjanen J (2007) Changes in habitat structure, benthic invertebrate diversity, trout populations and ecosystem processes in restored forest streams: a boreal perspective. Freshwater Biol 52: 724–737.

56. Jähnig SC, Lorenz AW (2008) Substrate-specific macroinvertebrate diversity patterns following stream restoration. Aquat Sci 70: 292–303.

57. Jähnig SC, Lorenz AW, Hering D (2008) Hydromorphological parameters indicating differences between single- and multiple-channel mountain rivers in Germany, in relation to their modification and recovery. Aquat Conserv: 1200–1216.

58. O'Connor NA (1991) The effects of habitat complexity on the macroinvertebrates colonizing wood substrates in a lowland stream. Oecologia 85: 504–512.

59. Jähnig SC, Lorenz AW, Hering D (2009) Restoration effort, habitat mosaics, and macroinvertebrates – does channel form determine community composition? Aquat Conserv 19: 157–169.

60. Lepori F, Palm D, Brannas E, Malmqvist B (2005) Does restoration of structural heterogeneity in streams enhance fish and macroinvertebrate diversity? Ecol Appl 15: 2060–2071.

61. Moerke AH, Lamberti GA (2003) Responses in fish community structure to restoration of two Indiana streams. N Am J Fish Manage 23: 748–759.

62. Tullos DD, Penrose DL, Jennings GD (2006) Development and application of a bioindicator for benthic habitat enhancement in North Carolina Piedmont. Ecol Eng 27: 228–241.

63. Tullos DD, Penrose DL, Jennings GD, Cope WG (2009) Analysis of functional traits in reconfigured channels: implications for the bioassessment and disturbance of river restoration. J N Am Benthol Soc 28: 80–92.

5

Using Uncertainty and Sensitivity Analyses in Socioecological Agent-Based Models to Improve Their Analytical Performance and Policy Relevance

Arika Ligmann-Zielinska[1]*, Daniel B. Kramer[2,3], Kendra Spence Cheruvelil[3,4], Patricia A. Soranno[3]

1 Department of Geography, Michigan State University, East Lansing, Michigan, United States of America, 2 James Madison College, Michigan State University, East Lansing, Michigan, United States of America, 3 Department of Fisheries and Wildlife, Michigan State University, East Lansing, Michigan, United States of America, 4 Lyman Briggs College, Michigan State University, East Lansing, Michigan, United States of America

Abstract

Agent-based models (ABMs) have been widely used to study socioecological systems. They are useful for studying such systems because of their ability to incorporate micro-level behaviors among interacting agents, and to understand emergent phenomena due to these interactions. However, ABMs are inherently stochastic and require proper handling of uncertainty. We propose a simulation framework based on quantitative uncertainty and sensitivity analyses to build parsimonious ABMs that serve two purposes: exploration of the outcome space to simulate low-probability but high-consequence events that may have significant policy implications, and explanation of model behavior to describe the system with higher accuracy. The proposed framework is applied to the problem of modeling farmland conservation resulting in land use change. We employ output variance decomposition based on quasi-random sampling of the input space and perform three computational experiments. First, we perform uncertainty analysis to improve model legitimacy, where the distribution of results informs us about the expected value that can be validated against independent data, and provides information on the variance around this mean as well as the extreme results. In our last two computational experiments, we employ sensitivity analysis to produce two simpler versions of the ABM. First, input space is reduced only to inputs that produced the variance of the initial ABM, resulting in a model with output distribution similar to the initial model. Second, we refine the value of the most influential input, producing a model that maintains the mean of the output of initial ABM but with less spread. These simplifications can be used to 1) efficiently explore model outcomes, including outliers that may be important considerations in the design of robust policies, and 2) conduct explanatory analysis that exposes the smallest number of inputs influencing the steady state of the modeled system.

Editor: Frederic Amblard, University Toulouse 1 Capitole, France

Funding: Financial support for this work was provided by the Center for Water Sciences and the Environmental Science and Policy Program at Michigan State University, as well as the National Science Foundation Geography and Spatial Sciences Program Grant No. BCS 1263477. Any opinion, findings, conclusions, and recommendations expressed in this material are those of the authors(s) and do not necessarily reflect the views of the National Science Foundation. The funders had no role in study design, data collection and analysis, decision to publish, or preparation of the manuscript.

Competing Interests: The authors have declared that no competing interests exist.

* Email: ligmannz@msu.edu

Introduction

Socioecological systems are perpetually dynamic and nonlinear [1,2,3,4,5,6,7]. To account for this complexity, researchers often employ agent-based models (ABMs). Socioecological ABMs are computational models composed of heterogeneous entities (called agents) that shape a common environment representing an integrated human and natural system [2,8,9]. ABM offers a robust vehicle for simulating socioecological systems, for example landscape dynamics, by providing means of representing autonomous and decentralized decision-making that results in emergent landscape-scale characteristics (e.g., land value, land use patterns) and phenomena (e.g., biodegradation, land conservation). For example, ABMs are often used to model agricultural land systems, in which the environment is operationalized by spatial layers (maps) including land use, soil productivity, vegetative cover, and precipitation. The agents, or actors, in the system may include farmers who cultivate their land, developers who buy and sell land parcels, residents who inhabit select locations, and authorities that adopt and enforce land-related policies [10,11,12,13,14,15,16]. As with all modeling of such complex systems, ABMs are inevitably prone to uncertainty reflecting insufficient knowledge of the processes driving these coupled human-natural systems. For example, we have incomplete knowledge on the relationships and feedbacks between crop market fluctuations, farmland management, and temporal dynamics in nutrient cycling affecting such systems, nor do we fully understand the inherent randomness of environmental and social events like the popularity of agritourism during a heat wave. Not surprisingly, comprehensive evaluation of uncertainty has emerged as an important topic of social-ecological research, including environmental modeling [17,18,19,20], land use and land cover change [5,21,22], and geographic information science [23,24,25,26]. The need for proper handling of uncertainty also has been widely recognized in ABMs of socioecological systems [27,28,29].

We argue that systematic evaluation of ABM uncertainty should comprise a joint quantification of model output variability and its sensitivity to inputs (called factors) - Figure 1 [5,21,28,29,30,31]. Consequently, we propose a new framework for evaluating uncertainty of ABMs. We demonstrate how an integrated quantitative uncertainty analysis and sensitivity analysis (UA-SA) can be employed in ABM development to meet three modeling objectives: 1) to evaluate the validity of simulation results (using uncertainty analysis - UA); 2) to generate a more parsimonious model (using sensitivity analysis - SA), and 3) to prioritize input data refinement by identifying the ABM factors that are mostly responsible for model output variability (using both UA and SA). Factors comprise various uncertain model components including variables, parameters, spatial data (maps), functions, and sub-models that jointly influence ABM results - Figure 1 [32]. Uncertainty analysis (UA) evaluates how the variability of factors propagates through the model and affects the variability of output values. The objective of UA is to quantify the distribution of results given uncertain factors. Conversely, sensitivity analysis (SA) evaluates how factor variability contributes to model output. Although ABMs are relatively common in socioecological research, studies rarely include the joint use of quantitative UA-SA, suggesting that these stages of ABM evaluation are perfunctorily, if at all, undertaken [33,34,35].

Recognized as important for scientific understanding, quantitative UA-SA has been employed in a number of non-ABM studies on socioecological systems. Examples include ecological modeling of ecosystem vulnerability to climate change [36], hydrology and water use [37,38,39,40], species interactions and community stability [41], changing human environmental attitudes [42], water eutrophication [43], and coral reef degradation [44]. In ABMs studies, the most common approach involves an UA that summarizes the results of Monte Carlo simulation based on simple random sampling or, in the case of SA, running the model with extreme values of its factors or using a limited number of values, with little or no quantification of the influence of these

factors on the variability of results [1,45,46,47,48]. One possible explanation for the lack of quantitative UA-SA in socioecological ABM is the relative infancy of the AB methodology coupled with the flexible protocol for executing AB simulations. There is a need for a well-defined UA-SA framework tailored to meet the specific needs of ABM, such as handling the very large number of heterogeneous factors (at least one per agent) of a highly nonlinear model.

Quantitative UA-SA in ABMs can serve many purposes [31,49,50,51]. It can be used to strengthen trust in model realism and to eliminate model factors that have negligible influence on the variability of the output, allowing for a simpler, easier to understand model. Therefore, UA-SA together provide information on influential factors that significantly affect the variability of model results. They allow scientists to gain a deeper understanding of the complexity of the model, its uncertainties, interrelationships, and its potential future scenarios. UA-SA provide a means of asking 'what if?' questions that help to validate or disqualify the results [52]. UA-SA should be included in all ABM exercises using methods that systematically examine model factors and outputs to build credible models necessary for addressing problems at the science/policy interface [53,54]. We argue that a reliable simulation-based policy analysis requires simplified yet practical models, and that integrated quantitative UA-SA associated with ABM of socioecological systems is essential for scientific understanding for two reasons. First, uncertainty is a fundamental property of complex systems that cannot be ignored. Moreover, because a portion of this uncertainty is irreducible, a socioecological model that generates results with little or no variability has little practical value. Second, a distribution of UA outputs, including the tails and means, provides an opportunity for *exploration* of extreme system behavior. Although highly unlikely, boundary cases may result in radical changes of significant consequence to society and/or the environment. On the other hand, scientific *explanation* requires considerable accuracy, which can be achieved through reducing output variability in order to improve model performance.

In the next sections, we describe how quantitative UA-SA can be used to build ABMs for policy analysis and exploration. While still nascent, a comprehensive UA-SA have been applied to ABMs in a few previous studies [28,30,35,55,56,57]. For example, Fonoberova et al. [35] use UA-SA to identify factors for model reduction in an ABM of criminal activity, while Parry et al. [55]

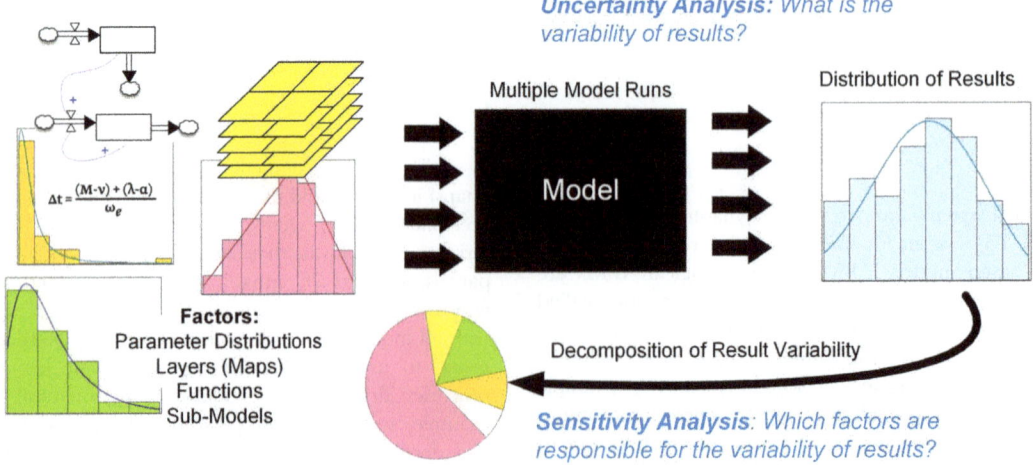

Figure 1. Uncertainty and sensitivity analyses of model output.

use UA-SA to identify the highly sensitive factors that need further refinement in an ABM of bird population. What sets our manuscript apart is the use of comprehensive UA-SA separately for model explanation and exploration, by focusing on the refinement of the most influential factors (explanation) and the reduction of the least influential factors (exploration) of a socioecological ABM. We employ variance decomposition of the ABM output - a method commonly used in ecological modeling outside of ABMs [58,59,60]. UA is applied to build a legitimate model, where the distribution of model results informs us not only about the expected value validated against independent data, but also provides information about the spread around the expected value and the extreme (boundary) results. SA is then employed to produce a parsimonious model. Two cases are examined. First, we build a practical *exploratory model* that allows scientists to simulate low-probability but high-consequence events that may be of high policy relevance. Second, we build a more *explanatory model* that provides the means of describing the system with higher accuracy. Specifically, we postulate that the explanatory power of a model lies in illuminating the core underlying processes [61] and exposing system-wide regularities [62], which manifest themselves through the mean of the output of interest. The proposed framework is applied to the problem of modeling farmland conservation and resulting land use change (from agriculture to fallow), demonstrating the utility of UA-SA for contributing to science and policy.

Materials and Methods

Uncertainty analysis

UA produces a distribution of model results (Figure 1). It requires multiple model runs, where factor values are randomly chosen from their respective distributions. Because the results of quantitative UA-SA are computationally expensive, the selection of the sampling method used to perform UA is essential. Following Saltelli et al. [63], we use quasi-random sampling that generates samples more uniformly over the entire factor space than simple random sampling. A sample is then used to execute the model, which produces an individual output value. In our case study, for example, the output is a measure of total area of land converted from agriculture to fallow. These results build a distribution of outputs that can be further analyzed using descriptive statistics. Two statistics are particularly useful: the mean that represents the central tendency of the stochastic process, and the variance that summarizes the variability of the results. Variance is then used as input to SA (UA is, therefore, a prerequisite to SA).

Sensitivity analysis using output variance decomposition

Commonly, SA involves modifying the value of one factor (while keeping the other factors constant) and observing the effects of this change on model results. This method, referred to as one-parameter-at-a-time (OAT) [33,64,65], is most often used by socioecological modelers. The prominent examples, closely related to socioecological ABMs, include the use of OAT in land use change cellular automata models to evaluate their sensitivity to map resolution and the size and configuration of neighborhoods [66,67] and the use of OAT to identify the most sensitive factors in an epidemiological ABM of the spread of measles among humans [68]. OAT popularity may be attributed to its simplicity, low computational cost, clear starting point in the form of a baseline parameter set, and the fact that the observed changes in outputs can be easily traced back to changes in specific factors [33]. Unfortunately, the utility of OAT for complex socioecological ABM is limited. The arbitrary choice of which factor to modify

and by what amount is problematic when the magnitude of key system drivers is hard to determine [28]. Also, OAT does not explore the variability of factors in combinations and, consequently, assumes a linear relationship between inputs and outputs. Finally, OAT is of limited use in exploratory modeling, because it does not test the full range of factor variability and therefore minimizes our ability to simulate extreme, but catastrophic, events. As an alternative to OAT, we utilize a global SA approach, which is based upon simultaneous perturbations of the entire model factor space, examining the factors both individually and in combinations [69,70].

Our global SA uses model output variance decomposition in which the variability of the area of fallow land (resulting from farmer agent decision making) is decomposed (partitioned) and distributed among model factors evaluated in various combinations [69,71]. Factor sensitivity is quantified using two measures referred to as a first order sensitivity index, S, and a total effects sensitivity index, ST [69,70,72]. Index S measures the independent, fractional contribution of each individual model factor to output variance. The ST index estimates the overall contribution of a given factor to output variance including its interactions with other factors. Assuming that model output Y has unconditional variance V, the indices of a given factor (i) are formalized as follows:

$$S_i = \frac{V_i}{V} \qquad \text{(Eq. 1)}$$

$$ST_i = \frac{V - VC_i}{V} \qquad \text{(Eq. 2)}$$

Where V_i is the variance of Y due to the variability of factor i alone, and VC_i is the conditional variance due to all model factors except i. The sum of all S indices (ΣS) is the fraction of output variance that can be explained by the individual factors alone. Therefore, the formula:

$$I = 1 - \sum S \qquad \text{(Eq. 3)}$$

gives the fraction of output variance due to the interactions (I) between the factors. This succinct measure of interactions can be further analyzed using the ST indices, which provide information about the total (first and higher order) influence of each factor on output variance. For more details on variance decomposition, the reader is referred to Saltelli et al. [70], Lilburne and Tarantola [69], and Homma and Saltelli [72], among others.

The (S,ST) pairs are quantified as ratios of the conditional output variances to the total variance and, thus, measure the relative contribution of each factor to output variance (Figure 2A). Factors with relatively high S (ST) values will have the greatest impact on the variability of model results. When these factors are refined or fixed to constant values, the result is a reduction in output variance. We use this property of the (S,ST) pairs to operationalize the *explanatory power* conception of modeling (Figure 2B). The major premise of model explanatory power is that additional observations used for estimating the most influential factors get us closer to an accurate representation of the underlying system. By better approximating values of the most influential factors, especially in cases where these factors dominate the output, we can unravel the interrelationships among other factors and expose model nonlinearities. Conversely, if we fix factors that have S (ST) values close to zero (i.e. the non-influential

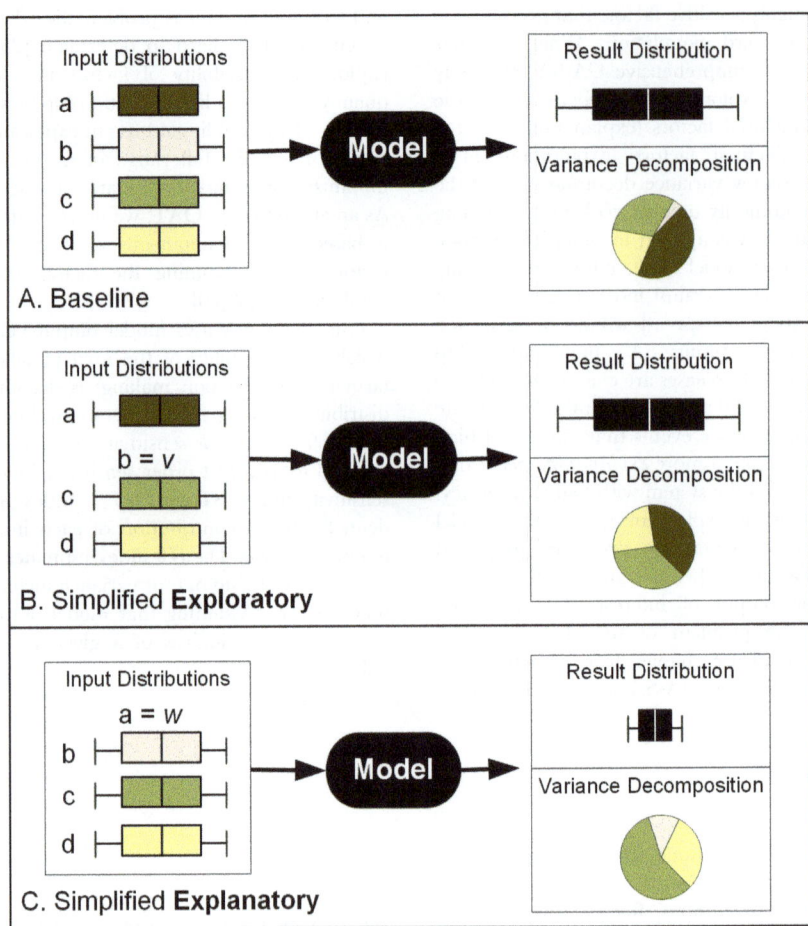

Figure 2. A framework for uncertainty and sensitivity analysis of ABMs of socioecological systems. Applying variance decomposition to simplify a stochastic model (A), and maintain its exploratory power embodied in outcome variability (B) or improving its explanatory power by reducing its outcome variability (C).

factors), we do not significantly change the variance of the results. Instead, we derive a simplified model with quantitative *exploratory power* (embodied in variance of a given output) equal to this model's baseline implementation (Figure 2C).

Case study: ABM of Michigan farmer enrollment in the Conservation Reserve Program

We use quantitative UA-SA for land use model simplification and factor prioritization. The goal is to build a simpler representation of an ABM with two distinct objectives: policy analysis that would benefit from exploratory modeling [73], and advancing science through explanatory modeling [74]. Our case study considers the participation of farmers in a land conservation program aimed at protecting ecologically valuable areas.

Published research suggests that farmers' decision to participate in a land conservation program is driven by both financial and nonmonetary drivers [75,76,77,78]. These findings are based on conventional statistical analyses of survey data. Few studies have explicitly modeled the decision processes and analyzed the resulting spatial configurations of conserved land [16]. Following this observation, we develop an ABM of agricultural land conservation decision making. The model simulates voluntary participation in the U.S.'s largest land protection program, the Conservation Reserve Program (CRP) [79]. We examine CRP enrollment in southwest Michigan, U.S. (Figure 3). The area

covers 985 square km, with 2687 farmland parcels. This area is characterized by large proportions of agricultural land with about 3% of farmland enrolled in CRP according to the U.S. 2007 agricultural census [80].

Model description

In the ABM reported herein, CRP enrollment is simulated based on well-defined federal regulations [79]. Two types of decision makers are involved in this process (Figure 4): [1] farmers who decide whether or not to participate in CRP and [2] the Farm Service Agency (FSA), which evaluates, selects, and accepts farmer enrollment offers. The basic spatial unit of CRP decision-making is a farmland parcel. During the model setup, a farmer agent (FA) is associated with various socio-demographic and economic factors (*land tenure*, operator's *retirement*, and the *value of production* on a farm; described in later sections). The FA is then assigned to a parcel.

As a first step, the FA calculates their willingness to enroll in the CRP based on decision criteria (factor values) including financial motives and nonmonetary drivers. To calculate the willingness to enroll, we apply a group of aggregation operations (aka *decision rules*) called Ordered Weighted Averaging - OWA [81,82]. OWA allows for simple representations of different conceptions of risk related to CRP enrollment, which, after acceptance, is mandatory for at least ten years. OWA decision rules range from the most

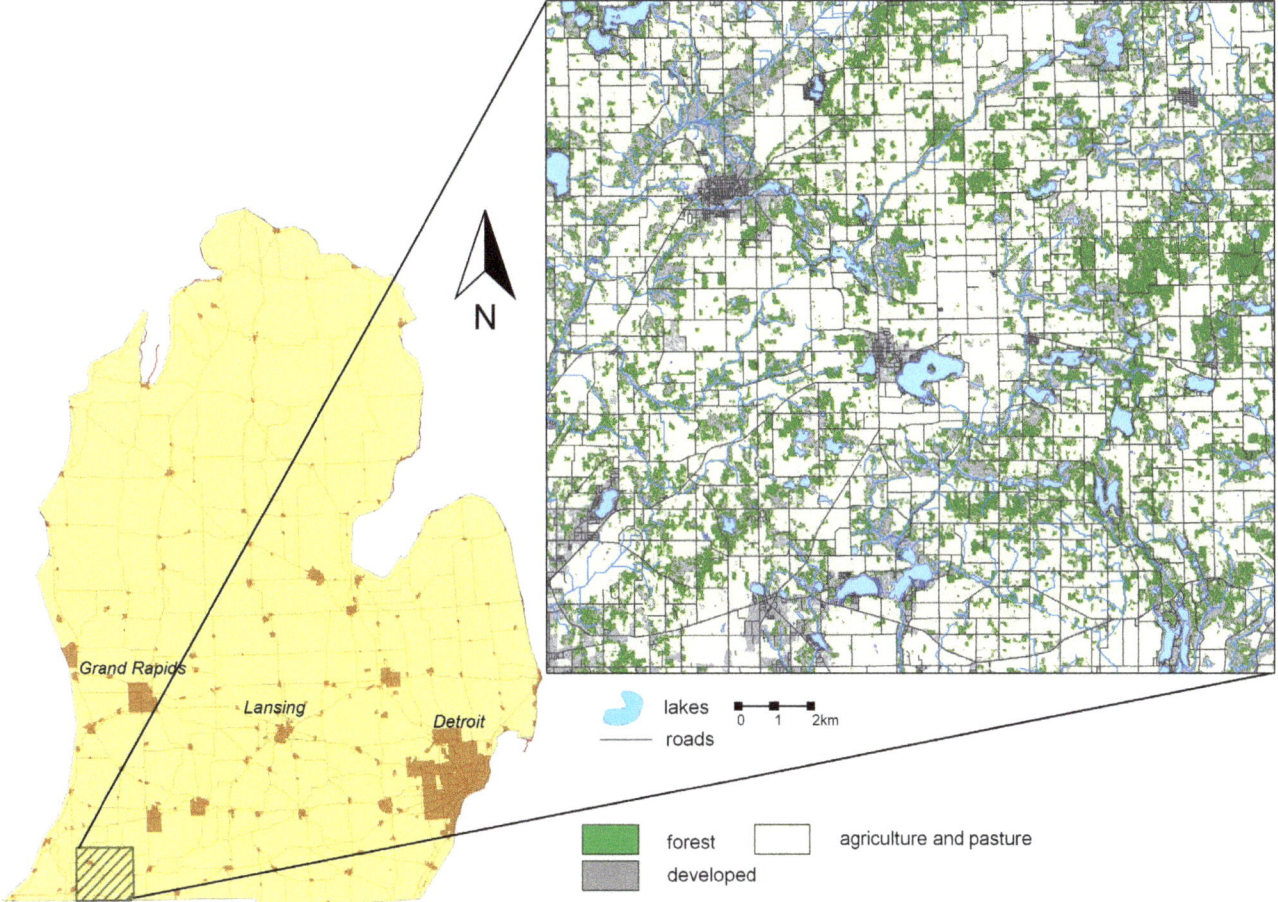

Figure 3. Study area in Michigan, U.S.

risk-averse, where values of all decision criteria must be positive, to the most risk-taking where only one decision criterion needs to be nonzero. For example, if an agent makes a decision to participate in the program based on low value of production AND retirement, that decision is risk-averse, whereas if the decision is made based on low value of production OR retirement, the agent is risk-taking. Agent's willingness to enroll is extended with a simple group interaction mechanism, in which farmers adopt imitative behavior [83] based on the decisions made by other proximal farmers. An FA incorporates into its decision mechanism the *density of enrollment* in its neighborhood, measured as the ratio of CRP-enrolled neighbors to the total number of neighbors within 0.5 km from the parcel.

If an FA's willingness to enroll exceeds an empirically derived threshold, the agent selects a *fraction* of its parcel for potential enrollment [77]. Eligible sites in the parcel (pasture and cropland) are rank ordered based on distance to water, distance to forest, and land slope, and the first *fraction* of sites is selected. Next, the FA builds an offer by calculating an expected annual payment based on soil rental rates [84]. To increase its offer competitiveness, the FA reduces the payment using a *bid* rate established by USDA [85] and estimates a discounted annual payment (DAP). The FSA agent collects offers from the FAs and selects a subset (*n*) of them based on the *environmental benefit index* (described in the following section), the signup budget, and the DAP. Next, FSA announces the signup results leading to land use change from agriculture to fallow. In sum, the location and area of fallow land

results both from the FAs' decisions to participate and the FSA's decision to accept their offers. The process of CRP signup is repeated annually for ten years (minimum CRP contract length). Land use change maps constitute the output of the model. They are summarized into the *total area of fallow land*. This scalar is used in the UA and SA.

Model input data

The ABM uses a number of factors, some that are readily available and some that we derived from auxiliary resources, including land uses obtained from 2010 cropland raster layers [86], freshwater ecosystems from a lakes and rivers geodatabase [87], soil data [88], and slope [89]. The major geoprocessing operations were mapping land eligible for CRP [79], deriving spatial layers that influence FA's choice of area to enroll (Figure 4 - distance to water, distance to forest, and land slope), and generating the soil rental rate (SRR) and the environmental benefits layers. The SRR layer (Figure 5) was derived from a soil productivity index map for the State of Michigan [90] and county cash rental rates [84].

Deriving environmental benefits

Ranking of CRP offers is based mainly on their Environmental Benefits Index (EBI) values. EBI is a composite index based on multiple rated criteria describing benefits for wildlife, water quality, soil erosion, long term maintenance of installed vegetation, and air quality [85]. To optimize environmental benefits per dollar

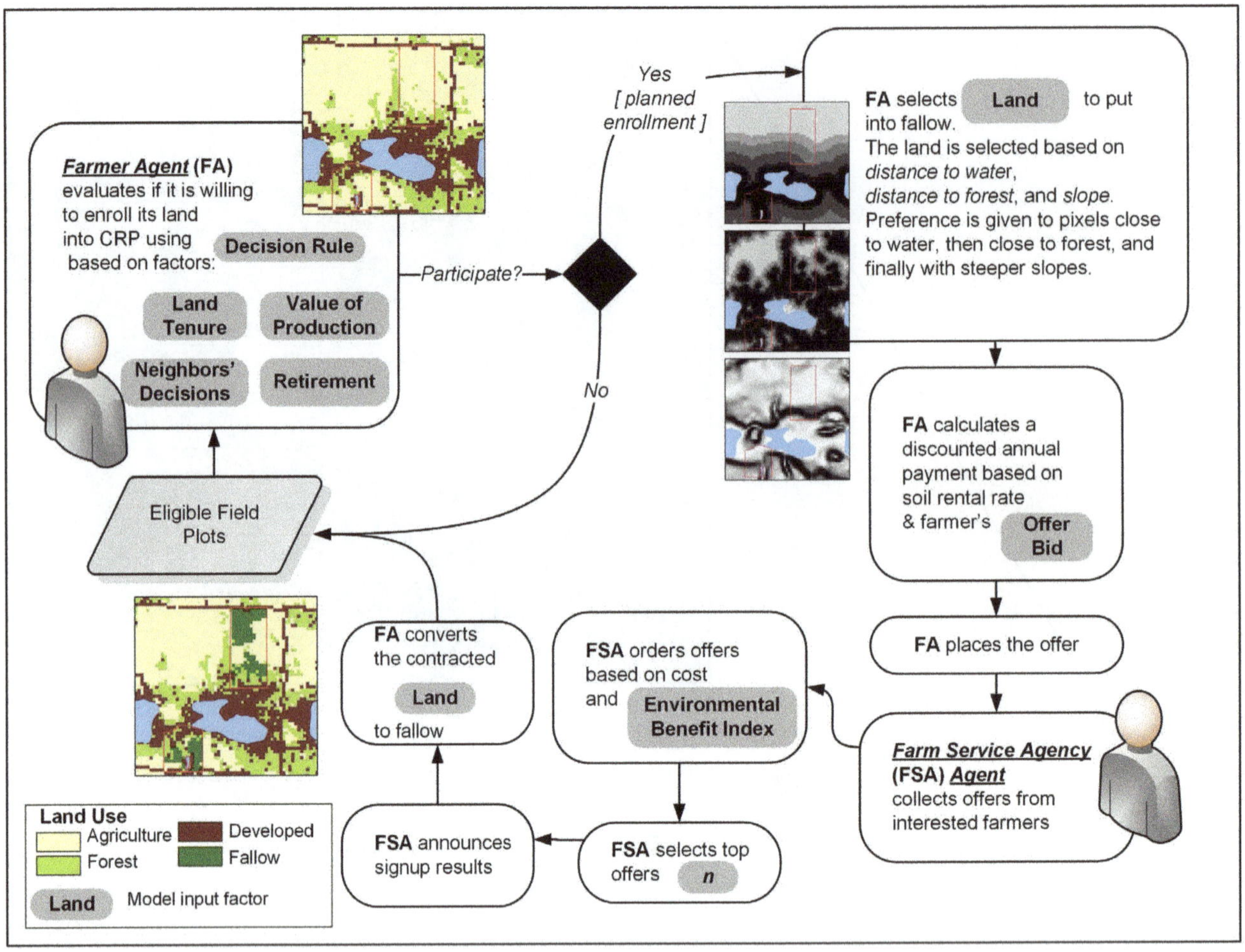

Figure 4. Agent-based model of enrollment in Conservation Reserve Program.

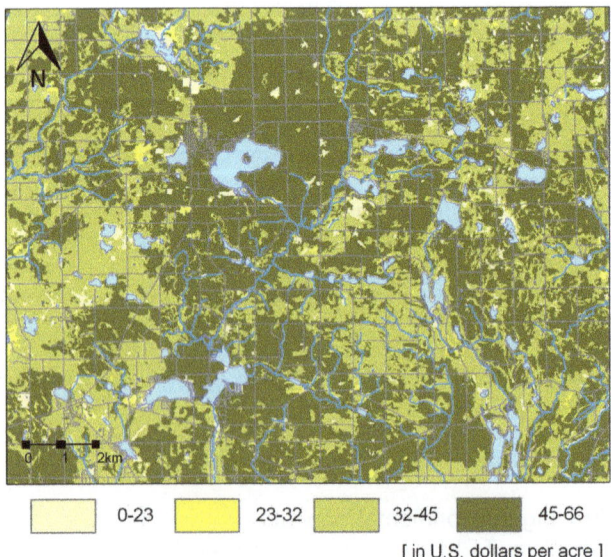

Figure 5. Soil rental rates (the southeast fragment of the study area).

Land Use
- Agriculture
- Forest
- Developed
- Fallow
- Land — Model input factor

0-23 | 23-32 | 32-45 | 45-66
[in U.S. dollars per acre]

expended for rental payments, the EBI is adjusted by a cost and bid rating scale. Offers with lower total annual payments and higher bids (voluntary reduction by the farmer of the offer value below the maximum payment) receive highest priority.

EBI can be quantified in many different ways, resulting in substantial uncertainty. Consequently, we used alternative conceptions of the benefits (Figure 6), weighed by their respective point scores [85] in various combinations to generate six different benefit layers used interchangeably in the ABM. The values of EBI range from 50 to 350 points and the six alternative EBI surfaces have moderate to high positive correlation (min Pearson's r = 0.35, max Pearson's r = 0.89).

Farmer's decision to participate

Statistical and econometric studies of CRP enrollment point to five major categories of independent variables used to predict participation in land conservation programs: farm, household, and environmental characteristics; government assistance; and farmers' attitude and perception [77,78,91,92,93,94]. We used USDA's Agricultural Resource Management Survey (ARMS) [95], a semi-annual survey of American farming businesses and households sponsored by USDA's Economic Research Service and the National Agricultural Statistical Service, as our data source for farm finances, production practices, and household characteristics.

We developed an *a priori* set of candidate multiple regression models to understand farmer participation in the CRP based on

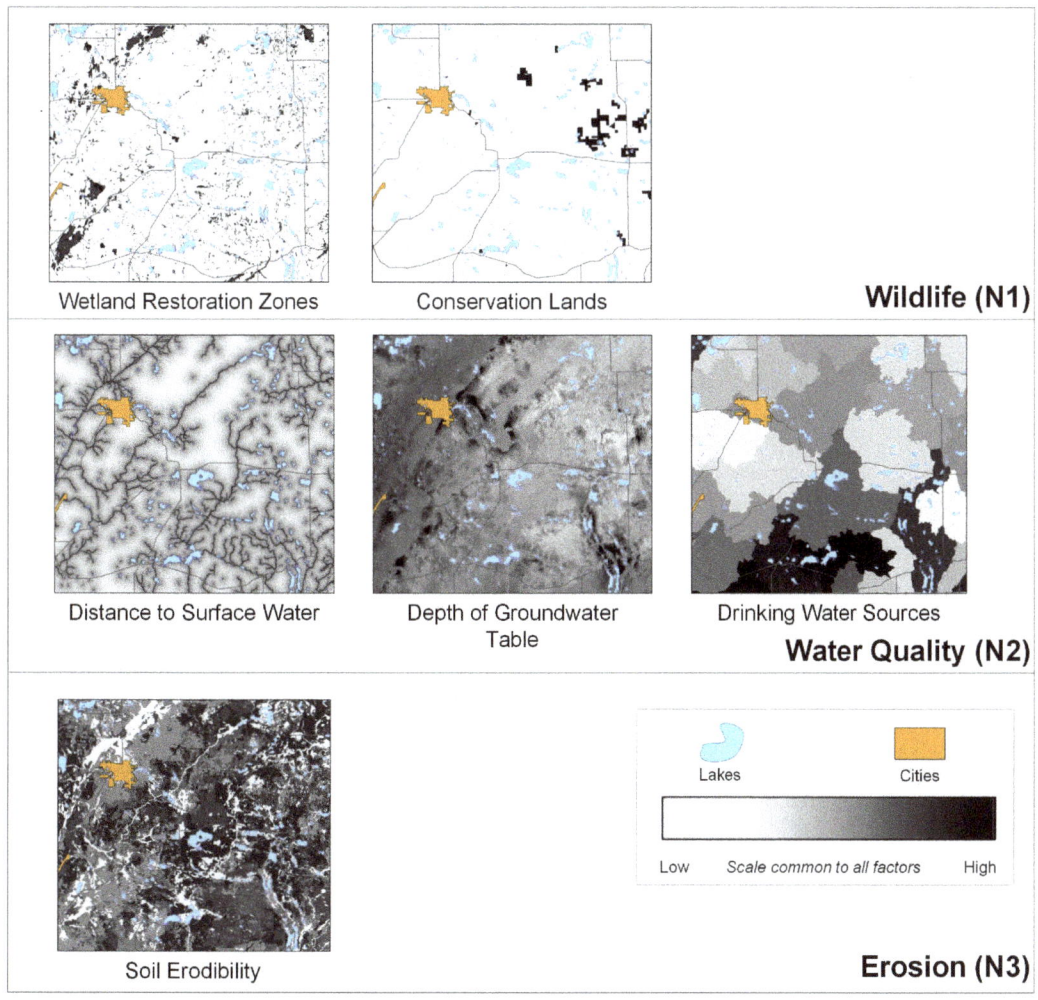

Figure 6. Benefit layers used to calculate six composite EBI surfaces. Each EBI surface is a sum of one of the N1 layers, one of the N2 layers, and the N3 layer. All N1, N2, and N3 layers are standardized based on their respective point scales [78]. The remaining benefit criteria used in EBI calculation (vegetation and air quality) were not used due to their negligible role in the area of study.

results of prior studies about farmer participation in land conservation programs. Using a multimodel inference analytical approach based on the Akaike information criterion (AIC), models with relatively low AIC values were considered the most parsimonious, balancing bias and variance of model predictions [96]. We assigned relative strengths of evidence to each candidate model according to AIC weights and evaluated explanatory variables in terms of deviance explained. From this process, farmers' retirement status (RETIREMENT), the total value of farm production (PRODUCTION), and the ratio of farmed to owned acres (TENURE) received the most support in terms of deviance explained (Table 1). Consequently, these three attributes became FA's decision criteria (Figure 4). Finally, the dependent variable used in the regression models (yes/no CRP participation decision) was used to estimate the threshold for FA's willingness to enroll equal to 0.87, which reflects the ratio of farmers enrolled in CRP to total farmers in 2010 in Michigan based on the ARMS data.

Factor distributions

Given the CRP enrollment procedure and the available data, we identified nine factors for the ABM. Seven factors are attributed to the FA and the remaining two to FSA (Figure 4). The three independent explanatory variables used by the FA in the enrollment decision (land tenure - TENURE, value of production - PRODUCTION, and operator's retirement status - RETIREMENT) were included as individual, farmer-level factors in the form of empirically-derived probability density functions - PDFs (Table 1). These functions epitomize the financial and nonmonetary drivers affecting the FA's land conservation decision.

Empirical data for other factors were only partially available. Consequently, we used a uniform PDF for the other model factors [97]. The density of enrollment in the FA's neighborhood depends on how the neighborhood is defined. In our model, we delineated neighborhood based on distance from FA's parcel, which varied from 500 to 1500 m. For OWA, we assumed various magnitudes of attitude to risk, where each level had an equal probability of selection. To select the fraction of land for potential enrollment (LAND in Figure 4), we assigned a uniform distribution from 1% to 100% of parcel area (partial to full land enrollment). The number of offers accepted by FSA was based on the budget allocated to CRP per county [98].

Table 1. Probability distributions for factors used in ABM simulations.

Factor Name	Factor Description	Probability Density Function
RETIREMENT	Primary operator retired from farming (0 -retired, 1- working).	D = {(0,.06), (1,.94)}
PRODUCTION	Total value of production on a farm (normalized).	D = {(0,0), (.2,.06), (.4,.06), (.6,.11), (.8,.15),(1,.62)}
TENURE	Ratio of owned to operated acres.	D = {(0,.04), (.2,.14), (.4,.18), (.6,.14), (.8,.15),(1,.35)}
DE	Extent of FA's neighborhood used to calculate the density of enrollment in FA's geographic vicinity.	U = {.5 km to 1.5 km with increments of 100 m, with equal probability of selection}
OWA	FA decision rule based on ordered weighted averaging, with varying attitudinal character i.e. the level of "orness" [81].	D = {17 combinations with equal probability}
LAND	Fraction of parcel to set aside for conservation.	U = (0, 1]
BID	Voluntary reduction by the farmer of the offer value below the maximum payment rate.	D = {0% to 16% of offer reduction with increments of 1, with equal probability of selection}
EBI	Environmental benefits index dataset.	D = {6 layers with equal probability}
n	Number of offers (contracts) accepted annually by FSA.	D = {18 to 28 with increments of 1, with equal probability of selection}

U - uniform distribution, D - discrete distribution (value, probability). All factors were normalized to [0.0, 1.0]. All data are for CRP sign-up 41 in 2010 [108].

Design of experiments

Our simulations include three computational experiments. In experiment one (EXP1, 2560 model runs), our base scenario (Figure 2), we run Monte Carlo simulations using all nine factors. In experiment 2 (EXP2, 1536 runs), the simplified exploratory scenario, we include only those factors that most influence the variability of the total area of fallow land (AREA), calculated at the end of model execution. The simplified explanatory scenario with variance reduction is implemented in experiment 3 (EXP3, 2304 runs), where we fix the value of the most influential factor from EXP1, leaving the remaining factors unchanged. All simulations were run using high performance computing at Michigan State University. Factor samples were produced using the quasi-random Sobol' experimental design [99], which is the most optimal method to approximate the values of the S and ST indices [63,69]. The ABM was implemented in the Python programming language (http://www.python.org/) and the (S, ST) indices were computed using the SimLab software package for uncertainty and sensitivity analysis (http://ipsc.jrc.ec.europa.eu/?id=756). Statistical regressions were completed utilizing R, version 3.0.2.

Results

The results of our ABM simulations are land use maps with one additional category, *fallow land*, when compared to our input land use maps. Example results from EXP1 are depicted in Figure 7A. Because FAs make decisions on a site-by-site basis, most of the parcels enrolled in CRP at the end of model execution have only a portion of their land enrolled in CRP. Figure 7B illustrates the frequency of site enrollment (number of times a site is enrolled for all ABM executions). Note the considerable spatial variability in site enrollment. Most sites are only selected 5–8% of the time. We hypothesize that this dispersed enrollment is caused by the complex interactions between the nine factors. We utilize UA-SA to illuminate these complexities and focus on the causes of CRP enrollment variability.

Uncertainty analysis

To explore the variability of CRP enrollment, we performed UA by examining the distribution of the total area enrolled in CRP (*total fallow land area* - AREA). Figure 8A summarizes the

distributions of AREA for each of the three experiments. We also calculated the mean and variance of AREA per experiment. The mean CRP AREA was between 5120 and 5150 acres, with results of no experiment being significantly different from any other (one-way ANOVA ($F(2,6397) = 0.961$, $p = 0.38$), confirming that all ABM representations are equivalent. Since our experimental design uses a more uniform (quasi-random) sampling compared to the typical ABM Monte Carlo simulations that are based on simple random sampling, we can infer that the calculated mean is indeed the true (accurate) measure of central tendency in AREA distribution. Consequently, we can use this value to validate the model against an independent dataset. The U.S. Agricultural Census [80] reported 5490 acres (~24,700 map units) of CRP land in the study area, which is about 7% more than the mean for the baseline EXP1, rendering the results plausible for further evaluation.

We used the variance to evaluate the degree of AREA variability. As expected, the variances of EXP1 and EXP2 are approximately equal. Consequently, the simplified model used in EXP2 can be used in *exploratory* analysis without the loss of variability necessary when evaluating the CRP policies. Conversely, EXP3 (in which data on the most sensitive factor was refined) produces a distribution much more centered around the mean when compared to the baseline. Consequently, the simplified and refined ABM used in EXP3 can be used in *explanatory* analysis of the social, economic, and ecological processes associated with CRP participation. The following section explains the details on how we arrived at these two ABM simplifications.

Sensitivity analysis and model simplification

UA alone does not provide any information about the influence of individual factors on AREA variability. Without SA it would be impossible to build the simpler yet equivalent versions of our ABM. By performing the decomposition of AREA variance, we can identify factors in the initial version of the ABM that can be either reduced without the loss of ABM exploratory power (EXP2), or refined if our objective is to explain the processes (EXP3).

Figure 8B shows pie charts of the S and ST indices for all three experiments. Because factors with relatively high values of S have the most effect on the of total fallow land area, we look for factors

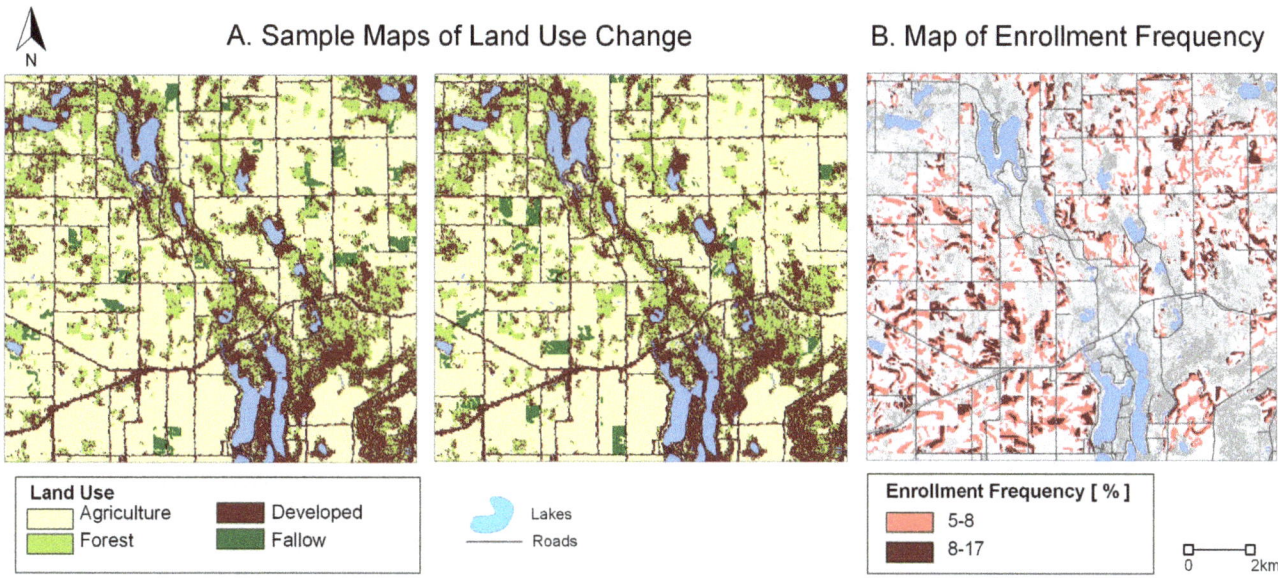

Figure 7. Example output land use maps (A), and the frequency of agriculture-to-fallow conversion (B). For clarity, only the southeast portion of the study area is shown.

that, if fixed singly, would most reduce the variance of AREA. In the baseline EXP1, the highest S is recorded for the number of offers accepted by the FSA (n). Trivially then, the extent of farmland conservation is first and foremost driven by the FSA signup choices. Given that CRP is competitive among farmers, the ABM confirms the observation that program participation depends on the federal budget allocated to annual payments. Only about 10% of AREA variance can be attributed to factor interactions, which occur between n, OWA, BID, LAND, and RETIREMENT. Due to their influence, these five factors were included in a simplified version of the model in EXP2 (central box plot and pie charts of Figure 8A and 8B). Because we only excluded factors that had negligible influence on the distribution of AREA (which were set to constant values - either their mean or median), the resulting and baseline distributions are nearly identical, including their means and variances. More importantly, variance decomposition generated S and ST indices consistent with the original model formulation. We can therefore conclude that our ABM formulation used in EXP2 meets the criteria of a simplified *exploratory* model (Figure 2). This simplified model is more efficient computationally - an indispensable feature for models used in policy analysis [73]. At the same time it maintains result variability, which can be of use when identifying the less probable but highly consequential policy scenarios.

In EXP3, we set n = 23 (its midpoint number of offers), to demonstrate how the behavior of our ABM changes when, instead of fixing the negligible factors, we do so for the most influential factor. This scenario imitates a situation in which we obtain more accurate data on the most sensitive factor of the model. There was a significant reduction in AREA (Figure 8A), and although the mean is roughly the same as its initial value, the spread around the mean decreased by 64% compared to EXP1. EXP3 is also characterized by a more complex behavior than the first two experiments. Only 35% of this reduced variance can be explained by individual factors (Figure 8B right). The total effect indices suggest that non-monetary motives (perception of risk and FA's retirement) are equally important in FA's decision as the financial drivers (BID, LAND, PRODUCTION, TENURE). We hypoth-

esize that a portion of these interactions can be attributed to the functional relationships between factors. For example, if the fraction of land to convert in a particular parcel has a relatively high value while the OWA rule is conjunctive (AND-only) [100], a large portion of land has the potential to become fallow. However, if the OWA rule is disjunctive (OR-only), an offer can be accepted (and the land can be set to fallow) even when the fraction of land to convert to fallow is relatively low, provided that the other factors (RETIREMENT, PRODUCTION, TENURE) compensate for LAND and encourage land conservation. In summary, while we reduced the range of the distribution for AREA in EXP3, we also exposed more complex dependency among the remaining factors than initially observed. By "improving" the most influential factor, we illuminated the complexity of FA's decision making. We can therefore postulate that this simplified ABM carries more *explanatory* power than the original model.

Limitations

Our combined quantitative UA-SA framework serves as a tool for better-informed ABM building. It leads to equivalent but simpler representations of a given socioecological system. Output uncertainty can be greatly reduced if more effort is put into improving the quality of data on the most influential factors (factor prioritization) through additional field studies, surveys, or auxiliary databases. However, the UA-SA framework also has limitations due to two design aspects: factor distributions and the type of output variables used (i.e., the way we measure or assess model results). A different output variable (e.g. the patchiness of fallow land, the cost of vegetation installation, or the long term reduction in nutrient loading to lakes) might point to a different set of influential factors. For example, the use of spatial metrics applied to output land use change maps [101] may lead to alternative explanations of model uncertainty [30]. Similarly, the type and characteristics of the probability distributions used for each factor (e.g. uniform versus normal distribution for LAND) could influence both the variability of outputs and the relative contribution of factors to this variability.

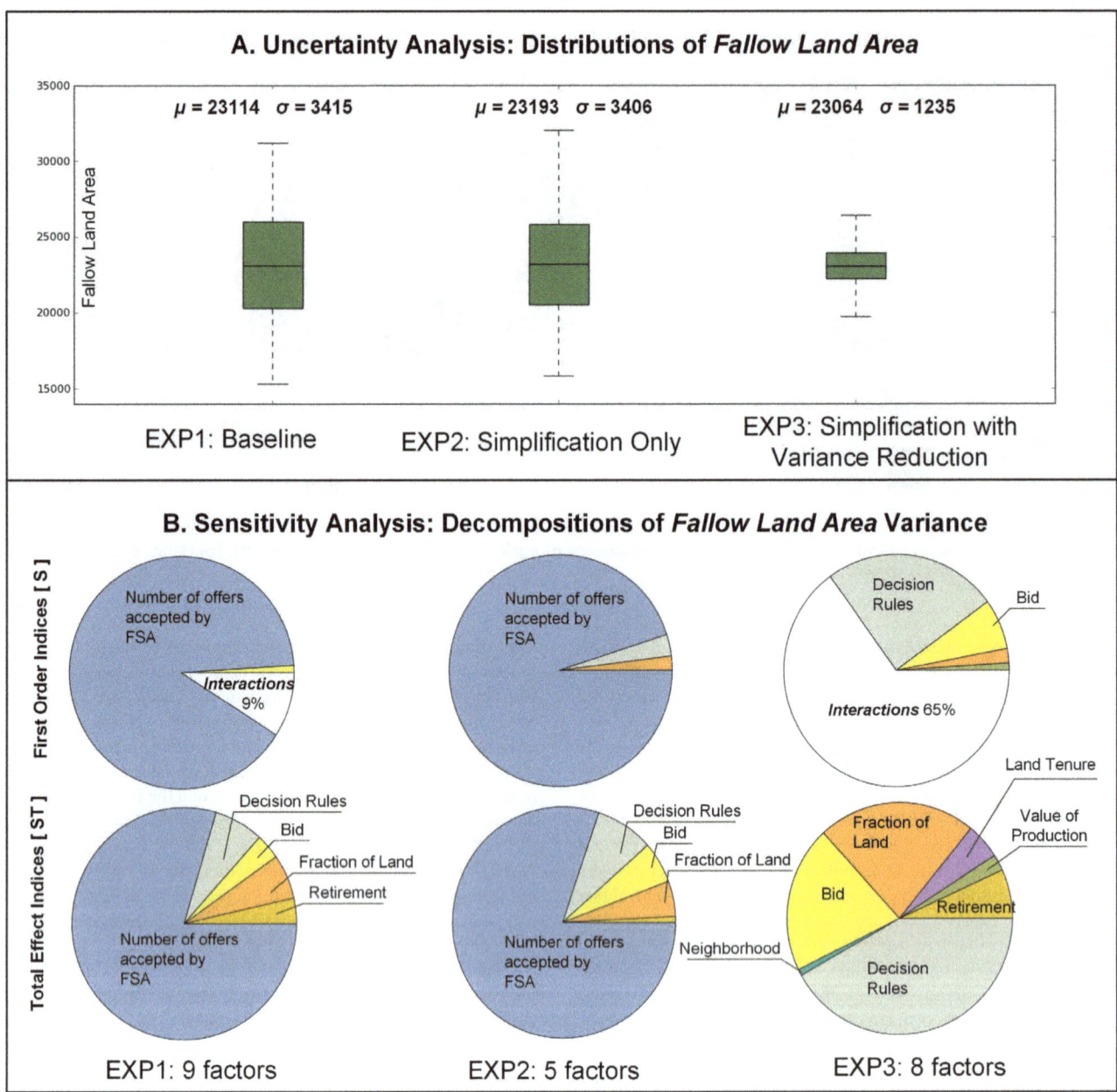

Figure 8. Results of uncertainty (A) and sensitivity (B) analysis for the output variable fallow land area. Fallow land area is reported in map units (equivalent of 30 m). Factor labels used in text: number of offers accepted by the Farm Service Agency - n, payment reduction used by the farmer agent to increase offer competitiveness - BID, FA's decision rule - OWA, fraction of farmland enrolled in CRP - LAND, FA's retirement status - RETIREMENT, FA's value of production - PRODUCTION, land tenure - TENURE, density of enrollment in the neighborhood - DE, measurement of environmental benefits - EBI, factor interactions - I (Equation 3).

The ABM presented here is of limited use for natural resource management practice. Data for most of the factors are either simulated or come from secondary sources and some of the mechanisms are poorly defined. Future model improvements will require surveys of and interviews with farmers and government officials. The recent decline in CRP enrollment suggests that increasing crop prices and government subsidies may play a significant role in the extent of land conservation [102], indicating the importance of such research. Finally, more insight into the spatial configuration of fallow land (connectivity, clustering, or dispersion of fallow land) may be necessary to better evaluate the ecological benefits of land conservation resulting in the prioritization of protected areas.

Discussion

ABMs have distinct advantages over other modeling approaches due to their abilities to couple human and natural systems, to incorporate micro-level behaviors among interacting agents, and to understand emergent phenomena due to these interactions. Their use thus far has been primarily by researchers for descriptive and predictive purposes [103]. This fact may explain their limited use in policy-making; ABMs' abilities to make accurate predictions have been questioned [62,104]. We have addressed this perceived limitation using our quantitative UA-SA approach by identifying and fixing the values of the most influential factors, thereby reducing the variance of model results. Doing so allows researchers to gain a greater understanding of the individual and interactive effects of different model factors. Further, by controlling the factors that explain the most variation in the output, researchers can expose the smallest number of factors that influence the steady state of a system. In our CRP example, we fixed the number of offers accepted by the FSA in our exploratory model (EXP 3), thereby reducing the number of factors by one as compared to the baseline model. Although the mean of our output variable, fallow land area, was essentially the same as that of the baseline, the variance decreased dramatically. Thus, this explanatory model revealed complex and important interactions among the remaining factors.

We also used the quantitative UA-SA approach to improve the ABM's policy relevance. Lempert [105] argued that ABM policy relevance might be improved if utilized for exploratory rather than predictive purposes, reflecting the fact that there is often great uncertainty and little agreement among stakeholders regarding complex, dynamic processes and corresponding decisions. Whereas his suggestion was to exercise large numbers of model runs and use various criteria including robustness, resilience, and stability to evaluate different policies, we have offered a more tractable approach. By identifying the most influential factors and ignoring others, we developed an ABM model for exploratory purposes; a simplified model with no loss in output that allows for the exploration of various policy scenarios, including rare but potentially catastrophic events. In our example, our exploratory model (EXP 2) used only five factors as compared to nine in the baseline model. Yet, the mean and variance of our output variable, fallow land in conservation, changed little from the baseline. Thus, by reducing model factors, we are able to efficiently explore different, policy-relevant scenarios.

Interest in the study of complex socioecological systems or coupled human and natural systems has risen concomitantly with the recognition of profound challenges in the Anthropocene including climate change, biodiversity loss, land use change, alteration of nitrogen and phosphorus cycles, and the depletion of freshwater [106]. Our ability to address these challenges depends greatly on how well we can make decisions despite great uncertainty. Although utilizing a variety of approaches is certainly of value [107], ABMs will likely play an important role in these efforts. Our intent, in utilizing a quantitative UA-SA approach, was to expand ABMs explanatory and exploratory potentials, contributing both to scientific efforts to increase our knowledge and predictive abilities and to policy requirements of making good decisions without complete knowledge.

Supporting Information

Code S1 Pseudo code of the main routine in the Agent-Based Model of Participation in the Conservation Reserve Program. CRP: Conservation Reserve Program, FALLOW: total area converted to fallow (in pixels), FSA: Farm Service Agency, OWA: ordered weighted averaging decision rule, SITE: a pixel belonging to a given farm parcel. Factors are presented using uppercase bold fonts.

Author Contributions

Conceived and designed the experiments: ALZ. Performed the experiments: ALZ. Analyzed the data: ALZ DBK. Contributed reagents/materials/analysis tools: ALZ DBK KSC PS. Wrote the paper: ALZ DBK KSC.

References

1. An L, Linderman M, Qi J, Shortridge A, Liu J (2005) Exploring complexity in a human-environment system: an agent-based spatial model for multidisciplinary and multiscale integration. Annals of the Association of American Geographers 95: 54–79.
2. Parker DC, Manson SM, Janssen MA, Hoffmann MJ, Deadman P (2003) Multi-agent systems for the simulation of land-use and land-cover change: A review. Annals of the Association of American Geographers 93: 314–337.
3. Parker DC, Entwisle B, Rindfuss RR, Vanwey LK, Manson SM, et al. (2008) Case studies, cross-site comparisons, and the challenge of generalization: comparing agent-based models of land-use change in frontier regions. Journal of Land Use Science 3: 41–72.
4. Rindfuss RR, Entwisle B, Walsh SJ, An L, Badenoch N, et al. (2008) Land use change: complexity and comparisons. Journal of Land Use Science 3: 1–10.
5. Verburg PH, Kok K, Pontius JRG, Veldkamp A (2006) Modeling Land-Use and Land-Cover Change. In: Lambin EF, Geist HJ, editors. Land-Use and Land-Cover Change: Local Processes and Global Impacts. Berlin: Springer. pp. 117–135.
6. Verburg PH (2006) Simulating feedbacks in land use and land cover change models. Landscape Ecology 21: 1171–1183.
7. NRC (2013) Advancing Land Change Modeling: Opportunities and Research Requirements: The National Academies Press.
8. Bousquet F, Le Page C (2004) Multi-agent simulations and ecosystem management: a review. Ecological Modelling 176: 313–332.
9. Ligmann-Zielinska A (2010) Agent-based models. In: Warf B, editor. Encyclopedia of Geography Sage Publications. Thousand Oaks, USA: Sage Publications.
10. Huang Q, Parker DC, Sun S, Filatova T (2013) Effects of agent heterogeneity in the presence of a land-market: A systematic test in an agent-based laboratory. Computers, Environment and Urban Systems 41: 188–203.
11. Berger T (2001) Agent-based spatial models applied to agriculture: a simulation tool for technology diffusion, resource use changes and policy analysis. Agricultural Economics 25: 245–260.
12. Bert FE, Podestá GP, Rovere SL, Menéndez ÁN, North M, et al. (2011) An agent based model to simulate structural and land use changes in agricultural systems of the argentine pampas. Ecological Modelling 222: 3486–3499.
13. Evans TP, Phanvilay K, Fox J, Vogler J (2011) An agent-based model of agricultural innovation, land-cover change and household inequality: the transition from swidden cultivation to rubber plantations in Laos PDR. Journal of Land Use Science 6: 151–173.
14. Happe K, Kellermann K, Balmann A (2006) Agent-based analysis of agricultural policies: an illustration of the agricultural policy simulator AgriPoliS, its adaptation, and behavior. Ecology and Society 11.
15. Schreinemachers P, Berger T (2011) An agent-based simulation model of human-environment interactions in agricultural systems. Environmental Modelling & Software 26: 845–859.
16. Sengupta R, Lant C, Kraft S, Beaulieu J, Peterson W, et al. (2005) Modeling enrollment in the Conservation Reserve Program by using agents within spatial decision support systems: an example from southern Illinois. Environment and Planning B: Planning and Design 32: 821–834.
17. Beven K (2008) Environmental Modelling: An Uncertain Future? New York: Routledge. 328 p.
18. CCSP (2009) Best Practice Approaches for Characterizing, Communicating, and Incorporating Scientific Uncertainty in Climate Decision Making. In: M. . Granger Morgan HD, Max Henrion, David Keith, Robert Lempert, Sandra McBride, Mitchell Small, and Thomas Wilbanks, editor. Washington DC: Climate Change Science Program and the Subcommittee on Global Change Research. National Oceanic and Atmospheric Administration. pp. 96 pages.
19. Larocque GR, Bhatti JS, Boutin R, Chertov O (2008) Uncertainty analysis in carbon cycle models of forest ecosystems: Research needs and development of a theoretical framework to estimate error propagation. Ecological Modelling 219: 400–412.
20. Warmink JJ, Janssen JAEB, Booij MJ, Krol MS (2010) Identification and classification of uncertainties in the application of environmental models. Environmental Modelling & Software 25: 1518–1527.
21. Pontius R, Neeti N (2010) Uncertainty in the difference between maps of future land change scenarios. Sustainability Science 5: 39–50.
22. Pontius RG (2000) Quantification error versus location error in comparison of categorical maps. Photogrammetric Engineering and Remote Sensing 66: 1011–1016.
23. Couclelis H (2003) The Certainty of Uncertainty: GIS and the Limits of Geographic Knowledge. Transactions in GIS 7: 165–175.
24. Fisher P (1991) Models of uncertainty in spatial data. Geographical information systems: Principles and applications 1: 191–205.

25. Goodchild MF (2008) Imprecision and Spatial Uncertainty. In: Shekhar S, Xiong H, editors. Encyclopedia of GIS. New York: Springer.

26. Zhang J, Goodchild MF (2002) Uncertainty in Geographical Information. London: Taylor & Francis.

27. Schindler J (2013) About the Uncertainties in Model Design and Their Effects: An Illustration with a Land-Use Model. Journal of Artificial Societies and Social Simulation 16: 6.

28. Ligmann-Zielinska A (2013) Spatially-explicit sensitivity analysis of an agent-based model of land use change. International Journal of Geographical Information Science 27: 1764–1781.

29. Brown DG, Verburg PH, Pontius Jr RG, Lange MD (2013) Opportunities to improve impact, integration, and evaluation of land change models. Current Opinion in Environmental Sustainability 5: 452–457.

30. Ligmann-Zielinska A, Sun L (2010) Applying Time Dependent Variance-Based Global Sensitivity Analysis to Represent the Dynamics of an Agent-Based Model of Land Use Change. International Journal of Geographical Information Science 24: 1829–1850.

31. Saltelli A, Chan K, Scott EM (2000) Sensitivity Analysis. Chichester, England: Wiley-Interscience. 475 p.

32. Campolongo F, Saltelli A, Sorensen T, Tarantola S (2000) Hitchhiker's Guide to Sensitivity Analysis. In: Saltelli A, Chan K, Scott EM, editors. Sensitivity Analysis. Chichester, England: Wiley-Interscience. pp. 15–47.

33. Saltelli A, Annoni P (2010) How to avoid a perfunctory sensitivity analysis. Environmental Modelling & Software 25: 1508–1517.

34. Saltelli A, D'Hombres B (2010) Sensitivity analysis didn't help. A practitioner's critique of the Stern review. Global Environmental Change 20: 298–302.

35. Fonoberova M, Fonoberov VA, Mezić I (2013) Global sensitivity/uncertainty analysis for agent-based models. Reliability Engineering & System Safety 118: 8–17.

36. Chu-Agor ML, Muñoz-Carpena R, Kiker G, Emanuelsson A, Linkov I (2011) Exploring vulnerability of coastal habitats to sea level rise through global sensitivity and uncertainty analyses. Environmental Modelling & Software 26: 593–604.

37. Makler-Pick V, Gal G, Gorfine M, Hipsey MR, Carmel Y (2011) Sensitivity analysis for complex ecological models - A new approach. Environmental Modelling & Software 26: 124–134.

38. Nossent J, Elsen P, Bauwens W (2011) Sobol' sensitivity analysis of a complex environmental model. Environmental Modelling & Software 26: 1515–1525.

39. Soboll A, Elbers M, Barthel R, Schmude J, Ernst A, et al. (2011) Integrated regional modelling and scenario development to evaluate future water demand under global change conditions. Mitigation and Adaptation Strategies for Global Change 16: 477–498.

40. Yang J (2011) Convergence and uncertainty analyses in Monte-Carlo based sensitivity analysis. Environmental Modelling & Software 26: 444–457.

41. Hosack GR, Li HW, Rossignol PA (2009) Sensitivity of system stability to model structure. Ecological Modelling 220: 1054–1062.

42. Mosler H-J, Martens T (2008) Designing environmental campaigns by using agent-based simulations: Strategies for changing environmental attitudes. Journal of Environmental Management 88: 805–816.

43. Estrada V, Diaz MS (2010) Global sensitivity analysis in the development of first principle-based eutrophication models. Environmental Modelling & Software 25: 1539–1551.

44. Melbourne-Thomas J, Johnson CR, Fulton EA (2011) Characterizing sensitivity and uncertainty in a multiscale model of a complex coral reef system. Ecological Modelling 222: 3320–3334.

45. Brown DG, Robinson DT (2006) Effects of heterogeneity in residential preferences on an agent-based model of urban sprawl. Ecology and Society 11.

46. Bennett D, McGinnis D (2008) Coupled and complex: Human-environment interaction in the Greater Yellowstone Ecosystem, USA. Geoforum 39: 833–845.

47. Schluter M, Pahl-Wostl C (2007) Mechanisms of resilience in common-pool resource management systems: an agent-based model of water use in a river basin. Ecology and Society 12.

48. Guzy MR, Smith CL, Bolte JP, Hulse DW, SV G (2008) Policy research using agent-based modeling to assess future impacts of urban expansion into farmlands and forests. Ecology and Society 13.

49. Pannell DJ (1997) Sensitivity analysis of normative economic models: theoretical framework and practical strategies. Agricultural Economics 16: 139–152.

50. Alexander ER (1989) SENSITIVITY ANALYSIS IN COMPLEX DECISION-MODELS. Journal of the American Planning Association 55: 323–333.

51. Insua DR, French S (1991) A FRAMEWORK FOR SENSITIVITY ANALYSIS IN DISCRETE MULTIOBJECTIVE DECISION-MAKING. European Journal of Operational Research 54: 176–190.

52. Ligmann-Zielinska A, Jankowski P (2008) A Framework for Sensitivity Analysis in Spatial Multiple Criteria Evaluation. In: Cova TJ, Miller HJ, Beard K, Frank AU, Goodchild MF, editors. Geographic Information Science Proceedings 5th International Conference, GIScience 2008, Park City, UT, USA, September 23–26, 2008. Berlin/Heidelberg: Springer. pp. 217–233.

53. Saltelli A, Funtowicz S (2014) When All Models Are Wrong. Issues in Science and Technology: 79–85.

54. Saltelli A, Guimarães Pereira Â, Van der Sluijs JP, Funtowicz S (2013) What do I make of your latinorum? Sensitivity auditing of mathematical modelling. International Journal of Foresight and Innovation Policy 9: 213–234.

55. Parry HR, Topping CJ, Kennedy MC, Boatman ND, Murray AWA (2013) A Bayesian sensitivity analysis applied to an Agent-based model of bird population response to landscape change. Environmental Modelling & Software 45: 104–115.

56. Segovia-Juarez JL, Ganguli S, Kirschner D (2004) Identifying control mechanisms of granuloma formation during M. tuberculosis infection using an agent-based model. Journal of Theoretical Biology 231: 357–376.

57. Dancik GM, Jones DE, Dorman KS (2010) Parameter estimation and sensitivity analysis in an agent-based model of Leishmania major infection. Journal of Theoretical Biology 262: 398–412.

58. Vanuytrecht E, Raes D, Willems P (2014) Global sensitivity analysis of yield output from the water productivity model. Environmental Modelling & Software 51: 323–332.

59. Convertino M, Muñoz-Carpena R, Chu-Agor ML, Kiker GA, Linkov I (2014) Untangling drivers of species distributions: Global sensitivity and uncertainty analyses of MaxEnt. Environmental Modelling & Software 51: 296–309.

60. Baroni G, Tarantola S (2014) A General Probabilistic Framework for uncertainty and global sensitivity analysis of deterministic models: A hydrological case study. Environmental Modelling & Software 51: 26–34.

61. Beven K (2002) Towards a Coherent Philosophy for Modelling the Environment. Proceedings: Mathematical, Physical and Engineering Sciences 458: 2465–2484.

62. Epstein JM (2008) Why Model? Journal of Artificial Societies and Social Simulation 11: 12.

63. Saltelli A, Annoni P, Azzini I, Campolongo F, Ratto M, et al. (2010) Variance based sensitivity analysis of model output. Design and estimator for the total sensitivity index. Computer Physics Communications 181: 259–270.

64. Chen Y, Yu J, Khan S (2013) The spatial framework for weight sensitivity analysis in AHP-based multi-criteria decision making. Environmental Modelling & Software 48: 129–140.

65. Daniel C (1958) 131 Note: On Varying One Factor at a Time. Biometrics 14: 2.

66. Menard A, Marceau DJ (2005) Exploration of spatial scale sensitivity in geographic cellular automata. Environment and Planning B: Planning and Design 32: 693–714.

67. Kocabas V, Dragicevic S (2006) Assessing cellular automata model behaviour using a sensitivity analysis approach. Computers, Environment and Urban Systems 30: 921–953.

68. Perez L, Dragicevic S (2009) An agent-based approach for modeling dynamics of contagious disease spread. International Journal of Health Geographics 8: 50.

69. Lilburne L, Tarantola S (2009) Sensitivity analysis of spatial models. International Journal of Geographical Information Science 23: 151–168.

70. Saltelli A, Ratto M, Andres T, Campolongo F, Cariboni J, et al. (2008) Global Sensitivity Analysis: The Primer. Chichester, England: Wiley-Interscience. 304 p.

71. Ligmann-Zielinska A (2013) Spatially-Explicit Sensitivity Analysis of an Agent-Based Model of Land Use Change. International Journal of Geographical Information Science online.

72. Homma T, Saltelli A (1996) Importance measures in global sensitivity analysis of nonlinear models. Reliability Engineering & System Safety 52: 1–17.

73. Lempert R, Popper S, Bankes S (2003) Shaping the Next One Hundred Years New Methods for Quantitative, Long-Term Policy Analysis. Santa Monica, CA: RAND. MR-1626-RPC MR-1626-RPC. 210 p.

74. Becker BJ, Schram CM (1994) Examining explanatory models through research synthesis. In: Cooper H, Hedges LV, editors. The handbook of research synthesis. New York: Russell Sage Foundation.

75. Lambert DM, Sullivan P, Claassen R (2007) Working Farm Participation and Acreage Enrollment in the Conservation Reserve Program. Journal of Agricultural and Applied Economics 39: 151–169.

76. Lambert DM, Sullivan P, Claassen R, Foreman L (2007) Profiles of US farm households adopting conservation-compatible practices. Land Use Policy 24: 72–88.

77. Lambert DM, Sullivan P, Claassen R, Foreman L (2006) Conservation-Compatible Practices and Programs: Who Participates?: United States Department of Agriculture. 48 p.

78. Wossink GAA, Wenum JHV (2003) Biodiversity conservation by farmers: analysis of actual and contingent participation. European Review of Agricultural Economics 30: 461.

79. USDA FSA (2012) Conservation Reserve Program Overview. United States Department of Agriculture Farm Service Agency, http://www.fsa.usda.gov/.

80. USDA (2013) United States Census of Agriculture, http://www.agcensus.usda.gov/.

81. Yager RR (1988) On ordered weighted averaging aggregation operators in multi-criteria decision making. IEEE Transactions on Systems, Man and Cybernetics 18: 183–190.

82. Rinner C, Malczewski J (2002) Web-Enabled Spatial Decision Analysis Using Ordered Weighted Averaging (OWA). Journal of Geographical Systems 4: 385–403.

83. Jager W, Janssen MA, De Vries HJM, De Greef J, Vlek CAJ (2000) Behaviour in commons dilemmas: Homo economicus and Homo psychologicus in an ecological-economic model. Ecological Economics 35: 357–379.

84. USDA FSA (2010) Notice CRP-665 Grouped Soil Productivity Factors for 2010 SRR's. Washington, DC United States Department of Agriculture Farm Service Agency.

85. USDA FSA (2011) Conservation Reserve Program Sign-up 41 Environmental Benefits Index (EBI) Fact Sheet. United States Department of Agriculture Farm Service Agency.

86. USDA NASS (2012) Cropland Data Layer (CDL). United States Department of Agriculture National Agricultural Statistics Service.

87. USGS (2012) National Hydrography Dataset (NHD). United States Geological Survey.

88. Soil Survey Staff (2013) The Gridded Soil Survey Geographic (gSSURGO) Database for Michigan. Natural Resources Conservation Service, United States Department of Agriculture.

89. USGS (2013) National Elevation Dataset. USGS.

90. Schaetzl RJ, Krist FJJ, Miller BA (2012) A Taxonomically Based, Ordinal Estimate of Soil Productivity for Landscape-Scale Analyses. Soil Science 177.

91. Kingsbury L, Boggess W (1999) An Economic Analysis of Riparian Landowners' Willingness to Participate in Oregon's Conservation Reserve Enhancement Program. The Annual Meeting of the American Agricultural Economics Association pp. 15.

92. Chang H-H, Lambert DM, Mishra AK (2008) Does participation in the conservation reserve program impact the economic well-being of farm households? Agricultural Economics 38: 201–212.

93. Greiner R, Patterson L, Miller O (2009) Motivations, risk perceptions and adoption of conservation practices by farmers. Agricultural Systems 99: 86–104.

94. Brady M, Nickerson C (2009) A Spatial Analysis of Conservation Reserve Program Participants: The Impact of Absenteeism on Participation Decisions. The Annual Meeting of the Agricultural and Applied Economics Association pp. 27.

95. USDA (2011) Agricultural Resource Management Survey (ARMS).

96. Burnham KP, Anderson DR (2002) Model selection and multimodel inference: A practical information-theoretic approach New York, USA: Springer-Verlag.

97. Saltelli A, Tarantola S, Campolongo F, Ratto M (2004) Sensitivity Analysis in Practice: A Guide to Assessing Scientific Models. Chichester, England: Wiley. 232 p.

98. USDA FSA (2013) Conservation Programs Reports and Statistics. United States Department of Agriculture Farm Service Agency.

99. Sobol' IM (1993) Sensitivity estimates for nonlinear mathematical models. Mathematical Modeling and Computational Experiment 1: 407–414.

100. Malczewski J (1999) GIS and Multicriteria Decision Analysis. New York: John Wiley & Sons, Inc.

101. McGarigal K, Marks B, J, (1995) FRAGSTATS: Spatial Pattern Analysis Program for Quantifying Landscape Structure. Portland, OR: USDA Forest Service, Pacific Northwest Research Station. PNW-GTR-351 PNW-GTR-351.

102. Hellerstein D (2010) Challenges facing the USDA's Conservation Reserve Program. USDA ERS.

103. Matthews RB, Gilbert NG, Roach A, Polhill JG, Gotts NM (2007) Agent-based land-use models: a review of applications. Landscape Ecology 22: 1447–1459.

104. Heppenstall AJ, Crooks AT, See LM, Batty M, editors(2012) Agent-Based Models of Geographical Systems. Dordrecht: Springer. 746 p.

105. Lempert RJ (2002) Agent-based modeling as organizational and public policy simulators. Proceedings of the National Academy of Sciences of the United States of America 99: 7195–7196.

106. Rockström J, Steffen W, Noone K, Persson Å, Chapin FS, et al. (2009) A safe operating space for humanity. Nature 461: 472–475.

107. Polasky S, Carpenter SR, Folke C, Keeler B (2011) Decision-making under great uncertainty: environmental management in an era of global change. Trends in ecology & evolution 26: 398–404.

108. USDA FSA (2010) Notice CRP-663 Sign-up 41 Revised Soil Rental Rates (SRR's) for 2010. Washington, DC United States Department of Agriculture Farm Service Agency.

6

The Occurrence, Sources and Spatial Characteristics of Soil Salt and Assessment of Soil Salinization Risk in Yanqi Basin, Northwest China

Zhang Zhaoyong[1,2], Jilili Abuduwaili[1]*, Hamid Yimit[3]

1 State Key Laboratory of Desert and Oasis Ecology, Xinjiang Institute of Ecology and Geography, Chinese Academy of Sciences, Urumqi, China, **2** University of the Chinese Academy of Sciences, Beijing, China, **3** Key Laboratory of Xingjiang Arid Land Lake Environment and Resource, Xinjiang Normal University, Urumqi, China

Abstract

In order to evaluate the soil salinization risk of the oases in arid land of northwest China, we chose a typical oasis-the Yanqi basin as the research area. Then, we collected soil samples from the area and made comprehensive assessment for soil salinization risk in this area. The result showed that: (1) In all soil samples, high variation was found for the amount of Ca^{2+} and K^+, while the other soil salt properties had moderate levels of variation. (2) The land use types and the soil parent material had a significant influence on the amount of salt ions within the soil. (3) Principle component (PC) analysis determined that all the salt ion values, potential of hydrogen (pHs) and ECs fell into four PCs. Among them, PC1 (Cl^-, Na^+, SO_4^{2-}, EC, and pH) and PC2 (Ca^{2+}, K^+, $Mg^{2+}$and total amount of salts) are considered to be mainly influenced by artificial sources, while PC3 and PC4 (CO_3^- and HCO_3^{2-}) are mainly influenced by natural sources. (4) From a geo-statistical point of view, it was ascertained that the pH and soil salt ions, such as Ca^{2+}, Mg^{2+} and HCO_3^-, had a strong spatial dependency. Meanwhile, Na^+ and Cl^- had only a weak spatial dependency in the soil. (5) Soil salinization indicators suggested that the entire area had a low risk of soil salinization, where the risk was mainly due to anthropogenic activities and climate variation. This study can be considered an early warning of soil salinization and alkalization in the Yanqi basin. It can also provide a reference for environmental protection policies and rational utilization of land resources in the arid region of Xinjiang, northwest China, as well as for other oases of arid regions in the world.

Editor: Andrew C. Singer, NERC Centre for Ecology & Hydrology, United Kingdom

Funding: This study was supported by the Knowledge Innovation Program of the Chinese Academy of Sciences (KZCX2-EW-308; KZCX2-YW-GJ04). The funders had no role in study design, data collection and analysis, decision to publish, or preparation of the manuscript.

Competing Interests: The authors have declared that no competing interests exist.

* Email: jilil@ms.xjb.ac.cn

Introduction

Soil salinization is a global problem and it is a potential environmental problem in all continents with the exception of unassessed Antarctica. Soil saline levels are found within a wide range, and soil salinization occurs in much of the waterfront, arid and semi-arid zones of more than 100 countries and regions [1–3]. According to statistics done by the United Nations Educational, Scientific and Cultural Organization (UNESCO), and the Food and Agriculture Organization (FAO), salinized soil covers an area of about 9.543×10^6 km^2 on Earth [4–6]. In China alone, the area of salinized soil is about 3.693×10^5 km^2, which accounts for about a third of the total arable land [7–8]. The area of salinized soil in the oasis basin of Xinjiang in northwest China is about 1.05×10^4 km^2, which accounts for 33.4% of the total land in this area and research has found that the salinity of this area is trending upwards [9,10]. Soil salinization restricts agricultural development, especially when sustainable agricultural development and environmental quality improvement strategies are being considered. Studies have found that when the salt ions in the soil attain 8 $g.kg^{-1}$, they can greatly harm and even kill crops in farmland [11,12].

In oases of arid regions of northwest china, the environment is so weak, including a lack of precipitation and high envapotion, that economic activities such as fishing, agriculture, forestry and grassland farming of the oases have been strongly limited, especially for agriculture [13–15]. Therefore, it is necessary to identify the distribution characteristics, sources of the soil properties, such as salt ions, potential of hydrogen (pH) and electrical conductivity (EC), and also the status and causes of salinization of the land of the oasis, in order to provide a scientific basis for protection of the soil that sustains land plants.

Multivariate analyses and other statistical methods have been widely applied in studies to determine the sources of elements found in soil, such as total soil salt content and heavy metals [16–18]. The spatial variation model and spatial distribution are used to make a hazard risk map of soil salt properties in regions of interest. Correlation analysis, principal component (PC) analysis and cluster analysis are classic methods used to identify the natural and man-made sources of salt ions and to simplify data. Additionally, use of the comprehensive index results in a class of data with high correlations that better reflect the associations between the data.

Table 1. Indicators used for risk assessment of soil salinization.

Indicators	Class limits and their ratings score				
	None	Slight	Moderate	Severe	Very severe
*EC (dS.m⁻¹) [49]	<4	4–8	8–16	16–32	>32
**SAR [50]	<8	8–13	13–30	30–70	>70
Total salt content (%) (0–20 cm) [51]	<1	2~3	3–4	4–8	>8

*EC is electrical conductivity; **SAR is sodium adsorption ratio.

The Geostatistical Analyst is based on GIS technology [19,20]. Among these, the ordinary kriging is the most widely used one in the study of soil salt distributions [21,22,29]. In recent years, the Geostatistical Analyst method has been used in the field of hydrology and water resources, including studies of groundwater pollution risk, water potential research and spatial distribution of soil salinization in arid land [23–25,28,32]. Since the 1990s, geostatistical methods have been widely used to study spatial variability characteristics of soil salt properties (salt ions, EC and pH). Sylla et al. [26] studied the spatial variation characteristics of soil salt content of an agricultural ecosystem under different scales in West Africa. Ammari et al. [27] studied the soil salinity changes in the Jordan Valley and the potential threat against the sustainable irrigated agriculture. In China, Bai et al. [30] researched the spatial variation characteristics and composition of soil salt content in Huang Huai Hai plain, northeast China and found that the average influential range of the soil salt content was higher than 200 km, indicating the salt content of the soil is mainly gathered in a large area of the regions. In Xinjiang in northwest China, Lin et al. [33] researched the spatial variation characteristics of soil salt in the Wei Gan He irrigated area and found the agricultural irrigation has resulted in serious soil salinization in this area and these deserve serious attention.

In arid regions, oases in basins are the main places where humans live and life can survive [31]. Therefore, it is important to understand the spatial distribution characteristics of the soil salt properties, including total salt content, salt ions, pH and EC. A quantitative grasp of soil salinity levels could serve as a reference and a basis for maintaining soil quality, which would help to effectively control the human pollution and develop the regional economy in a reasonable and orderly fashion [34,35]. However, previous research has focused on rapidly developing areas, such as coastal plains and large irrigation areas in eastern china and elsewhere of the world with the purpose of assessing land usability, environmental effects and soil salinization risk [36–38]. Since the 1990s, implementation of the "western development policy of China" has led to prodigious economic development in many oases in Xinjiang, and the agriculture in these regions has undergone rapid progress. However, the rational irrigation of the agriculture, lack of precipitation and high envapotion of these regions have negatively influenced soil salt properties, resulting in increased soil salt contents, ECs and pHs, which can result in serious soil salinization [39]. Unfortunately, research on the soil salt property distribution characteristics and soil salinization risk assessment in the oases of arid regions of northwest of China is lacking.

The Yanqi basin is a typical oasis in a basin in the southern Tianshan Mountains, Xinjiang in northwest China. Since the 1990s, both the implementation of the "western development policy of China" and the development policy made by the Xinjiang Province, China, have led to prodigious economic

development in the Yanqi basin, but regional economic development and associated human activity have left the current ecological environment fragile [39,40]. Together with economic development, the blind expansion of farmland and unrestrained surface water irrigation led to a rise in groundwater and an increase in soil salinization in the basin oasis. 64.12% of the area experienced mild soil salinization, 8.25% had moderate salinization, and 27.07% had severe salinization. Research has shown that excessive use of water resources by agriculture has made the soil salinization status severe and decreased agricultural production [41].

After a basic analysis of land use and soil parent materials types in the area, we created land use and soil type geological maps, and, using ArcGIS 10.0 software and combining the grid sampling method with 3S technology, we made sampling points to get soil samples across the whole area. We evaluated the soil salt properties in different land use types in the laboratory, and assessed the soil salinization status and the cause in the Yanqi basin. Then ordinary kriging of Geostatistical Analyst method was used to reveal the spatial distribution characteristics of the soil salt properties in this region. Then by combining these properties with the climate, precipitation, evaporation and temperature, we assessed the soil salinization risk of this area. From this we can provide helpful proposals to prevent the environmental risks that could lead to soil salinization in this area. This research can serve as a helpful reference for environmental protection in this region and for soil salinization prevention in arid regions of northwest China.

Materials and Methods

Study area

The area studied in this work is a desert basin oasis in the arid region of northwest China including four counties in the Yanqi basin: Yanqi County, Hejing County, Bohu County and Heshuo County. This region lies within the geographical coordinates of 85°50′–87°50′E and 41°40′–42°20′N with a length of about 85 km from north to south and width of 130 km from east to west, totaling an area of about 723100 km². The terrain slopes up from the northwest down to the southeast. The northwest is mountainous and the south is low-lying desert that is 1050–2000 m above sea level. The western area has extensive intrusive rock and metamorphic rock from the *Proterozoic era, Neoproterozoic* and *Cenozoic*. Weathering of this rock results in brown earth soil, acidic rocky soil and an acidic soil skeleton in the west. The east is primarily made up of quaternary sediments, which form *Takyic* (Calcisols), *Chemic* (Phaeozems), *Stagnic* (Gleysols), *Irragric* (Anthrosols), *Fragic* (Arenosols), *Eutric* (Gleysols) and *Yemic* (Solonchaks) [42]. The area researched is in a continental desert climate temperate zone with an annual mean temperature of 14.6°C, 186 frost-free days per year, and 50.7–79.9 mm of annual

Table 2. Descriptive statistics of the soil salt properties from Yanqi basin.

Elements	Ranges (g.kg⁻¹)	Contributions(%)	Median (g.kg⁻¹);EC (dS.m⁻¹)	Average (g.kg⁻¹)	Standard deviation (%)	Coefficient of variation (%)	Kurtosis (%)	Skewness (%)
HCO$_3^-$	0.13-0.98	3.51	0.171	0.19	12.25	35.37	0.14	0.58
CO$_3^-$	0.18-0.85	4.08	0.252	0.46	10.35	32.08	1.63	8.76
Ca^{2+}	0.59-1.95	6.54	0.681	0.75	9.83	191.67	31.38	42.43
Na$^+$	0.69-2.43	13.18	0.955	1.19	12.38	23.01	2.45	16.22
Mg^{2+}	0.47-1.89	7.75	0.987	0.58	21.02	20.07	1.10	10.88
K$^+$	0.49-2.13	12.36	0.855	0.69	23.04	226.25	23.47	51.81
SO$_4^{2-}$	0.93-1.58	18.95	1.245	1.14	12.56	19.63	1.46	0.76
Cl⁻	0.75-2.36	25.63	1.167	0.98	25.7	12.94	2.30	14.79
SAR	3.41-33.241	-	22.417	10.51	121.45	148.56	11.54	15.78
EC	0.7-1.39	-	0.981	0.95	15.09	21.27	1.46	10.99
pH	7.85-8.55	-	8.141	8.15	16.39	30.32	1.17	0.73
Total salt	1.16-14.77	-	8.56	9.73	15.36	126.73	16.84	18.46

precipitation. By calculating the potential evaporation (ET_0) by the method of Hargreaves (1985) [43], we then got the annual average potential evaporation of this area as 2438.9 mm, the $\geq 10°C$ active accumulated temperature 3414.4–3694.1°C and an annual average relative humidity of 72%.

Soil sampling and analyses

In order to perform a basic analysis of the land use and soil type, geological maps were made of the study area using ArcGIS 10.0 software to lay out a grid of soil sampling points on a digital map of the Yanqi basin. All samples were acquired in July 2012 or July 2013 from a collection area. In order to best assess the ecological risk in the Yanqi basin, diverse land use types were encompassed in our study of salt ion distribution. Soil samples were collected at depths of 0–20 cm, where a hard plastic shovel was used to dig a vertical 20×20 cm soil profile. 1 kg uniform samples were collected, and then they were put into a clean cloth, numbered and sealed. The collection position, date, sample vegetation types and surrounding vegetation conditions of each sampling area were recorded. After the soil samples were taken back to the laboratory, they were air dried and impurities, such as plant residues and rocks, were removed. The samples were then pushed through a 20 mesh nylon sieve (0.84 mm) to eliminate the plant residue and stones. We then used agate to grind the soil samples through 100 mesh nylon sieves (0.25 mm) to prevent contamination and then stored the samples in plastic bottles [46].

Total soil salt content, soil salt ions, pH and EC tested are as follows: 50 g of ground sample were removed from the plastic bottles, dissolved in 250 ml of deionized water (CO_2 has been removed) (1:5, soil:water) for 2 hours to fully dissolve salt ions contained in the soil. The samples were then put in a centrifuge tube, vibrated for 3 min with an oscillator and then centrifuged at a speed of 4,500–5,000 r·min⁻¹. To get the supernatant prepared for analysis of total soil salt content, salt ions content, pH and EC, the method described by Lu (2000) was followed [44].

The total salt content of the soil was determined by gravimetry of the evaporation residue. First, the supernatant was absorbed in a porcelain dish, and hydrogen peroxide (H_2O_2) was used to oxidize organic matter. Then, the samples were boiled in a water bath at 105–110°C until it dried, and weighed. The drying quality of the residue is expressed as total salt content of the soil. The pH of the extracted supernatant was tested using a Potentiometric Titrimeter (G20, METTLER, and TOLEDO). Burette drive resolution was 1/20000. Mv/pH electrode measurement range was ±2000 mv. The ECs were tested using a Conductivity Meter (DDSJ-308A, Shanghai, China) with a measurement range of $0–1.999 \times 10^5$ µs/cm and a test error of ±0.5% (FS) ±1.

The supernatant was run through a 0.45 µm drainage cellulose acetate membrane. Then, the cation content (K^+, Na^+, Ca^{2+}, Mg^{2+}) of the solutions was determined using an inductively coupled plasma atomic emission spectrometer (Vista MPX, Varian, USA). The anion content (Cl^-, SO_4^{2-}, CO_3^{2-}, and HCO_3^-) of the solutions was determined using an Ion Chromatograph (ICS-90, Dionex, USA). All tests were conducted using the following protocol: a standard solution was prepared for Na^+, K^+, Ca^{2+}, Mg^{2+}, Cl^-, SO_4^{2-}, CO_3^{2-} and HCO_3^-. The salt ion content was determined by comparing each sample to the standard solution of known concentration. The standard solutions used for the salt ions in this study were national level standard material (Gss series, China). The coefficient of the best fitting curve was determined by the testing equipment based on the standard material and then the amount of the salt ions (Na^+, K^+, Ca^{2+}, Mg^{2+}, Cl^-, SO_4^{2-}, CO_3^{2-}, and HCO_3^-) in the liquid supernatant was tested. After all the samples had been tested for their salt ion

Table 3. Statistical parameters of the soil salt ions found at 0–20 cm depth within the investigated land use and land cover categories of the study area.

LUCC SPM	Parameters	HCO₃⁻ (g.kg⁻¹)	CO₃⁻ (g.kg⁻¹)	Na⁺ (g.kg⁻¹)	Mg²⁺ (g.kg⁻¹)	K⁺ (g.kg⁻¹)	Ca²⁺ (g.kg⁻¹)	SO₄²⁻ (g.kg⁻¹)	Cl⁻ (g.kg⁻¹)	Total salt (g.kg⁻¹)	EC (dS.m⁻¹)	pH
Farmland (n = 51)	Ranges	0.18–0.98	0.27–0.85	0.69–2.43	1.05–1.89	1.31–2.13	1.17–1.95	1.05–1.51	1.47–2.36	8.56–14.77	0.96–1.39	8.05–8.55
	Average	0.32a	0.62a	1.36a	1.15a	1.83a	1.71a	1.21a	1.68a	11.2a	1.02a	8.15a
	SD	22.45	15.23	31.35	23.3	27.57	37.73	24.57	19.19	23.97	18.64	14.77
Forest (n = 46)	Ranges	0.13–0.53	0.18–0.78	0.69–0.98	0.47–0.98	0.49–0.97	0.59–1.02	1.09–1.58	0.75–0.97	9.13–13.35	0.7–1.13	7.85–8.45
	Average	0.33a	0.41b	0.75b	0.87b	0.75a	0.77a	1.18b	0.82a	9.98b	0.94b	8.17b
	SD	14.75	17.73	23.53	25.55	23.74	23.73	17.86	28.33	16.79	18.71	14.37
Grassland (n = 63)	Ranges	0.21–0.73	0.25–0.82	1.03–1.63	0.54–1.02	0.61–1.31	0.89–1.14	0.93–1.19	0.96–1.61	9.08–11.24	0.76–1.25	7.92–8.42
	Average	0.42a	0.65b	1.32b	0.68b	1.18b	0.98b	1.04b	1.28b	9.45a	0.95b	8.27b
	SD	17.54	23.77	27.52	23.13	11.51	15.23	12.22	22.14	13.75	16.71	21.37
Desert (n = 71)	Ranges	0.26–0.83	0.19–0.76	0.98–1.57	0.47–1.44	0.95–1.56	0.99–1.19	1.05–1.24	1.08–2.29	9.31–13.89	0.78–1.34	8.06–8.44
	Average	0.58b	0.32b	1.31b	1.13b	1.21b	1.02b	1.11.6b	1.57b	10.56a	1.08b	8.28a
	SD	17.33	15.65	17.52	25.55	18.85	13.43	17.33	19.25	23.75	12.52	15.31
Urban construction areas (n = 40)	Ranges	0.49–0.93	0.33–0.83	0.91–1.54	0.65–1.03	0.51–1.51	0.88–1.21	0.94–1.18	0.79–1.31	6.16–14.13	0.75–1.33	7.94–8.51
	Average	0.53c	0.39c	1.21c	0.91c	0.97c	0.97c	1.06c	0.94c	10.52a	1.11c	8.31c
	SD	15.75	16.51	13.25	15.52	12.35	19.52	23.37	21.54	32.24	22.31	24.87
Sandy shale of weathered material (n = 66)	Ranges	0.13–0.86	0.18–0.38	1.19–2.39	0.47–1.75	0.58–2.11	0.59–1.36	0.93–1.53	0.75–2.36	8.69–14.77	0.79–1.26	7.86–8.55
	Average	0.45a	0.42a	1.51a	0.95a	0.91a	0.82a	1.01a	1.21a	10.88a	0.3a	7.93a
	SD	25.55	17.53	13.52	22.31	22.35	23.35	15.75	18.65	11.54	12.25	21.35
Coarse crystalline rock weathered material (n = 74)	Ranges	0.18–0.52	0.22–0.79	0.69–2.43	0.51–1.54	0.49–2.13	0.69–1.45	0.97–1.56	0.78–2.12	6.16–12.41	0.7–1.39	7.85–8.32
	Average	0.32b	0.38b	1.72b	0.81b	0.87b	0.95b	1.13b	1.46b	8.72a	1.14b	8.14b
	SD	13.35	11.41	21.25	23.74	11.29	12.52	21.57	22.53	15.54	12.57	24.58
Diluvial material (n = 68)	Ranges	0.21–0.93	0.24–0.85	0.81–2.23	0.65–1.89	0.64–1.72	0.71–1.56	0.95–1.58	0.85–2.26	7.89–13.25	0.84–1.28	7.89–8.19
	Average	0.67b	0.51b	1.42b	0.75b	0.98b	1.01b	1.05b	1.62b	9.82a	0.91b	8.01b
	SD	12.15	12.25	23.73	27.35	11.37	12.75	12.73	22.36	21.57	23.37	11.52
Lacustrine deposits (n = 63)	Ranges	0.33–0.98	0.23–0.79	0.85–2.35	0.52–1.49	0.71–1.98	0.62–1.95	0.96–1.51	0.96–2.19	8.98–11.51	0.85–1.32	7.97–8.37
	Average	0.74a	0.44b	1.16a	0.91c	1.02c	0.85c	1.06a	1.31b	9.24a	1.24a	8.23a
	SD	13.51	22.35	21.26	23.59	15.34	15.62	21.94	19.31	26.49	23.77	12.35
R²	LUCC (%)	31.21	9.74	13.49	8.4	71.45	37.85	11.3	10.8	56.71	12.64	12.43
	SPM (%)	58.79	34.59	16.47	28.93	32.7	11.4	17.9	15.9	23.51	17.53	15.2

Different small letters represent a significance of 0.05; LUCC represent land use types; SPM represent soil parent material types.

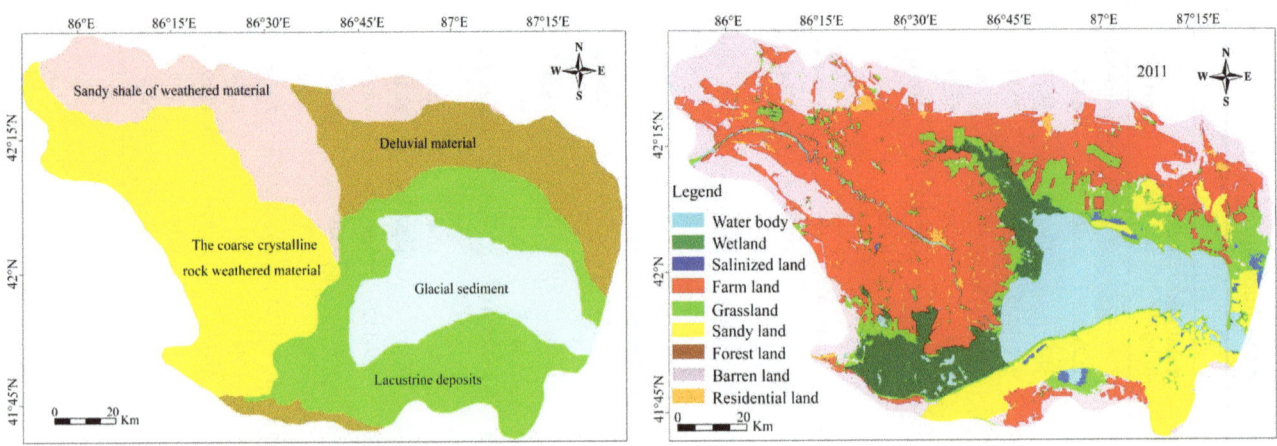

Figure 1. Land use types and parental material pattern in Yanqi basin.

content (Na^+, K^+, Ca^{2+}, Mg^{2+}, Cl^-, SO_4^{2-}, CO_3^{2-}, and HCO_3^-), we chose approximately 20% for retesting and found that 97.3% of the results were repeatable, inspiring confidence in the original data. After all the total salt ion contents (Na^+, K^+, Ca^{2+}, Mg^{2+}, Cl^-, SO_4^{2-}, CO_3^{2-}, and HCO_3^-) of the solution were determined, we recalculated them from unit of $\mu g/ml$ into mg/g (g/kg) using the method described by Bao (2005) [45]. To prevent contamination during the testing process, all glassware was soaked in 5% HNO_3 for 24 hours, rinsed and then dried.

Statistical analyses

Descriptive and multivariate statistical analysis. Descriptive statistical methods were used to analyze the range, mean, median, standard deviation, coefficient of variation, kurtosis and skewness of the total salt content, each salt ion, SAR, pH and EC of the soil samples. Correlation analysis, PC analysis and cluster analysis of the classic multivariate statistical method were used to process data and identify the soil salinity. Single factor analysis of variance (ANOVA) was used to analyze the differences in the amount of salt ions between different land use types. These analyses were all processed using the software SPSS 19.0.

Ordinary kriging method. Ordinary kriging (OK) is a commonly used linear spatial interpolation method that estimates variables at unsampled locations by using information from neighboring points and assigning weights to these points based on their distance from the point and the spatial variability structure. The OK method can be formulated as

$$Z_{OK}^*(x_0) = \sum_{i=1}^{n} w_i Z(x_i) \tag{1}$$

where $Z_{OK}^*(x_0)$ is the OK estimation at an unsampled location (x_0), n is the number of samples in a search neighborhood, and w_i are the weights assigned to the ith observation $Z(x_i)$. Weights are assigned to each sample such that the estimation or kriging variance $E\left[\{Z^*(x_0) - Z(x_0)\}^2\right]$ is minimized and the estimates are unbiased [47]. Weights are determined after computing a semivariogram that models spatial correlation and covariance structure between data points for each variable using Eq. 1 [48].

$$\gamma(h) = \frac{1}{2N(h)} \sum_{i=1}^{N} [Z(x_i + h) - Z(x_i)]^2 \tag{2}$$

where $\gamma(h)$ is the semivariance between two observation points $Z(x_i)$ and $Z(x_i + h)$ separated by a distance h, and N is number of observation pairs at the distance h.

Soil salinization evaluation criteria used in this research as under below (Table 1).

Results and Discussion

Descriptive statistical analysis of soil salinity

The descriptive statistics concerning the soil properties in Yanqi basin in Table 2 show that the maximum and average values of HCO_3^-, Ca^{2+}, CO_3^-, Na^+, Mg^{2+}, K^+, SO_4^{2-}, Cl^-, SAR, total amount of salts, EC, and pH were 0.98(0.19) g.kg^{-1}, 1.95(0.75) g.kg^{-1}, 0.85 (0.46) g.kg^{-1}, 2.43(1.19) g.kg^{-1}, 1.89(0.58) g.kg^{-1}, 2.13(0.69) g.kg^{-1}, 1.58(1.14) g.kg^{-1}, 2.36(0.98) g.kg^{-1}, 33.241(10.51), 14.77(9.73) g.kg^{-1}, 1.39(0.95) dS.m^{-1}, and 8.55(8.15), respectively. Within the analysis of the soil samples, a large amount of variation occurred, suggesting that the sources and influencing factors of the salt properties in Yanqi basin are complex. This work found that the main salt ions accounted for 84.41% of the total salt content and were K^+, Ca^{2+}, Na^+, Cl^-, Mg^{2+} and SO_4^{2-}. Meanwhile the amount of HCO_3^- and CO_3^- was very low. The pH of the soil of the study areas ranged from 7.85 to 8.55, which had only a small variation. However, there was a dramatic change within the total salt content of the soil, ranging from 1.16 to 14.77 g.kg^{-1}.

The coefficient of variation is the ratio between the standard deviation and average, and it can be used to compare different dimensions of indicators. The coefficients of variation of HCO_3^-, CO_3^-, Na^+, Mg^{2+}, SO_4^{2-}, Cl^-, EC, and pH were 35.37%, 32.08%, 23.01%, 20.07%, 19.63%, 12.94%, 21.27%, and 30.32%, respectively, and were of medium variation (10%<CV<100%). However, the coefficients of variation of SAR, the total amount of salts, Ca^{2+} and K^+ were 148.56%, 126.73%, 191.67% and 226.25%, and, therefore, had high levels of variation (CV> 100%)[52]. In particular, Ca^{2+} and K^+ had higher coefficients of variation as compared to the other elements. From the perspective of skewness, the values of these ten soil properties are ordered as K^+>Ca^{2+}>total amount of salts>Na^+>SAR>Cl^->EC>Mg^{2+}> CO_3^->SO_4^{2-}>pH>HCO_3^-.

The differences in soil salt properties from different land use types and soil parent materials

Land utilization types and soil parent materials are the main examples of human activity and geological background that

Table 4. The correction matrix of soil salt properties in Yanqi basin.

	EC	TSA	TSC	TS	Mg^{2+}	Na$^+$	K$^+$	SO$_4^{2-}$	Cl$^-$	CO$_3^-$	Ca^{2+}	HCO$_3^-$	pH
EC	1												
TSA	0.58	1											
TSC	0.40	0.30	1										
TS	0.41	0.52**	0.38**	1									
Mg^{2+}	0.10	0.04	0.11	0.06**	1								
Na$^+$	0.98**	0.34	0.96**	0.95**	0.01	1							
K$^+$	0.65	0.48	0.78*	0.71**	0.51	0.64	1						
SO$_4^{2-}$	0.87**	0.94**	0.42	0.93**	0.001	0.82**	−0.24	1					
Cl$^-$	0.66**	0.55*	0.12	0.57**	0.13	0.69**	−0.15	0.24	1				
CO$_3^-$	0.48*	0.49	0.27	0.48**	−0.24	0.57	0.12	0.27	0.17	1			
Ca^{2+}	0.42	0.55	0.51*	0.54**	0.22	0.25	0.57*	0.23	0.03	−0.14	1		
HCO$_3^-$	−0.25	−0.22	−0.20	0.21**	0.03	−0.16	−0.19	−0.42	0.28	0.16	−0.21	1	
pH	0.72**	0.62	0.14	0.82	0.42	0.32*	0.27	0.67**	0.58**	0.10	0.41	0.01	1

EC is electrical conductivity, TSA is the total anionic salt, TSC is the total cationic salt and TS is total salt content.

influence the salt ion content of soil. We analyzed the relationship between soil salt properties, human activity and geological background to further explore the distribution characteristics and sources of the soil salt properties of the Yanqi basin. The salt properties of the soil from each land use type and soil parent material of Yanqi basin are in Table 3. The land use types and soil parent materials of Yanqi basin are shown in Fig. 1. The analysis of these data suggests that the manner in which land is used has a significant influence on the amount of Ca^{2+} and K$^+$. In farmland, the average content of Ca^{2+} was 1.83 g.kg^{-1}, K$^+$ was 1.71 g.kg^{-1}, and the total amount of salts was 10.88 g.kg^{-1}. This was significantly higher than in the other land use types including areas of urban construction, forest, grassland, urban construction areas and desert. Meanwhile the maximum average values of Na$^+$, Mg^{2+}, SO$_4^{2-}$, Ca^{2+}, the total amount of salts, and K$^+$ found in farmland were higher than in grassland, desert, and areas of urban construction. The variance test attained a significant level of 0.05, indicating these elements and their distribution are mainly controlled by human activity. The activities that had a significant influence on these soil salt properties include agricultural activities, such as irrigating, fertilizing and farming. The CV calculated indicates that the pH, EC, total amount of salts and all the soil salt ions measured belong to medium variability categories (10–100%). Among the five land use types, the differences between these groups were small and, therefore, the classes of element enrichment were not obvious.

This research also found differences in the amount of organic soil salt ions. For example, HCO$_3^-$ reached a maximum in lacustrine deposits of 0.74 mg.kg^{-1}. The maximum average values of Na$^+$, Mg^{2+}, K$^+$, SO$_4^{2-}$, total amount of salts, and Cl$^-$ were found in coarse crystalline rock weathered material, sandy shale of weathered material and lacustrine deposits, while the maximum average values of CO$_3^-$, Ca^{2+}, and EC were found to be significantly higher in diluvial material than in sandy shale or coarse crystalline rock from weathered material or lacustrine deposits. For the five land use types, the class of element enrichment was not obvious as the differences between these groups were small.

R^2 represents the ratio of the sum of squares in groups and the total error of the sum of squares. It reflects the contribution of different factors on the soil salt properties [53]. For this study, the land use types explained the variances in Ca^{2+} (37.85%), K$^+$ (71.45%) and total amount of salts (56.71%), which were higher than that of the soil parent material. This demonstrates that the way land was used played a major role in the accumulation of Ca^{2+}, K$^+$ and total amount of salts in Yanqi basin. This analysis also found that the variances of HCO$_3^-$, CO$_3^-$, Mg^{2+}, Na$^+$, SO$_4^{2-}$, and Cl$^-$ in the soil parent material are higher than those of the land use factors (Table 4), indicating that the soil parent material played a major role in the accumulation of these elements. However, there was little difference in the variances of EC and pH, indicating that these soil salt properties were mainly influenced by land use and soil parent material. Overall, the R^2 analysis fits well with the results of the multivariate statistical analysis.

Multivariable statistics

Correlation analysis. Table 4 shows the Pearson correlation coefficients between the soil salinity variables. There is a significant correlation of 0.96 (P<0.01) between the total amount of salt cations and Na$^+$, as well as the total amount of salt cations and Ca^{2+} at 0.51 (P<0.05) in the soil in the Yanqi basin, but there is no significant correlation with other salt cations. Furthermore, we found a significant correlation between the amount of salt

Table 5. Factors matrix of soil salt properties from Yanqi basin.

Soil salt properties	Principal components			
	PC 1	PC 2	PC 3	PC 4
K^+	−0.18	0.65	0.37	0.56
Mg^{2+}	0.37	0.80	−0.07	0.14
Ca^{2+}	0.63	0.74	0.50	0.45
CO_3^{2-}	−0.05	−0.02	0.59	−0.10
SO_4^{2-}	0.61	0.54	0.42	0.17
pH	0.49	0.15	0.06	−0.79
Cl^-	0.91	−0.04	0.29	0.13
Na^+	0.68	−0.08	0.56	0.31
EC	0.46	−0.01	−0.73	0.13
HCO_3^-	−0.09	−0.09	0.29	0.37
Total salt	0.21	0.78	0.14	0.62
Percentage of variance (%)	33.75	28.54	18.25	15.18
Percentage of cumulative variance (%)	33.75	62.29	80.54	95.72

anions and SO_4^{2-} of 0.94 (P<0.01), and the correlation coefficients between the total amount of salt anions and Cl^- or CO_3^{2-} are 0.55 and 0.49 (P<0.05), respectively. This indicates that SO_4^{2-} was the primary salt anion, Cl^- was the secondary and CO_3^{2-} was the tertiary. Together, the correlation coefficients between the total amount of salts and Cl^-, and the total amount of salts and SO_4^{2-} are 0.57 and 0.92, respectively (P<0.01), indicating that the main types of salt in the soil were sulfate and chloride. This study also found that there are close correlation coefficients between the EC and Na^+, the EC and Cl^-, and the EC and SO_4^{2-} content, where the coefficients are 0.98, 0.66, and 0.87, respectively (P<0.01). Additionally, the correlation coefficient between the soil EC and CO_3^{2-} is 0.48, which is also significant (P<0.05). Overall, a significant correlation was found between pH and EC, and total amount of salts (TS) and salt anions (Na^+, Ca^{2+}, Ca^{2+}, Ca^{2+}, Cl^-, SO_4^{2-}, HCO_3^-, and CO_3^{2-}) (P<0.01). Further analysis shows that the correlation coefficient between the pH and SO_4^{2-} is 0.67 and between pH and Cl^- is 0.58 (P<0.01), indicating the pH of the soil is primarily influenced by the SO_4^{2-} and Cl^- content.

Principal component analysis and clustering analysis. All of the elements studied were found to fall into four PCs (Table 5) with a cumulative variance of 95.72%. This is a reflection of their sources and main influences, particularly for the ten indicators of Cl^-, Na^+, EC, pH, K^+, Mg^{2+}, Ca^{2+}, HCO_3^-, CO_3^{2-}, total amount of salts, and SO_4^{2-}. Among these, the variance contribution rate of the first PC (Cl^-, Na^+, SO_4^{2-}, EC and pH) was 33.75% and the second PC (Ca^{2+}, K^+, Mg^{2+}, and total amount of salts) was 28.54%. The primary contribution was by agricultural development, in particular from herbicide application and acid salt fertilizer [54,55]. The variance contribution rates of the third (CO_3^-) and fourth PCs (HCO_3^{2-}) were at 18.25% and at 15.18%, respectively. Upon combining the sampling sites within their respective land use types, it was found that the samples that had a high content of salts were often taken from the desert, grassland and forest, and appear to have originated from natural sources. This analysis also showed there were larger loads of Ca^{2+} in the first PC, SO_4^{2-} in the second PC and K^+, total salt content in the fourth PC, indicating these elements were influenced by both artificial and natural sources. We used clustering analysis to see if the results are consistent with the results from the PC analysis

(Fig. 2). All the elements were classified into four categories: the first category consisted of Cl^-, Na^+, SO_4^{2-}, pH, and EC; the second of Mg^{2+}, Ca^{2+}, K^+, and the total amount of salts; the third of HCO_3; and the fourth of CO_3^{2-}.

Spatial distribution of soil salt ions in Yanqi basin. The spatial dependence of the soil salt properties was determined by semivariance analysis in order to quantitate the spatial variability. The parameters of the semivariogram included model type, nugget, sill and effective range. The nugget value (C_0) represents the random variation derived from measurement inaccuracy or variations in properties that cannot be detected in the sample range [54]. The sill value is the upper limit of the fitted semivariogram model [47]. The ratio of nugget to sill was a criterion to classify the spatial dependence of soil properties and it reflects the influence of regional factors (nature) and the role of the non-regional factors (human factors). The range of the semivariogram (A_0) represents the average distance through which the variable semivariance reaches its peak value. A small effective

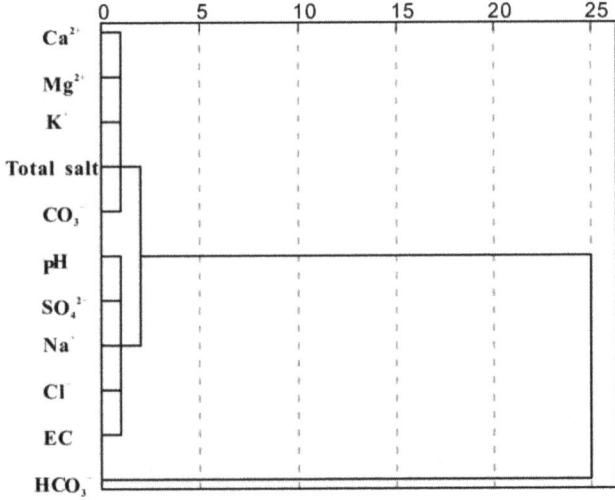

Figure 2. Clustering tree of soil salt properties of Yanqi basin.

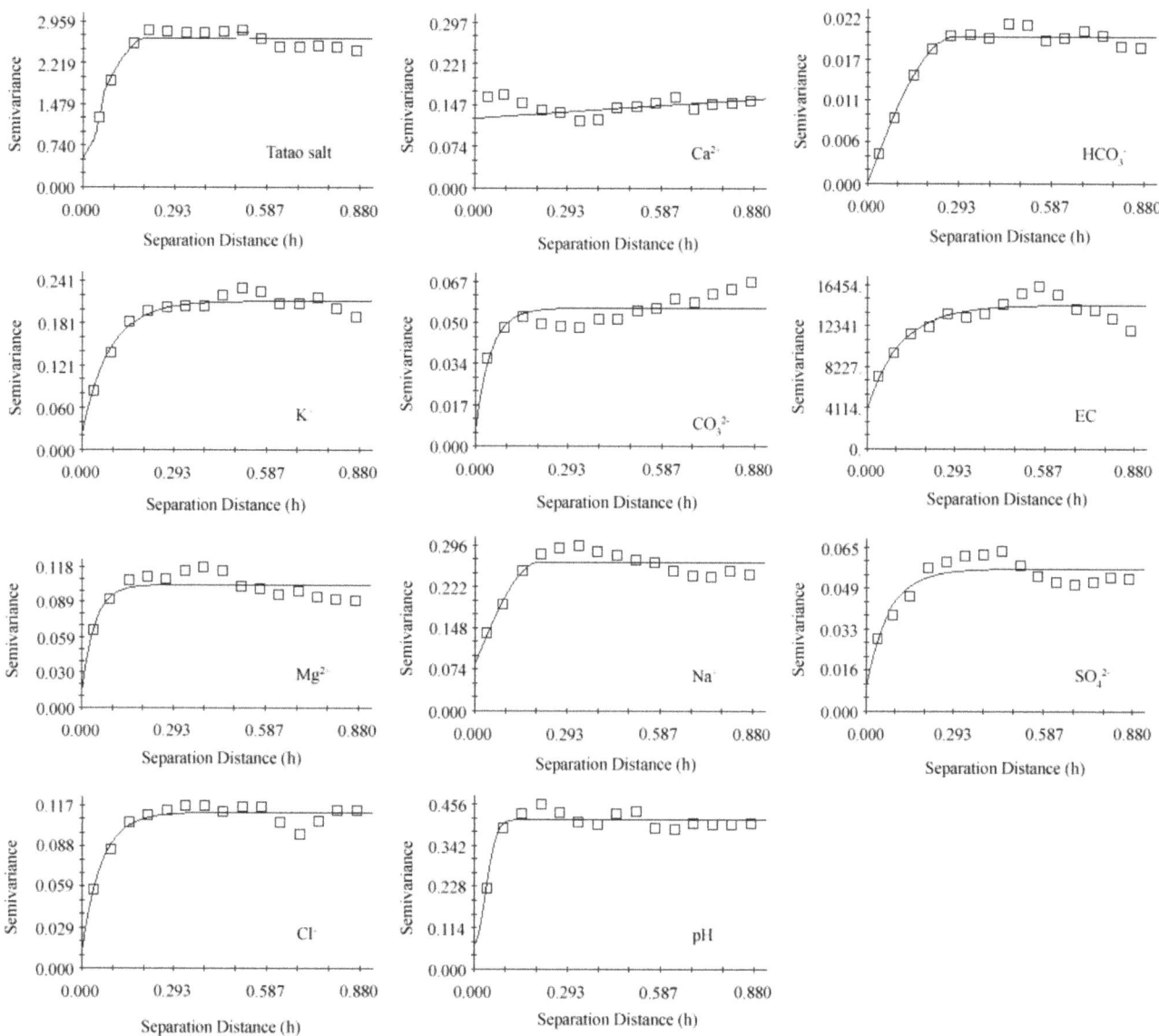

Figure 3. The semivariance function diagram of the soil salt properties of Yanqi basin.

range implies a distribution pattern composed of small patches. The cross-validation value is the coefficient of determination (R^2) of the correlation between the measured values and the cross-validation values, which were predicted based on the semivariogram and neighbor values [53]. Previous studies have shown that the semivariogram often differs considerably from its regional counterpart [33,53]. Fig. 3 and Table 6 show the tested variables over the study area that were modeled using spherical, gaussian and exponential semivariograms with the lower nugget effect based on the coefficients of determination (R^2) and residual sum of square (RSS). In this respect, a low ratio (less than 25% as was found for pH, Ca^{2+}, Mg^{2+}, and HCO_3^-) means that a large part of the variance is introduced spatially. This implies a strong spatial dependency of the variable, most likely due to intrinsic factors, including soil formation factors, such as soil parent materials, topography, and/or climate. A high ratio (more than 75% was found with Na^+ and Cl^-) often indicates a weak spatial dependency in the present sampling that was most likely due to extrinsic factors, including contamination, irrigation, and soil management

practices, where the fertilization and the soil chemical properties were continuously shifting. In this research the Nug/Sill ratios of CO_3^{2-}, and total salt content are 67.27 and 67.53% in the range between 25% and 75%, indicating the total salt content of the soil in Yanqi basin was influenced by the above factors. Additionally, all the other variables have a moderate spatial dependency for both intrinsic and extrinsic factors. The effective range calculated for variograms of different soil salt properties was 50–280 m, indicating that the sample distance was adequate for the characterization of the spatial variability of these properties.

The main application of geostatistics in soil science has been estimating and mapping the chemical properties in the soil of unsampled areas. Maps for each of the soil properties can be obtained using an ordinary kriging interpolation based on the best-fit semivariogram model. The skewness, defined as more than +1 or less than -1, indicates that some soil properties were not normally distributed. For these properties, it is difficult to estimate the semivariogram and doing so would result in a high value of kriging standard deviations [53]. Lognormal kriging with non-

Table 6. The spatial variation parameters of the soil salt properties of Yanqi basin.

Soil salt properties	Model	Nugget (Co)	Sill (Co+C)	Nug/Sill ratios $C_0/(C_0+C)$ (%)	Range A_0 (m)	R^2	RSS
Na^+	Spherical	0.084	0.102	82.35	200	0.921	4.053E-03
K^+	Exponential	0.068	0.212	32.08	90	0.861	2.900E-03
Ca^{2+}	Exponential	0.034	0.217	15.67	101	0.954	1.131E-02
Mg^{2+}	Spherical	0.023	0.103	22.33	100	0.733	1.288E-03
Cl^-	Gaussian	0.087	0.111	78.38	100	0.891	4.104E-04
CO_3^{2-}	Exponential	0.037	0.055	67.27	50	0.914	4.014E-04
HCO_3^-	Spherical	0.004	0.021	19.05	280	0.927	2.446E-05
SO_4^{2-}	Exponential	0.012	0.041	29.27	80	0.823	1.826E-04
pH	Gaussian	0.071	0.415	17.11	50	0.872	5.215E-03
EC	Exponential	0.082	0.106	77.36	120	0.791	2.081E-05
Total salt	Spherical	0.497	0.836	59.45	180	0.833	1.472E-02

linear transformations is an alternative method for dealing with a data set with outliers or a non-normal distribution [31,33]. Since the concentration of the variables with the non-normal distribution had a lognormal distribution, their concentrations were log-transformed, resulting in more regular variograms. The kriging interpolations were performed on the log-concentrations and the estimated values were back-transformed by an exponential function. As seen in the results, total amounts of salt, pH, EC, and the concentrations of Na^+, K^+, Mg^{2+}, CO_3^{2-}, and Cl^- were relatively higher when from farmland and grassland. The spatial patterns of these variables had a significant geographical distribution with their primary occurrence in the western, north and central areas, which were higher than in other areas. This further proves the effect of the intrinsic factors of topography, soil forming factors and soil type, and extrinsic factors of soil management practices, such as fertilization, use of organic fertilizer and land management in farms, on the spatial distribution of soil chemical properties [33,53]. The soil salinity tended to increase from the margin to the center across the study area, where agricultural wells are denser. It is clear that substantial soil salinization has taken place in these areas due to the effects of land management, farming and climate conditions, such as rainfall and high evaporation. This shows that more attention should be paid to these areas to prevent future problems. Soil pH in the study area ranged from 7.95 to 8.55 in most parts of the Yanqi basin. This pH range falls into the middle level of values that meets the fundamental conditions for plant growth and fertility. The ECs of the soil in all sections were within the acceptable value of <4 dS/m as based on the soil quality standard given by Bao et al. [45]. Areas presently not affected by salinity, but near saline areas, are potential areas for the development of salinity in the future, especially if they are also low-lying. In this respect, it is important to take necessary precautions and implement proper land use plans and cultivation practices.

Assessment of soil salinization risk in Yanqi basin. Soil salinization occurs mainly in arid and semi-arid regions. It may arise due to climate, but is more likely to occur when irrigation practices alter the natural salt balance. Irrigation promotes soil salinization by raising the water table of the underlying aquifer, thus carrying salts upwards. Salts, unlike water, remain in the soil as evaporation and plant transpiration take place, thereby amplifying the salinity. The accumulation of salts in the surface and near-surface zones of soil is a major issue of environmental degeneration and is one of the main causes of low crop yields, loss of land and decreased production. The risk of soil salinization in the Yanqi basin is presented in Table 7. We observed that the whole area has a low salinization risk, as previously mentioned, mainly due to anthropogenic activities and climatic variation [39,40]. The overall salinization risks for farmland, grassland, forest, urban construction areas and desert were none, none, none, low, and moderate, respectively. The mean values of the salinization risk indicators, except the SAR in farmland and desert, are lower than the corresponding maximum limitations that are predicted for the risk (Table 7). This indicates that farmland and desert in the study area have a low salinization risk, but they have mean values of 15.2 and 32.7 for SAR, which showed a moderate grade of salinization (Table 1, Table 7). This indicates a potential risk for environment declination. This research showed the grassland and forest have no risk of salinization, while the areas of urban construction have a moderate risk of salinization. In terms of spatial distribution, the soil salinization risk and the soil salt ions present at higher concentrations in the study area were similar. Environment declination due to soil salinity within the study area almost all

Table 7. Classification of soil salinization risk taken at 0–20 cm depth within the investigated land use and land cover categories in Yanqi basin.

LUCC categories	EC	Total salt content	SAR	Levels of risk
Farmland	1.981	0.552	15.2	Low
Grassland	0.901	0.715	7.5	None
Forest	0.857	0.382	6.8	None
Urban construction areas	1.035	2.761	8.2	Low
Desert	1.024	3.657	32.7	Moderate

took place on the edge of a lake, pond or river. These areas are heavily affected by climate warming, which in turn results in an increase in the amount of evapotranspiration that exceeds 2438.9 mm/a and in the average annual rainfall <100 mm (the drought and waterlog) (Fig. 4). Apart from natural factors, the main driving factors that jointly determined how local dwellers changed the landscape pattern were land use policies, economic systems and population growth. Human activity increases salinization through excessive application of irrigation water without adequate drainage. In addition, the cultivation of grassland is another major cause of inland salinization [40,41]. On one side of the study area, plenty of grass landscape suffered damage, and, in some regions, the land salinization and desertification problem were serious enough to destroy the harmony between the material cycle and energy flow of the ecosystems [42]. On the other side of the study area, the habitat for wildlife was deteriorating, thus seriously threatening the biological diversity. Therefore, establishing and modifying policy, adjusting the irrigation system, improving drainage, dredging the surface water system, promoting circulation between surface water and underground water, and setting up wetland resource monitoring systems are necessary in order to restore damaged wetland and grassland [49,50,55].

Conclusions

(1) From all the soil samples taken in this study, it can be gathered that there are numerous salt properties that vary largely between different samples. This analysis shows that the main salt ions in the soil were K^+, Ca^{2+}, Na^+, Cl^-, Mg^{2+}, and SO_4^{2-}, which accounted for 84.41% of the total salt content of the soil samples. Conversely, the concentrations of HCO_3^- and CO_3^- were very low. Except for the high variation found for the amounts of Ca^{2+} and K^+ (191.67% and 226.25%, respectively), the other soil salt properties of Yanqi basin had moderate levels of variation (10%< CV<100%).

(2) From the analysis shown that the average values of Na^+, Mg^{2+}, SO_4^{2-}, Ca^{2+}, total amounts of salt and K^+ being higher in farmland than grassland, desert, and urban construction areas. This work also determined that the maximum average values of Na^+, Mg^{2+}, K^+, SO_4^{2-}, total amount of salts, and Cl^- were found in coarse crystalline rock weathered material, sandy shale of weathered material and lacustrine deposits. Within the five land use types examined, they were not obvious which classes of elements were enriched and the differences between these groups were small.

(3) PC analysis determined that PC1 (Cl^-, Na^+, SO_4^{2-}, EC, and pH) and PC2 (Ca^{2+}, K^+, Mg^{2+}, and total amount of salts) originated from artificial sources, while PC3 and PC4 (CO_3^- and HCO_3^{2-}) originated from natural sources. Together, this research

shows that Ca^{2+}, K^+, SO_4^2 and the total amount of salts were influenced by both artificial and natural sources. Clustering analysis is consistent with the results from the PC analysis.

(4) From the geo-statistical point of view, it can be speculated that pH and soil salt ions, such as Ca^{2+}, Mg^{2+} and HCO_3^-, had a strong spatial dependency. Meanwhile, Na^+ and Cl^- had only a weak spatial dependency, which was probably due to extrinsic factors, such as contamination, irrigation, and current soil management practices. We evaluated the EC, SAR and total salt content standard to reveal the risk of soil surface salinization. Soil salinization indicators suggest that the entire area had a low risk of salinization as mentioned previously, and this risk was mainly due to anthropogenic activities and climate variation. It is recommended that management of salinized land be preceded by an assessment of local factors and processes that may affect land composition.

Although the overall soil environment was healthy in the Yanqi basin, human activity, such as excessive groundwater pumping, have negatively impacted conditions by inducing soil salinization in the oasis. This matter deserves increased attention. This study can be considered an early warning of soil salinization and alkalization in the Yanqi basin. It can also provide a reference for environmental protection policies and for rational utilization of land resources in the arid region of Xinjiang, northwest China, as well as for other oases of arid regions in the world.

Acknowledgments

Many thanks to the members from the Institute of Geographical Science and Tourism of Xinjiang Normal University, Urumqi, China, and the

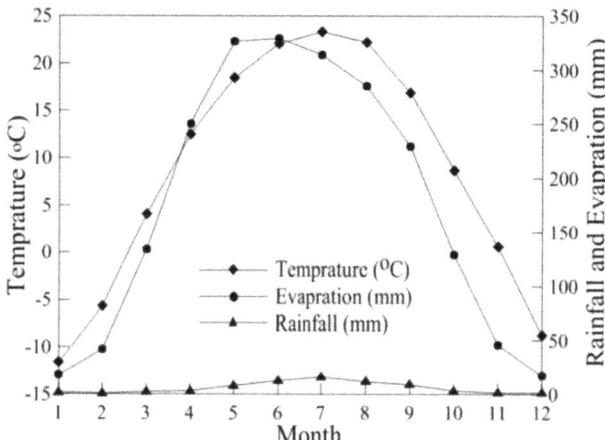

Figure 4. The monthly evaporation, precipitation and temperature of Yanqi basin in 2011.

members form College of Resources and Environment Sciences of Xinjiang University, Urumqi, China for the data collection and assistance in processing.

Author Contributions

Conceived and designed the experiments: ZZ JA. Performed the experiments: ZZ JA HY. Analyzed the data: ZZ JA. Contributed reagents/materials/analysis tools: ZZ JA HY. Contributed to the writing of the manuscript: ZZ.

References

1. Yang YJ, Yang SJ, Liu GM, Yang XY (2005) Space-Time Variability and Prognosis of Soil Salinization. Pedosphet 15(6):797–804.

2. Amezketa E (2006) An integrated methodology for assessing soil salinization, a pre-condition for land desertification. Journal of Arid Environments 67(4):594–606.

3. Masoud AA, Koike K (2006) Arid land salinization detected by remotely-sensed landcover changes: A case study in the Siwa region, NW Egypt. Journal of Arid Environments 66(1):151–167.

4. Yang JS (2008) Development and prospect of the research on salt-affected soils in China. Acta Pedologica Sinica 45(5):837–845. doi:10.3321/j.issn:0564-3929.2008.05.010

5. Stirzaker RJ, Cook FJ, Knight JH (1999) Where to plant trees on cropping land for control of dryland salinity: some approximate solutions. Agricultural Water Management 39(2):115–133.

6. Ghassemi F, Jakeman AJ, Nix HA (1995) Salinisation of land and water resources: human causes, extent, management and case studies. CAB international.

7. Chai S X, YangBZ, Wang XY, Wei L, Wang P, et al. (2008) Analysis of salinization of saline soil in west coast area of Bohai gulf. Rock and Soil Mechanics 29(5):1217–1221. (In Chinese).

8. Wang XL, Zhang FR, Wang YP, Feng T, Lian XJ, et al. (2013) Effect of irrigation and drainage engineering control on improvement of soil salinity in Tianjin. Transactions of the Chinese Society of Agricultural Engineering (20):82–88. (In Chinese).

9. Ren JG, Heng XL, Xi JM, Li JL (2005) Salinization characteristics of the soil in Yeerqiang river valley, Xinjiang. Soils 37 (6): 635–639. (In Chinese).

10. Chen XB, Yang JS, Liu CQ, Hu SJ (2007) Soil Salinization Under Integrated Agriculture and Its Countermeasures in Xinjiang. Soils 39(3):347–353. (In Chinese).

11. Wang YG, Xiao DN, Li Y (2008) Spatial and Temporal Dynamics of Oasis Soil Salinization in Upper and Middle Reaches of Sangonghe River, Northwest China. Journal of Desert Research 28(3):478–484. (In Chinese).

12. Jiang L, Li PC, Hu AY, Yi X (2009) Analysis and evaluation of soil salinization in oasis of arid region. Arid Land Geography 32(2):234–239. (In Chinese).

13. Sawut M, Eziz M, Tiyip T (2013) The effects of land-use change on ecosystem service value of desert oasis: a case study in Ugan-Kuqa River Delta Oasis, China. Canadian Journal of Soil Science 93(1): 99–108.

14. Ling H, Xu H, Fu J, Fan Z, Xu X (2013) Suitable oasis scale in a typical continental river basin in an arid region of China: A case study of the Manas River Basin. Quaternary International 286:116–125.

15. Eziz M, Yimit H, Mamat Z, Li JT (2013) Driving forces of farmland dynamics and its ecological effects in Keriya Oasis in recent 60 years. Agricultural Research in the Arid Areas 3:033. (In Chinese).

16. Triki I, Trabelsi N, Zairi M, Dhia HB (2014) Multivariate statistical and geostatistical techniques for assessing groundwater salinization in Sfax, a coastal region of eastern Tunisia. Desalination and Water Treatment 52(10-12):1980–1989.

17. Fu S, Wei CY (2013) Multivariate and spatial analysis of heavy metal sources and variations in a large old antimony mine, China. Journal of Soils and Sediments 13(1): 106–116.

18. Gil PM, Saavedra J, Schaffer B, Navarro R, Fuentealba C, et al. (2014) Quantifying effects of irrigation and soil water content on electrical potentials in grapevines (*Vitis vinifera*) using multivariate statistical methods.Scientia Horticulturae173:71–78.

19. Hu W, Shao MA, Wan L, Si BC (2014) Spatial variability of soil electrical conductivity in a small watershed on the Loess Plateau of China. Geoderma 230: 212–220.

20. Oliver MA, Webster R (2014) A tutorial guide to geostatistics: Computing and modelling variograms and kriging.Catena 113:56–69.

21. Emadi M, Baghernejad M (2014) Comparison of spatial interpolation techniques for mapping soil pH and salinity in agricultural coastal areas, northern Iran. Archives of Agronomy and Soil Science 60(9):1315–1327.

22. Elbasiouny H, Abowaly M, Abu-Alkheir A, Gad A (2014) Spatial variation of soil carbon and nitrogen pools by using ordinary Kriging method in an area of north Nile Delta, Egypt. Catena 113:70–78.

23. Wu Y, Wang Y, Xie X (2014) Spatial occurrence and geochemistry of soil salinity in Datong basin, northern China. Journal of Soils and Sediments 1–11.

24. Li SJ, Sun YN, Wang HB, Chen ZW (2013) Impact of Soil Nutrient Contents and Spatial Variability in Different Sampling Schemes under the Conservation Tillage.Advanced Materials Research 718:316–320.

25. Bilgili AV (2013) Spatial assessment of soil salinity in the Harran Plain using multiple kriging techniques. Environmental monitoring and assessment 185(1):777–795.

26. Sylla M, Stein A, Van Breemen N, Fresco LO (1995) Spatial variability of soil salinity at different scales in the mangrove rice agro-ecosystem in West Africa. Agriculture, ecosystems & environment 54(1), 1–15.

27. Ammari TG, Tahhan R, Abubaker S, Al-Zu'Bi Y, Tahboub A, et al. (2013) Soil salinity changes in the Jordan Valley potentially threaten sustainable irrigated agriculture.Pedosphere 23(3):376–384.

28. Walter C, McBratney AB, Douaoui A, Minasny B (2001) Spatial prediction of topsoil salinity in the Chelif Valley, Algeria, using local ordinary kriging with local variograms versus whole-area variogram. Soil Research 39(2), 259–272.

29. Jordán MM, Navarro-Pedreno J, García-Sánchez E, Mateu J, Juan P (2004) Spatial dynamics of soil salinity under arid and semi-arid conditions: geological and environmental implications. Environmental Geology 45(4): 448–456.

30. Bai YL, Li BG, Hu KL (1999) Spatial variability of soil salt and its composing ions in salt-affected soil in Huang-Huai-Hai plain. Soil and fertilizer (3): 22–26. (In Chinese).

31. Hu Kl, Li BG, Chen DL (2001) Spatial variability of soil water and salt in field and estimating soil salt using CoKriging. Advances in water science 12(4): 460–466. (In Chinese).

32. Xu Y, Chen YX, Shi HB (2004) Scale effect of spatial variability of soil water-salt. Transactions of the Chinese Society of Agricultural Engineering 20(2): 1–5. (In Chinese).

33. Lin J, Anwar M, Dilbar S (2007) Investigation of the spatial variability of soil salts in saline soil in Xinjiang. Research of Soil and Water Conservation 14(6): 189–192.

34. Aragüés R, Medina ET, Clavería I, Martínez-Cob A, Faci J (2014) Regulated deficit irrigation, soil salinization and soil sodification in a table grape vineyard drip-irrigated with moderately saline waters. Agricultural Water Management 134, 84–93.

35. Bouksila F, Bahri A, Berndtsson R, Persson M, Rozema J, et al. (2013) Assessment of soil salinization risks under irrigation with brackish water in semiarid Tunisia. Environmental and Experimental Botany 92, 176–185.

36. Liang SY, Wu TN, Wu YS, Chou YC, Lee CH (2013) Assessment of Aquifer Salinization Beneath an Offshore Industrial Park Based on Solute Transport Calculation. Advanced Materials Research 779, 1285–1288.

37. Chen L, Feng Q (2013) Geostatistical analysis of temporal and spatial variations in groundwater levels and quality in the Minqin oasis, Northwest China.Environmental Earth Sciences 70(3):1367–1378.

38. Wang LC, Wu RW, Gao J (2014) Spatial coupling relationship between settlement and land and water resources–based on irrigation scale–A case study of Zhangye Oasis.Advanced Engineering and Technology 225.

39. Mamat Z, Yimit H, Eziz M, Ablimit A (2013) Analysis of the Ecology-Economy Coordination Degree in Yanqi Basin, Xinjiang, China. Asian Journal of Chemistry 25(16): 9034–9040.

40. Mamat Z, Yimit H, Eziz A (2014) Oasis land-use change and its effects on the eco-environment in Yanqi Basin, Xinjiang, China. Environment Monitoring and Assessment 186(1): 335–348.

41. Wu JS, Zhang YQ, Liu ZH, Peng J, He JF (2010) Land salnization monitorng with remote sensing on Yanqi County, Xin jiang. Arid Land Geography 32(7):251–257. (In Chinese).

42. IUSS Working Group, WRB (2006). World reference base for soil resources. World Soil Resources Report,103.

43. Hargreaves G H, Allen R G (2003) History and evaluation of hargreaves evapotranspiration equation. Journal of Irrigation and Drainage Engineering 129(1): 53–63.

44. Lu RK (2000) Soil and agricultural chemistry analysis. Beijing/China Agricultural.

45. Bao SD (2005) Soil Agricultural Chemistry Analysis. Beijing/China Agriculture Press. pp: 17–200.

46. Carter M R (1993) Soil sampling and methods of analysis. CRC Press.

47. Lark RM (2001) Geostatistics for environmental scientists.European Journal of Soil Science 52(3), 526–526.

48. Cressie N (1988) Spatial prediction and ordinary kriging. Mathematical Geology 20(4): 405–421.

49. Fan LQ, Yang JG, Xu X, Sun ZJ (2012) Salinity characteristics and correlation analysis of saline soil in irrigation area of Ningxia. Soil and Fertilizer Sciences in China 6: 003. (In Chinese).

50. Metternicht G, Zinck JA (1997) Spatial discrimination of salt-and sodium-affected soil surfaces. International Journal Remote Sensing 18(12): 2571–2586.

51. Agriculture Department of Xinjiang (1996) Soil Survey Office in Xinjiang. Xinjiang Soil. Beijing/Science press 458–464.

52. Wilding LP (1984) Spatial variability: Its documentation, accommodation and implication to soil surveys//MNielson D R, Bouma J. Soil Spatial Variability. Purdoc, Wageningen: 166–193.

53. Fan XM, Liu GH, Liu HG (2014) Evaluating the spatial distribution of soil salinity in the Yellow river delta based on Kriging and Cokriging Methods. Resources Science 36(2):0321–0327. (In Chinese).

54. Wang SX, Dong XG, Liu YF (2009) Spatio-Temporal variation of subsurface hydrology and groundwater and salt evolution of the oasis area of Yanqi Basin in 50 Years recently. Geological Science and Technology Information 28(5):101–108. (In Chinese).

55. Li XG, Lai N, Chen SJ, Mamattursun E (2014) Spatial variability of soil salt based on geostatistics and GIS in the oasis of the lower reaches of Kaidu River: a case study on Yanqi County. Gco-graphy and Gco-information Science 30(1):105–109. (In Chinese).

7

Density and Stability of Soil Organic Carbon beneath Impervious Surfaces in Urban Areas

Zongqiang Wei[1,2], Shaohua Wu[1]*, Xiao Yan[2], Shenglu Zhou[1]

1 School of Geographic and Oceanographic Science, Nanjing University, Nanjing, China, **2** School of Environmental and Land Resource Management, Jiangxi Agricultural University, Nanchang, China

Abstract

Installation of impervious surfaces in urban areas has attracted increasing attention due to its potential hazard to urban ecosystems. Urban soils are suggested to have robust carbon (C) sequestration capacity; however, the C stocks and dynamics in the soils covered by impervious surfaces that dominate urban areas are still not well characterized. We compared soil organic C (SOC) densities and their stabilities under impervious surface, determined by a 28-d incubation experiment, with those in open areas in Yixing City, China. The SOC density (0–20 cm) under impervious surfaces was, on average, 68% lower than that in open areas. Furthermore, there was a significantly ($P<0.05$) positive correlation between the densities of SOC and total nitrogen (N) in the open soils, whereas the correlation was not apparent for the impervious-covered soils, suggesting that the artificial soil sealing in urban areas decoupled the cycle of C and N. Cumulative CO_2-C evolved during the 28-d incubation was lower from the impervious-covered soils than from the open soils, and agreed well with a first-order decay model ($C_t = C_1 + C_0(1 - e^{-kt})$). The model results indicated that the SOC underlying capped surfaces had weaker decomposability and lower turnover rate. Our results confirm the unique character of urban SOC, especially that beneath impervious surface, and suggest that scientific and management views on regional SOC assessment may need to consider the role of urban carbon stocks.

Editor: Dafeng Hui, Tennessee State University, United States of America

Funding: This work was funded by the National Natural Science Foundation of China (41001047). The funders had no role in study design, data collection and analysis, decision to publish, or preparation of the manuscript.

Competing Interests: The authors have declared that no competing interests exist.

* Email: shaohuawu@126.com

Introduction

At present more than half of the world's population resides in cities and towns, and the percentage of urban population is projected to increase to 70% by 2050 [1]. As a result, urban areas are increasing in extent at a greater pace than any other land use type [2]. The rapid expanding of urban areas have caused large areas of agricultural, pasture, or forest soils to be changed to urban soils [3,4]. To date, urban land is estimated to cover 9% of the continent [5], in which the impervious surface (e.g., buildings, roads, and other pavements) is estimated to cover nearly 580 000 km² globally, an area larger than France [6]. Presently, the resulting impacts of the impervious surfaces on urban soil function, including soil organic carbon (SOC) storage and dynamics, remain largely unknown.

Yet a growing body of literature suggest that urban soils still have robust C storage capacity [7–10], especially in the areas covered by green vegetation (e.g., meadow, forest) which could provide fairly C sequestration [11,12]. Generally, the SOC storage capacity of urban open soils (without impervious surfaces) is comparable to that of adjacent agricultural soils and varies highly amongst different cities/regions, which is might be controlled by several factors, such as urbanization histories, land use types, soil parent materials, topography, and climate (Table 1). However, the storage and turnover of SOC beneath impervious surfaces are still poorly characterized due to their inaccessibility. The impervious

surface in urban area is still rapidly expanding due to urbanization, thus it is critical to investigate the SOC stocks and dynamics beneath impervious, to provide accurate inventories in estimates of the entire SOC storage in urban areas, and to promote our understanding of the net impact of urbanization on terrestrial C pools. The relatively limited previous studies suggest that, in urban areas, the SOC density, an important parameter to calculate SOC storage, is significantly lower under impervious surfaces than in open sites [8,15]. These studies mainly focus on the amount or stocks of SOC in urban areas, whereas few attempts are made to study the SOC dynamics in the impervious-covered soils.

Here we collected some urban soil samples from typical impervious-covered and open areas in Yixing City, China, an area experienced rapid urbanization these decades. The main objectives of this study were to investigate the SOC density underlying impervious surfaces in urban areas, and to further study the stability or decomposability of SOC in the sealing environment.

Materials and Methods

Ethics Statement

This study was conducted in Yixing, China. The impervious-covered and open soils were collected from part of Yixing urban area (31°20'–31°25'N, 119°45'–119°50'E). The selected sampling

Table 1. Soil organic carbon densities (kg m^{-2} at a depth where indicated) for urban soils located in different cities.

City	Latitude/longitude	Urban SOC density kg m^{-2}	Suburban SOC density	Adjacent agricultural SOC density	Literature cited
Liverpool, UK	53.4° N/2.98° W	4[a]	–	–	[13]
Boston, U.S.A.	42.35° N/71.06° W	4.02–4.24[b]	3.33–3.99	2.83–4.29	[14]
New York, U.S.A.	42.34° N/75.19° W	5.67[a]	–	–	[15]
New York, U.S.A.	42.34° N/75.19° W	5.1[c]	3.5	3.4	[7]
Baltimore, U.S.A.	39.28° N/76.62° W	4–7[a]	–	–	[7]
Chuncheon, South Korea	37.87° N/127.73° E	2.48[d]	–	3.16	[16]
Phoenix, U.S.A.	33.45° N/112.06° W	0.5–1.1[b]	–	0.75	[17]
Nanjing, China	32.05° N/118.77° E	4.52[e]	–	–	[18]
Chongqing, China	29.56° N/106.57° E	2–3.6[e]	–	–	[19]

a, 0–15 cm; b, 0–10 cm; c, 0–100 cm; d, 0–60 cm; e, 0–20 cm; –, not determined.
Cities were ranked by latitude.

areas did not involve endangered or protected species, and no specific permissions were required for the soil sampling.

Study area

The soils in this study were sampled in January 2008 from Yixing City (31°07′–31°37′N, 119°31′–120°03′E), which has an urban area of 13.4 km^2 in 2000 [3]. Yixing which locates on the plain of the lower reaches of the Yangtze River has had rapid urbanization in recent decades mainly at the expense of agricultural land. Average annual rainfall in Yixing is 1177 mm, and average annual temperature is 15.7°C. The soils in the studied area were formed on the alluvium of the Yangtze River, and the dominant soil type in the agricultural lands around Yixing City is Hydragric Anthrosols [20].

Soil sampling and analysis

Seven sites were selected for impervious-covered soils, and six sites with similar soil parent materials were selected for urban open soils (Figure 1). Soil in our study area had SOC content of 1.48 (±0.19 SE, $n = 9$) g kg^{-1}, TN content of 1.38 (±0.11 SE, $n = 9$) g kg^{-1}, and bulk density of 1.3 (±0.13 SE, $n = 61$) g cm^{-3} in 1982, which these parameters had relatively small variation as their CVs ranged from 8% to 13% (Table 2) [21]. The impervious sites consisted of road pavements, and paved residential squares and alleys, whereas the open area reference sites mainly comprised residential and commercial lawns and gardens, and public greenspaces. Outline description (0–20 cm) for the impervious-covered and open soils selected in our study is shown in Table 3. The soils sampled in our study area primarily had a slit loam texture but varied in the contents of soil admixtures. Compared with the soils in the open areas, the impervious-covered soils had little root penetration and greater amounts of artifacts, including asphalt, cement, brick, tiles, and gypsum. Generally, soil physico-chemical properties were more variable in impervious areas than in the grass areas, and more severely affected by human activities. The mean soil pH and particle fraction larger than 2 mm beneath impervious surfaces (0–20 cm) were 7.72 (range, 7.08–8.33) and 4.28% (range, 6.4%–7.74%), respectively (Table 3).

Each sample, consisting of 3 separate soil cores (5 cm in diameter), was taken from 0–20 cm depth, as SOC storage was supposed to be mainly allocated in the upper soil layers. To make the impervious-covered and open soils more comparable, in the impervious areas, only soils having roots penetrations and little mixture of artifacts were selected. We assumed that the selected soil horizon having these properties is of significance in sequestrating SOC. Specific soil horizon from which the soil was sampled was not determined, but the sampled soils had similar soil texture and chemical properties (Table 3). At the impervious sites, we punched holes on the hard ground near the central position of the selected sealing area (2 m×2 m) and removed the padding to make the covered soil accessible for sampling. After collection, soil samples were transported to the laboratory and air dried at room temperature. Then, the air-dried soils were grounded and sieved through a 2-mm nylon mesh; stones, artifacts, and coarse roots greater than 2 mm in size were weighed.

Soil pH was measured in water (1 soil: 2.5 water, w/v) using a glass electrode, and soil particle size analysis was determined using the hydrometer method [22]. Soil organic C concentrations in the samples were determined using the potassium dichromate sulfuric acid oxidation method [23]. Total nitrogen (N) concentrations were measured by Kjeldahl digestion [24]. Soil bulk density was measured by automated three-dimensional laser scanning [25] (NextEngine Desktop 3D Scanner, NextEngine, Inc., US). The densities of SOC and total N in a horizon of unit area (1 m^2) in each site were calculated as:

$$c = c_c \times BD \times (1 - \delta_{2mm}) \times H \qquad (1)$$

where c (kg m^{-2}) represents SOC density or total N density, c_c (g kg^{-1}) represents SOC concentration or total N concentration, BD (g cm^{-3}) represents soil bulk density, δ_{2mm} (g g^{-1}) represents the fraction of material lager than 2 mm diameter, and H (m) represents the soil sampling depth. In our study, the densities of SOC and total N in each site were calculated based on 1 m^2 square to a 0.2 m depth.

To investigate the stability of SOC in urban areas, a thermostatic soil incubation experiment was conducted. Before the incubation, we first determined soil full water holding capacity. The sieved soil (<2 mm) was packed to the same bulk density (~1.5 g cm^3) in cutting ring (100 cm^3), saturated with water, weighed and then dried at 105°C for 48 hours to determine soil water content. Soil full water holding capacity was the water

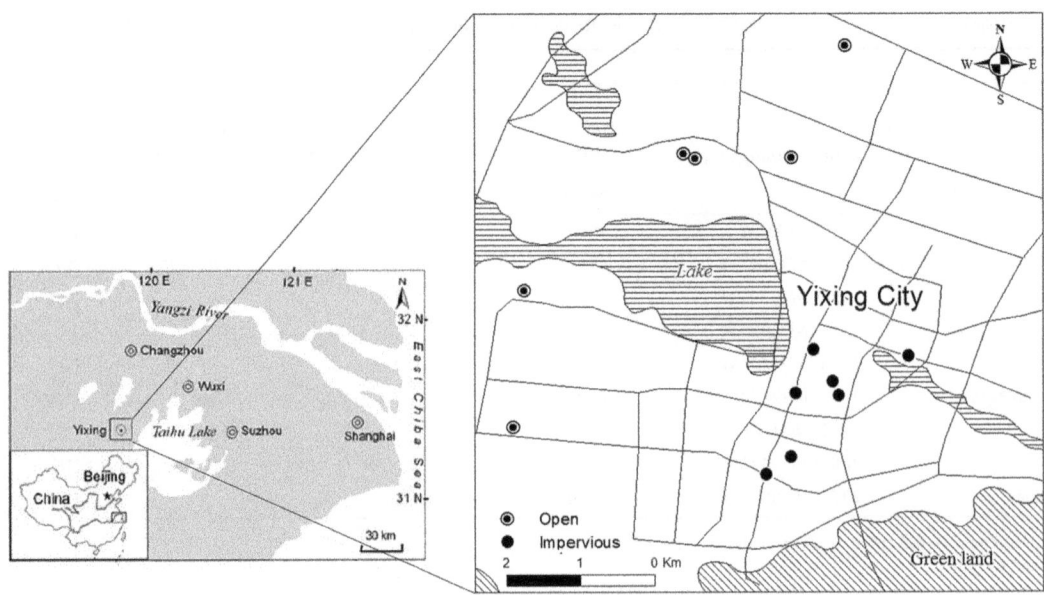

Figure 1. Soil sampling points in Yixing city. Seven sites were selected for impervious-covered soils, and six sites with similar soil parent materials were selected for open soils. Soil samples were collected at 0–20 cm depth.

content of the saturated soil. Fifty grams of each urban soil (< 2 mm) were placed into 500-mL capacity plastic bottles and incubated simultaneously in one incubator (SP-300B, Hengyu, China) for 28 days. The soil water content of each sample in the bottle was then adjusted to 60% of full water holding capacity, which the actual amount of water needed was calculated based on the preliminary test. Each plastic bottle was sealed airtight and incubated in the dark at 25°C. Each sample was pre-incubated without CO_2 absorption for 5 days [18]. Then, a beaker with 5 mL of 0.6 mol L^{-1} NaOH was placed in each jar to absorb the evolved CO_2 during the incubation. Control jars (without soil) were used to measure the background CO_2 concentration. The NaOH was renewed after 1, 3, 5, 7, 14, 21, and 28 days. Any loss of water from the cylinders (checked by weighing) was corrected by a mist sprayer. The CO_2 trapped in the NaOH was determined by back-titration of excess NaOH with 1.5 mol L^{-1} H_2SO_4 after precipitation with 1 mol L^{-1} $BaCl_2$ [26].

A first-order decay equation was used to describe the organic C mineralization in the samples studied [27]:

$$C_t = C_1 + C_0(1 - e^{-kt}) \qquad (2)$$

where C_t (mg C g^{-1} C) is the cumulative amount of SOC mineralized during time t (day), C_1 (mg C g^{-1} C) is the rapidly mineralizable SOC pool, C_0 (mg C g^{-1} C) is the potentially mineralizable SOC pool, and k (mg C g^{-1} C d^{-1}) is the corresponding mineralization rate constant.

The SOC mineralization half-time (i.e., time required to mineralize half of the potentially mineralizable SOC) was calculated as follows [28]:

Table 2. Initial properties of the urban soil in Yixing obtained from the second Chinese soil survey [a] (n = 9 except where noted).

Soil property	0–14 cm		14–25 cm	
	Mean ±SE	CV, %	Mean ±SE	CV, %
Bulk density [b], g cm^{-3}	1.3±0.13	10	1.44±0.15	10.4
pH	6.6±0.3	4.5	6.9±0.6	8.7
TOC [c], %	1.48±0.19	13.3	1.14±0.30	26.4
N, g kg^{-1}	1.38±0.11	8.0	1.13±0.28	24.8
C/N	10.6±1.2	11.3	9.9±1.5	15.2
P, %	0.095±0.01	10.5	0.09±0.125	16.7
K, %	1.67±0.09	5.4	1.68±0.11	6.7
Olsen P, mg kg^{-1}	5.8±2.8	48.3	5±3.8	76.0
Available K [d], mg kg^{-1}	85±30	35.3	96±30	31.2
CEC, cmol kg^{-1}	19.51±2.36	12.1	20.11±2.02	10.2

a, Office of Soil Survey in Yixing County, conducted in 1979–1982; b, n = 61; c, determined using the dichromate oxidation method; d, CH_3COONH_4 extractable K.

Table 3. Outline description for the urban soils (0–20 cm) selected in our study.

Soil sample	Texture	pH	artifacts, rock fragments, roots	> 2 mm fraction
				%
Impervious soils				
1	Silt loam	7.08	coarse angular weathered brick, roots	1.60
2	Silt	7.42	coarse angular freshly weathered rock fragment, roots	9.01
3	Silt loam	7.9	plastic, roots	5.19
4	Silt loam	8.15	coarse angular highly weathered brick and tiles	2.67
5	Silt loam	7.56		8.53
6	Silt	7.58	gypsum, roots	1.44
7	Silt loam	8.33		1.51
Open soils				
8	Silt loam	7.21	heavily roots	5.61
9	Silt loam	6.4		0.72
10	Silt loam	7.12	coarse angular weathered rock fragment, roots	4.25
11	Silt loam	7.58	heavily roots	2.78
12	Silt loam	7.74	heavily roots	2.33
13	Silt loam	6.94	heavily roots	2.35

$$t_{1/2} = \ln 2 / k \qquad (3)$$

Statistical analysis

Model fit was conducted using the Global curve fit wizard in SigmaPlot 12.0 software package (Systat Software, Inc., Chicago, IL, USA). Linear regression analysis and the comparisons in the densities of SOC and total N and the model parameters between the impervious and open land uses (group t-test) were conducted using SAS 8.2 software (SAS Institute, Cary, NC).

Results and Discussion

SOC density under impervious surfaces

The impervious-covered sites had lower SOC and TN concentrations (0–20 cm) compared with the open sites and those in 1982 (Figure 2a, b). The SOC and TN concentrations of the open soils were comparable with those in 1982. The mean SOC density beneath impervious surfaces was 68% (±7.7% SE, $P <$ 0.05) lower than that in open areas (2.46 versus 7.59 kg m^{-2}, respectively, Figure 2c). Similar results were also reported by Pouyat *et al.* (2006) and Raciti *et al.* (2012) that SOC densities under impermeable surfaces were lower than that of open soils [8,15]. It is interesting to note that the SOC densities observed at the impervious sites were comparable to those in New York City (Raciti *et al.* 2012) although the samples were distributed at different cities [15], suggesting that there might be an equilibrium value for the depletion of SOC densities under impervious surfaces. The total N densities were 0.25 kg m^{-2} and 0.32 kg m^{-2}, respectively, for the impervious-covered and open soils in Yixing City, but no significant difference was found between them (Figure 2d), probably due to the relatively small sample size in our study.

The exact mechanisms of SOC loss from the impervious areas are unknown, but likely possibilities include gaseous losses, aqueous losses as dissolved organic and inorganic C, and physical removal of topsoil during the construction process. The removal of surface soil may readily result in a significant depletion of organic C stock due to enhanced mineralization and re-use. The situation could be worse when topsoil is not re-used and is left slowly to decompose. If the potential loss of SOC inferred by this study hold true for other impervious-covered soils, this would suggest that roughly 0.45 Pg (1 Pg = 10^{15} g) of SOC would be potentially lost in China associated with 87128 km^2 of impervious surface [6]. The potential loss of SOC stock could be more significant in magnitude when taking into account the large vegetable C losses in response to initial land use change to urban.

Although the SOC densities beneath impermeable surfaces were relatively lower, it is not adequate to exclude these C pools for estimates of ecosystem C stocks in national scale as required by Kyoto Protocol signatories [10]. In fact, investigation of SOC stocks and dynamics in urban impervious areas are getting more important to the C study, since there is ongoing significant expanding of constructed impervious surfaces around the world [5,6]. In our study, soil samples were collected at 0–20 cm soil layer and therefore SOC densities were calculated to 0.2 m depth. Urban subsoil horizons may also contain considerable amounts of SOC since physical disturbance (e.g., mixing, burying) can result in translocation of topsoil that is rich in OC to deeper soil layer [29]. Thus, the SOC storage underlying capped surfaces in urban areas could be larger than we estimated when it was calculated to 1 m soil depth as usually did. The SOC inventory involving subsoil horizons beneath impervious surfaces in urban areas will strengthen our understanding on the impact of urbanization on the ecosystem. Furthermore, the SOC inventory in urban area should consider the definition and built-up age of the urban used by the investigators. Urban areas, regardless of definition, are rapidly expanding at unprecedented rates; inconsistent definitions of 'urban' will result in different conclusions about the size of urban C stocks [30]. Moreover, it was suggested that the built-up age could modify urban C stocks [31].

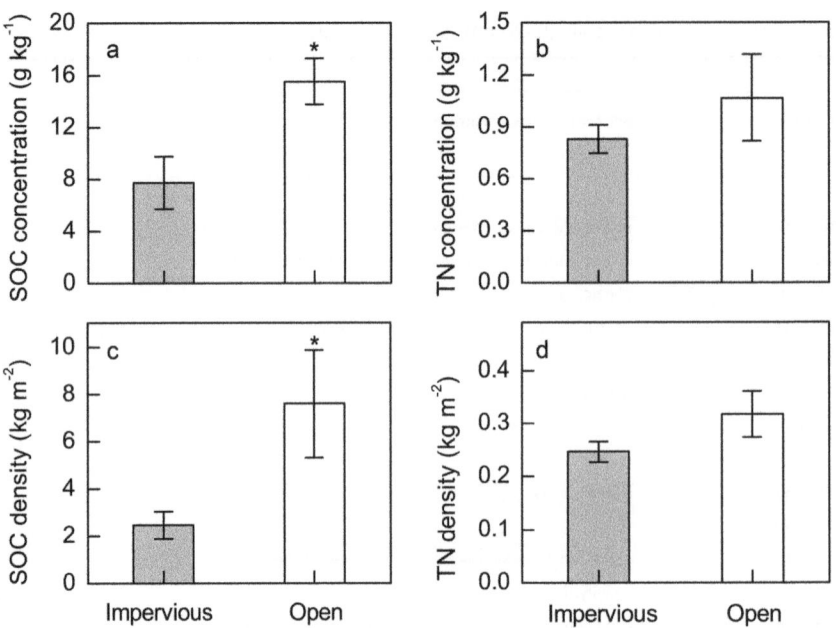

Figure 2. The concentrations and densities of SOC and TN for the impervious-covered and open soils in Yixing city. Values are means ± SE, SOC represents soil organic carbon, and TN represents total nitrogen, * $P<0.05$ ($n=7$ for urban impervious-covered soils, and $n=6$ for urban open soils).

In most soils, there is a tight coupling between the stocks and fluxes of C and N. The mean C/N ratio in the soils covered by impervious surfaces was significantly lower than the ratio in open soils (10.8 versus 22.1, respectively, $P<0.05$, data not shown), indicating that the microorganisms had a relative C deficit state in the impervious-covered soils. The higher C/N ratios in the open soils were probably due to the abundant input of organic materials (e.g., leaves, branches) from plant as these materials always had relatively higher C/N ratios [32]. Regression of SOC density as a function of total N density in the open soils revealed a strong linear relationship between the two variables at the 0–20 cm depth ($r^2=0.73$, $P<0.05$, Figure 3). In contrast, there was no clear relationship between SOC and total N for the soils beneath impervious surfaces, suggesting that the paving in urban areas decoupled the cycle of C and N in the soils [15].

Installation of impervious surfaces in urban areas had negative impact on urban ecosystem, which was indicated by large amount of SOC loss that can offset the C stored in trees or other green vegetation in urban areas, and perturbed soil C and N cycling in impervious area. More greenspaces or semi-pervious pavement systems were, therefore, recommended in future urban construction to mitigate the negative consequences of urban artificial soil sealing [5]. In addition, soil management (e.g., fertilization, tillage, irrigation) for the open sites could be optimized to enhance the SOC sequestration in urban soils.

SOC stability under impervious surfaces

The laboratory incubation can provide an insight of SOC dynamics (e.g., [33]). The amount of SOC mineralized (mg C g^{-1} C) over the 28-d incubation was calculated based on the total (whole soil) SOC content in each sample to enable us directly compare the potential C loss (i.e., CO_2 emissions) from the impervious-covered and open soils, because the portion of SOC is comparable amongst urban soils although the initial SOC content for the urban soils were unequal (Figure 4). It was found that less CO_2 was emitted from the soils underlying capped surfaces than

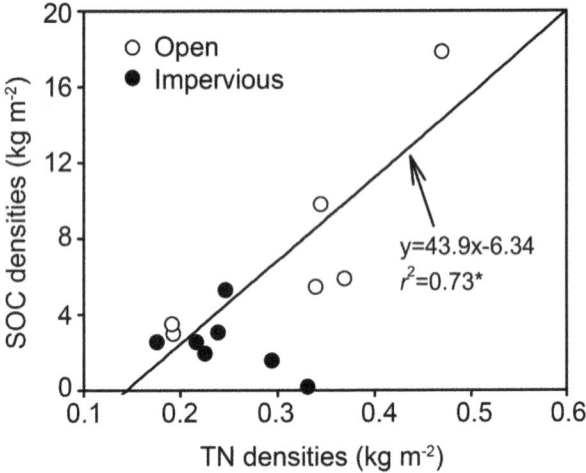

Figure 3. The correlations between the densities of SOC and TN for urban soils in Yixing city. SOC represents soil organic carbon, and TN represents total nitrogen, * $P<0.05$ ($n=7$ for urban impervious-covered soils, and $n=6$ for urban open soils).

from the open soils during the incubation, indicating that the SOC was more stable beneath impervious surfaces than in open areas. The lower transformation of organic C in the covered soils can be partially explained by a low microbial activity in these soils [34].

The first-order decay model ($C_t=C_1+C_0(1-e^{-kt})$) used in this study fitted the data well (r^2 were 0.97 and 0.96, Table 4). The easily decomposable SOC pool (C_1) was small relative to the potentially decomposable SOC pool (C_0) in all samples. The C_1 and C_0 of SOC under impervious surfaces were 47% and 27%, respectively, lower than those in open areas, indicating that there is a more severe depletion of readily decomposable SOC pool after

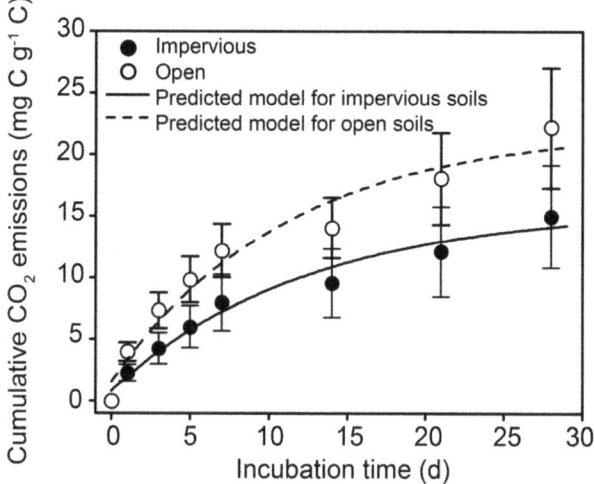

Figure 4. Cumulative carbon release (as CO_2) from the impervious-covered and open soils during the 28-d incubation. Data were fitted by the first-order decay model. The bars indicate standard errors ($n = 7$ for urban impervious-covered soils, and $n = 6$ for urban open soils).

paving. The impervious-covered and open soils had similar mineralization rate constants (k) of the potentially decomposable organic C. However, the lower value of C_0k, a parameter that could be comparable with the initial potential rate of SOC mineralization [35], together with the longer mineralization half-life time ($t_{1/2}$) of SOC under impervious surfaces revealed these soils had weaker organic C decomposability and lower turnover rate.

Edmondson *et al.* (2012) suggested that the turnover of SOC beneath impervious surface likely depend upon the type of capping and the extent of impervious surface [10]. In our study, the covered soils were collected conformably from areas with more than 95% impervious surface, in which the exchanges of gas and water between the soil and atmosphere were supposed to be rare. In fact, there are some urban impervious-covered soils (e.g., patio, garden path) which were distributed in areas dominated by vegetation. The soils underlying capped surfaces in these areas could be colonized by the root systems of lawn grasses and garden trees and shrubs. Thus, it is likely that below these smaller patches of impervious surfaces the soil remains active potentially accumu-

lating SOC and has more robust organic C transformation. Accordingly, more systematic research and soil samples including different land use types and sealing degrees in urban areas are required to better understand the SOC turnover under impervious surfaces.

Additionally, some impervious surfaces in urban area may be removed due to urban land use change. The differences in soil biochemical properties between the impervious and open sites are supposed to be minimized, since the removal of impervious surface. Investigation on the temporal dynamic of SOC after removal of sealing will be beneficial for comprehensively understanding the ecological effects of urbanization. Therefore, management and time, which were needed to rebuild organic C sequestration capacity for the soil after removal of sealing, need to be studied.

Conclusions

Our data demonstrated that the SOC density decreased at 0–20 cm depth in urban areas (regardless of the definition) after the installation of impervious surfaces, although the precise mechanism of organic C loss from these soils was uncertain. The artificial soil sealing in urban areas also decoupled the cycle of C and N, since SOC density correlated positively with total N density in the open soils ($P < 0.05$), but they did not exhibit an apparent relation in the impervious-covered soils. The SOC underlying capped surfaces had weaker decomposability and lower turnover rate compared with that in open areas, which was indicated by a smaller readily decomposable SOC pool, a longer half-time, and a smaller amount of CO_2-C emission during the 28-d incubation. More greenspaces or semi-pervious pavement systems in future urban construction will mitigate the negative consequences of urban artificial soil sealing.

Acknowledgments

We are grateful to the Center of Modern Analysis Nanjing University for supporting soil analysis in our study. We thank the three anonymous reviewers for their very helpful comments and revisions of the manuscript.

Author Contributions

Conceived and designed the experiments: ZQW SHW SLZ. Performed the experiments: ZQW SHW XY. Analyzed the data: ZQW SHW. Contributed reagents/materials/analysis tools: SHW XY. Wrote the paper: ZQW SHW.

Table 4. The first-order decay model (Eq. 2) parameters and coefficients of determination (r^2) for carbon mineralization in urban impervious-covered and open soils.

Parameters	Impervious	Open	
C_1 (mg C g^{-1} C)	0.84	1.59	n.s.
C_0 (mg C g^{-1} C)	14.9	20.53	n.s.
k (mg C g^{-1} C d^{-1})	0.08	0.09	n.s.
C_0k	1.17	1.87	n.s.
$t_{1/2}$ (d)	8.84	7.62	n.s.
r^2	0.97	0.96	n.s.
p	0.01	0.01	n.s.

C_1, rapidly mineralizable SOC pool; C_0, potentially mineralizable SOC pool; k, SOC mineralization rate constant; C_0k, a parameter that could be comparable with the initial potential rate of SOC mineralization [35]; $t_{1/2}$, SOC mineralization half-time.
n.s., not significant.

References

1. United Nations (2008) World urbanization prospects: the 2007 revision.
2. Hansen AJ, Knight RL, Marzluff JM, Powell S, Brown K, et al. (2005) Effects of exurban development on biodiversity: patterns, mechanisms, and research needs. Ecol Appl 15: 1893–1905. doi:10.1890/05-5221.
3. Pan XZ, Zhao QG (2007) Measurement of urbanization process and the paddy soil loss in Yixing city, China between 1949 and 2000. Catena 69: 65–73. doi:10.1016/j.catena.2006.04.016.
4. Su SL, Zhang Q, Zhang ZH, Zhi JJ, Wu JP (2011) Rural settlement expansion and paddy soil loss across an ex-urbanizing watershed in eastern coastal China during market transition. Reg Environ Change 11: 651–662. doi:10.1007/s10113-010-0197-2.
5. Scalenghe R, Ajmone-Marsan F (2009) The anthropogenic sealing of soils in urban areas. Lanscape Urban Plan 90: 1–10. doi:10.1016/j.landurbplan.2008.10.011.
6. Elvidge CD, Tuttle BT, Sutton PC, Baugh PC, Howard AT, et al. (2007) Global distribution and density of constructed impervious surfaces. Sensors 7: 1962–1979. doi:10.3390/s7091962.
7. Pouyat R, Groffman P, Yesilonis I, Hernandez L (2002) Soil carbon pools and fluxes in urban ecosystems. Environ Pollut 116: S107–S118. doi:10.1016/S0269-7491(01)00263-9.
8. Pouyat RV, Yesilonis ID, Nowak DJ (2006) Carbon storage by urban soils in the United States. J Environ Qual 35: 1566–1575. doi:10.2134/jeq2005.0215.
9. Churkina G, Brown DG, Keoleian G (2010) Carbon stored in human settlement: the conterminous United States. Global Change Biol 16: 135–143. doi:10.1111/j.1365-2486.2009.02002.x.
10. Edmondson JL, Davies ZG, McHugh N, Gaston KJ, Leake JR (2012) Organic carbon hidden in urban ecosystems. Sci Rep 2: 963. doi:10.1038/srep00963.
11. Townsend-Small A, Czimczik C (2010) Carbon sequestration and greenhouse gas emissions in urban turf. Geophys Res Lett 37: L02707. doi:10.1029/2009GL041675.
12. Raciti SM, Groffman PM, Jenkins JC, Pouyat RV, Fahey TJ, et al. (2011) Accumulation of carbon and nitrogen in residential soils with different land-use histories. Ecosystems 14: 287–297. doi:10.1007/s10021-010-9409-3.
13. Beesley L (2012) Carbon storage and fluxes in existing and newly created urban soils. J Environ Manage 104: 158–165. doi:10.1016/j.jenvman.
14. Rao P, Hutyra LR, Raciti SM, Finzi AC (2013) Field and remotely sensed measures of soil and vegetation carbon and nitrogen across an urbanization gradient in the Boston metropolitan area. Urban Ecosystems 16: 593–616. doi:10.1007/s11252-013-0291-6.
15. Raciti SM, Hutyra LR, Finzi AC (2012) Depleted soil carbon and nitrogen pools beneath impervious surfaces. Environ Pollut 164: 248–251. doi:10.1016/j.envpol.2012.01.046.
16. Jo HK (2002) Impacts of urban greenspace on offsetting carbon emissions for Middle Korea. J Environ Manage 64: 115–126. doi:10.1006/jema.2001.0491.
17. Kaye JP, Majumdar A, Gries C, Buyantuyev A, Grimm NB, et al. (2008) Hierarchical Bayesian scaling of soil properties across urban, agricultural, and desert ecosystems. Ecol Appl 18: 132–145. doi:10.1890/06-1952.1.
18. Wei ZQ, Wu SH, Zhou SL, Li JT, Zhao QG (2014) Soil organic carbon transformation and related properties in urban soil under impervious surfaces. Pedosphere 24: 56–64. doi:10.1016/S1002-0160(13)60080-6.
19. Liu Y, Wang C, Yue W, Hu Y (2013) Storage and density of soil organic carbon in urban topsoil of hilly cities: a case study of Chongqing Municipality of China. Chin Geogra Sci 23: 26–34. doi:10.1007/s11769-013-0585-x.
20. IUSS Working Group WRB (2007) World reference base for soil resources 2006, first update 2007. World Soil Resources Reports No. 103. FAO, Rome.
21. Office of Soil Survey in Yixing County (1988) The soils of Yixing County. Office of Soil Survey in Jiangsu Province, Nanjing. (in Chinese).
22. Gee GW, Or D (2002) Particle-size analysis. In: Dame JH, Topp GC, editors. Methods of Soil Analysis, Part 4. Physical Methods.pp.255–293.
23. Nelson DE, Sommers LE (1982) Total carbon, organic carbon, and organic matter. In: Page AL, Miller RH, Keeney DR, editors. Methods of soil analysis, Part 2. Chemical and microbiological properties. pp.539–580.
24. Bremner JM, Mulvaney CS (1982) Nitrogen total. In: Page AL, Miller RH, Keeney DR, editors. Methods of soil analysis. Part 2. Chemical and microbiological properties. pp.595–624.
25. Rossi AM, Hirmas DR, Graham RC, Sternberg PD (2008) Bulk density determination by automated three-dimensional laser scanning. Soil Sci Sco Am J 72: 1591–1593. doi:10.2136/sssaj2008.0072N.
26. Goyal S, Chander K, Mundra MC, Kapoor KK (1999) Influence of inorganic fertilizers and organic amendments on soil organic matter and soil microbial properties under tropical conditions. Biol Fertil Soils 29: 196–200. doi:10.1007/s003740050544.
27. Jones CA (1984) Estimation of an active fraction of soil nitrogen. Commun Soil Sci Plant Anal 15: 23–32. doi:10.1080/00103628409367451.
28. Zhang ZD, Yang XM, Drury CF, Reynolds WD, Zhao LP (2010) Mineralization of active soil organic carbon in particle size fractions of a Brookston clay soil under no-tillage and mouldboard plough tillage. Can J Soil Sci 90: 551–557. doi:10.4141/cjss09081.
29. Lorenz K, Kandeler E (2005) Biochemical characterization of urban soil profiles from Stuttgart, Germany. Soil Biol Biochem 37: 1373–1385. doi:10.1016/j.soilbio.2004.12.009.
30. Raciti SM, Hutyra LR, Rao P, Finzi AC (2012) Inconsistent definitions of "urban" result in different conclusions about the size of urban carbon and nitrogen stocks. Ecol Appl 22: 1015–1035. doi:10.1890/11-1250.1.
31. Scalenghe R, Malucelli F, Ungaro F, Perazzone L, Filippi N, et al. (2011) Influence of 150 years of land use on anthropogenic and natural carbon stocks in Emilia-Romagna region (Italy). Environ Sci Technol 45: 5112–5117. doi:10.1021/es1039437.
32. McGroddy ME, Daufresne T, Hedin LO (2004) Scaling of C:N:P stoichiometry in forests worldwide: implications of terrestrial Redfield-type ratios. Ecology 85: 2390–2401. doi:10.1890/03-0351.
33. Haile-Mariam S, Collins HP, Wright S, Paul EA (2008) Fractionation and long-term laboratory incubation to measure soil organic matter dynamics. Soil Sci Soc Am J 72: 370–378. doi:10.2136/sssaj2007.0126.
34. Wei ZQ, Wu SH, Zhou SL, Lin C (2013) Installation of impervious surfaces in urban areas affects microbial biomass, activity (potential C mineralization), and functional diversity of the fine earth. Soil Res 51: 59–67. doi:10.1071/SR12089
35. Fernández JM, Plaza C, Hernández D, Polo A (2007) Carbon mineralization in an arid soil amended with thermally-dried and composted sewage sludges. Geoderma 137: 497–503. doi:10.1016/j.geoderma.2006.10.013.

8

Through the Eyes of Children: Perceptions of Environmental Change in Tropical Forests

Anne-Sophie Pellier[1,4]*, Jessie A. Wells[2,4], Nicola K. Abram[3,4], David Gaveau[1,4], Erik Meijaard[1,2,4]

1 Center for International Forestry Research (CIFOR), Bogor, Indonesia, **2** ARC Centre of Excellence for Environmental Decisions, Centre for Biodiversity & Conservation Science, University of Queensland, Brisbane, Australia, **3** Durrell Institute for Conservation and Ecology, School of Anthropology and Conservation, Marlowe Building, University of Kent, Canterbury, Kent, United Kingdom, **4** Borneo Futures initiative, People and Nature Consulting International, Jakarta, Indonesia

Abstract

This study seeks to understand children's perceptions of their present and future environments in the highly biodiverse and rapidly changing landscapes of Kalimantan, Indonesian Borneo. We analyzed drawings by children (target age 10–15 years) from 22 villages, which show how children perceive the present conditions of forests and wildlife surrounding their villages and how they expect conditions to change over the next 15 years. Analyses of picture elements and their relationships to current landscape variables indicate that children have a sophisticated understanding of their environment and how different environmental factors interact, either positively or negatively. Children appear to have landscape-dependent environmental perceptions, showing awareness of past environmental conditions and many aspects of recent trends, and translating these into predictions for future environmental conditions. The further removed their present landscape is from the originally forested one, the more environmental change they expect in the future, particularly declines in forest cover, rivers, animal diversity and increases in temperature and natural disasters. This suggests that loss of past perceptions and associated "shifting environmental baselines" do not feature strongly among children on Borneo, at least not for the perceptions we investigated here. Our findings that children have negative expectations of their future environmental conditions have important political implications. More than other generations, children have a stake in ensuring that future environmental conditions support their long-term well-being. Understanding what drives environmental views among children, and how they consider trade-offs between economic development and social and environmental change, should inform optimal policies on land use. Our study illuminates part of the complex interplay between perceptions of land cover and land use change. Capturing the views of children through artistic expressions provides a potentially powerful tool to influence public and political opinions, as well as a valuable approach for developing localized education and nature conservation programs.

Editor: Nicolas Chaline, Universidade de São Paulo, Brazil

Funding: The study was financially supported by the United States Agency for International Development (USAID) (www.usaid.gov/) and the Arcus Foundation (http://www.arcusfoundation.org/). The CGIAR Research Program on Forests, Trees and Agroforestry, led by the Center for International Forestry Research (CIFOR), provided additional funds. The funders had no role in study design, data collection and analysis, decision to publish, or preparation of the manuscript.

Competing Interests: The authors have declared that no competing interests exist.

* Email: pellier.annesophie@gmail.com

Introduction

Environmental change typically occurs over long, decadal, time frames. Deforestation, pollution of water ways, or declines in species numbers often take place gradually, and notable changes are only apparent once certain thresholds are passed, for example, when species become extinct, or pollution creates significant health problems [1]. Gradual environmental changes can lead to inter-generational differences in perceptions of environmental conditions, often referred to as the shifting baseline syndrome [2,3]. Shifting baselines have been extensively demonstrated in perceptions of fisheries [3,4,5,6], but this phenomenon also occurs in terrestrial environments, impacting conservation management and influencing environmental policy-making [7]. For example, if perceptions of the former abundance of threatened species, such as the orangutan (*Pongo pygmaeus*), are lost over time, there may be little public or governmental support to enable populations to recover to ecologically optimal densities. Similarly, people's perceptions of what constitute forests can change, if people forget what forests were once like, or have adapted their perception of forests to their present reality [8]. These altered perceptions can impact conservation efforts, if they result in lack of public support for conserving forest ecosystems or protecting areas of remnant, intact forest from loss or degradation.

Shifting baseline perceptions are of particular concern among children and the younger generations of a community, as it is they who will ultimately play the biggest role in determining future use and management of the natural environment [2,9,10]. Compared to older generations, however, children have a shorter term association with past environmental conditions and are thus more likely to use present-day contexts as their reference framework. For example, a child who has only experienced current degraded environments might be unaware of earlier, more natural conditions, or of the potential for recovery to pre-disturbance conditions. This shift is more likely to occur with a lack of communication between generations and/or of information on past (and current) environment conditions [4,7]. Most children have longer lives ahead of them than adults and thus greater stakes in future environmental and social conditions, which will influence their long-term well-being. Children are also engaged in rapid learning and have fewer pre-conceptions than older generations. They may therefore be the ideal group for targeting conservation education and awareness programs that aim to increase consciousness of how natural ecosystems and biodiversity are changing, and motivate more sustainable interactions with their environment.

In this study we focus on Indonesian Borneo, South-East Asia, a hotspot of endemism and biodiversity [11]. These biologically rich environments are, however, under increasing threat, as Borneo has some of the highest rates of deforestation and forest degradation in the world [12,13]. Between 1973 and 2010, at least 168,493 km^2 (30.2%) of closed-canopy forests were cleared across the island, and a further 179,917 km2 (32%) were logged [14]. By 2010, Borneo's land cover consisted of 28% intact forests, 24% logged forests, 10% industrial plantations, and the remaining 38% included secondary forests, agroforests, grasslands, croplands and other land uses [14].

Governments in Borneo are principally democratically elected and political decision-making can be influenced by peoples' opinions. Therefore, capturing perceptions about land use and forest values, and understanding what drives spatial variation within these perceptions, offers important knowledge for spatial land use planning and natural resource policies. The Borneo Futures initiative, under which the present study was conducted, was established to perform these assessments, and has led to new insights about how people view their relationships with forests and other land uses, how this varies across the large and culturally diverse island, and what environmental and social factors underlie these perceptions [8,15,16,17]. These findings are vital if land use conflicts are to be minimized, environmental health restored, biological diversity conserved, and sustainable economic development enhanced.

In this paper, we build an additional dimension to the Borneo Futures initiative by focusing on children's perceptions of forest landscapes and wildlife surrounding villages in Kalimantan, Indonesian Borneo. Specifically, we assess children's perceptions of the current and future state of their environment, and whether these perceptions differ among children who are growing up in different landscape contexts: forested landscapes, recently deforested landscapes, or old agro-forestry landscapes. To assess this, we employ a novel approach by using children's drawings to gain insights into their perceptions of environmental change. Although other studies have used similar techniques for studying children's thoughts [18,19,20], to our knowledge this is the first study to use this approach to assess perceptions of environmental change across dynamic forest environments.

Materials and Methods

Ethics statement

Institutional and authority level. The surveys of children's drawings were conducted in parallel with a broader set of interview-based surveys of adult villagers regarding ecosystem services and wildlife [8,15,16]. Written approval for our survey was given by the Indonesian State Ministry of Research and Technology (RISTEK) under their research agreement with the Center for International Forestry Research (CIFOR). Prior to commencing field work in each district, the researcher registered the project's methods and purpose with district government officials. At the village level, approval by local authorities was given verbally by the village head or other senior village leader at the time of the research. As CIFOR has neither an ethics

committee nor institutional review board, the proposed methodology was discussed with CIFOR's senior social scientists who gave us verbal approval to proceed, pending each potential research subject's free, prior and informed consent to participate in the research. This consent required us to give villagers, teachers and children sufficient information about the study's design and objectives so that they could make a free and informed decision about whether they agreed with the research occurring in the village, and specifically whether they wished to participate.

Village level. In each village, the consent process involved meetings with the village head and village leader, followed by a community meeting with school teachers and other adult members of the village, including the parents of potential participants. During these meetings a standard statement in Indonesian was read out (and provided in hard copy) which specified the identities of the researcher and her field assistant, the purpose and value of the research, and that each child's participation was to be voluntary and anonymous. We outlined that children would be asked to draw the environment around their village, with features such as the forest and animals. (More specific details of the drawing activity and its request to consider the environmental conditions in 15 years' time were given on the day of the activity, not during the village meeting, because we did not want the children's views to be influenced by ourselves or by others in the village before the time of the activity). We specified that we were seeking children between 10 and 15 years old to participate, and that younger children (7–9 years) would be invited to participate if fewer children in the target range were present. We requested verbal consent to use the drawings in research, workshops, magazines and any research papers, and on a web platform. We assured villagers that the data would be treated confidentially, and neither the names of villages nor the names of participants would be publicized with these drawings. An additional meeting with the school head and teachers on the morning of the research activity was held providing information on the aims of the study and requesting their verbal confirmation of their consent. In the classroom, the children were clearly informed of the identity of the researcher and the details of the drawing activity (for example, that they would each join a group of children and draw a picture together, which would show either how they saw their present environment, or their expectations for their environment 15 years in the future). We emphasized the voluntary nature of participation and that non-participation would not affect them negatively, and assured anonymity of their names and their village. Finally, children were asked to verbally confirm that they understood this explanation, then asked to freely choose to participate or not. Those who chose not to participate were free to leave the classroom. After completion of the drawings, each participant was asked to write his or her first name on the reverse side of the drawing sheet (or the first letter of his/her first name if he/she was not able to write), to indicate their voluntary participation in the activity, noting that their name would remain anonymous and would not be visible in any display or in any data sheet analysis of the research.

Reasons for verbal consent. We did not obtain written consent for two reasons. Firstly, there is a common reluctance by people in Indonesia to sign documents due to fears of their use by authorities to their detriment. Secondly, the resulting lag time between introducing the research topic to villagers and the actual implementation of the drawing studies would have created significant opportunities for children's perceptions to be influenced by their parents, teachers or each other. As the primary goal of the study was to capture the children's own perceptions, it was essential to minimize such influences. Formal written consent would have required parents to sign on behalf of their children, resulting in time lags of several days because of the logistical difficulty of visiting people who are often working in fields away from the village.

Data Collection

We chose three broadly different environmental settings for the surveys: 1) an area dominated by old agroforestry and plantations (mostly for palm oil and rubber production) in the lower Kapuas River region in West Kalimantan Province; 2) an area dominated by old growth forests and slash-and-burn agriculture in the upper Kapuas River region of West Kalimantan; and 3) a remote area dominated by selectively logged natural forest in the upper Kelay River area in East Kalimantan (Figure 1). Without having prior knowledge about the villages, we randomly selected 22 villages, consisting of 20 in West Kalimantan and two in East Kalimantan.

In each village, classes were composed of children from a range of ages, where often the age of a child could not easily be linked with their level of school education, especially in the more remote villages. If there was no school in the village or if the visit occurred during school holidays, children were asked by the village head whether they wanted to participate. In the classroom, children were asked to join one of two groups to work together on one drawing per group. These groups consisted of both boys and girls mostly aged between 10 and 15 years old (see Table 1). We targeted this age range because our pilot study had revealed that the aim of the activity was clearly understood by children in this age class, whereas younger children sometimes found it difficult to understand what they were asked to draw. In some villages, where the number of children in the target age range was less than six per group, younger children aged 7, 8 and 9 years old joined the activity. The children of different ages were allocated evenly between groups, to ensure that the age ranges for the groups drawing the present and future were as similar as possible. The ages of the children were then confirmed by their teacher or head of school. Groups were allocated in different parts of the classroom to minimize interactions between them.

Drawing surveys

In each village, we asked one group to draw the current condition of the forest and the animals they would expect to find in the forest. The second group was asked to draw their vision of the forest and animal life in the forest in 15 years from now. To clarify the 15 year time frame, we referred to a period when participating children would be adults and may have their own children. We reminded the children that it was not important how skillful they were at drawing and that the important thing was to draw what their group really thought about present or future conditions of forests. Children were encouraged to discuss ideas within their group before committing them to paper. Sometimes children of the same group had different ideas, and emphasis was given for them to either draw a compromise idea or to include all the different ideas in their drawing. No time restrictions were imposed. The average time spent was two hours, and in all cases the children reported that they completed their drawing. We requested that during the activity the teacher would not interact with the children, discuss their drawings or provide additional explanation. Also, no books or images were allowed for the activity, to ensure that we captured the children's unbiased perceptions. During the drawing exercises, author ASP also observed the discussion and drawing processes and took notes about particular decisions the children made.

Drawing interpretations

Once the drawings were complete, each group was asked to describe their drawings and these explanations were recorded on video and later used to verify particular features of the drawings and gain further insights into their perceptions. These interviews lasted until the children had nothing more to say about their drawing. If some elements were still unclear, we asked children for clarification through some specific questions regarding the amount of forest left in their landscapes, and its current and expected future conditions. Additionally, we asked whether the forests they had drawn contained many, few or no animals, and which animals occurred or had disappeared. Communication during the surveys was primarily in Indonesian. In remote areas, children understood Indonesian but were more comfortable in their local language. Many items in the drawings are captioned in these different local languages and we asked for input from both children and the school teachers to translate them into Indonesian.

Survey data management

All of the elements depicted by children in their drawings and described in their writings, and all oral comments were entered in a database. We assigned features to eight categories (from which six were used for analysis, see Table 2), with some categories following Bowker [18]: 1) Broader landscape features (e.g., mountains, sun, sky, clouds, rainbow, rain, temperature condition, rivers, waterfalls, and lakes); 2) Botanical diversity (e.g., frequency and number of each different type of trees and plants in either natural forest or agro-forest: fruit trees, hardwood trees, medicinal plants, resinous trees, other tree types, shrubs, flowers, vines, fungi and grassland, and total of this vegetation diversity (sum of different types)); 3) Faunal diversity (e.g., the frequency and number of different animal species and general condition of animal wildlife); 4) Forest condition features (e.g. including village-forest distance, areas of continuous-canopy forest (not being cleared) and disturbed forest areas, and the presence of felled trees); 5) Agricultural features (e.g., oil palm trees or plantations, rice fields, and rubber plantations); 6) Features relating to other environmental issues (natural and anthropogenic) (e.g., natural disasters, such as floods, landslides, erosion, and forest fires; condition and pollution of rivers; and elements related to development, including roads, vehicles, hotel buildings, large housing, factories, electricity poles/towers, etc.); 7) People in the forest or village (e.g., the number of rural people in the forest or in the village, their location, their activity, boat types, and equipment); and 8) Village features (e.g., village occurrence in the drawing, number of houses, and any other building types such as churches, schools, and traditional monuments).

We coded each individual element either as binary values (absence/presence of each element, or good/bad for the element's condition), as ordinal values (e.g., few, several, many; or, beside, close, far, far away), or as numerical counts of elements or element types such as particular species. In addition, some variables were assigned a set of further 'conditions' only as nominal values (e.g., 'undisturbed forest areas'; 'threats to animals'; etc.). Each of the measures or individual elements described above constitutes an 'art variable' for statistical analysis.

Spatial variables

We assessed relationships between children's perceptions of forests and their village landscape contexts represented by 13 spatial variables consisting of land use and land cover (LULC), climate, topographical and socio-economic variables (see Table S1 for descriptions and abbreviations of spatial variables; sources and

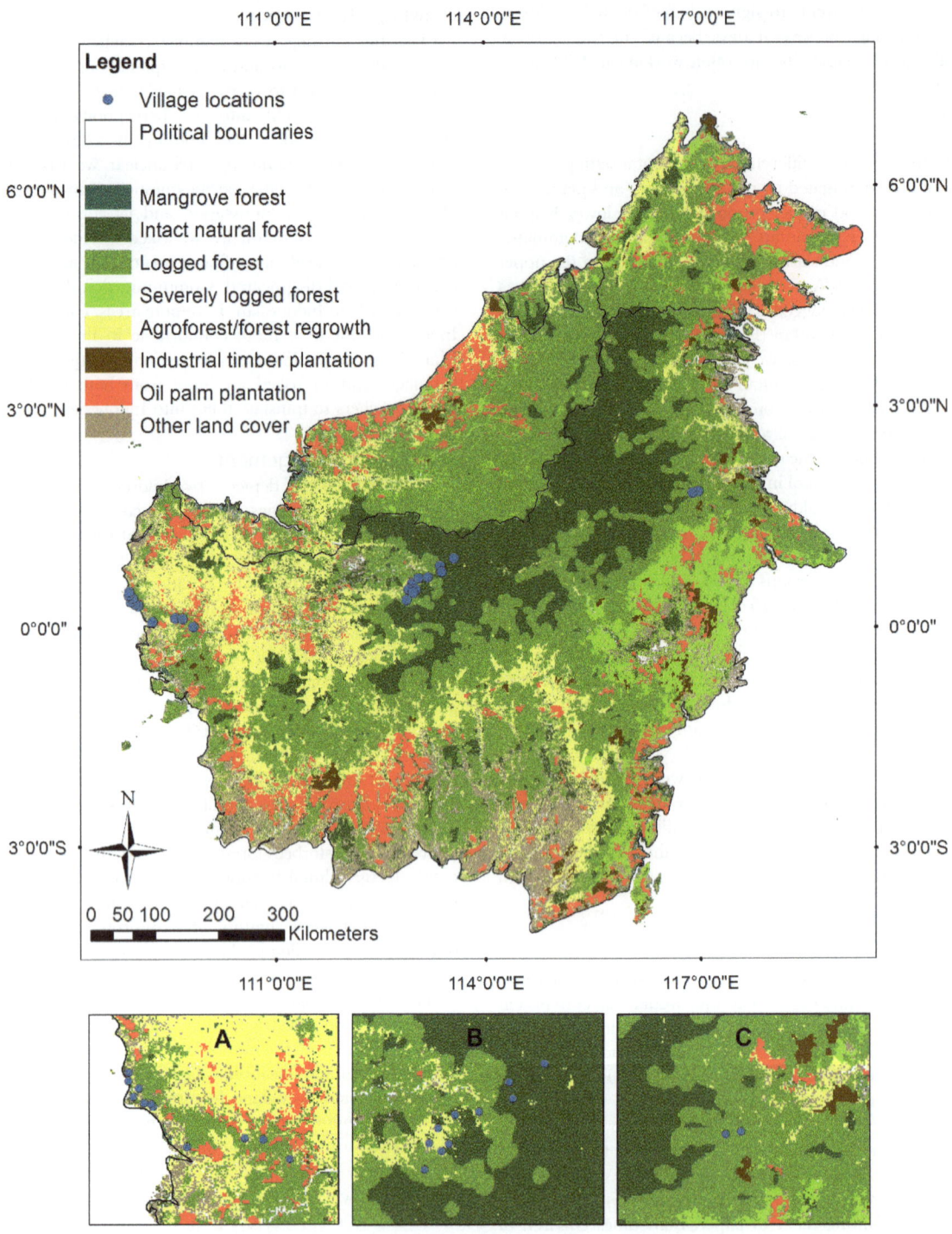

Figure 1. Surveyed villages in Kalimantan, Indonesian Borneo with land cover (2010) classes. The map shows the land cover and land use across Borneo island in the year 2010 (see Figure legend, and methods for sources), and the locations where the children's drawing surveys were conducted in twenty-two villages (blue dots) in three different parts of Kalimantan, Indonesian Borneo. Insets: (A) Ten villages at the lower Kapuas in West Kalimantan Province; (B) Ten villages at the upper Kapuas in West Kalimantan Province; (C) Two villages at the upper Kapuas in East Kalimantan Province.

processing steps are outlined below, and are described in full in reference [8]).

We extracted values of these spatial variables for each sampled village by mapping village locations within a Geographic Information System (GIS), using the Global Positioning System (GPS) way points taken in the centre of each village.

Analysis of land use and land cover considered five LULC classes: intact natural forest; agro-forests/forest re-growth; oil palm plantations; other land cover types; and a general 'forest cover' layer developed by combining the intact forest and logged forest layers. The villages surveyed were surrounded mainly by 'logged forest', whereas intact forest was less extensive. The overall 'forest cover' was strongly positively correlated with the cover of 'logged forest', and weakly negatively correlated with 'intact forest' cover. The LULC spatial variables were derived from the integration of three datasets: firstly, a SarVision PALSAR 2010 (50 m resolution) classification, where classes were either used separately, or aggregated to form more general classes; secondly, logging road data networks (digitized from Landsat imagery from the years 1990, 2000 and 2002, [21]) were used to distinguish intact and logged forest classes, and to calculate a road density index (the length of logging roads and major roads within a 5 km radius of the village); and thirdly, digitized data sets of oil palm plantations developed through onscreen digitising of >150 Landsat images from the 1990-, 2000-, and 2010-eras, downloaded from the Global Land Survey database (http://earthexplorer.usgs.gov/) (for details see [21,22]). For each LULC class, we calculated the Euclidean distance from the village to the nearest example of that class (in meters), such as the distance to the nearest oil palm plantations or to logged forest (see Table S1). For each of the five LULC classes, and for an additional class indicating protected area status [21], we also calculated the summed values of neighbouring cells within a 10 km radius (using focal statistics in ArcGIS 10.0). This gave a set of values for each village, detailing the summed cover of each LULC class and of protected areas in the landscape around the village. Secondly, the climatic variable 'annual precipitation' was included, along with elevation grid data from the WorldClim, ver. 1.4 dataset (www.worldclim.org) at 30 arc-seconds resolution. A river density index was generated by applying a kernel density tool in ArcGIS 10 to the Hydrosheds rivers dataset sourced from http://hydrosheds.cr.usgs.gov/index. php [23]. Thirdly, a human population density layer (i.e., estimated number of people per 1 km^2) was derived from the LandScan 2007 dataset [24] and used to calculate 'settlement density' using a kernel density function for cells with 10 or more

people per km^2. All spatial data were developed at 30 arc-seconds (approximately 1 km^2) resolution and projected. We then extracted all spatial data for each village location to use within the analyses outlined below.

Statistical analysis

Art variables were first summarized using descriptive statistics. To compare the art variables between the present and future drawings we used contingency table tests (contingency coefficient (Cc)) for statistical differences in frequencies for each binary or nominal variable, and we used non-parametric Mann-Whitney (U) tests for differences in numerical (ratio-scale) and ordinal variables. To understand the relationships between variables, correlation coefficients were estimated for each pair of art variables and spatial variables using Pearson's correlation coefficient (r_P) for ratio scale variables, and Spearman's rank correlation (r_s) for pairs involving binary or ordinal variables. To summarize the current physical landscape of the villages, we conducted a principal component analysis on the 13 spatial variables in SPSS 17.0 (SPSS 2010) using the factor analysis function. We next conducted a categorical principal component analysis (CatPCA) of the art variables as a combination of numerical (ratio scale), ordinal, binary and nominal variables, and compared the multivariate axis scores between drawings of the present and the future. We also explored the relationships between the current land cover of the villages and the art variables describing the children's perceptions of current and future environmental conditions. In the CatPCA analyses, numeric variables that contain decimal values were converted to rank variables, to enable handling of values <1 (these variables consisted of the physical landscape layers and one art variable 'Vegetation diversity'). All variables were given equal weight in the multivariate analyses.

Results

Drawings summary

A total of 247 children participated in the drawing study (131 girls and 116 boys) from 22 different villages, resulting in 44 drawings (see Figure 2 for an example). The four animal taxa with highest representation in the drawings were: birds (22% of wild animals), mammals (21%), reptiles (21%), and fish (19%). Children often drew people in the forest engaged in some activity, such as fishermen in boats or farmers working in their field and gave several examples of different trees (fruit, resinous and timber trees), animal species, and forest products (e.g. rattan and honey).

Table 1. Ages and numbers of children who participated in drawing surveys in 22 villages in Kalimantan.

Drawing time period	Number of villages	Average number of children	Average age of children	Male	Female	Total of children*
Present	22	6	11.4	73	72	145
Future	22	6	12.0	65	76	141
Totals or Mean	44 drawings, 2 per village	6	11.7	138	148	286

*Note: In four of the villages, 39 children participated in both the present and the forest future drawing sessions, however, if a child had already drawn, then he or she only participated in giving ideas for the second drawing. Consequently, a total of 247 children participated in the drawing study with 131 girls and 116 boys from 22 different villages.

Within each village, the average age of children in the 'present' and 'future' drawing groups were as similar as possible.

Table 2. Environmental art variables depicted in drawings by children from 22 villages in Kalimantan.

Category	Environmental variables from drawings	Code used in figures
Landscape	**Condition of mountain forests**	**For.Mtain**
	Temperature condition	**Temp**
	River condition	**RiverC**
Botanical diversity	**Vegetation diversity** (Total number of vegetation types, excluding grassland)	**VegE**
Faunal diversity	**Faunal condition - Abundance and diversity of wildlife**	**FaunaC**
	Threats to animal wildlife (9 individual threats coded from 1 to 9: a single code value was assigned to each drawing; if several threats occurred in a given drawing, a single code value was assigned as the addition of the specific threats)	**ThreatA**
Forest condition	**Village-Forest Distance** (areas of non-felled trees and agro-forest trees if represented in forests)	**ViForD**
	Disturbed Forest areas (Shown as location of logged, burned or degraded forests; e.g. 'only in mountain part'; 'surrounded village'; 'both mountainous areas and surrounding the village')	**Defo**
	Undisturbed Forest areas (i.e. Location of forests not being cleared or degraded at this time; can be primary and/or secondary forests)	**NoDefo**
Agriculture fields	**Oil palm area cover**	**Poil**
Other environmental issues (natural and anthropogenic)	**Presence of industries**	**Ind**
	Presence of main roads	**Mroad**
	People clearing the forest (no clearance; few; a lot)	**MenClear**
	Flood occurrence	**FL**
	Non-flood disasters occurrence	**NFLdis**

Sometimes children, especially those living in more densely forested areas, represented their cultural and spiritual beliefs relating to environmental features, for example by drawing a '*rumah hantu*' (house of ghosts) in the forest, or a dragon in the river. Overall we distinguished 180 different elements in the eight different categories. Fifteen art variables were selected for further analysis (see Table 2), consisting of individual elements or their summaries (e.g., total 'vegetation diversity' as the sum of plant types) (see below). These art variables were selected as the elements that appeared frequently in drawings from multiple groups, and had been specifically mentioned by children during their explanation of their drawings.

Correlations between art variables within each time period

In the 'present day' drawings, children depicted a wide range of environmental conditions for most variables, such as the condition of forest near the village, faunal diversity, and vegetation diversity. Of the 15 variables, three were represented as being in 'good' conditions in all villages ('temperature', 'forest mountain', and 'river' conditions), and hence showed no statistical variation (in contrast, these variables did show variation across drawings of the future).

Several of the art variables from 'present' drawings are strongly correlated (Table S2). For instance, 'main road' and 'flood occurrence', which are positively correlated to each other ($r = 1.0$, $p < 0.05$ in all cases mentioned), are both similarly and positively correlated with the 'presence of industries' ($r = 0.69$) and 'threats to animals' ($r = 0.65$). The latest variable is also negatively correlated with 'faunal condition' ($r = -0.79$), while 'vegetation diversity' shows a negative correlation with 'disturbed forest areas' ($r = -0.64$).

The drawings and discussion of the 'future' showed variation across villages in the condition of almost all environmental features, with generally higher variance than across the 'present' drawings. All art variables showed variation, except for 'people clearing the forest', which was categorized as 'many' in the future for all villages. Correlations among 'future' art variables are shown in Table S3. Interestingly, 'flood occurrence' is significantly correlated with all other variables, except 'oil palm cover' and 'threats to animals'. Though 'flood occurrence' is only weakly correlated with 'main road' presence ($r = 0.42$), they both show similar negative correlations with other art variables (conditions of 'mountain forests' (respectively, $r = -0.65$ and -0.68), of 'rivers' (resp., $r = -0.47$ and -0.46), and of 'animals' (resp., $r = -0.48$ and -0.50), 'undisturbed forest areas' (resp., $r = -0.46$ and -0.64) and the 'vegetation diversity' (resp., $r = -0.45$ and -0.49)); and positive correlations with 'village-forest distance' (resp., $r = 0.46$ and 0.56) and 'disturbed forest areas' (resp., $r = 0.48$ and 0.56). 'Main road' presence is also positively correlated with 'oil palm cover' ($r = 0.43$).

Differences between present and future environments

Comparison between drawings of present and future conditions showed the following differences: local temperatures are expected to be higher in the future ($p < 0.001$, $U = 99.0$); future conditions of mountain forests and rivers are expected to deteriorate (respectively: $p < 0.001$, $U = 19.0$; and $p < 0.001$, $U = 50.0$, for example 100% showing 'good' river condition in the present, compared to 58% 'bad' in the future), and the number of people clearing the forest is expected to increase ($p < 0.001$, $U = 66$). A closer analysis of forest conditions indicates that distance between villages and the nearest forest will be much greater in the future ($p < 0.001$, $U = 54.5$), with 22.7% of 'future' drawings showing the highest distance class, and 68.2% placing villages in areas that were previously, but no longer, forested. In the present, the 'undisturbed' forest areas occur around villages and on mountains (68.2%), though some villages showed this undisturbed forest only on the mountains (27.3%). In the future, undisturbed forests are

Figure 2. Drawings of the present and future by children in lower Kapuas, West Kalimantan. Both the drawings represent respectively the present (A) and future conditions (B) of the forest and animal wildlife. The children provided explanations in the drawing as the following: A) Present conditions: "The forest is an oxygen giver which is used to make a living from forest trees: rattan, rubber trees, durian, bamboo, etc. Forest is home to wild animals (e.g. snakes, wild cats, monkeys, pigs, birds, deer, squirrels, etc.) and among those there are several which live in the river (e.g. crocodile, fish and turtles). Nevertheless, wild animals are being hunted by humans and are thus disappearing (e.g. snakes, monkeys, deer, pigs, pangolins, birds, etc.). Animals that live around the village are chickens, pigs, dogs, and cats. The forest is still good, especially in the mountains." B) Future conditions: "Forest is currently home to many different animals and to human livelihoods. In the future, the air that was fresh in the present time is getting warmer, including the mountainous regions. Nothing is green anymore and generally forest is replaced by oil palm even in the mountain where it only remains as few small trees. The number of trees decreases due to palm oil companies, factories, and other human activities such as illegal logging to build houses, mining for materials, and others for building highways. These changes will cause natural disasters such as erosion, landslides, and floods. The number of animals decreases, and little by little they will disappear, except dogs and chickens. The rivers (with water boats) are being polluted by scattered waste, and fish will be less abundant and will then go extinct."

expected to be more limited, either restricted to smaller areas (40.9%), or disappearing completely (40.9%) (p<0.001, Cc = 0.635). In contrast, the disturbed forest areas are much more widespread in future drawings. For example, the majority of 'present' drawings had few felled trees or deforested or degraded areas (63.6%), whereas all of the 'future' drawings contained deforested or degraded areas (p<0.001, Cc = 0.651). In the future drawings, disturbed forest was indicated by cleared and degraded areas drawn around the village and even on mountains (summed value: 72.7%), or by the complete absence of forest areas from the drawing (27.3%). Furthermore, felled trees appeared in 59.1% of present drawings compared to 95.5% of the future drawings (p = 0.004; Cc = 0.398). The presence or absence of oil palm plantations did not differ significantly across the two time periods, however, the size of oil palm areas drawn in the 'future' pictures was much greater than in any drawings of the 'present' (p = 0.01, U = 0.0).

Analysis of biodiversity features indicated that stark changes were expected in wild animal taxa, with declines in richness for all species groups (Figure 3) and in animal abundances between the present and future environments (p≤0.001, U = 7.5). (The presence of domestic animals did not show significant changes between present and future drawings.)

Similarly, children expected declines in botanical diversity (p≤ 0.05, U = 148.0), and in specific vegetation elements such as the number of resinous trees and number of different types of fruit trees. Grasslands (considered separately from other vegetation diversity) were often drawn in former forest areas in the future. Children often paid specific attention to the condition of wild animals: in the present they depicted 'many animal species' (77.3% of present drawings) and in some cases they indicated the animals were already threatened or very threatened (22.7%). In the future, the majority of the species disappear, with either few species persisting (22.7%) or no wild animals at all (40.9%). The drawings indicate that children expect an increase in the range of threats to

animals (p<0.001, Cc = 0.676), with only two threats mentioned in the present (hunting and buildings, each in 5% of the drawings), but nine threats perceived for the future. For the future, three major threats were identified as arising from hunting, agriculture and forest fire, while several other threats also contributed to the disappearance of animals. The threats with highest frequencies in 'future' drawings consisted of urban development and logging for buildings (25%), agricultural development (20%), hunting (15%), agriculture and logging activities (15%), and a mix of these different threats (10%). Illegal logging was not drawn in the present, but was expected to occur in the future (35% of drawings, p≤0.05, Cc = 0.386).

The drawings indicate that environmental disasters are expected to become more frequent, with flood and landslide occurrence being significantly higher in the future (respectively: p = 0.019, Cc = 0.333; p = 0.008, Cc = 0.369; for the overall non-flood/landslide disasters: p = 0.019, Cc = 0.333). All elements involving development are also expected to increase, including the number of factories (timber and others, p = 0.019, U = 187.0), number of large buildings (p = 0.019, U = 187.0), presence and number of main roads (p = 0.004, respectively Cc = 0.398 and U = 152.5), and presence of industries (logging, mining, others; p = 0.031, Cc = 0.309). Finally, there was a clear expectation that active forest clearance will increase dramatically (p<0.001, U = 66.0). In the 'present' drawings, 40.9% showed 'no clearing', 31.8% showed 'a few' and 27.3% showed 'many' people clearing the forest. In contrast, all of the future drawings showed 'many' people clearing the forest.

Physical landscape of villages

Figure 4 shows the relationships among land use and land cover variables for the areas surrounding the villages, as summarized by the principal component analysis (PCA). Principal component 1 captures 51.3% of the variation in the LULC data and the second principal component 15.9%. The PCA separates the villages into 3

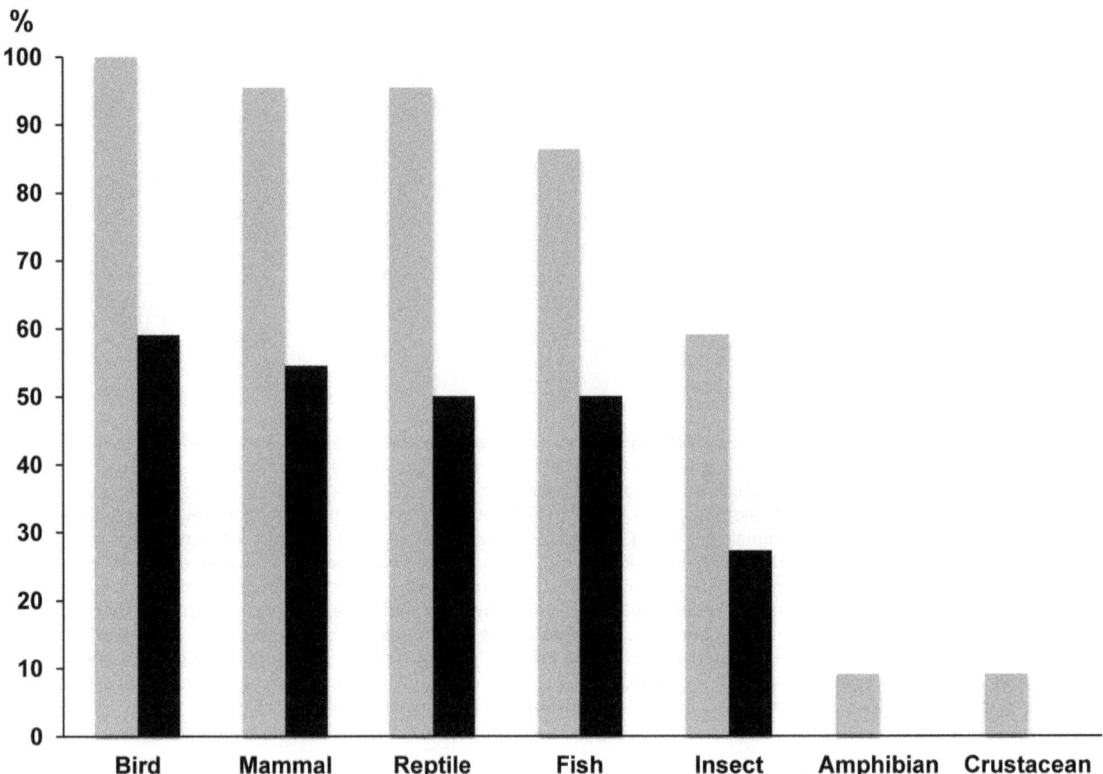

Figure 3. Relative frequency of wild animal taxa (%) between the present and future drawings. Frequency of presence of animal wildlife in drawings by children from 22 villages in Indonesian Borneo. Frequency of presence for each animal taxon is shown for present-day drawings (grey bars), and the future drawings (black bars).

groups: 1) a group consisting of eight villages from the upper Kapuas, particularly Kapuas Hulu District; 2) a group of eight villages from the lower Kapuas, particularly Pontianak District; and 3) a group consisting of two villages in East Kalimantan (further along component 2), two in upper Kapuas, and two in the lower Kapuas close to Pontianak district.

The first group is related to intact forest, protected area cover, annual precipitation, distances to logged forest and to oil palm, as well as river density and elevation. Four villages of this group are close to large protected areas (higher along component 1), with the latter three of these villages being close to a large national park and remote (especially 19 and 20). The second group is associated with other (non-forest) land covers, agroforest/regrowth and higher settlement density. Village 10 is rather set apart from this group along the component 1 due to its high values for oil palm plantations (oil palm land cover) and road density, and its lower values for distances to logged forests and to oil palm. The final group is composed of villages characterized by total forest cover (consisting mainly of logged rather than intact forest) and road density. The two villages from the lower Kapuas are also correlated with high oil palm coverage in their surroundings, while the villages from the upper Kapuas are slightly set apart in the PCA due to their relatively high values for river density and elevation.

Multivariate analysis of art variables

A clear direction of change through time is visible in the positions of 'present' and 'future' variables in the drawings on the first two principal components in the CatPCA of the art variables (Figure 5), indicating consistent expectations about the direction of

change in environmental conditions. The present drawings show better 'conditions of mountain forests, river, fauna', higher 'vegetation diversity', and less 'deforestation and degradation'. In the future, children expect all these conditions to deteriorate, as shown by shifts along the first axis in a direction of more 'floods and non-flood disasters', higher 'temperature', more 'disturbed areas' with more 'people clearing the forest', and increases in 'roads', 'industries', 'oil palm cover', 'threats to wildlife' and 'distance of the village to forest'. Two villages have 'present' conditions that are more degraded than other villages (visible as two grey dots outside the compact group of other villages, where grey represents conditions in the 'present' drawings). This separation from other villages is mainly due to the 'presence of industries' (varying along axis 2). Nonetheless, the children's drawings from these villages show similar directions of future changes compared to other villages. There are no villages in which the children expected improved environmental conditions in the future (i.e., no black dots to the right of corresponding grey ones). Only three villages had 'future' conditions that were close to the main group of 'present' conditions which showed relatively good current conditions (high on axis 2) and relatively small declines into the future. All three were in the upper Kapuas.

It appears that the existing variation in landscape context across villages is expected to increase over time. Some patterns correspond to geographic regions, with villages (in the top quadrant) from the lower Kapuas (except one, no. 11, in the upper Kapuas), all being expected to suffer from higher 'temperatures', more environmental 'disasters', and to see more 'industrial', 'main road', and 'oil palm' developments. However, there is a very broad geographic spread among the villages (one

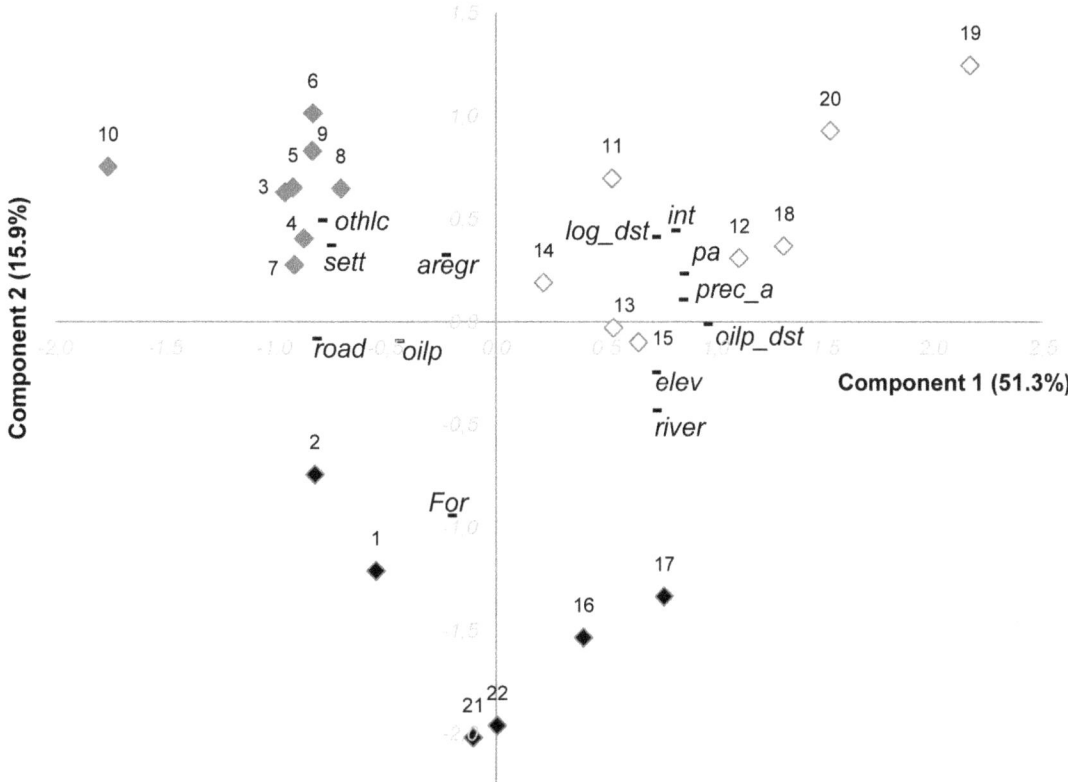

Figure 4. Principal component of the current land cover and land use data related to the villages. The figure shows the current physical landscapes of the villages. The first two components explain 67.2% of the variability of the data (respectively, component 1: 51.3% and component 2: 15.9%). Villages are numbered points (diamonds) and variables are shown as a short line (black). Villages with similar physical landscapes form a group characterized in the figure by a specific color (white: villages 11, 13, 14, 15, 12, 18, 19 and 20; grey: villages 3, 4, 5, 6, 7, 8, 9 and 10; black: villages 1, 2, 16, 17, 21, and 22).

from East Kalimantan, and the others from the two areas in West Kalimantan) that show future expectations of increased areas of 'disturbed forest', 'people clearing the forest', and increased 'village-forest distance' and 'threats to animals'.

The three villages with the largest expected changes between the present and future (as measured in the PCA by the Euclidean distance between the two periods for each village), are villages from the lower Kapuas area characterized by relatively high road and settlement densities, agro-forest/regrowth cover, other non-forested and oil palm land covers (i.e. high oil palm cover for village 1 and 2; and more rubber, coconut, cacao plantations, and oil palm plantations planning during 2012 for village 3). The two villages with the smallest expected changes between present and future consist of one village in a presently degraded area, where children expected slight further degradation into the future, and one village where children perceived good environmental conditions in the present only degrading slightly into the future. The first is a village in the lower Kapuas (village 10), which shows small changes from the degraded conditions of the present, to slightly further degraded conditions in the future. Current levels of degradation are also indicated in the drawing, and by the landscape variables for which it has the highest values of settlement and road densities, highest oil palm and other non-forested land covers, and lower distances to logged forests and to oil palm. The second village shows small changes from the generally good conditions of the present, and this is a village in East Kalimantan with the largest forest area (selectively logged natural forest), and lower settlement density (village 21).

Correlation between land cover data and art variables

Several land cover variables are correlated with art variables in the present (see Table S4). Children living in landscapes with higher oil palm or other non-forest land covers drew pictures from the 'present' with larger 'distance of the village to forest' (respectively, $r = 0.43$ and 0.44, and $p < 0.05$ in all cases mentioned). The land cover of oil palm plantations is also correlated with drawing of 'main roads' and 'flood disasters' (both resp., $r = 0.52$). Both oil palm land cover and road density are positively correlated with 'threat to animals' (resp., $r = 1.0$ and 0.52) and negatively correlated with 'faunal condition' ($r = -0.73$ for oil palm; -0.47 for road density). Total forest cover (characterized by more logged than intact forest) is positively correlated with 'presence of industries' ($r = 0.44$). Logged forest distance is negatively correlated with the number of 'people clearing the forest' ($r = -0.47$) and with the 'distance of the village to forest' ($r = -0.58$). The three physical layers of current river density, elevation and annual precipitation are negatively correlated with the art variable of 'oil palm cover' (resp., $r = -0.59$; -0.54 and -0.60).

Correlations between land cover variables and art variables in the future (see Table 3) show current intact forest cover and distance to oil palm negatively correlated with future 'presence of main roads' in drawings. On the contrary, this future art variable is positively correlated with current cover of oil palm plantations and road density. Children living in villages with lower forest cover drew greater 'non-flood disasters' while children living with larger cover of agroforest/regrowth surrounding their village drew more

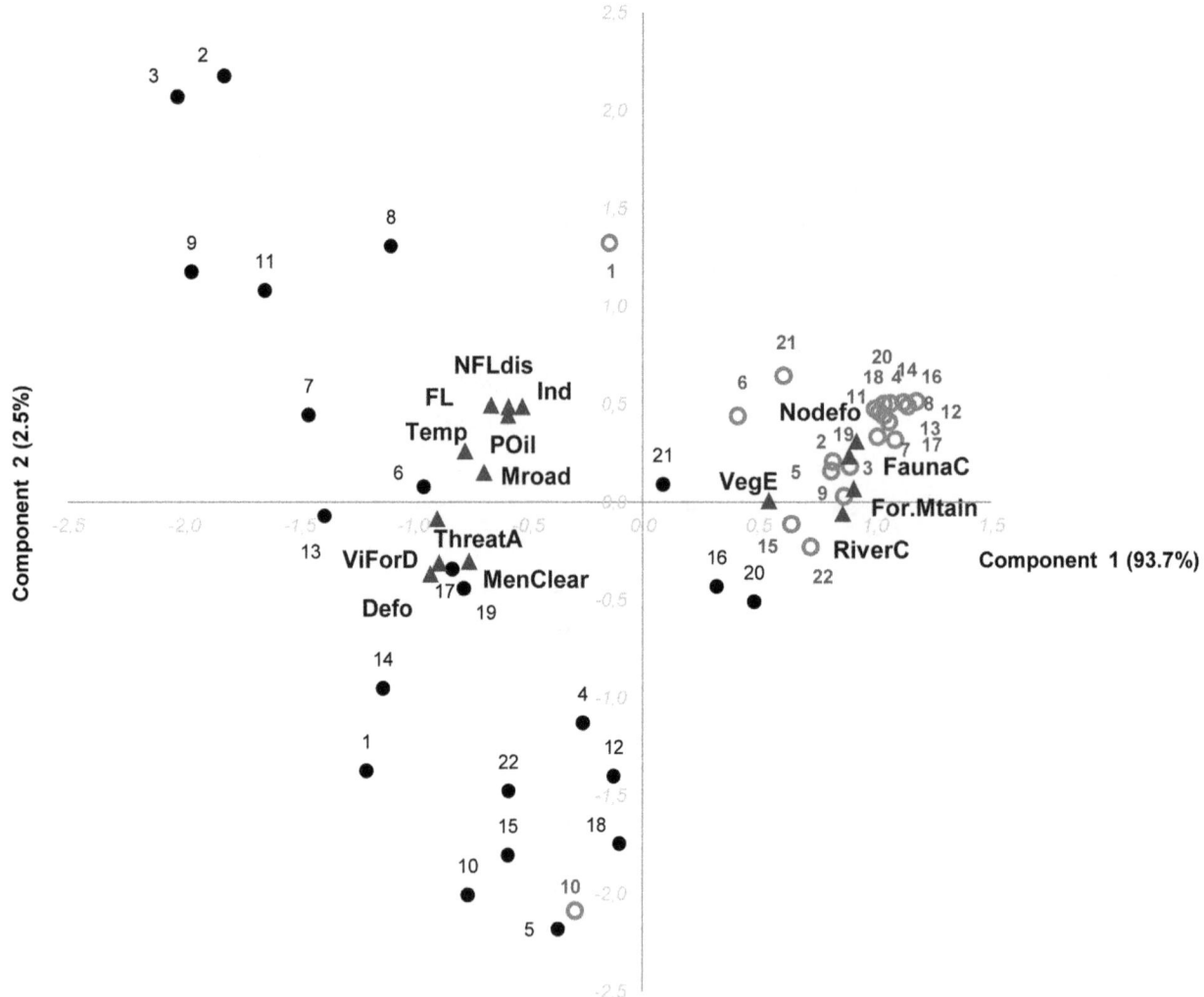

Figure 5. Categorical principal component of 15 art variables from 'present' and 'future' drawings. The figure shows the relationships between the art variables (green triangles; see Table 2 for their names and codes) and the villages from different landscapes (see Fig. 4) describing the children's perceptions of environmental change between the present (grey circles) and the future (black dots). The two first components explain 96.2% of the variability of the data (respectively, component 1: 93.7% and component 2: 2.5%).

'flood disasters'. Other land cover, settlement density and agroforest/regrowth cover are negatively correlated with 'undisturbed forest' and 'condition of mountain forests'. These first two landscape variables were also negatively correlated with the future 'river condition' and positively with 'oil palm cover' in drawings. Agroforest/regrowth and other land covers are also positively correlated with 'disturbed forest areas'. Agroforest/regrowth cover is also associated with lower 'faunal condition'. Distance to oil palm is negatively correlated with the art variable 'oil palm cover' in drawings of the future (though oil palm plantation cover is not). Other current landscape variables are correlated with 'oil palm cover' in future drawings, including negative correlations with protected area cover, river density and annual precipitation.

Multivariate analysis of art and land cover variables: present-day context

Figure 6 shows the multivariate relationships between the current landscape context of the villages where the children are living, and their perceptions of current environmental conditions (bivariate correlations between land cover and art variables were

presented previously in Table S4). The two first components explain 85.2% of the variability of the data (respectively, component 1: 64.6% and component 2: 20.6%). The villages separate into 3 groups (Figure 6). Group 1 consists of ten villages with high values on axis 1; Group 2 of seven villages with low values on axis 2; and Group 3 of five villages with relatively low values on axis 1.

Group 1 consists of ten villages from the upper Kapuas, including the three remote villages close to a national park, associated with high values for several physical landscape layers: protected area cover, intact forest cover, larger distances to oil palm plantations and to logged forest, higher river density, elevation and annual precipitation. Drawings from these villages showed more 'undisturbed forest areas', higher 'total vegetation diversity', and better 'faunal condition'.

Group 2 consists of seven villages in the lower Kapuas region (Pontianak district), that have higher settlement densities and other non-forest land cover, and in which children drew more 'oil palm areas', and 'non-flood disasters'.

Table 3. Correlations between current land cover surrounding the village, and art variables representing children's drawings of their future environment.

		Intact forest	Logged forest distance	Forest cover	Agroforest / regrowth	Oil palm plantations	Oil palm distance	Other land cover	Protected areas	River density	Elevation	Settlements	Road density	Annual precipitation
Temperature condition	Corr.	-.23	-.02	.23	.01	.07	-.37	.12	-.33	-.10	-.18	.13	.29	-.17
Forest Mountain condition	Corr.	.18	.08	.05	-.68**	-.28	.60**	-.68**	.58**	-.09	.65**	-.65**	-.36	.12
River Condition	Corr.	.11	.07	.11	-.45	-.32	.44	-.59**	.37	.03	.38	-.54*	-.21	.22
Village-Forest Distance	Corr.	-.05	.01	-.05	.26	.27	-.34	.32	-.31	.09	-.35	.26	.24	-.01
Undisturbed forest	Corr.	.09	.19	.11	-.44*	-.24	.52*	-.65**	.55**	.10	.63**	-.53*	-.35	.17
Disturbed forest	Corr.	.06	.02	-.15	.46*	.35	-.40	.47*	-.35	.23	-.41	.35	.22	.09
Oil palm area cover	Corr.	-.22	-.19	-.01	.22	.09	-.47*	.57**	-.43*	-.53*	-.40	.47*	.38	-.51*
Industries	Corr.	-.20	-.11	.22	.08	-.04	-.25	.04	-.28	-.10	-.05	.04	.22	-.13
Floods	Corr.	.06	.28	-.22	.46*	.00	-.30	.31	-.39	-.02	-.31	.24	.08	.02
Main road	Corr.	-.46*	-.24	.17	.26	.48*	-.55**	.39	-.40	-.02	-.31	.34	.47*	-.16
Faunal Condition	Corr.	.14	.09	-.11	-.65**	-.26	.37	-.39	.28	-.27	.08	-.29	-.26	-.06
Vegetation diversity	Corr.	.03	-.13	.15	-.28	-.28	.32	-.29	.25	-.11	.19	-.22	-.28	.09
Non–Flood disasters	Corr.	.18	.25	-.45*	.17	.00	-.22	.32	-.17	-.27	-.28	.30	.07	-.25
Threats to Animals	Corr.	-.08	-.21	.06	.24	-.27	-.05	.34	-.10	-.37	-.08	.20	.02	-.34

For descriptions of land cover variables and of art variables, see respectively Table S1 and Table 2. The art variable 'People clearing the forest' has no variance for the future, with the highest number of men clearing the forest for each village, its correlation with the land cover is thus not calculated. Correlation coefficients shown in bold font are statistically significant (with *p<0.05 and **p<0.005).

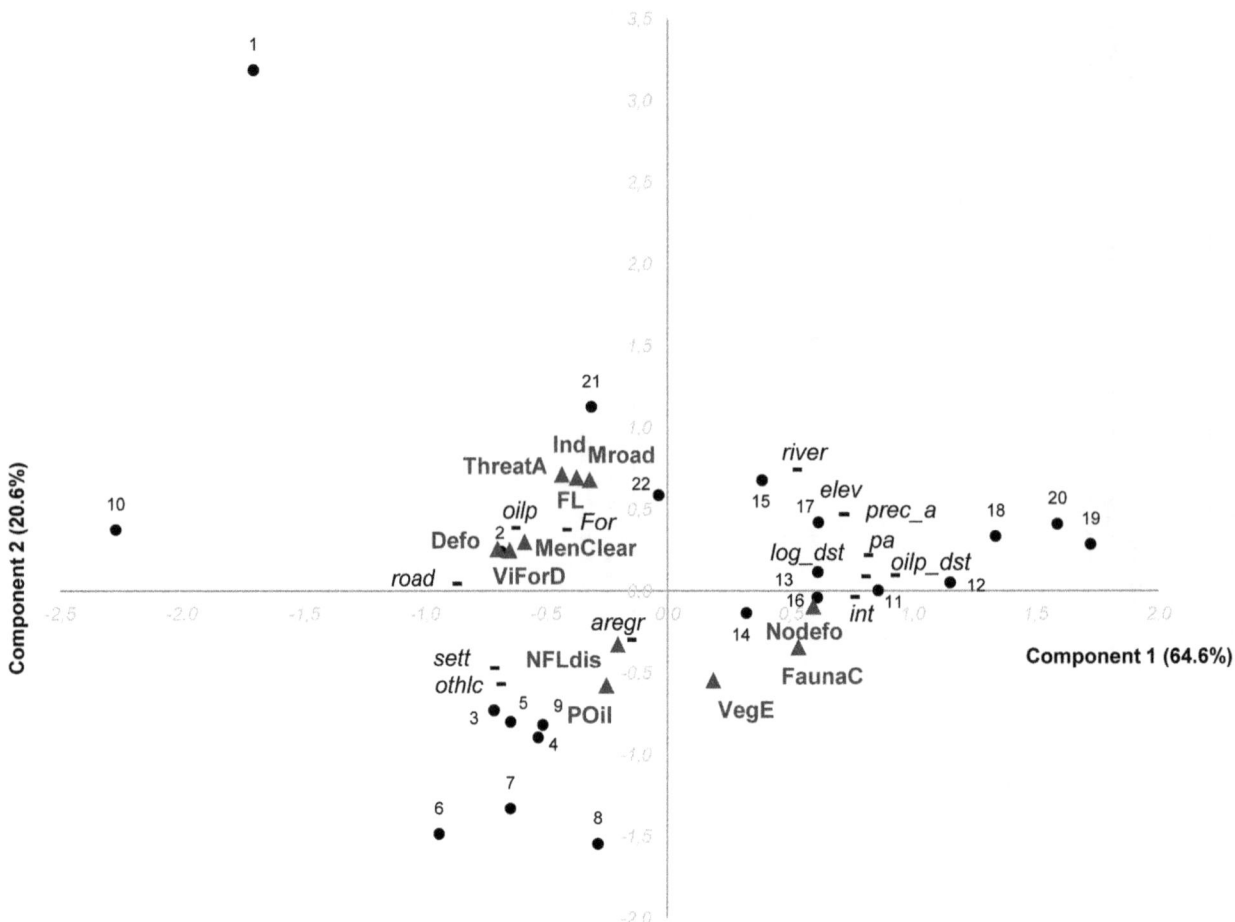

Figure 6. Categorical principal component of the current land cover, and art variables of the 'present' environment. The figure shows the relationships among the current land cover (black short line) of the villages (black dots) and the art variables from drawings (green triangles; see Table 2 for their names and codes) describing the children's perceptions of current environmental conditions. The two first components explain 85.2% of the variability of the data (respectively, component 1: 64.6% and component 2: 20.6%) and separate three village groups (Group 1: high values along axis 1, consisting of villages 11, 13, 14, 15, 16, 17, 12, 18, 19 and 20; Group 2: low values on axis2, consisting of villages 3, 4, 5, 6, 7, 8 and 9; Group 3: relatively low values on axis 1, consisting of villages 1, 2, 21, 22 and 10). The art variables 'temperature, forest mountain, and river conditions' have no variance in the present-day context across villages (perceived in 'good condition'), and were thus not included in the PCA.

Group 3 consists of two remote villages in East Kalimantan, along with three villages in the lower Kapuas. These villages show moderate forest cover (dominated by logged forest), and the three villages in the lower Kapuas also have higher oil palm land cover and road density (showing the highest values for these variables). The village from Pontianak district is further apart along axis 1 also due to its high value for other non-forest land cover and highest value for settlement density. Drawings from children in this group of villages showed more frequent occurrences of 'industries', 'main roads' and of 'floods', and higher 'disturbed forest areas', 'village-forest distance' and greater 'threats to animals' and 'people clearing the forest'.

Multivariate analysis of art and land cover variables: current landscape, future drawings

Figure 7 shows the multivariate relationships between the current landscape context of the villages where the children are living, and their perceptions of environmental conditions in the future (bivariate correlations were presented in the Table 3). The two first components explain 86.8% of the variability of the data (respectively, component 1: 72.0% and component 2: 14.7%).

Generally the villages separate into 3 groups: 1) a group consisting of the two remote villages in East Kalimantan, along with seven villages in the upper Kapuas (including the three close to a national park (18, 19 and 20)); 2) a group of three villages from the lower Kapuas and three from the upper Kapuas; and 3) a group of seven villages in the lower Kapuas, within or near Pontianak district.

Village group 1 is correlated with the art variables for future conditions of 'river', 'animals', 'mountain forests', the 'undisturbed forest areas' and 'vegetation diversity'. This group is correlated with the physical land cover of intact forest, protected areas, river density, elevation, and annual precipitation, and, to some extent, distances to logged forest and to oil palm plantations.

Village group 2 is largely associated with future drawings of worse 'temperature condition', greater occurrences of 'floods' and other 'natural disasters', 'presence of industries' and 'main roads', higher 'village-forest distance' and areas of 'disturbed forest'. Three villages in the upper Kapuas are higher along axis 2 due to relatively highest values of agroforest/regrowth cover and their association with 'disturbed forest'. This group is correlated with higher current agroforest/regrowth cover, explained in both components.

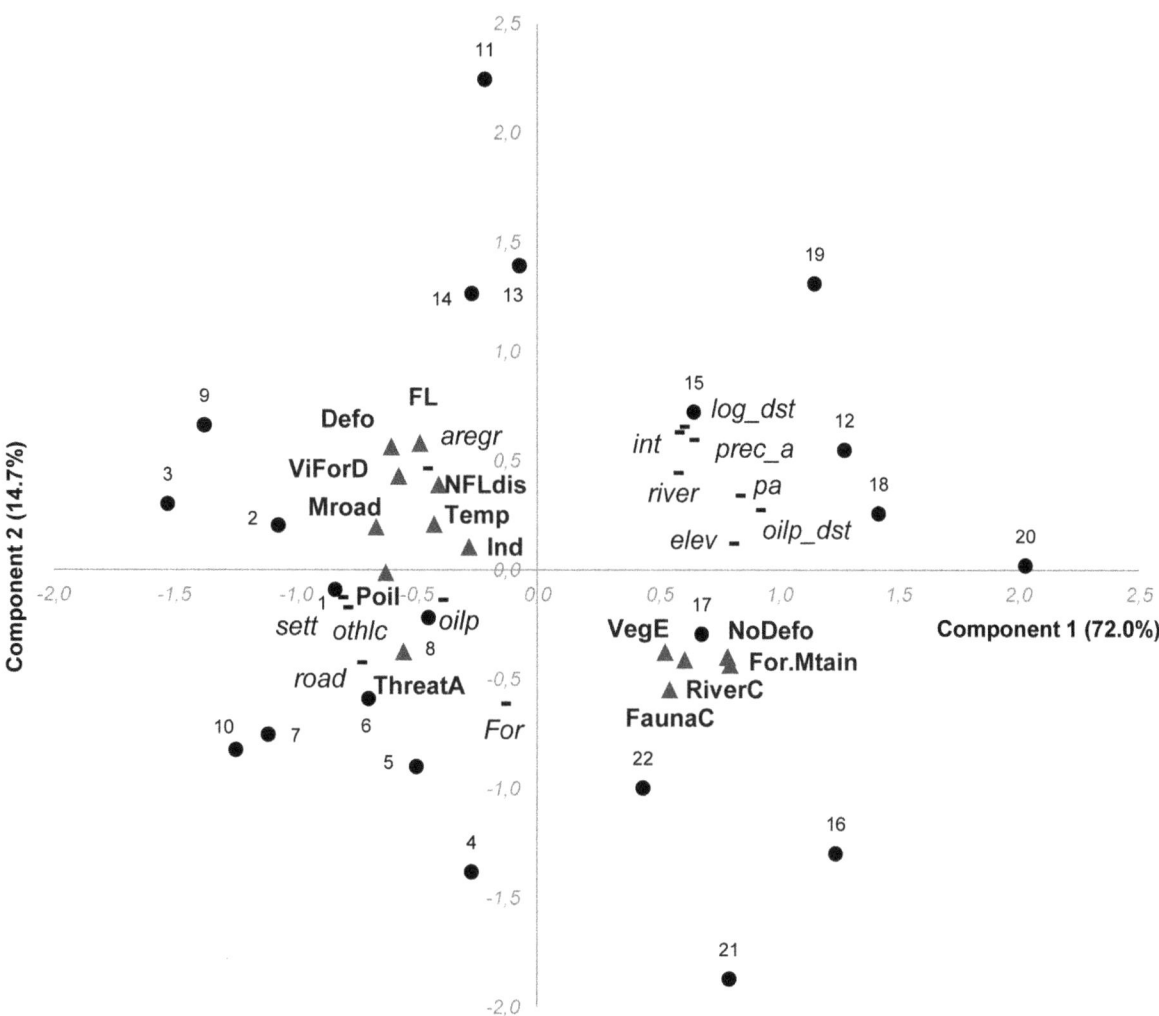

Figure 7. Categorical principal component of current land cover, and art variables of the 'future' environment. The figure shows the relationships between the current land cover (black short line) of the villages (black dots) and the art variables from drawings (green triangles; see Table 2 for their names and codes) describing the children's perceptions of environmental conditions for the future, in 15 years' time. The two first components explain 86.8% of the variability of the data (respectively, component 1: 72.0% and component 2: 14.7%). The villages fall into three broad groups (Group 1: high values on axis 1, consisting of villages 15, 16, 17, 12, 18, 19, 20 and 21, 22; Group 2: low values on axis 1, consisting of villages 2, 3, 9, 11, 13 and 14; Group 3: low values on axis 2, consisting of villages 1, 4, 5, 6, 7, 8, and 10. The art variable 'People clearing the forest' was high for all villages, and thus has no variance across villages and was not included in the PCA.

Village group 3 is correlated with the art variables of future 'oil palm cover', and 'threats to animals', and the current physical land covers of oil palm plantations, other non-forest land cover, settlements and road densities (along axis 1) and total forest cover (along axis 2).

Discussion

This study confirms that drawings provide powerful tools to gain insights into children's thoughts about forests, wildlife and environmental change [25,26]. Verbal explanations from children about their drawings were an important complement to the drawings themselves, as suggested by Malchiodi [27]. Similar to findings by Alerby [25], in the present study young children were actively engaged in thinking about their environment, and were able to convey their thoughts through the symbolic language of their drawings.

Our results indicate that children have sophisticated perceptions of their direct environments, including general conditions of forests

and rivers, fauna and flora, and human activities; and have an understanding of how these variables influence each other. Indonesia's rich biodiversity is being rapidly degraded by landscape change, pollution and overharvesting. The children in this study show a strong awareness of these issues, and generally share similar perceptions of their environment and the causes and consequences of environmental degradation, along with an understanding of human needs and environmental protection. They comprehend the impacts of human activities on wildlife and other natural resources, via land use change and landscape fragmentation, and also through increasing natural disasters and temperature.

Our results indicate that children have landscape context-dependent perceptions of current and future environmental conditions, which may be influenced by awareness about past environmental conditions. Children's perceptions often depend strongly on their previous and current experiences in specific contextual conditions [18]. Our study proves that this is the case for rural children in Borneo who have consistent perceptions of their current environment and expectations of deteriorating future environmental conditions that both reflect the landscape that currently surrounds them. As long as forest cover remains high, the views of the current environment and its expectation are that the atmosphere, rivers and wildlife continue to be in relatively good condition, although some other aspects of the environment are still expected to decline in the future. In contrast, in areas where limited forest remains or forests are heavily degraded due to human activities, the current environment is perceived with similarly high awareness, but the expectation is of much faster, 'run-away' transition to non-forest landscapes and a collapse of natural environmental functions. Moreover, it appears that the more the children have been exposed to land use and environmental change, the more they expect such trends to continue and to worsen in the future (e.g. increase in infrastructure, climate change and natural disasters).

The strong associations that our study found between present and future environmental conditions indicate that there is no obvious shifting baseline among Kalimantan's children regarding perceptions of their environment. Children expressed strong perceptions of how present and future land uses are entwined. Most children predicted future environmental conditions to be worse than the present, suggesting an insight into past environmental trends. If either generational [2] or personal amnesia [28] existed, a more likely outcome would have been that children would make similar predictions about their future environmental conditions, irrespective of how current conditions had been caused by past processes. In our present study, this was not the case. Clearly, a proper test of shifting baselines requires consideration of older generations as well as children. The shifting baseline syndrome would assume that adults have a more accurate understanding of past environmental conditions and trends than children. Because these trends are negative (from a forest conservation point of view), adults would be expected to have more negative environmental expectations than children who would then perceive current and probable future land use patterns as a normal condition [7]. As part of our broader research program, we did assess the perceptions of adults about the past, present and future condition of their village environments, but using a different approach, namely the pebble distribution method [29]. Preliminary analysis of the adult's perceptions indicate that, similar to children, adults perceive deteriorating environmental conditions in their present village environments, and see the causes of present-day conditions as likely to continue and worsen into the future (ASP, unpubl. data). Children and adults expressed qualitatively similar predictions of future environmental conditions (especially an increase in temperature, pollution and flood frequencies (for adult perceptions of floods, see [17]). However, we cannot assess these similarities quantitatively, due to the different methods of collecting information.

The apparent absence of a shifting environmental baseline among Kalimantan's children may be explained by diverse informal and formal learning processes that shape children's perception on current and future environmental conditions. Testing of these factors is beyond the scope of this paper, however we note that children in the most remote and forested areas, where conservation education at school was noted to be rare during the present study, often attend village meetings that involve discussion of natural resources and decisions on land use management. We noted that during such meetings, experiences from other nearby villages were often discussed to inform the villagers' own decision-making. For example, if oil palm plantations are expanding in neighbouring village areas, many stories are discussed about land prices, labour opportunities, local access to forest resources, or conflicts with companies or local governments. Children's participation in such informal meetings likely forms a strong basis for their environmental awareness. In addition, it is likely that the more communities depend on the use of forests, forest gardens, and other natural resources, and the more children within such communities engage in these processes, the more they understand about their environment and connection to nature, and the more aware they would be of the changes that might occur in their direct environment. For example, communities that depend for much of their animal protein on fish and wild meat such as pigs and deer are highly aware of the temporal and spatial trends in populations of their target species. Communities that instead have livelihoods based mainly on income from agriculture or from employment in timber, mining, or plantation industries would be expected to have lower awareness or reduced sensitivity to environmental trends. However, this was not apparent in the current study, since children in already-degraded areas did express awareness of current environmental problems, and greater expectations of negative changes in the future than did children in more intact forest landscapes.

Finally, children's perceptions on deteriorating environmental conditions across different landscapes may be shaped by the provision of environmental courses at school, or the availability of a range of media (including books, television and radio). National and international information on environmental degradation is generally negative and focuses on declining populations of endangered species (e.g., orangutans), deforestation, or increases in environmental disasters (e.g., flooding). Children living in more remote areas (usually with more intact forests) may have less access to such information compared to children in more degraded landscapes and so may hold views of their current and future environments that are less influenced by the generally negative trends portrayed in media. These different forms of learning and sources of information are not mutually exclusive and are likely to interact, depending on the natural and social context in which the children are living. Further studies would be needed to differentiate how diverse sources of information affect environmental thinking among children in different parts of tropical landscapes. This would involve assessing differences in the accessibility of environmental education and other information sources, and the time children spend in forests in different multifunctional landscapes. Understanding which factors and information sources most strongly affect environmental opinions would help design more effective methods for increasing environmental awareness.

A further message from our study for forest conservation and sustainable resource management is that a large shift in practices and perceptions is needed to alter the current trajectory of change, and avert severe losses of forest and wildlife. As most children foresee a future of further environmental degradation, it is essential to develop approaches to show these children that past trends do not necessarily determine future trends, and that positive changes are possible if development actively considers biodiversity and dependence of communities on ecosystem services. The strong environmental awareness among children in Kalimantan provides a good starting point for educational programs and their support for conservation of biodiversity and ecosystem services. Awareness of trade-offs involved in forest conservation and conversion to different land uses will enable children to value the benefits from the environment and to make more informed decisions later in their lives when they participate in decision making. In that regard, it is important for children to start learning about nature as early as possible, especially in view of the high sensitivity of children to stereotypical imaging such as they encounter in the media and literature [30]. Education, environmental experience, media and other types of communication could thus play major roles in promoting support among younger generations for more sustainable environmental management, and the children's drawings themselves could play a useful role in this. Indeed, the strong story-telling potential of the drawings could provide a powerful tool by orienting NGOs and educational practitioners in educational initiatives to regions where conservation awareness is the most needed. Also the drawings could be used to influence political thinking and decision-making on land use, particularly in democratic governments where citizens' views hold considerable influence in public policy. Children have the highest stake in ensuring that land use maximizes their long-term well-being, and their views and aspirations should thus be seriously considered by local and international political decision-makers elected to represent the citizens' voices and fulfill efficient natural resource management.

References

1. Liu JG, Dietz T, Carpenter SR, Alberti M, Folke C, et al. (2007) Complexity of coupled human and natural systems. Science 317: 1513–1516.
2. Kahn PH, Friedman B (1995) Environmental views and values of children in an inner-city black community. Child Devel 66: 1403–1417.
3. Pauly D (1995) Anecdotes and the shifting baseline syndrome of fisheries. Trends Ecol Evol 10: 430.
4. Sáenz-Arroyo A, Roberts C, Torre J, Cariño-Olvera M, Enríquez-Andrade R (2005) Rapidly shifting environmental baselines among fishers of the Gulf of California. Proc Royal SocB: Biol Sc 272: 1957–1962.
5. Pinnegar JK, Engelhard GH (2008) The 'shifting baseline' phenomenon: a global perspective. Rev Fish Biol Fisheries 18: 1–16.
6. Turvey ST, Barrett LA, Yujiang HAO, Lei Z, Xinqiao Z, et al. (2010) Rapidly Shifting Baselines in Yangtze Fishing Communities and Local Memory of Extinct Species. Conserv Biol 24: 778–787.
7. Papworth SK, Rist J, Coad L, Milner-Gulland EJ (2009) Evidence for shifting baseline syndrome in conservation. Conserv Lett 2: 93–100.
8. Meijaard E, Abram NK, Wells JA, Pellier A-S, Ancrenaz M, et al. (2013) People's perceptions on the importance of forests on Borneo. PLOS ONE 8: e73008.
9. Miller JR (2005) Biodiversity conservation and the extinction of experience. Trends Ecol Evol 20: 430–434.
10. Kareiva P (2008) Ominous trends in nature recreation. Proc Nat Acad Sc USA 105 (8): 2757–2758.
11. Whitten T, van Dijk PP, Curran L, Meijaard E, Supriatna J, et al. (2004) Sundaland. In: Mittermeier RA, Gil PR, Hoffmann M, Pilgrim J, Brooks T, et al., editors.Hotspots revisited: Another look at Earth's richest and most endangered terrestrial ecoregions. Mexico: Cemex.
12. FAO (2010) Global forest resources assessment 2010. Progress towards sustainable forest management. FAO Forest Paper 163. Rome, Italy: Food and Agricultural Organization of the United Nations.
13. Miettinen J, Shi C, Liew SC (2011) Deforestation rates in insular Southeast Asia between 2000 and 2010. Glob Chang Biol 17: 2261–2270.
14. Gaveau D, Sloan S, Molidena E, Husnayaen H, Sheil D, et al. (2014) Four decades of forest persistence, clearance and logging on Borneo. PLOS ONE. 9 (7): e101654.
15. Meijaard E, Buchori D, Hadiprakoso Y, Utami-Atmoko SS, Tjiu A, et al. (2011) Quantifying killing of orangutans and human-orangutan conflict in Kalimantan, Indonesia. PLOS ONE 6: e27491.
16. Meijaard E, Mengersen K, Buchori D, Nurcahyo A, Ancrenaz M, et al. (2011) Why don't we ask? A complementary method for assessing the status of great apes. PLOS ONE 6: e18008.

17. Wells J, Meijaard E, Abram NK, Wich SA (2013) Forests, floods, people and wildlife on Borneo.A review of flooding and analysis of local perceptions of flooding frequencies and trends, and the roles of forests and deforestation in flood regimes, with a view to informing government decision-making on flood monitoring, forest management and biodiversity conservation. Bangkok, Thailand: UNEP.
18. Bowker R (2007) Children's perceptions and learning about tropical rainforests: an analysis of their drawings. Environ Educ Res 13: 75–96.
19. Dove J (2012) Tropical rainforests: a case study of UK, 13-year-olds' knowledge and understanding of these environments. Int Res Geog Environ Educ 21: 59–70.
20. Snaddon JL, Turner EC, Foster WA (2008) Children's Perceptions of Rainforest Biodiversity: Which Animals Have the Lion's Share of Environmental Awareness? PLOS ONE 3: e2579.
21. Wich SA, Gaveau D, Abram N, Ancrenaz M, Baccini A, et al. (2012) Understanding the Impacts of Land-Use Policies on a Threatened Species: Is There a Future for the Bornean Orang-utan? PLOS ONE 7: e49142.
22. Carlson KM, Curran LM, Asner GP, Pittman AM, Trigg SN, et al. (2013) Carbon emissions from forest conversion by Kalimantan oil palm plantations. Nature Clim Change 3: 283–287.
23. Lehner B, Verdin K, Jarvis A (2006) HydroSHEDS Technical documentation. World Wildlife Fund US, Washington, DC.
24. Bright EA, Coleman PR, King AL, Rose AN (2008) LandScan 2007. 2007 ed. Oak Ridge, TN: Oak Ridge National Laboratory.
25. Alerby E (2000) A Way of Visualising Children's and Young People's Thoughts about the Environment: A study of drawings. Environ Educ Res 6: 205–222.
26. Barraza L (1999) Children's Drawings About the Environment. Environ Educ Res 5: 49–66.
27. Malchiodi CA (1998) Understanding children's drawings. London: Jessica Kingsley Publishers.
28. Simons DJ, Rensink RA (2005) Change blindness: past, present, and future. Trends Cogn Sci 9: 16–20.
29. Colfer CJP, Brocklesby MA, Diaw C, Etuge P, Günter M, et al. (1999) The grab bag: supplementary methods for assessing human well-being. The Criteria & Indicators Toolbox Series, Number 6. Bogor, Indonesia: Center for International Forestry Research (CIFOR).
30. Anderson S & Moss B (1993) How wetland habitats are perceived by children: consequences for children's education and wetland conservation. Int J Sci Educ: Routledge. pp. 473–485.

Acknowledgments

We give particular thanks to the children, the village heads and other villagers for their participation and for giving us their time. We also thank the Center for International Forestry Research (with the CGIAR Research Program on Forest, Trees and Agroforestry; and the CoLUPSIA project led by CIRAD), the United States Agency for International Development, and the Arcus Foundation for providing logistical and financial support to our study. We extend our sincere gratitude to the field assistants and our NGO partners: Riak Bumi and The Nature Conservancy for their logistic support; as well as WWF-Indonesia and Fauna and Flora International for providing useful information to the study. We would like to thank the reviewers of this paper for their helpful and supportive comments.

Author Contributions

Conceived and designed the experiments: ASP EM. Performed the experiments: ASP EM JAW NKA. Analyzed the data: ASP JAW NKA. Contributed reagents/materials/analysis tools: ASP EM JAW NKA DG. Wrote the paper: ASP EM JAW NKA DG.

9

Value Assessment of Ecosystem Services in Nature Reserves in Ningxia, China: A Response to Ecological Restoration

Yan Wang[1,2], **Jixi Gao**[1,2]*, **Jinsheng Wang**[1], **Jie Qiu**[2]

1 College of Water Science, Beijing Normal University, Beijing, China, **2** Nanjing Institute of Environmental Science, Ministry of Environmental Protection, Nanjing, China

Abstract

Changes in land use can cause significant changes in the ecosystem structure and process variation of ecosystem services. This study presents a detailed spatial, quantitative assessment of the variation in the value of ecosystem services based on land use change in national nature reserves of the Ningxia autonomous region in China. We used areas of land use types calculated from the remote sensing data and the adjusted value coefficients to assess the value of ecosystem services for the years 2000, 2005, and 2010, analyzing the fluctuations in the valuation of ecosystem services in response to land use change. With increases in the areas of forest land and water bodies, the value of ecosystem services increased from 182.3×10^7 to 223.8×10^7 US$ during 2000–2010. Grassland and forest land accounted for 90% of this increase. The values of all ecosystem services increased during this period, especially the value of ecosystem services for biodiversity protection and soil formation and protection. Ecological restoration in the reserves had a positive effect on the value of ecosystem services during 2000–2010.

Editor: Ben Bond-Lamberty, DOE Pacific Northwest National Laboratory, United States of America

Funding: The National Special Scientific Research of Environmental Protection Public Welfare Project (No. 201209027), the National Science and Technology Support Project (No. 2012BAC01B00), and the National Ecological Environment Change Assessment by Remote Sensing Survey Project (2000–2010). The funders had no role in study design, data collection and analysis, decision to publish, or preparation of the manuscript

Competing Interests: The authors have declared that no competing interests exist.

* E-mail: gaoeco@163.com

Introduction

Ecosystem services (ESs) refer to the benefits people obtain from ecosystems or to aspects of ecosystems used (actively and passively) to produce human well-being [1,2]. ESs emphasize not only provisioning services (marketable goods), but also supporting (e.g. nutrient cycling), regulating (e.g. soil and water conservation), and cultural services (e.g. aesthetic values). As the Millennium Ecosystem Assessment [1,3] reminded us, our lives, much less our societies, economies, or well-being, depend on ESs [4]. Ecosystem service values (ESVs) are the values of the contributions of ESs to the sustainability of human well-being [5]. The growing body of literature on ESV includes studies on the value generated by ecosystems [6,7], changes in ESVs in response to changes in land use [8,9], climate change [10,11], approaches and models for assessing ESV [12,13], and other factors [14,15].

Ecosystems have been substantially exploited, degraded, and destroyed in the last century as a consequence of the global increase in economic and societal prosperity [1]. Humans have changed ecosystems more rapidly and extensively in the last 50 years than in any comparable period of human history; some 60% of the ESs studied have been degraded during this period [1,16]. The concept of ESs has become a central issue in conservation planning and the assessment of environmental impacts for preventing further abatement of the quality of ecosystems [17,18]. Many case studies on ESs have been performed, but too few have paid enough attention to long-term fluctuations of ESs, even though long-term study is necessary for detecting the

response of ESs to land use change, climate change, or other factors. Monitoring the fluctuations in ESVs would benefit the management and maintenance of ecosystem sustainability, e.g. the identification and measurement of variations in ESs as a consequence of land use changes appears to be an adequate means of evaluating the environmental costs and benefits of decisions affecting the planning of land use [19].

The autonomous region of Ningxia is a typical region of western China, and harsh natural conditions and the over-exploitation of resources have deteriorated the already fragile ecological environment in the past few decades. The Chinese government has recently taken measures to improve the ecological environment, such as the conversion of cropland to forest, prohibition of enclosures and grazing, and sand prevention and control. Estimating the effects of these measures is thus essential for restoring ecologically fragile regions. The effects of the restoration efforts during 2000–2010 in western China, the first decade of such efforts, have generated concern. We have assessed the variation of ESVs based on land use change in the national nature reserves in Ningxia to determine the effects of ecological restoration.

The unit prices of ESVs in global biosphere ecosystems were estimated by Costanza et al. [20]. ESVs are now commonly estimated by integrating the use of the Geographic Information System (GIS) and remote sensing (RS) [21,22,23]. The estimates of Costanza et al. [20] have been criticized because they overestimate ESVs for wetland and underestimate ESVs for farmland [24,25].

Value Assessment of Ecosystem Services in Nature Reserves in Ningxia, China...

85

A survey of Chinese ecologists led to an improved approach suitable for the situation in China [26]. This approach includes merging some ES functions, as suggested by Costanza et al. [20,27], and extracting the equivalent weighting factors for ESs per hectare for terrestrial ecosystems [28]. Much of the research on ESVs based on the equivalent weighting factors and land use data suggest that land use type can be a proxy for ESs by matching the types to the equivalent biomes, which is convenient and simple for ESVs of large areas [29,30,31].

Based on a decade of RS data combined with equivalent weighting factors and the use of ArcGIS, a package of programs for working with GIS data, this study assessed the ESVs of national nature reserves in Ningxia for the years 2000, 2005, and 2010. Our objectives were to: (1) analyze the changes in land use in the reserves during the past 10 years, (2) assess the variation in ESVs in the reserves during this period, and (3) discuss the effects of ESVs on ecological restoration.

Materials and Methods

Study Site

The study area included six national nature reserves in Ningxia, which is located in western China at 35°25′N–39°22′N and 104°49′E–107°40′E (Figure 1). The area is in a temperate arid/semi-arid region that has a continental monsoon climate, four distinct seasons, and abundant sunshine. The area is geographically diverse with annual average temperatures of −0.9 to 9.6°C and an annual precipitation of 186–800 mm. The multi-type nature reserves in the study site are classified in China as forest, wetland, and desert ecosystems and contain approximately 1000 species of plants and vertebrates in the six nature reserves. The abundant variety of natural resources provides a variety of ESs.

Data Collection and Land Use Classification

An integrated approach utilizing a Geographic Information System (GIS) and Remote Sensing (RS) was used to extract a data set for the changes in land use. The data set was extracted from GIS and RS data from Landsat Thematic Mapper imagery for 2000 and 2005 and from environmental satellite data (http://www.secmep.cn/secPortal/portal/index.faces) for 2010. To maintain consistency of classification with the remote sensing data, all the remote sensing images with good image quality in July were selected for analysis. After converting the data to the unified coordinate system and projection, we used the Krasovsky ellipsoid and the Transverse Mercator projection of ENVI 4.8 to perform RS image radiation correction and geometry correction, respectively.

After completing the pretreatments, including image mosaics, grooming, and data fusion, we used ArcGIS10.0 to consolidate and analyze the land use data with a background of raster images. The maximum likelihood classifier of the supervised classification method was used for classification with ENVI 4.8. According to the confusion matrix for classification accuracy, qualitative errors of precision for deciphering the remote sensing data for different years were controlled at the 90% level. Comparing the results of the interpretation with those of the field survey of typical points, the total classification accuracies were all higher than 90%, and the total Kappa coefficients were all greater than 0.8, which were higher than the minimum acceptable (0.7). The accuracy could thus meet the monitoring accuracy of the demands for land use change. Our data set included seven classified land use types listed in the resource and environmental database established by the Chinese Academy of Sciences (Table 1). ArcGIS 10.0 and SPSS 19.0 were used for the statistical analysis.

We obtained data sets for the normalized difference vegetation index (NDVI) from the Goddard Space Flight Center (NASA)

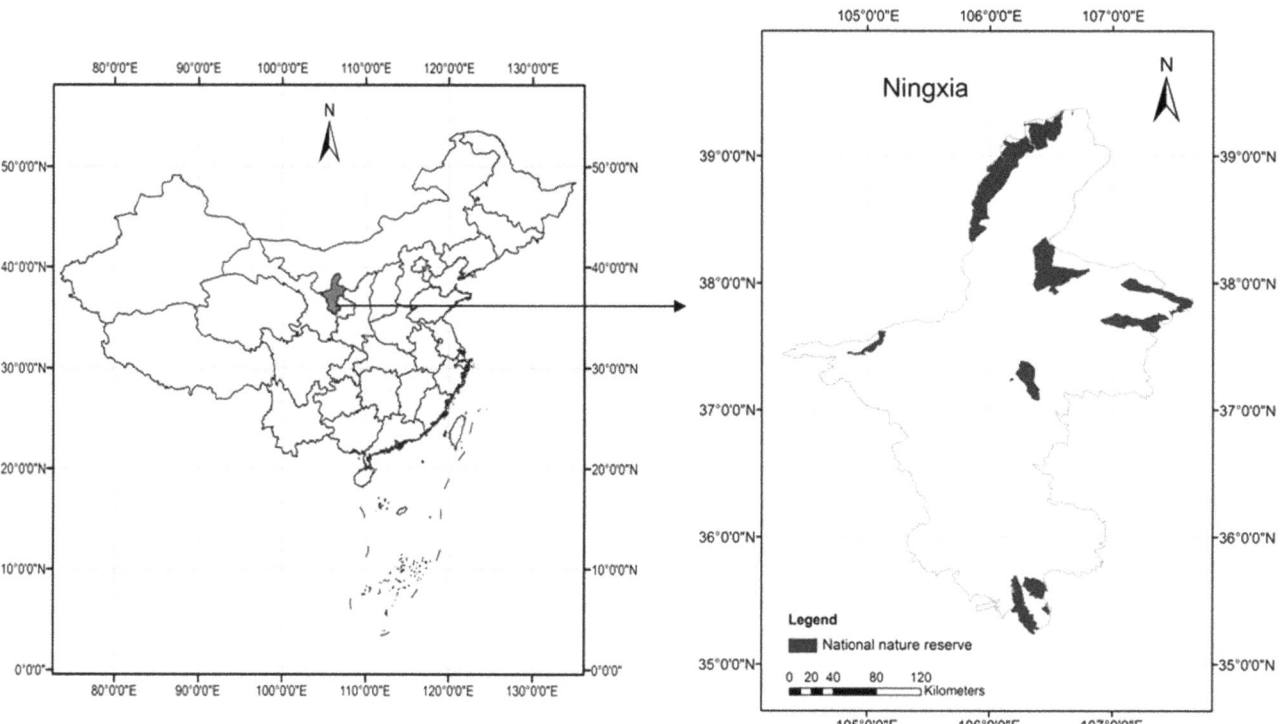

Figure 1. Location of the study site.

Table 1. Definitions of land use type in the national nature reserves in Ningxia.

Type	Definition
Forest land	Arbor, bush forest, broad-leaved forest, coniferous forest, and mixed forest
Grassland	Meadow and steppe
Farmland	Dry land, irrigable land, and crop fields
Wetland	Herbaceous swamp and thicket swamp
Water body	Rivers, ponds, reservoirs, and lakes
Construction land	Land used for industry, residences, and transportation
Unused land	Bare soil, bare rock, and saline-alkali soil

(http://ladsweb.nascom.nasa.gov/data/search.html). We obtained climatic data sets, including the monthly records from 122 radiation stations and 756 ground-based meteorological and automatic stations, from the web of China meteorological data sharing service (Figure 2, http://cdc.cma.gov.cn/index.jsp). Data for vegetation types were obtained from GLC2000 China regional land cover data (http://solargis.info/doc/32).

Land Use Dynamics

ESVs were determined by the value coefficients and areas of land use types. The changes of areas directly caused the variation in ESVs. The areas were statistical calculations of the areas of various land use types from the remote sensing data, so the remote sensing data were the basis of the calculated ESV data and had a direct relation with the ESV calculation. Images of the national nature reserves in Ningxia from 2000, 2005, and 2010 were used to estimate the land use changes in the past decade. The Map Algebra function of ArcGIS10.0 was used to calculate the figures of land use type for 2000, 2005, and 2010 and the dynamics of land use. The rate (R) of change of land use was calculated as:

$$R = \frac{A_1 - A_0}{A_0} \times 100\% \tag{1}$$

where A_0 and A_1 represent the initial and final areas of a given land use, respectively.

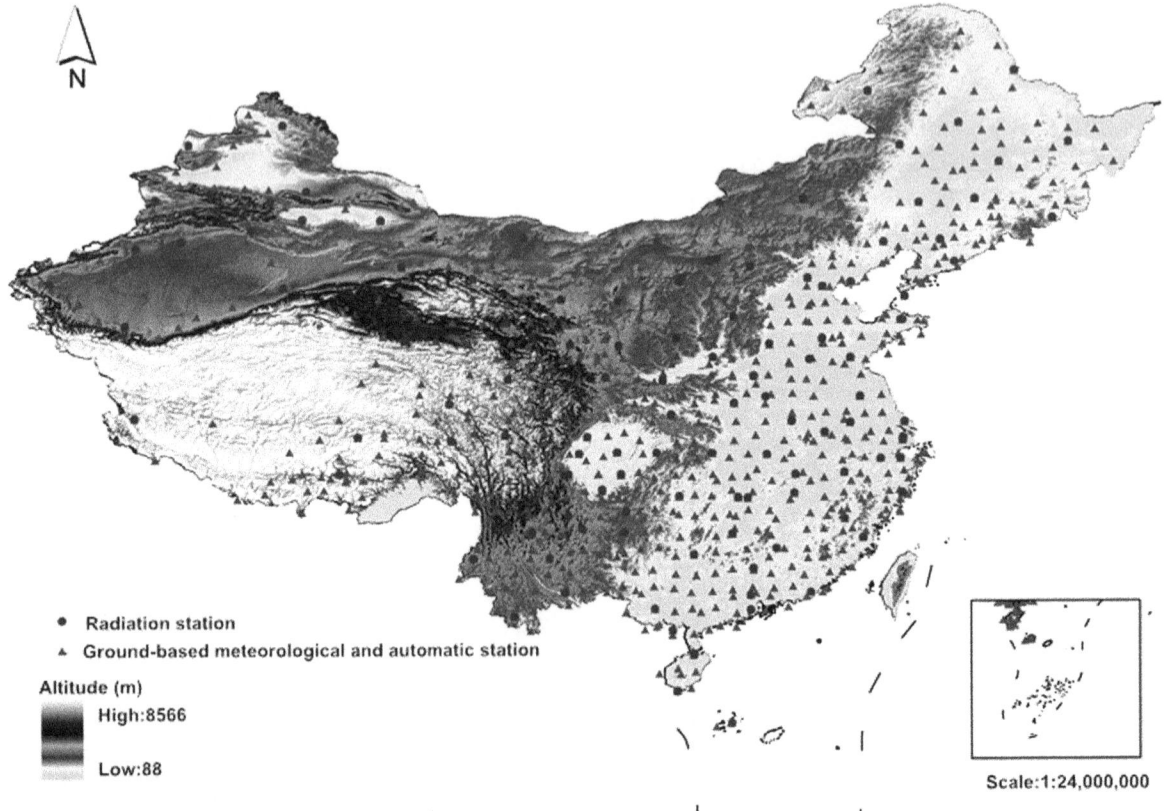

Figure 2. Distribution of weather stations.

ESV Assessment

Costanza's [20] theory and the survey of 500 Chinese ecologists [26] suggested that the equivalent value per unit area (Table 2) was practicable in China, and it has been widely used to assess ES [28,32,33]. The ESV of one equivalent weight factor was calculated as [34]:

$$VC_0 = \frac{1}{7} \times q_i \times \frac{1}{10} \sum_{i=1}^{10} p_i \qquad (2)$$

where VC_0 is the economic value of one equivalent weight factor (Yuan$\cdot$ha$^{-1}\cdot$yr^{-1})(1 Yuan = 0.16 US$), q_i is the average grain price nationwide (Yuan$\cdot$kg^{-1}), p_i is the yield of per unit area of crops in year i (kg$\cdot$ha$^{-1}\cdot$yr^{-1}), and i is the year.

Species resources, especially for rare species, are much more abundant inside than outside nature reserves, so adjusting the equivalent value per unit area of biodiversity protection is essential. A database of 3337 rare and endangered species in China was established from lists of national key protected wildlife species, the CITES (Convention on International Trade in Endangered Species) appendix, IUCN (the International Union for Conservation of Nature) endangered species level 3.1, a list of China's endemic species, and the IUCN Red List of Threatened Species. Based on information of species protection in 861 nature reserves in China, the distributions of 2157 rare and endangered species were determined for the nature reserves. Information about protection in national nature reserves accounted for 96.2% of the objects and 99.7% of the area, indicating that the calculated value per unit area was better representative of the important species. We then used the density of important species as a parameter for correction, calculated as:

$$V_b = \frac{d}{D} \times V_{b0} \qquad (3)$$

where V_b is the equivalent value per unit area of biodiversity protection in nature reserves, d is the density of important species in nature reserves (species$\cdot$ha^{-1}), D is the average distribution density of important species on a national scale (species$\cdot$ha^{-1}), and V_{b0} is the equivalent value per unit area of biodiversity protection in China.

The equivalent value per unit area of ES was based on the national average, so we calibrated for regional differences when assessing the ESV of a local area. Because ESV is closely related to ecosystem productivity, we used a correction factor based on ecosystem productivity to adjust the calculation of ESV. The regional correction factor reflected the difference in net primary production (NPP) caused by the variations in climate between local areas and the country as a whole. The ESV of one unit area for a year was calculated as:

$$VC_i = \frac{b_i}{B_i} \times VC_0 \qquad (4)$$

where b_i and B_i are the average NPPs of ecosystems in the study areas and the country in year I, respectively, and i is 2000, 2005, or 2010.

The CASA (Carnegie-Ames-Stanford Approach) model is based on the principle that plant productivity is correlated with the amount of photosynthetically active radiation absorbed or intercepted by green foliage [35,36]. Transformation to regional scales is easily achieved by the model, which is valuable for estimating the annual dynamic change in NPP on regional scales. RS data can provide information on many parameters of the vegetation. We obtained the parameter $FPAR$ (see below) required for calculating NPP from time-series data for NDVI from the MODIS spectroradiometer aboard the EOS satellites. The climatic data, including total solar radiation, average temperature, and duration of sunshine, were obtained from 122 radiation stations and 756 ground-based meteorological and automatic stations in China. The equations we used were:

$$NPP(x,t) = APAR(x,t) \times \varepsilon(x,t) \qquad (5)$$

$$APAR(x,t) = FPAR(x,t) \times PAR(x,t) \qquad (6)$$

$$\varepsilon(x,t) = \varepsilon_{max} \times T_{\varepsilon1}(x,t) \times T_{\varepsilon2}(x,t) \times W_\varepsilon(x,t) \qquad (7)$$

where $APAR$ is photosynthetically active radiation (mJ$\cdot$m^{-2}); ε is the actual light use efficiency of vegetation (g$\cdot$mJ^{-1}); x and t refer to location and time, respectively; PAR is the total incident photosynthetically active radiation (mJ$\cdot$m^{-2}); $FPAR$ is the fraction of PAR absorbed by the vegetation canopy; ε_{max} is the maximum light use efficiency under ideal conditions (g$\cdot$mJ^{-1}); $T_{\varepsilon1}$ and $T_{\varepsilon2}$ refer to stress effects of low and high temperatures on the use efficiency of light energy, respectively; and W_ε is the water stress

Table 2. Equivalent value per unit area of ecosystem services in China [26].

	Forest land	Grassland	Farmland	Wetland	Water body	Unused land
Gas regulation	4.32	1.50	0.72	2.41	0.51	0.06
Climate regulation	4.07	1.56	0.97	13.55	2.06	0.13
Water supply	4.09	1.52	0.77	13.44	18.77	0.07
Soil formation and protection	4.02	2.24	1.47	1.99	0.41	0.17
Waste treatment	1.72	1.32	1.39	14.40	14.85	0.26
Biodiversity protection	4.51	1.87	1.02	3.69	3.43	0.40
Food production	0.33	0.43	1.00	0.36	0.53	0.02
Raw material	2.98	0.36	0.39	0.24	0.35	0.04
Recreation and culture	2.08	0.87	0.17	4.69	4.44	0.24
Total	28.12	11.67	7.90	54.77	45.35	1.39

Table 3. Areas of land use types in the national nature reserves in Ningxia in 2000, 2005, and 2010.

Land use type	2000		2005		2010	
	Area(ha)	Percentage(%)	Area(ha)	Percentage(%)	Area(ha)	Percentage(%)
Forest land	59809.5	14.01	61652.5	14.44	62374.9	14.61
Grassland	254377.2	59.58	254469.9	59.61	251606.6	58.94
Farmland	43310.2	10.14	41412.9	9.70	42941.3	10.06
Wetland	65.9	0.02	49.8	0.01	33.2	0.01
Water body	1989.5	0.47	2137.1	0.50	2701.3	0.63
Unused land	60476.3	14.17	59360.9	13.90	57013.7	13.35
Construction land	6887.4	1.61	7833.0	1.83	10245.0	2.40
Total	426916	100.00	426916	100.00	426916	100.00

influence coefficient, which represents the influence of moisture conditions.

The ESVs of each land use type and service function and the total ESV were then calculated as:

$$ESV_k = \sum_f A_k \times VC_{ikf} \tag{8}$$

$$ESV_f = \sum_k A_k \times VC_{ikf} \tag{9}$$

$$ESV = \sum_k \sum_f A_k \times VC_{ikf} \tag{10}$$

where ESV_k, ESV_f, and ESV refer to the ESVs of land use type k, service function f, and the ecosystem (Yuan·ha^{-1}), respectively; A_k is the area of land use type k (ha); and VC_{ikf} is the value coefficient for land use type k with ES function type f (Yuan·ha^{-1}) in year i.

The contribution rate used to assess the effect of ESV variation on land use change was calculated as [20]:

$$S_{kt} = \frac{|\Delta ESV_{kt}|}{\sum\limits_{k=1}^{n} |\Delta ESV_{kt}|} \times 100\% \tag{11}$$

where S_{kt} is the proportion of the absolute value of ESV variation of land use type k in period t to the total amount of ESV variation of land use type k in period t.

Sensitivity Analysis of ESV

The coefficient of sensitivity (CS) validates the land use types representative of ecosystem type and certainties in the value coefficients [8,37,38]. CS takes the response of ESV to the ecological value of changes in unit price as a measure of the degree of sensitivity of ESV to a coefficient. CS was calculated as:

$$CS = \left| \frac{(ESV_j - ESV_i)/ESV_i}{(VC_{jk} - VC_{ik})/VC_{ik}} \right| \tag{12}$$

where ESV_i and ESV_j are the initial and adjusted total ESVs, respectively, and VC_{ik} and VC_{jk} are the initial and adjusted value coefficients, respectively. ESV is considered to be unaffected by the

coefficient, and the results will be reliable when CS<1, and ESV is considered elastic relative to the coefficient when CS>1. Larger values of CS will define VCs more accurately. Regardless of how the value coefficients change, the sensitivity of ESV to changes in the value coefficients must be kept low to ensure the reliability of our results. To verify CS, a 50% adjustment in the value coefficients was made to estimate the percent changes in the calculated total ESV and the CSs.

Results

Changes of Land Use

Table 3 and Figure 3 show the land use changes in the national nature reserves in Ningxia during 2000–2010. The area of grassland was an important factor in the reserves, accounting for approximately 60% of the total land area. The areas of wetland and unused land decreased during this period by 49.7% and 5.7%, respectively (Figure 4). The area of grassland increased during 2000–2005, decreased during 2005–2010, and had decreased by 1.1% by the end of the decade. The area of farmland decreased during 2000–2005, increased during 2005–2010, and had eventually decreased by 0.58% by the end of the decade. The areas of forest land, water bodies, and construction land continuously increased throughout the decade, by 4.3, 35.8, and 48.8%, respectively. The amount of wetland in the reserves declined more sharply than did the other land use types. The amount of construction land rose sharply and had the highest rate of increase due to the increasing encroachment of human activities.

Changes of ESV

The range of annual mean NPP was determined based on the observed NPP data of different vegetation types, including evergreen broad-leaved forest, evergreen coniferous forest, deciduous broad-leaved forest, deciduous coniferous forest, mixed forest, grassland and farmland [39,40,41]. The modeled NPP values were all within the range of observed values (Figure 5 A), indicating that the modeled result was consistent with the actual NPP. The modeled NPP values between 2000 and 2010 from other studies were compared with this study through correlation analysis [41,42]. The correlation coefficient was $R^2 = 0.81$ (Figure 5 B), showing that the modeled NPP in this study was in agreement with other studies. Through the above validation, we concluded that the NPP calculation by CASA in this study was reliable. The NPP values of terrestrial ecosystems in China calculated with the CASA model were 689, 711, and 692 gC·m^{-2}·yr^{-1} in 2000,

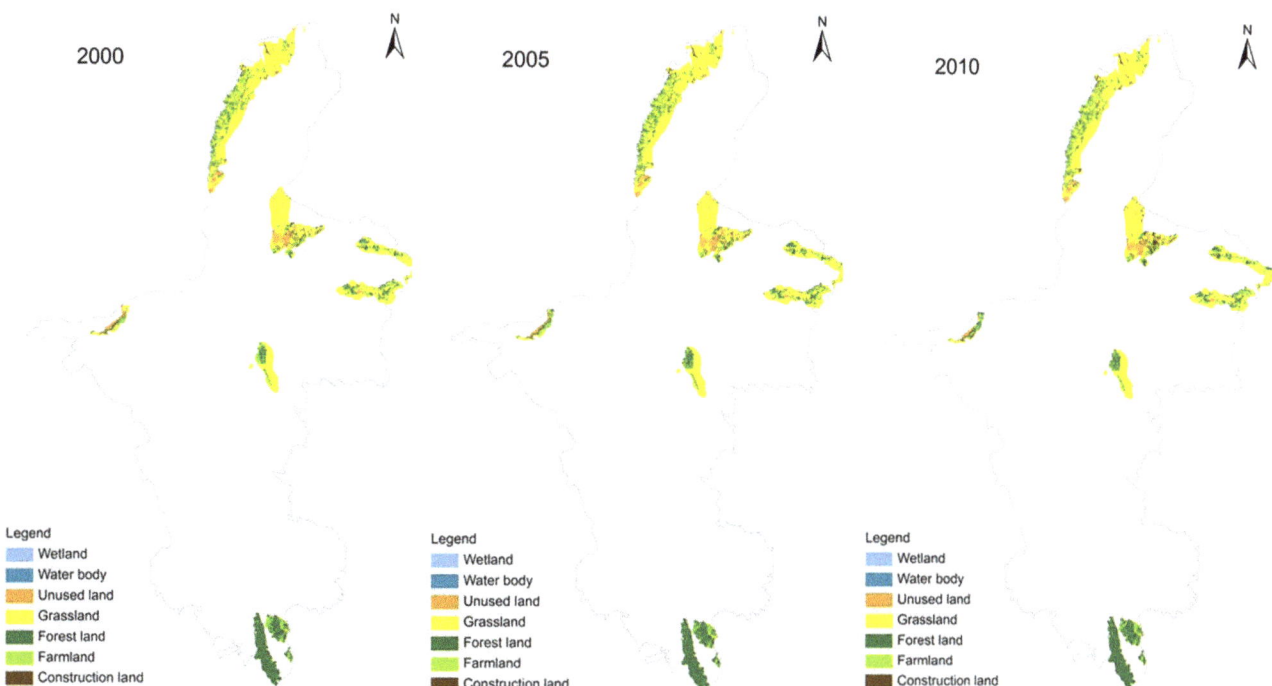

Figure 3. Distribution of land use types during 2000–2010.

2005, and 2010, respectively. The NPP values of terrestrial ecosystems in Ningxia were 384, 441, and 468 $gC \cdot m^{-2} \cdot yr^{-1}$ in the three years, respectively. The ESVs of each land use type in 2000, 2005, and 2010 are shown in Table 4 and Figure 6 (calculated with Eqs. 1–4, 8–10). The total ESV of the reserves in Ningxia increased throughout 2000–2010 by 22.7%. Total ESV increased at a higher percentage during 2000–2005 than during 2005–2010, by 12.3 and 9.2%, respectively.

The ESV of grassland, with such a large area, contributed most to the total ESV, accounting for approximately 57%. The ESV of forest land, accounting for approximately 33% of the total, continually increased during the decade. The ESVs of forest land and grassland thus constituted a substantial portion of the total ESV. The contribution rates used to assess the effect of ESV

variation on land use change are shown in Figure 7. Grassland contributed most in nearly every period, indicating that changes in the area of grassland had the strongest influence on the variation in ESV. The areas and influences of forest land and farmland were also relatively large. The area of water bodies was much lower than the area of unused land, but the ESV of water bodies was nearly that of unused land, with a high value coefficient. The contribution rates of water bodies were higher than those of unused land in every period.

The ESVs of the ESs in the reserves are shown in Figure 8. The ESVs of each ES increased throughout 2000–2010. Biodiversity protection contributed most to the total ESV, accounting for more than 50% of the total value in the decade, indicating the effect of nature reserves on biodiversity protection. The value of food

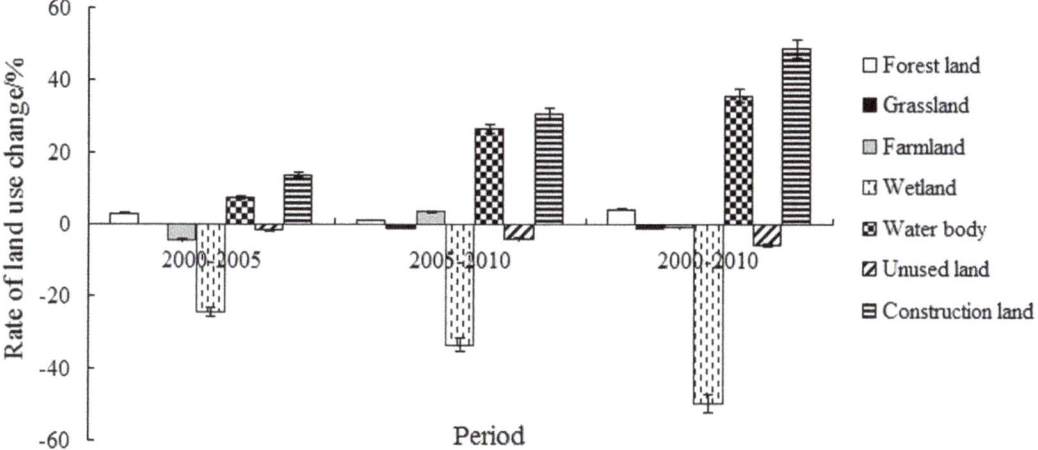

Figure 4. Dynamic rates of each land use type during 2000–2010.

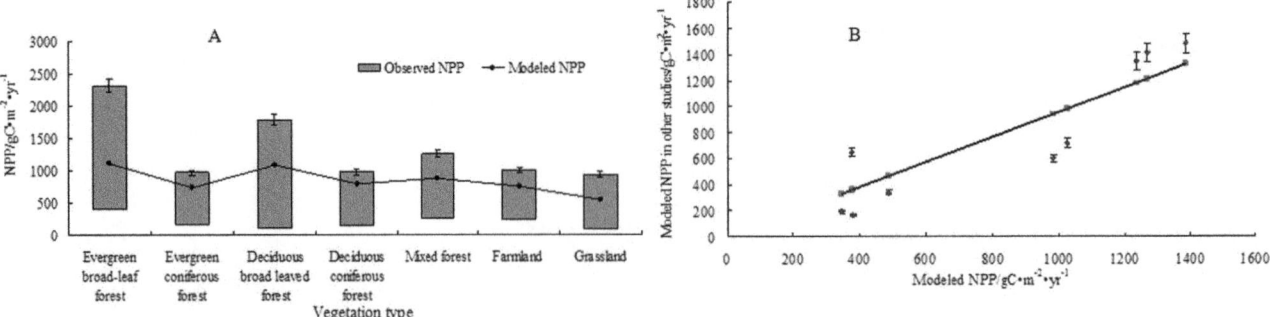

Figure 5. A. Relationship between modeled NPP and observed NPP. B. Relationship between modeled NPP and other evaluations.

production was the lowest, accounting for less than 2% of the total, and demonstrated the importance of the policies to protect nature reserves from commercial exploitation. The value of biodiversity protection increased most during the period. The areas of forest land and water bodies increased sharply with high value coefficients, showing good effects on habitat supply. The ESV of soil formation and protection increased during this period, indicating that conditions of soil desertification in the reserves in Ningxia had improved. The Wilcoxon Signed Ranks Test of SPSS 19.0 showed that the ESVs of various land use types increased significantly in different periods, and the ESVs of each ecosystem service increased quite remarkably in different periods (Table 5). These increases indicated that the ESVs of the study area increased significantly during 2000–2010.

Ecosystem Sensitivity Analysis

As is shown in Table 6, the CS in each case was less than 1, indicating that the estimated ESV was not affected by changes in the value coefficients. The CSs of grassland and forest land, with large areas and high value coefficients, were higher than those of the other land use types, with values of approximately 0.6 and 0.3, respectively. The areas of farmland and unused land were relatively large, but these types had low CSs and value coefficients. We can conclude from the estimated CSs that the calculated ESVs were responsible for the uncertainties in the value coefficients.

Discussion

The recent enthusiasm for analyzing the concepts and methods for ES valuation appears to have been mostly initiated by the needs of conservationists to recommend broader and better-founded policies for protecting natural resources [43]. The responses of ecosystem quality to changes in land use and other factors are apparent in ESV variation. Economic analysis of ESs is an adequate framework for timely and effectively improving decisions involving various aspects of nature conservation and ecological restoration. Valuation may be a first step toward a "commodification" of nature and is not an end in itself but rather a conceptual and methodological framework for organizing information as a guide to making decisions and for managing ecological restoration and nature conservation.

Ecological restoration has been practiced in western China since the turn of the century. Through a wide range of comprehensive measures including policy, projects, technology, and capital, the trends of ecological degradation in the western region have been relieved. Ecological function is gradually being restored through the efforts of ecological restoration in weatern China. Variations in ESVs can reflect changes to the health of ecosystems and can thus evaluate the effects of these efforts at restoration. The evaluation of ESVs conducted in this study is a fast and effective way to assess the results of ecological restoration in western China. The increase of ESVs of the national nature reserves in Ningxia reflected the effects of ecological restoration to some extent. For example, the

Table 4. Ecosystem service values (ESVs) and their variation for each land use type in the national nature reserves in Ningxia in 2000, 2005, and 2010.

Land use type	2000		2005		2010	
	ESV (10^7 US$)	Percentage(%)	ESV (10^7 US$)	Percentage(%)	ESV (10^7 US$)	Percentage(%)
Forest land	59.57	32.67	68.42	33.40	75.43	33.70
Grassland	105.11	57.64	117.16	57.20	126.22	56.40
Farmland	11.13	6.11	11.86	5.79	13.40	5.99
Wetland	0.06	0.03	0.08	0.04	0.10	0.04
Water body	2.49	1.37	2.98	1.46	4.11	1.84
Unused land	3.96	2.17	4.34	2.12	4.54	2.03
Construction land	0	0	0	0	0	0
Total	182.3	100.00	204.8	100.00	223.8	100.00

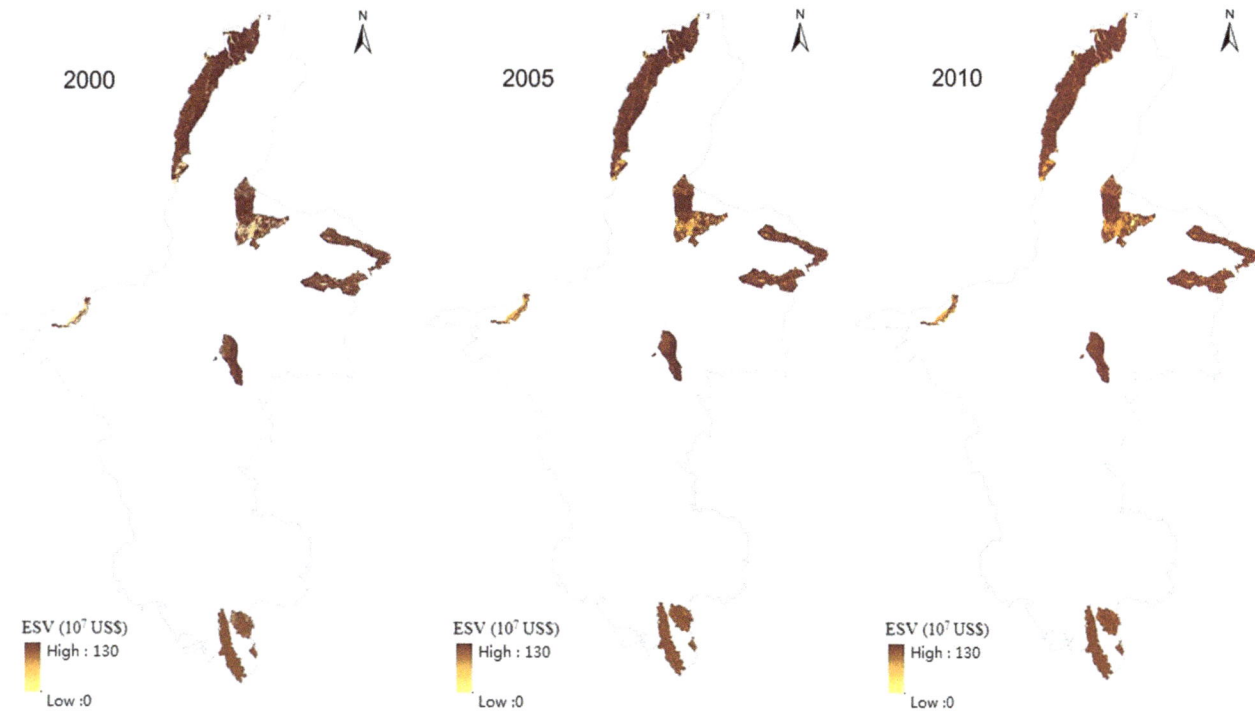

Figure 6. Ecosystem services values of land use types during 2000–2010.

ESVs of forest land and water bodies increased with expanded areas and improved ecosystem quality (e.g. ecosystem productivity), and the ESVs of grassland, farmland, wetland, and unused land increased with improved ecosystem quality, even when the areas decreased. Water bodies, with a high value coefficient, had great potential for ESs. We thus recommend that more attention be paid to these land use types (e.g. water bodies) for ecological restoration and construction in the near future.

The method of deriving ESVs by multiplying the area of land use types and the value coefficients was proposed by Costanza et al. [20]. Their value coefficients were based on the average of global ecosystems, not tallied from the Chinese situation. Xie et al. [26] modified the value coefficients to apply to China. The accuracy and reliability of the evaluated results are mainly determined by the accuracy of the value coefficients. More

accurate value coefficients are thus necessary. ESs have considerable spatial heterogeneity, so RS and GIS technology must be used to conduct ESs assessments to improve the reliability of ESVs at the regional scale. For our study, the protection of ecosystems in nature reserves confers advantages in ecosystem productivity and biodiversity, so parameter corrections to the value coefficients are needed for accurate estimation of the ESVs in nature reserves. With NPP and biodiversity parameter corrections on value coefficients, the ESV of biodiversity protection was the highest, and the ESV of food production was the lowest, in agreement with the actual situation of nature reserves in China.

Land use can be used as a proxy measure of ESs through matching land use types with equivalent biomasses and ESVs can be easily conducted based on land use data. Using the average value coefficients, however, may not be precise enough, because

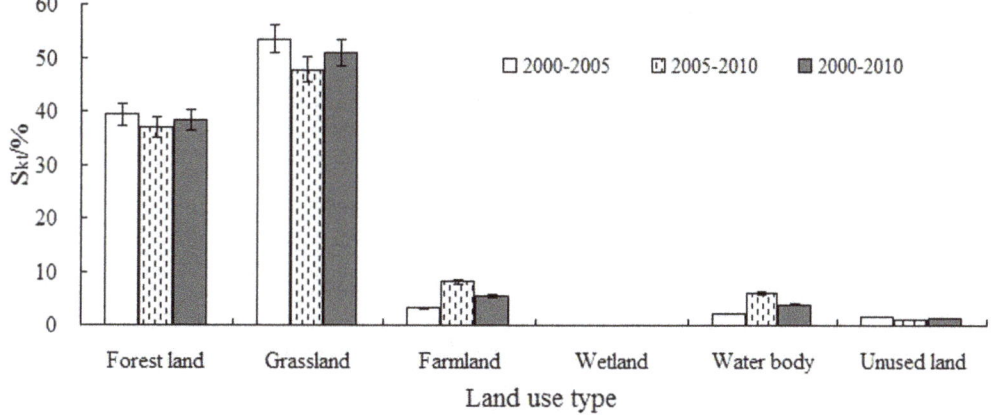

Figure 7. Contribution rate of each land use type during 2000–2010.

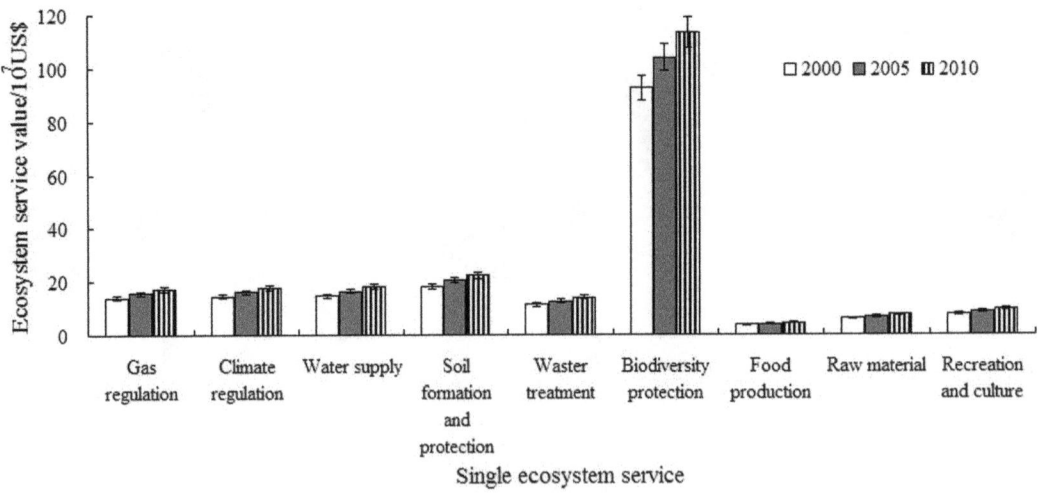

Figure 8. Values of ecosystem services during 2000–2010.

Table 5. Difference analysis of ecosystem service values (ESVs) in the national nature reserves in Ningxia in 2000, 2005 and 2010.

		2000	2005	2010
ESVs of the land use types	2000	—	Z = −2.201 (P = 0.028<0.05)	Z = −1.992 (P = 0.046<0.05)
	2005	Z = −2.201 (P = 0.028<0.05)	—	Z = −1.992 (P = 0.046<0.05)
	2010	Z = −1.992 (P = 0.046<0.05)	Z = −1.992 (P = 0.046<0.05)	—
ESVs of the land use types	2000	—	Z = −2.666 (0.008<0.01)	Z = −2.666 (0.008<0.01)
	2005	Z = −2.666 (0.008<0.01)	—	Z = −2.666 (0.008<0.01)
	2010	Z = −2.666 (0.008<0.01)	Z = −2.666 (0.008<0.01)	—

the land use classification was only applied to the first-level classification (e.g. forest land and grassland). The structural and functional differences of different ecosystems at the same level, e.g. the forest land type includes broad-leaved forests, coniferous forests, bush forests, etc., may lead to uncertain ESVs. Further studies, then, must apply the detailed land use classification and

Table 6. Variation of the estimated total ecosystem service value (ESV) and the coefficient of sensitivity (CS) resulting from a 50% adjustment in the value coefficient in the national nature reserves in Ningxia in 2000, 2005, and 2010.

Land use type	Variation of ESV (%)			CS		
	2000	2005	2010	2000	2005	2010
Forest land	±16.33	±16.70	±16.85	0.33	0.33	0.34
Grassland	±28.82	±28.60	±28.21	0.58	0.57	0.56
Farmland	±3.05	±2.90	±2.99	0.06	0.06	0.06
Wetland	±0.03	±0.02	±0.01	0.00	0.00	0.00
Water body	±0.68	±0.73	±0.92	0.01	0.01	0.02
Unused land	±1.09	±1.06	±1.01	0.02	0.02	0.02
Construction land	±0.00	±0.00	±0.00	0.00	0.00	0.00

value coefficients to improve the understanding of the distributed characteristics of ESs.

Different methods of evaluation can provide different results. For example, both Li et al. [32] and Peng et al. [44] evaluated the ESV of Shenzhen for the same year, giving estimates of 2.9 billion and 126.5 billion Yuan, respectively. Absolute numbers of ESVs have less meaning, and the dynamics of ESVs are commonly indicating ecological problems. Despite the residual uncertainties due to the complex, dynamic, and nonlinear nature of ecosystems [45,46,47], accurate coefficients are often less critical for time-series than for cross-sectional analyses because value coefficients tend to affect estimates of directional change less than estimates of ecosystem values at specific points in time [32]. Our evaluation is thus valid for calculating ESVs over extended periods as a means of assessing the changes in ESVs in response to changes in land use.

Conclusions

We assessed ESVs and their changes for national nature reserves in Ningxia from land use data obtained during 2000–2010. The areas of forest land, water bodies, and construction land increased, while the areas of grassland, farmland, wetland, and unused land decreased. Ecological restoration helped to increase the total ESV in the reserves during 2000–2010 from 182.3 million to 223.8 million US$, an increase of 22.7%. Grassland and forest land contributed approximately 90% of the total ESV. The ESVs of all ESs increased throughout the decade, especially biodiversity protection and soil formation and protection. We can thus conclude that ecological restoration in the national nature reserves in Ningxia during 2000–2010 had achieved good results.

By matching the land use types to equivalent biomes, ESV can be estimated by the land use data and the value coefficients using GIS and RS data. The value coefficients are the essential issue for the accuracy and reliability of ESV estimation. The coefficients of sensitivity indicated that the estimated ESVs were relatively rigid relative to the changes in the value coefficients. Our analysis of the variation in the ESVs in the national nature reserves in Ningxia will be able to serve as a reference for future analyses.

Acknowledgments

We are grateful to the staff of the report of the key techniques for monitoring and conserving important biological species and for demonstrating its application in China (No. 2008BAC439B00).

Author Contributions

Conceived and designed the experiments: YW JG JW. Performed the experiments: YW. Analyzed the data: YW JQ. Contributed reagents/materials/analysis tools: YW JG. Wrote the paper: YW.

References

1. MEA (2005) Millennium Ecosystem Assessment - Ecosystems and wellbeing: A framework for assessment. Washington, DC: Island Press.
2. Fisher B, Turner RK, Morling P (2009) Defining and classifying ecosystem services for decision making. Ecological Economics 68: 643–653.
3. MEA (2003) Millennium Ecosystem Assessment - Ecosystems and wellbeing: A framework for assessment. Washington, DC: World Resources Institute.
4. Summers JK, Smith LM, Case JL, Linthurst RA (2012) A review of the elements of human well-being with an emphasis on the contribution of ecosystem services. Ambio 41: 327–340.
5. Costanza R, Folke C (1997) Valuing ecosystem services with efficiency, fairness, and sustainability as goals. In: Daily GC, editor. Nature's Services: Societal Dependence on Natural Ecosystems. Washington D.C: Island Press. pp. 49–69.
6. Kubiszewski I, Costanza R, Dorji L, Thoennes P, Tshering K (2013) An initial estimate of the value of ecosystem services in Bhutan. Ecosystem Services 3: e11–e21.
7. Yang W, Li F, Wang R, Hu D (2011) Ecological benefits assessment and spatial modeling of urban ecosystem for controlling urban sprawl in Eastern Beijing, China. Ecological Complexity 8: 153–160.
8. Kreuter UP, Harris HG, Matlock MD, Lacey RE (2001) Change in ecosystem service values in the San Antonio area, Texas. Ecological Economics 39: 333–346.
9. Sawut M, Eziz M, Tiyip T (2013) The effects of land-use change on ecosystem service value of desert oasis: a case study in Ugan-Kuqa River Delta Oasis, China. Canadian Journal of Soil Science 93: 99–108.
10. Lorencová E, Frélichová J, Nelson E, Vačkář D (2013) Past and future impacts of land use and climate change on agricultural ecosystem services in the Czech Republic. Land Use Policy 33: 183–194.
11. Su C, Fu B (2013) Evolution of ecosystem services in the Chinese Loess Plateau under climatic and land use changes. Global and Planetary Change 101: 119–128.
12. Kaplowitz MD (2000) Identifying ecosystem services using multiple methods: Lessons from the mangrove wetlands of Yucatan, Mexico. Agriculture and Human Values 17: 169–179.
13. Koschke L, Fürst C, Frank S, Makeschin F (2012) A multi-criteria approach for an integrated land-cover-based assessment of ecosystem services provision to support landscape planning. Ecological Indicators 21: 54–66.
14. Su CH, Fu BJ, He CS, Lü YH (2012) Variation of ecosystem services and human activities: A case study in the Yanhe Watershed of China. Acta Oecologica 44: 46–57.
15. Wang X, Chen W, Zhang L, Jin D, Lu C (2010) Estimating the ecosystem service losses from proposed land reclamation projects: A case study in Xiamen. Ecological Economics 69: 2549–2556.
16. Daily GC (1997) Nature's Services: Societal Dependence on Natural Ecosystems. Washington, DC: Island Press.
17. Burkhard B, Petrosillo I, Costanza R (2010) Ecosystem services - Bridging ecology, economy and social sciences. Ecological Complexity 7: 257–259.
18. Fisher B, Kerry Turner R (2008) Ecosystem services: Classification for valuation. Biological Conservation 141: 1167–1169.
19. Barral MP, Oscar MN (2012) Land-use planning based on ecosystem service assessment: A case study in the Southeast Pampas of Argentina. Agriculture, Ecosystems & Environment 154: 34–43.
20. Costanza R, d'Arge R, Groot Rd, Farber S, Grasso M, et al. (1997) The value of the world's ecosystem services and natural capital. Nature 387: 253–260.
21. Troy A, Wilson MA (2006) Mapping ecosystem services: Practical challenges and opportunities in linking GIS and value transfer. Ecological Economics 60: 435–449.
22. Wang ZM, Zhang B, Zhang SQ (2004) Study on the effects of land use change on ecosystem service values of Jilin province. Journal of Natural Resources 19: 55–61.
23. Zhao J, Wei L, Chen S (2010) Dynamics of the ecosystem service values along the upper reaches of Shiyanghe River Basin. Journal of Arid Resources and Environment 1: 36–40.
24. Heal G (2000) Valuing ecosystem services. Ecosystems 3: 24–30.
25. Wilson MA, Howarth RB (2002) Discourse-based valuation of ecosystem services: establishing fair outcomes through group deliberation. Ecological Economics 41: 441–443.
26. Xie GD, Zhen L, Lu CX, Xiao Y, Chen C (2008) Expert knowledge based valuation method of ecosystem services in China. Journal of Natural Resources 23: 911–919.
27. Costanza R, Cumberland J, Daly H, Goodland R, Norgaard R (1997) An introduction to ecological economics. Delray Beach Fla USA: St. Lucie Press.
28. Liu Y, Li J, Zhang H (2012) An ecosystem service valuation of land use change in Taiyuan City, China. Ecological Modelling 225: 127–132.
29. Li H, Wang S, Ji G, Zhang L (2011) Changes in land use and ecosystem service values in Jinan, China. Energy Procedia 5: 1109–1115.
30. Li J, Wang W, Hu G, Wei Z (2010) Changes in ecosystem service values in Zoige Plateau, China. Agriculture, Ecosystems & Environment 139: 766–770.
31. Song G, Fu C, E Y (2011) The Analysis of Ecosystem Service Value's Change in Yueqing Bay Wetland Based on RS and GIS. Procedia Environmental Sciences 11: 1365–1370.
32. Li T, Li W, Qian Z (2010) Variations in ecosystem service value in response to land use changes in Shenzhen. Ecological Economics 69: 1427–1435.
33. Wu KY, Ye XY, Qi ZF, Zhang H (2013) Impacts of land use/land cover change and socioeconomic development on regional ecosystem services: The case of fast-growing Hangzhou metropolitan area, China. Cities 31: 276–284.
34. Xie GD, Lu CX, Leng YF, Zheng D (2003) Ecological assets valuation of the Tibetan Plateau. Journal of Natural Resources 18: 189–196.
35. Monteith JL, Moss CJ (1977) Climate and the Efficiency of Crop Production in Britain [and Discussion]. Philosophical Transactions of the Royal Society of London B, Biological Sciences 281: 277–294.
36. Potter CS, Randerson JT, Field CB, Matson PA, Vitousek PM, et al. (1993) Terrestrial ecosystem production: A process model based on global satellite and surface data. Global Biogeochemical Cycles 7: 811–841.
37. Li RQ, Dong M, Cui JY, Zhang L, Cui QG, et al. (2007) Quantification of the Impact of Land-Use Changes on Ecosystem Services: A Case Study in Pingbian County, China. Environmental Monitoring & Assessment 128: 503–510.
38. Wang ZM, Zhang SQ, Zhang B (2004) Effects of land use change on value of ecosystem services of Sanjiang Plain. China Environmental Science 24: 125–128.
39. Gao ZQ, Liu JY (2008) A comparative study of Chinese vegetation net productivity. Chinese Science Bulletin 53: 317–326.
40. Chen YM, Zhang WQ, Yang TX, Zhao G, Wang SB (2012) The change characteristics of net primary production in different vegetation types in China. Journal of Fudan University 51: 377–381.
41. Zhang FM, Ju WM, Chen JM, Wang SQ, Yu GR, et al. (2012) Characteristics of terrestrial ecosystem primary productivity in East Asia based on remote sensing and process-based model. Chinese Journal of Applied Ecology, 23: 307–318.
42. Gao YN, Yu GR, Zhang L, Liu M, Huang M (2012) The changes of net primary productivity in Chinese terrestrial ecosystem: based on process and parameter models. Progress in Geography 31: 109–117.
43. Balmford A, Bruner A, Cooper P, Costanza R, Farber S, et al. (2002) Economic reasons for conserving wild nature. Science 297: 950–953.
44. Peng J, Wang YL, Chen YF, Li WF, Jiang YY (2005) Economic value of urban ecosystem services: a case study in Shenzhen. Acta Scientiarum Naturalium Universitatis Pekinensis 41: 594–604.
45. Hein L, van Koppen K, de Groot RS, van Ierland EC (2006) Spatial scales, stakeholders and the valuation of ecosystem services. Ecological Economics 57: 209–228.
46. Limburg KE, O'Neill RV, Costanza R, Farber S (2002) Complex systems and valuation. Ecological Economics 41: 409–420.
47. Turner RK, Paavola J, Cooper P, Farber S, Jessamy V, et al. (2003) Valuing nature: lessons learned and future research directions. Ecological Economics 46: 493–510.

10

Streamflow Impacts of Biofuel Policy-Driven Landscape Change

Sami Khanal[1]*, Robert P. Anex[2]*, Christopher J. Anderson[3], Daryl E. Herzmann[3]

1 School of Environment and Natural Resources, Ohio State University, Wooster, OH, United States of America, **2** Dept. of Biological Systems Engineering, University of Wisconsin-Madison, Madison, Wisconsin, United States of America, **3** Dept. of Agronomy, Iowa State University, Ames, Iowa, United States of America

Abstract

Likely changes in precipitation (P) and potential evapotranspiration (PET) resulting from policy-driven expansion of bioenergy crops in the United States are shown to create significant changes in streamflow volumes and increase water stress in the High Plains. Regional climate simulations for current and biofuel cropping system scenarios are evaluated using the same atmospheric forcing data over the period 1979–2004 using the Weather Research Forecast (WRF) model coupled to the NOAH land surface model. PET is projected to increase under the biofuel crop production scenario. The magnitude of the mean annual increase in PET is larger than the inter-annual variability of change in PET, indicating that PET increase is a forced response to the biofuel cropping system land use. Across the conterminous U.S., the change in mean streamflow volume under the biofuel scenario is estimated to range from negative 56% to positive 20% relative to a business-as-usual baseline scenario. In Kansas and Oklahoma, annual streamflow volume is reduced by an average of 20%, and this reduction in streamflow volume is due primarily to increased PET. Predicted increase in mean annual P under the biofuel crop production scenario is lower than its inter-annual variability, indicating that additional simulations would be necessary to determine conclusively whether predicted change in P is a response to biofuel crop production. Although estimated changes in streamflow volume include the influence of P change, sensitivity results show that PET change is the significantly dominant factor causing streamflow change. Higher PET and lower streamflow due to biofuel feedstock production are likely to increase water stress in the High Plains. When pursuing sustainable biofuels policy, decision-makers should consider the impacts of feedstock production on water scarcity.

Editor: Ben Bond-Lamberty, DOE Pacific Northwest National Laboratory, United States of America

Funding: This work is supported by the National Science Foundation Grant CBET-1137677. The funders had no role in study design, data collection and analysis, decision to publish, or preparation of the manuscript.

Competing Interests: The authors have declared that no competing interests exist.

* Email: khanal.3@osu.edu (SK); anex@wisc.edu (RA)

Introduction

As demand for renewable fuels grows, biofuels from lignocellulosic feedstock are considered a promising alternative to corn-based ethanol [1–2]. Cellulosic biofuels are expected to be both environmentally and energetically superior to grain-based biofuels [3–6]. The mandate set by the Renewable Fuel Standard [7] to use 16 billion gallons of cellulosic biofuel per year by 2022 is projected to have significant impact on agricultural land use in the U.S. as lands are converted for the production of bioenergy crops [8]. Prior studies [6,9] have investigated yields, land use, economics and greenhouse gas emissions of bioenergy crops, but one key factor often overlooked is the hydrologic balance associated with bioenergy crop production.

There is strong coupling between the land surface and atmosphere that is heavily influenced by the vegetative land cover [10–13]. Change in land cover thus has the potential to impact local and regional climate through alteration of the energy and moisture balances of the land surface [14–18]. The longer growing season and greener vegetative cover of biofuel crops result in higher water loss to the atmosphere through evapotranspiration (ET), decline in soil water depth [17,19] and reduced surface runoff [20] relative to annual cropping systems. Changes in soil moisture and runoff determine streamflow, groundwater recharge and influence water quality.

Bioenergy crops, e.g., switchgrass and miscanthus, can transpire as much as 38% more than corn over a growing season [20]. Replacing traditional annual cropping systems with switchgrass in the Midwest and High Plains may cause additional stress to water resources because the agricultural crop production in large portions of these areas (e.g., Kansas and Nebraska) is dependent upon irrigation water from already stressed local resources [21]. Streamflow volume (Q) is responsive to changes in both climate and land cover [22–23], and changes in Q have important biological and socioeconomic implications [24–26]. Anthropogenic alteration of Q has been shown to impair aquatic communities and ecosystems, and the likelihood of impairment rises rapidly with increasing severity of reduced Q [25].

Water may be a significant limiting factor for biofuel crop production in many agricultural regions. It is important that we develop projections of future water use for agricultural crop production under climate change induced by land use change and to account for the impact of that water use on critical water resources. Prior studies have examined the potential for biofuel crops to affect regional climate [15]. The climate feedback of biofuel crops examined in these studies, however, is based on hypothetical scenarios that do not account for the socio-economic responses of land managers and thus do not represent plausible land use patterns that might result from current biofuel policies. To our knowledge, no prior studies have explored the changes in streamflow volume in response to climate change induced by land use/land cover (LULC) change; certainly none have examined this under the constraints of enacted legislation. Future projections of climate and climate-driven streamflow under plausible landscape scenarios will aid state and federal agencies in assessing the local cost of adaptation, increase public awareness, and guide the development of new mitigation programs related to water resources.

In this study, we examine changes in hydrologic processes including precipitation (P), ET, PET, runoff and Q that result from modification of local/regional climate driven by switchgrass cropping systems predicted to replace current cropping systems in the High Plains (hereafter referred to as the "biofuel scenario"). A regional climate model coupled to a land surface model is used to capture feedback between changes in the vegetation canopy due to switchgrass planting and regional climate processes. The change in Q to climate change under the biofuel scenario compared to the current cropping system scenario (hereafter referred to as the "baseline scenario") is estimated based on widely used non-parametric approaches [22–23,27]. These non-parametric approaches utilize the concept of elasticity of Q that is usually derived using the historic relationship between Q, P and PET. Following the similar approach, we first derived the elasticity of Q to climate, and later used it in combination with projected changes in P and PET to derive changes in Q under the biofuel scenario relative to baseline across the conterminous U.S.

Materials and Methods

Regional climate modeling framework

The Weather Research Forecast (WRF) model version 3.1.1 [28–29] and NOAH land surface model (NOAH LSM) are used for regional climate simulations. The simulation domain covers the continental U.S. at a resolution of 0.25 degree (i.e., 24 km). The NOAH LSM coupled to the WRF model is used to represent the interaction of soil and vegetation with the atmosphere [30]. Regional climate simulations were produced for the period of 1979–2004 under the two land use scenarios. The choice of the simulation period is constrained by 1) availability of data describing the baseline scenario; 2) the accuracy of extrapolating land use categories derived from 1991–1995 satellite data further into the future; and 3) the large computational burden associated with longer simulations. The baseline scenario represents land use categories and monthly phenology based on satellite derived data from 1991–1995, and the biofuel scenario represents projected alternative (i.e., switchgrass) land use categories (Figure 1) with identical atmospheric forcing data. Following the recent studies [16,31–32] that have used 2-years as a minimum length for spin-up, we discarded the first two years (i.e., 1979 and 1980) of each simulation to allow for adjustment of the land surface with the atmosphere. Details of the model configuration are provided in the Anderson et al [19], and thus are not described here.

Land use scenarios

The baseline scenario uses the NOAH LSM default settings of land use and vegetation parameters, including 24 vegetation classes, a vegetation parameter table and satellite-based (1991–1995) monthly vegetation fraction from which leaf area index (LAI) and albedo are derived (Table S1). The projection of LULC change produced by the Policy Analysis System (POLYSYS) model [33–34] in support of the DOE study report "U.S. Billion-

Ton Update" [7] is used to create the biofuel land use scenario. The Billion Ton Update study examines the feasibility of attaining annual production of one billion dry tons of biomass feedstock by 2030 based on projections of future biomass demand, inventory, production capacity, availability, and technology. In the 2011 update, the greatest conversion of traditional cropping system to switchgrass occurs in the Great Plains, with 20–30% in Kansas and 30–45% in northern and northwest Oklahoma [35]. The POLYSYS simulation contained county level switchgrass and crop production for 2022, the first year Renewable Fuel Standard (RFS) goals reach maximum levels, at a $60/dry ton farm gate switchgrass price with the Billion-Ton Study baseline assumptions, including an extension of the USDA 10-year yield forecast for major food and forage crops to 2022. An area weighted method was used to resample county-level POLYSIS estimates of switchgrass conversion to the WRF grid [19]. The NOAH LSM uses a single vegetation category for each grid cell, and vegetation parameters are homogenous within each grid cell. Thus, a grid cell that contains a mixture of vegetation types does not explicitly account for each type, but represents an average of vegetation parameters over all vegetation types present in the grid cell.

For the biofuel scenario, four new vegetation classes, their related vegetation parameters and monthly vegetation fraction are introduced in the WRF model based on default land use categories (Figure 1A). The new vegetation classes are modified to represent a mixture of the default land use categories and switchgrass (Figure 1B). This reflects that a mixture of biofuel and conventional crops are expected in regions where biofuel crops are adopted rather than complete replacement of conventional crops with switchgrass. Two of the new classes (Switchgrass/Grassland Mosaic and Switchgrass/Cropland Mosaic) are used for grid cells in which the switchgrass fraction exceeds 30%, and two (Grassland/Switchgrass Mosaic and Cropland/Switchgrass Mosaic) are used for grid cells in which the switchgrass fraction is below 30% [36] (Figures 1B and 1C). We prevented land use change to switchgrass in regions beyond that projected by the Billion ton update by only changing the parameters based on latitude/latitude. For example, land cover changes were mostly in Oklahoma and Kansas (Figure 1B); we did not change parameters in California.

As the new land use categories under the biofuel scenario reflect a mixture of switchgrass and conventional crops rather than complete replacement of switchgrass, the phenology for new land use categories is characterized in our simulations by adjusting the satellite-based monthly vegetation fraction based on discussion with scientists working on field trials of switchgrass. The monthly greenness fraction is the basis for LAI and albedo calculations in NOAH LSM, and both LAI and albedo increase with vegetation fraction. During the growing season, spectrally weighted albedo (which is required in WRF) increases as a crop is greening up. To represent change in phenology consistent with managed plots of switchgrass stands, vegetation fraction is increased during February–October to simulate earlier greening, denser foliage at peak LAI, and later senesce of switchgrass (a perennial grass) compared to annual crops and rangeland. Increase of vegetation fraction ranges 10–20%, except north central Oklahoma where it is increased by 90% to offset low LAI due to winter wheat harvest [19].

The maximum and minimum values of LAI and albedo are adjusted to reflect changes in vegetation under the biofuel scenario (Table S1). And, the same monthly phenology is imposed for each simulated year. When the vegetation fraction is at its peak, the LAI and albedo are as well. Maximum LAI is based upon observations of field stands of managed perennial grass that grows a denser

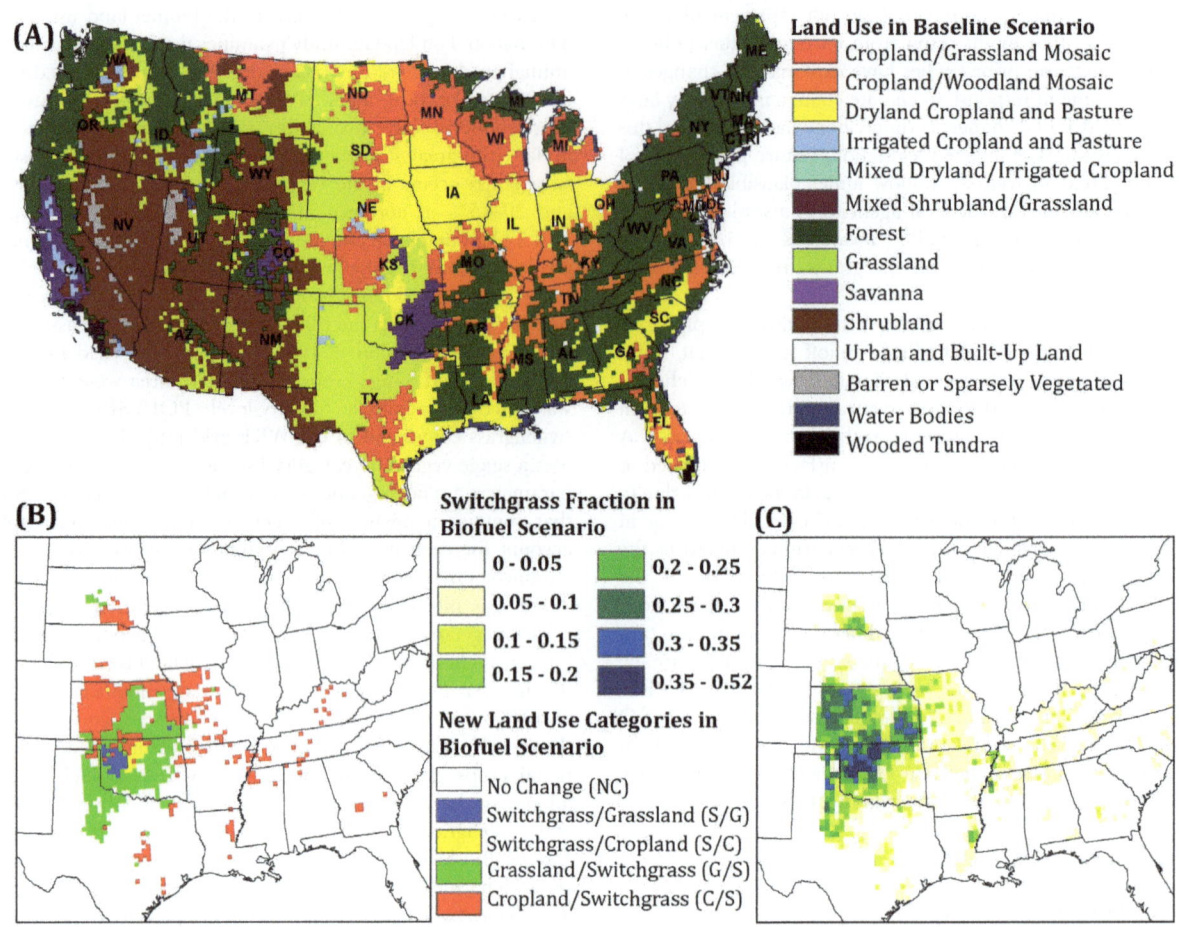

Figure 1. Default land use categories in the WRF model (A); new land use categories defined for the biofuel scenario (B); and fraction of land use that is switchgrass in the biofuel scenario (C).

canopy than prairie grass. Although LAI>6 as measured in field trials of managed switchgrass is used in simulations of switchgrass production [37–38], maximum LAI<6 is set to reflect a regional vegetation mixture. This approach is consistent in simulations with Van Loocke *et al* [17].

Climate elasticity of streamflow and changes in streamflow

In this version of WRF model, surface runoff is computed as the excess of precipitation that does not infiltrate into the soil [39]. Although NOAH-LSM describes the canopy and root zone in detail, the interactions between groundwater, the root zone, and surface water were not yet included at the time our project was undertaken and completed. This version parameterizes surface runoff with a simple infiltration-excess scheme rather than terrain slope channel routing, and it treats baseflow as a linear function of bottom soil-layer drainage [40]. Thus, runoff estimates from WRF are not representative of the changes in streamflow. For the purpose, we used non-parametric approaches that are demonstrated as or more robust than complex and detailed hydrologic models for evaluating the sensitivity of streamflow to climate [22,27,41]. These non-parametric approaches use the concept of climate elasticity of streamflow (ε_x), computed based on historic P, PET and Q data. The climate elasticity of streamflow is defined by the proportional change in Q to the change in a climate variable (x), such as P or PET [42]. It is an index commonly used to

quantify the sensitivity of Q to changes in climate. Often this index (i.e., ε_x) is derived from the historic climate and hydrologic data (i.e., P, PET and Q) [23,42]. Streamflow in unimpaired watersheds (i.e., watersheds in which streamflow are not subject to regulation or diversion, and defined as reference watershed in this study) can be modeled as a function of P and PET [23]. The changes in Q due to changes in P and PET can be approximated as:

$$\Delta Q/Q = \varepsilon_p \Delta P/P + \varepsilon_{pet}\, \Delta PET/PET \qquad (1)$$

In equation 1, ΔQ, ΔP and ΔPET are changes in Q, P and PET, respectively; ε_p and ε_{pet} are the elasticities of streamflow with respect to P and PET. Prior studies [23,27,43] have proposed nonparametric approaches to estimate ε_x from observed climatic data. Of the various approaches (Table S2, Figure S2) found in the literature, we have no reason to favor one over the others, and thus use the average of ε_p estimated from all available non-parametric approaches to predict average changes in Q under the alternative LULC scenario. Details about the non-parametric approaches used in this study and hydro-climatology of the conterminous U.S are discussed in Figure S3.

To compute elasticity estimates, we used historical annual streamflow, precipitation and PET information for 1,845 reference watersheds across the conterminous United States (see Figure S1).

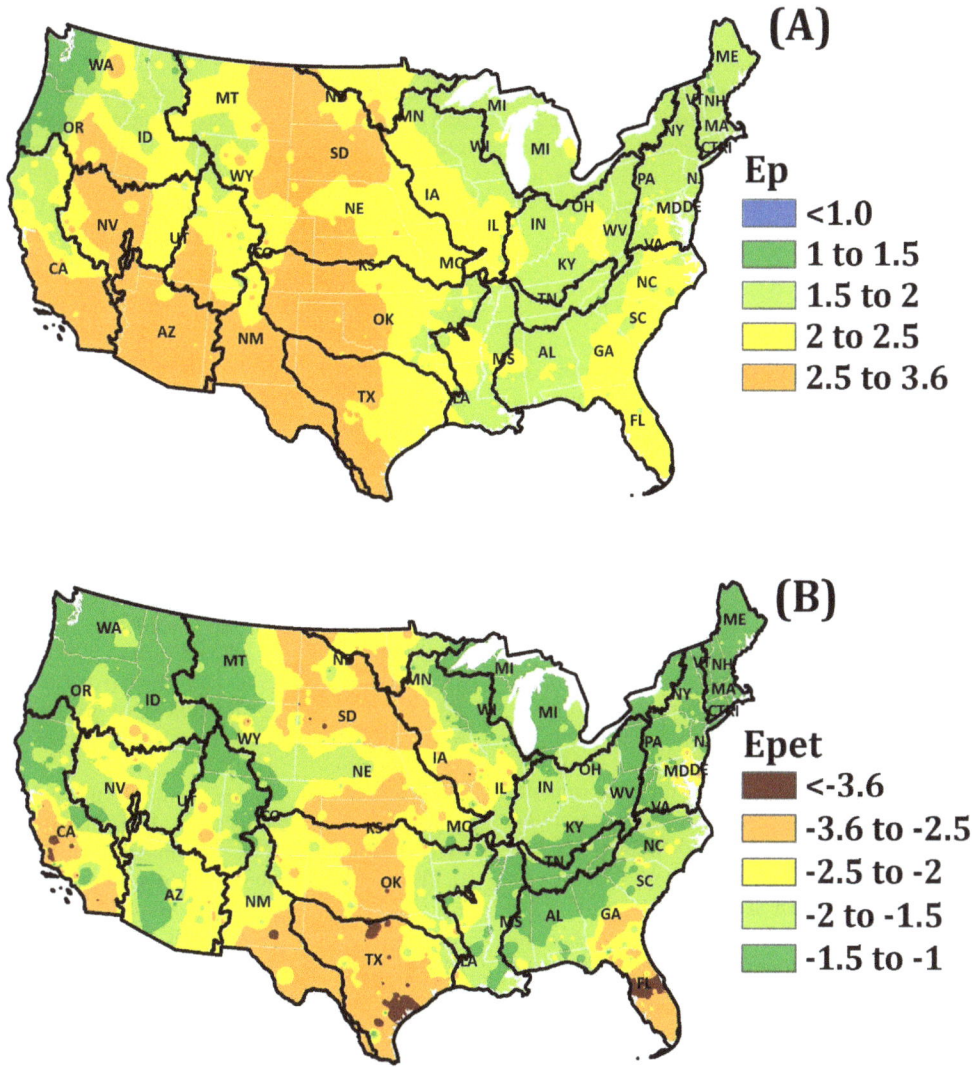

Figure 2. Hydro-climatology of the conterminous US; (A) Precipitation elasticity of streamflow (ε_p) and (B) Evapotranspiration elasticity of streamflow (ε_{pet}).

To compute ε_x estimates from the historic climate record, we used PET instead of ET similar to prior studies [22–23] due to data limitations associated with the estimation of actual ET from 1950–2009. Further, as ε_{pet} was computed using historic PET estimates, we used PET instead of ET from the regional climate model simulations to estimate changes in Q to maintain consistency in the methodology. These climate elasticity values were combined with the differences in mean annual P and PET between the biofuel and baseline scenarios, expressed as a percentage of the baseline, to compute the relative change in Q across the nation under the biofuel scenario (equation 1).

Sensitivity of streamflow change

Uncertainty in the estimated change in Q under the biofuel scenario is evaluated based on: 1) difference in elasticity estimates computed from various non-parametric approaches (discussed in the supporting material), and 2) year to year changes in simulated P and PET for the period 1981–2004. To evaluate the sensitivity of changes in streamflow to changes in both climate and elasticity estimates, we calculated the standard deviation (std) of percent annual change in P and PET between the biofuel and baseline

scenarios, and the std of elasticity estimates from the seven different non-parametric methods as shown in equations 2 and 3.

$$\Delta Q = (P \pm std\ P) \times \varepsilon_p + (PET \pm std\ PET) \times \varepsilon_{pet} \quad (2)$$

$$\Delta Q = P \times (\varepsilon_p \pm std\ \varepsilon_p) + PET \times (\varepsilon_{pet} \pm std\ \varepsilon_{pet}) \quad (3)$$

$$\Delta Q = (P \pm std\ P) \times \varepsilon_p + PET \times \varepsilon_{pet} \quad (4)$$

In equations 2, 3 and 4, P and PET indicate the percent change in precipitation and evapotranspiration under the biofuel scenario relative to the baseline. ε_p and ε_{pet} indicate the mean values of ε_p and ε_{pet} from the seven empirical methods.

The sensitivity of estimated change in Q is evaluated by varying the percent change in mean annual P and PET under the biofuel scenario (equation 2; Figure S6A and S6B), and the mean estimates of ε_{pet} and ε_p (equation 3; Figures S6C and S6D) by ±

their standard deviation. We estimated the sensitivity of predicted change in Q by varying the percent change in mean annual P and PET simultaneously because P observed in many regions including High Plains are correlated with PET [44] and ET [45]. Sensitivity measures are also computed varying only P (equation 4; Figure S7) because the impacts of varying P and PET simultaneously will tend to cancel each other in the regions where impacts are inversely correlated, thus not reflecting the full contribution of P or PET to Q (equation 4).

Results

Climate elasticity of streamflow

Precipitation elasticity of streamflow is estimated in the range of 1–3.6 with a mean of 2.2 for watersheds across the U.S., implying that a 1% change in P will result in more than a 1% change in Q. The relationship between P and Q is generally non-linear, and this non-linearity is influenced by catchment properties including storage processes, ET and vegetation properties, and these factors are implicitly factored into elasticity estimates. Approximately 46% of all watersheds examined have ε_p higher than 2, and these watersheds are clustered in the Southwest, Midwest and Southeastern parts of the nation (Figure 2A) where PET usually exceeds P. Only a few basins in the Northwest have ε_p less than 1.5.

Evapotranspiration elasticity of streamflow for watersheds across the U.S. is estimated to be in the range of negative 3.6 to negative 1, with a mean of negative 1.9, indicating that a 1% reduction in PET would result in about a 1.9% increase in Q. The geographical distribution of ε_{pet} (Figure 2B) in the conterminous U.S. is similar to the distribution of ε_p with lower values ε_{pet} in the arid and semiarid regions of the Southwest and Midwest.

Variability in climate elasticity of streamflow

The variability in elasticity estimates computed using seven different non-parametric approaches (expressed as a standard deviation) is high in the arid and semi-arid regions of the Midwest and Southwest, and lower in the humid and semi-humid regions (Figures S4A and S4B). Also, the standard deviation of ε_{pet} estimates is higher than the standard deviation of ε_p estimates. This is due to differences among the seven different approaches; five of seven approaches [46–51] depend upon aridity index (PET/P) to estimate elasticity estimates, while the other two depend on Q and P or PET [23,27] (see Table S2 for details).

Projected climate change under biofuel scenario

Change in annual precipitation and evapotranspiration. Under the biofuel scenario, the mean annual change in P relative to the baseline is projected to be in the range of negative 10% to positive 10% (Figure 3A). About 84% of the conterminous U.S. is predicted to experience changes in P in the range of negative 5% to positive 5% under the biofuel scenario; and a general decrease in P is predicted over 52% of the area. The magnitude of projected change in mean annual P in the main biofuel crop producing region (i.e., Kansas and Oklahoma) is between 2.5% to 5% relative to the baseline scenario. Under the biofuel scenario, western Kansas and Oklahoma show 5 to 15 mm higher annual P than under the baseline scenario.

The projected change in mean annual PET under the biofuel scenario is estimated to be in the range of negative 6% to positive 16% relative to the baseline (Figure 3C). Higher PET in the switchgrass planted region is consistent with lower mean temperatures due to earlier green-up and the higher LAI of switchgrass compared to current vegetation, and higher net radiation under the biofuel scenario.

In addition to the observed changes in climate in switchgrass planted region, we observed changes in climate patterns in areas away from the switchgrass concentrated area. Under the biofuel scenario, southern regions including parts of Arizona, New Mexico and Texas, the High Plains including western Kansas and Oklahoma, the Midwest including eastern part of Nebraska and Iowa, and a large region of the eastern states show an increase in annual P of between 2.5% to 10% relative to the baseline scenario. A large decline in annual P (i.e., between 2.5% to 10% relative to the baseline) is predicted across much of the northern (i.e., northern Minnesota, South Dakota, Wisconsin), Midwestern (i.e., Wyoming, Idaho, northern Colorado) and the High Plains (i.e., Missouri) regions. Due to the internal non-linear climate dynamics, a single simulation is insufficient to conclude that they are systematically caused by land use change in the Great Plains. They could, in fact, be an artifact of the initial atmospheric conditions.

PET is usually a good representation of actual ET when there is no plant water stress, and is thus commonly used in precipitation-runoff modeling applications [41]. We observed similar trends in PET and ET in the switchgrass perturbed region (Figures 4B and 4C) although they differed in magnitude. Increases in ET increase low-level humidity and the potential for more P [19]. Change in mean annual P, PET, ET and runoff when examined by land use (i.e., switchgrass altered and unaltered) categories in the High Plains, suggests that the difference in ET, PET and runoff represent a change induced by land cover perturbation (Figures 4 and S5).

Differences in climate between the baseline and biofuel projections vary annually in both magnitude and direction between 1981 and 2004. The magnitude of mean annual change in PET and ET is higher than the year to year changes in PET and ET in switchgrass planted regions under the biofuel scenario. Also, the mean annual runoff in the switchgrass dominated region is lower than other regions in Kansas and Oklahoma. Compared to land cover with a lower fraction of switchgrass (i.e., cropland/switchgrass and grassland/switchgrass), the land cover with a large fraction (i.e., >30%) of switchgrass (i.e., switchgrass/grassland and switchgrass/cropland) demonstrated a higher magnitude of change in PET, ET and runoff in the biofuel scenario (Figure 4). This indicates that the observed changes in PET, ET and runoff associated with biofuel feedstock production are large and significant. Decrease in runoff results from lower soil moisture levels due to higher evapotranspiration of switchgrass during the growing season [19]. Conversely, the inter-annual variability of change in P is as large as or larger than the magnitude of mean annual change in P (Figure 4). Thus it is hard to conclude that P is changed under the biofuel scenario.

Streamflow response to projected climate change. Mean annual change in Q in response to changes in P and PET under the biofuel scenario is shown in Figure 5. Across the conterminous U.S., the change in mean Q under the biofuel scenario is estimated to be in the range of negative 56% to positive 20% relative to the baseline scenario. An increase in Q with magnitude greater than 5% is predicted over 12% of the area. Lower PET but higher P in New Mexico and Arizona are estimated to increase Q under the biofuel scenario. However, a decrease in Q with magnitude greater than 5% is predicted over 30% of the area. The increase in P is smaller than the increase in PET under the biofuel scenario, and this causes a net decline in Q in the High Plains. In the High Plains, Q is predicted to be about 20% lower than the baseline. Streamflow in the biofuel crop region within the High Plains is 18% lower relative to the baseline (Figure 5). The switchgrass areas in the biofuel crop region show decreases in

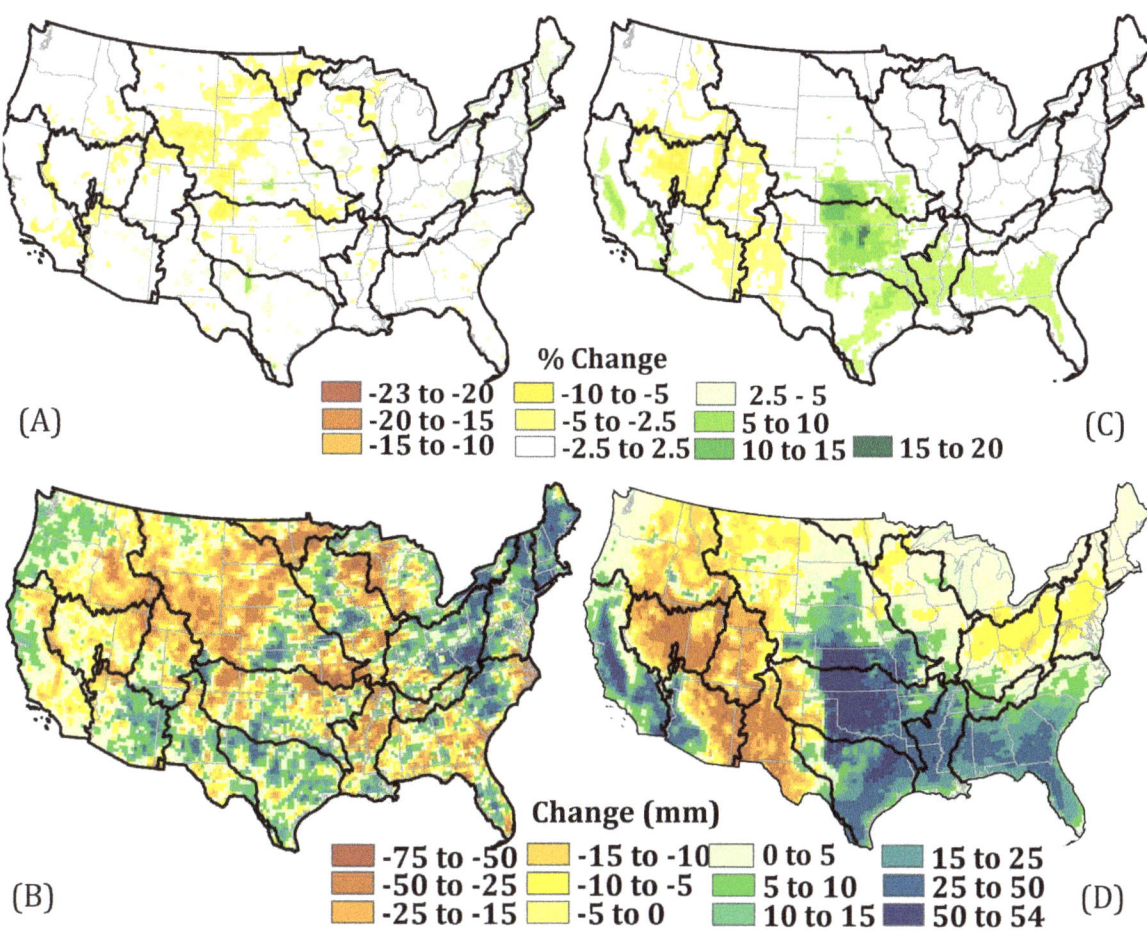

Figure 3. Change in mean annual precipitation and potential evapotranspiration for 1981–2004 expressed in A) percentage change in mean annual precipitation; B) change in mean annual precipitation (millimeters); C) percentage change in mean annual PET; and D) change in PET (millimeters) under the biofuel scenario. Percent change under the biofuel scenario relative to the baseline scenario for a given location is estimated as the mean of: $\frac{Baseline_{xi} - Biofuel_{xi}}{Baseline_{xi}} \times 100$ where x is either P or PET and i is year between 1981 and 2004.

streamflow that are twice as large as the decrease in the unperturbed area.

Sensitivity of streamflow. The change in Q under the biofuel scenario is observed to be highly sensitive to mean annual P and PET for large parts of the conterminous U.S., however, it is less sensitive in the switchgrass planted region where change in PET is larger than P (Figure S6). As parameters associated with precipitation and evapotranspiration are very likely to be correlated with each other, PET and P were changed independently in the sensitivity test to examine the relative significance of changes in P and PET on streamflow estimates (see the section 'Materials and Method' for details). Under these sensitivity analyses, Q always decreases under the biofuel scenario, even in the case when PET is held constant and P is increased by one standard deviation (Figure S7). Sensitivity analysis also showed that predicted changes in streamflow are robust to differences in ε_p and ε_{pet} stimation approaches. Change in Q across the range of ε_p and ε_{pet} estimation approach (Figure S6C and S6D) is less than the change in Q resulting from changes in P and PET equal to their inter-annual variability (Figure S6A and S6B).

Discussion

Climate change is a likely result of a policy that encourages expansion of energy crop production in the High Plains and

Midwest U.S [19]. This climate change is caused by alteration of the surface energy and moisture balances induced by changes in land cover when current cropping systems are replaced by energy crops. Notable changes include lower temperature, higher PET, lower runoff and a decline in Q in the region where switchgrass is predicted to replace current vegetation. Our analyses show changes in the mean annual PET, ET and streamflow to be a stronger climate change impact of biofuel crop production than changes in mean annual P.

The main goal of this study is to use a simple and robust model to communicate to policy makers and analysts the potential implications of biofuels policy-induced climate change on streamflow to inform the search for a sustainable renewable fuel production system. While adopting the simple but robust model for estimating streamflow changes, we made several assumptions, and these assumptions are likely to introduce uncertainties in our streamflow estimates. Despite some of the limitations (as discussed below), our conclusion that increase cellulosic feedstock production are likely to reduce water yield is found to be in agreement with other studies conducted [52,20] in other parts of the nation.

Weather conditions simulated with regional and global climate models over short periods are sensitive to their initial conditions. A change in the vegetation type at the initial condition will result in a different sequence of weather conditions. Thus, differences

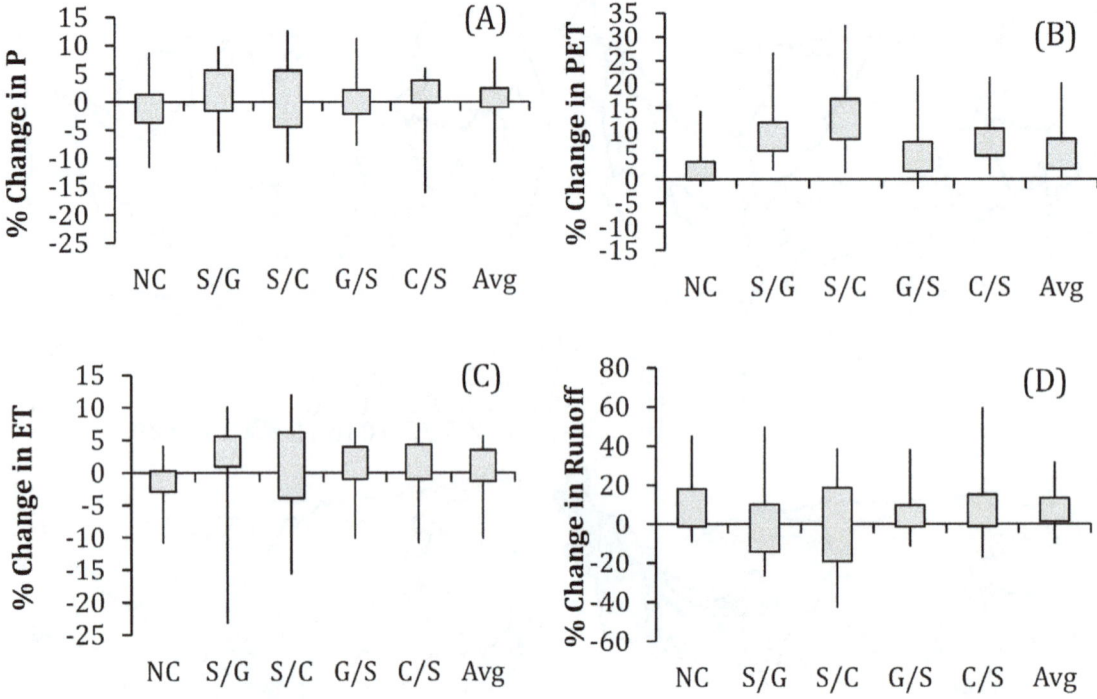

Figure 4. Percent difference in (A) annual precipitation; (B) PET averaged by land use categories in switchgrass altered regions in Kansas and Oklahoma as shown in Figure 1. Box top and bottom edges are the interquartile range of percent difference for each year, and whiskers are maximum and minimum annual values. X-axis labels are land use categories: No Change (NC), Switchgrass/Grassland (S/G), Switchgrass/Cropland (S/C), Grassland/Switchgrass (G/S), Cropland/Switchgrass (C/S), and average over all categories (Avg).

between simulations may be a combination of a transient response from unpredictable nonlinear dynamics acting upon a different initial state as well as a systematic response from a structural change in forcing of local climate, in this case land use change [53–54]. The average change for a transient response is expected to be zero given either a long data record or multiple simulations

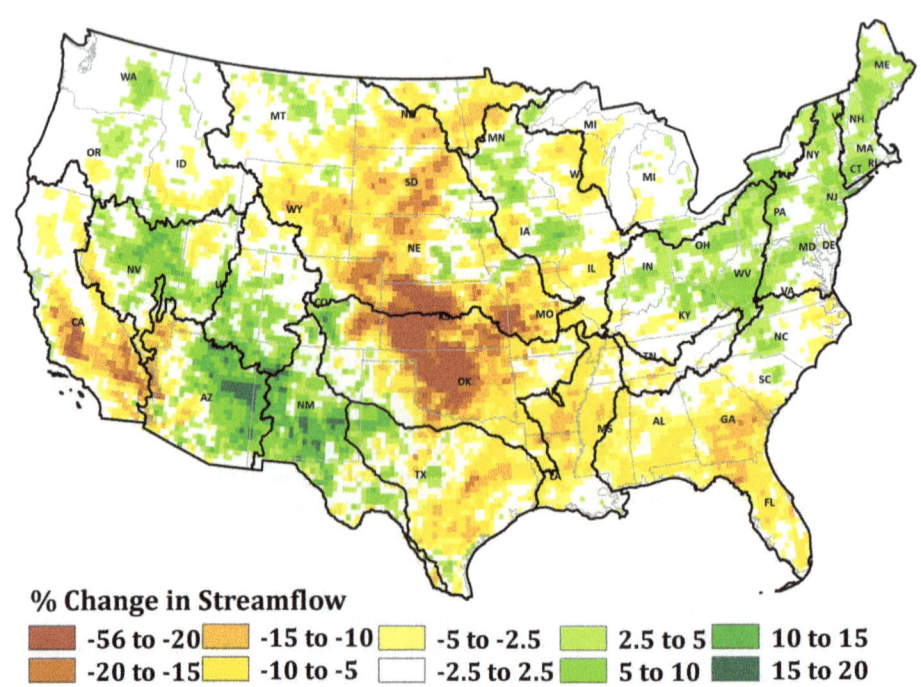

Figure 5. Percent change in mean annual streamflow as a function of change in annual precipitation and potential evapotranspiration under the biofuel scenario.

from alternative initial conditions. Changes in Q estimated in this study thus include uncertainties inherent in P and PET estimated over a relatively short time series (in this case 24 years). Anderson et al [19], using one-way ANOVA analysis, found statistically significant change of monthly values for ET and P. However, the small change in annual ET resulted from offsetting statistically significant monthly changes. To examine further whether the ET and P responses were possibly transient, they performed a second simulation of a single year of the control design beginning from a different initial condition. The difference of ET in the two control simulations was much smaller than the difference between the control and LULC scenario. Precipitation, however, showed substantial sensitivity to the initial condition. This sensitivity of P to initial conditions indicates a larger number of simulations would be needed to identify a forced response in P, if one exists at all. Therefore, we are unable to state that a forced change in P exists anywhere in our simulation domain. Since P, ET and PET are correlated, we conclude that outside of the switchgrass planted region, change in PET and Q should not be considered a forced response. Within the switchgrass planted region, however, change of PET and Q is a forced response. Another source of uncertainty in our estimates can be driven by the use of temperature based approach to estimate PET instead of approach like Penman-Monteith (see the supporting material). As temperature-based PET estimates are likely to overestimate the impacts of changes in temperature, our streamflow estimates might have been overestimated to some extent.

Another possible area of limitation includes elasticity estimates. In the study, we interpolated the elasticity estimates based on reference watershed for the conterminous US. This means, our elasticity estimates may not accurately reflect the connection of climate to Q in non-reference watersheds where streamflow is heavily influenced by land use practices and ground water pumping. Predicted change in Q in non-reference watersheds thus might be over or under estimated in the biofuel scenario. Despite this limitation, this study is useful in providing the direction of change in streamflow under the biofuel scenario without requiring the use of detailed hydrologic models that are computationally complex and often provide ambiguous results when compared [27].In the study we show a potential change in annual streamflow volume as an outcome of a landscape influenced climatic system. Our analyses also suggest that under the biofuel scenario, there is a change in seasonal P and ET. In the growing seasons (i.e., April–June), P decreases and ET increases. Evapotranspiration increases until soil moisture nears wilting point, eliminating transpiration and inhibiting further decline in soil moisture [19]. Decrease in P and increase in ET suggests the possibility for a higher magnitude of change in Q seasonally than annually, and we recommend that this possibility should be explored further with detailed hydrologic models.

This study examines landscape induced climate change and ignores projected climate change due to atmospheric concentration of greenhouse gases (GHGs). There is a strong coupling among landscape processes, atmospheric GHGs and climate, since landscape acts as a sink or source to GHGs like CO_2 which affect the distribution of heat over land and in the atmosphere and feedback to the climate [55]. However, these processes are often uncoupled when we make future projections, and this is likely to introduce biases in the projection of future changes in climate. Thus, we recommend the direct coupling of landscape and other climate forcing factors (e.g., GHGs) to better predict the range of future climate change and its impacts on environmental resources [19]. In this study, we examine landscape induced climate change and its impact on Q on an annual basis. Changing landscape and

climate alter not only Q, but also stream quality due to changes in the nutrient and sediment content of runoff, particularly from agricultural fields. Evaluation of stream quality under landscape driven climate change is thus recommended as a topic for future research.

In this study, we find interaction between land use change and climate change in Q that is not considered in previous studies. Although processes related to landscape characteristics, such as soil moisture, infiltration, and surface roughness could affect Q, but are not considered in estimation in Q in this study, we assume that the effect on streamflow of changing climate is larger than that resulting from change in landscape characteristics [56–58]. For example, Tu [56] examined change in streamflow in eastern Massachusetts under three different climate change scenarios under IPCC and land use scenarios relative to current conditions, individually and in combination. Tu [56] found that the change in streamflow under both climate and land use scenarios is similar to the streamflow changes examined under a climate change scenario only. Although the process-based hydrologic models (e.g., Soil and Water Assessment Tool (SWAT) and the Variable Infiltration Capacity (VIC)) have the ability to integrate the effects of land use change and climate change on Q and streamflow quality, they are computationally expensive given that they operate at the field- or sub-field scale and need to be calibrated with large amounts of empirical data. Using the method developed here, areas of concern can be identified and then detailed hydrologic models can be used to examine in more detail the specific trade-offs between land use and management options and streamflow.

The results here suggest that changes induced in the climate system by biofuel crop production may increase stress on the water resources of the High Plains. Under the biofuel scenario, increased ET reduces soil moisture [19], and lower soil moisture during the growing season can cause plant water stress and reduce crop yield. Irrigation can reduce water stress but additional irrigation will increase pressure on already strained water resources in arid agricultural regions such as the High Plains. The predicted 20% decline in Q in the biofuel crop producing region under the biofuel scenario would exacerbate on-going conflicts over water allocation between agriculture and other uses. As we develop and implement policies to pursue more sustainable cellulosic biofuel production, we should carefully consider potential water limitations and other impacts to the hydrologic cycle.

Supporting Information

Figure S1 Location of stream gauges used in this study and major water resource regions in the U.S.

Figure S2 Relationship between climate elasticity of streamflow and aridity index in the U.S. watersheds. Precipitation elasticity of streamflow versus aridity index (left); Evapotranspiration elasticity of streamflow versus aridity index (right).

Figure S3 Hydro-climatology of the conterminous US; (A) Aridity Index (Ø); (B) Runoff Coefficient; (C) Correlation coefficient between precipitation and streamflow; and (D) Dominant land cover of the watersheds in year 2006.

Figure S4 Standard deviation of (A) ε_p; (B) ε_{pet}; (C) mean annual change (%) in precipitation; and (D) mean annual change (%) in PET under the biofuel scenario relative to baseline scenarios. Standard deviation of elasticity estimates here reflects the variability in elasticity estimates among seven non-parametric approaches from their mean estimate, computed as $\sqrt{\frac{1}{N-1}\sum_{i=1}^{N}(x_i-\bar{x})^2}$ where $N=7$, x_i is $\varepsilon_{pet(i)}$ or $\varepsilon_{p(i)}$ where i represents a non-parametric approach and $\bar{x}$ is the mean of $\varepsilon_{pet\ (i)}$ or $\varepsilon_{p\ (i)}$ from seven non-parametric approaches for a given location. Standard deviation of precipitation and PET change under the biofuel scenario here reflects the inter-annual variability in changes in P or PET, and is computed as $\sqrt{\frac{1}{N-1}\sum_{i=1}^{N}(x_i-\bar{x})^2}$ where $N=24$, the number of years between 1981 and 2004, x_i is $\frac{Baseline_{x_i}-Biofuel_{x_i}}{Baseline_{x_i}}\times 100$ where $Baseline_{x_i}$ and $Biofuel_{x_i}$ represent P or PET in a year i in a given location under the baseline and biofuel scenarios, respectively, and $\bar{x}$ is the mean of x_i.

Figure S5 Change in mean annual runoff over 1981–2004 expressed in A) percentage change; B) absolute change (millimeters).

Figure S6 Change in streamflow volume when mean annual P and PET are varied by (A) adding and (B) subtracting a standard deviation of P and PET; (C) adding a standard deviation of elasticity (ε_p and ε_{pet}) estimates; and (D) subtracting a standard deviation of elasticity estimates among seven non-parametric approaches.

Figure S7 Streamflow prediction when mean annual P and PET are varied by (A) adding, and (B) subtracting a standard deviation of P holding PET constant at its mean annual value.

Table S1 Values of land use parameters in the NOAH land surface model as coupled within the WRF regional climate model. Text in bold face are the new land use categories and the associated parameterization of these categories used in the switchgrass scenario. Meanings of the parameters are listed below the table (from Anderson et al. [s19]).

Table S2 Mathematical expression of f(∅) and f'(∅) across the studies.

Text S1 Climate Elasticity of Streamflow.

Acknowledgments

We would like to acknowledge the valuable suggestions of Drs. Chris Kucharik and Anita Thompson. This material is based upon work supported by the National Science Foundation Grant CBET-1137677. Any opinions, findings, and conclusions or recommendations expressed in this material are those of the author(s) and do not necessarily reflects the views of the National Science Foundation.

Author Contributions

Conceived and designed the experiments: RA CA SK. Performed the experiments: CA SK. Analyzed the data: SK DH. Contributed reagents/materials/analysis tools: SK. Wrote the paper: SK RA CA.

References

1. Hill J (2007) Environmental costs and benefits of transportation biofuel production from food-and lignocellulose-based energy crops. A review. Agronomy for Sustainable Development 27: 1–12.
2. Hess JR, Foust TD, Hoskinson R, Thompson D (2003) Roadmap for agriculture biomass feedstock supply in the United States. DOE/NE-ID-11129, US Department of Energy, Washington DC, USA.
3. Costello C, Griffin WM, Landis AE, Matthews HS (2009) Impact of biofuel crop production on the formation of hypoxia in the Gulf of Mexico. Environ Sci Technol 43: 7985–7991.
4. Donner SD (2007) Surf or turf: A shift from feed to food cultivation could reduce nutrient flux to the Gulf of Mexico. Global Environ Change 17: 105–113.
5. Schnoor J, Doering III OC, Entekhabi D, Hiler EA, Hullar TL, et al. (2008) Water implications of biofuels production in the United States. National Academy of Sciences, Washington DC, USA.
6. Tilman D, Hill J, Lehman C (2006) Carbon-negative biofuels from low-input high-diversity grassland biomass. Science 314: 1598–1600.
7. U.S. Department of Energy (2011) U.S. Billion-Ton Update: Biomass Supply for a Bioenergy and Bioproducts Industry. Perlack, R.D., Stokes, B.J.; Leads, ORNL/TM-2011/224. Oak Ridge National Laboratory, Oak Ridge, TN. 227.
8. Marshall E, Caswell M, Malcom S, Motamed M, Hrubovcak J, et al. (2011) Measuring the indirect land-use change associated with increased biofuel feedstock production. A review of modeling efforts. US Department of Agriculture-ERS, AP054. Washington DC, USA.
9. Schmer MR, Vogel KP, Mitchell RB, Perrin RK (2008) Net energy of cellulosic ethanol from switchgrass. PNAS 105: 464–469.
10. Diffenbaugh N (2009) Influence of modern land cover on the climate of the United States. Climate Dynamics 33: 945–958.
11. Pielke Sr RA, Adegoke J, Beltran-Przekurat A, Hiemstra CA, Lin J, et al. (2007) An overview of regional land-use and land-cover impacts on precipitation. Tellus B 59: 587–601.
12. Twine TE, Kucharik CJ, Foley JA (2004) Effects of land cover change on the energy and water balance of the mississippi river basin. J Hydrometeor 5: 640–655.
13. Bonan GB (2001) Observational evidence for reduction of daily maximum temperature by croplands in the Midwestern United States. J Clim 14: 2430–2442.
14. Halgreen W, Schlosser CA, Monier E, Kicklighter D, Sokolov A, et al. (2013) Climate impacts of a large-scale biofuels expansion. Geophys Res Lett 40: 1624–1630.
15. Georgescu M, Lobell D, Field C (2011) Direct climate effects of perennial bioenergy crops in the United States. PNAS 108(11): 4307–4312.
16. Mishra V, Cherkauer KA, Niyogi D, Lei M, Pijanowski BC, et al. (2010) A regional scale assessment of land use/land cover and climatic changes on water and energy cycle in the upper Midwest United States. Int. J. of Climatology 30(13): 2025–2044.
17. Vanloocke A, Bernacchi CJ. Twine TE (2010) The impacts of Miscanthus× giganteus production on the Midwest US hydrologic cycle. GCB Bioenergy, 2, 180–191. Skamarock WC, Klemp JB (2008) A time-split nonhydrostatic atmospheric model for weather research and forecasting applications. Journal of Computational Physics 227: 3465–3485.
18. Pielke Sr RA (2005) Land use and climate change. Science 310: 1625–1626.
19. Anderson CJ, Anex RP, Arritt RW, Gelder BK, Khanal S, et al. (2013) Regional climate impact of a biofuels policy projection. Geophys Res Lett 40: 1217–1222. doi:10.1002/grl.50179.
20. Le PVV, Kumar P, Drewry DT (2011) Implications for the hydrologic cycle under climate change due to the expansion of bioenergy crops in the Midwestern United States. PNAS 108: 15085–15090.
21. Stone K, Hunt P, Cantrell K, Ro K (2010) The potential impacts of biomass feedstock production on water resource availability. Bioresource Technology 101: 2014–2025.
22. Chiew F, Teng J, Vaze J, Kirono D (2009) Influence of global climate model selection on runoff impact assessment. Journal of Hydrology 379: 172–180.
23. Zheng H, Zhang L, Zhu R, Liu C, Sato Y, et al. (2009) Responses of streamflow to climate and land surface change in the headwaters of the Yellow River Basin. Water Resources Research 45 W00A19.
24. Anderson CJ (2012) Local adaption to changing flood vulnerability in the Midwest. Climate change in the Midwest; Pryor, S.C., Eds.; Indiana University Press, ISBN: 978-0-253-00682-0.
25. Carlisle DM, Wolock DM, Meador MR (2011) Alteration of streamflow magnitudes and potential ecological consequences: a multiregional assessment. Front Ecol Environ 9(5): 264–270.
26. Vogel RM, Wilson I, Daly C (1999) Regional regression models of annual streamflow for the United States. Journal of Irrigation and Drainage Engineering 125: 148–157.
27. Sankarasubramanian A, Vogel RM, Limbrunner JF (2001) Climate elasticity of streamflow in the United States. Water Resources Research 37: 1771–1781.
28. Skamarock W, Klemp JB (2008) A time-split nonhydrostatic atmospheric model for weather research and forecasting applications. J. of Computational Physics 227(7): 3465–3485.

29. Skamarock W, Klemp JB, Dudhia J, Gill DO, Barker DM, et al. (2008) A description of the Advanced Research WRF version 3. NCAR Technical Note NCAR/TN-475 STR.

30. Ek M, Mitchell K, Lin Y, Rogers E, Grunmann P, et al. (2003) Implementation of NOAH land surface model advances in the National Centers for Environmental Prediction operational mesoscale Eta model. J Geophys Res 108: doi:10.1029/2002JD003296.

31. Steiner AL, Pal JS, Rauscher SA, Bell JL, Diffenbaugh NS, et al. (2009) Land surface coupling in regional climate simulations of the West African monsoon, Climate Dynamics, 33(6):869–892.

32. Elía R, Caya D, Côté H, Frigon A, Biner S, et al. (2008) Evaluation of uncertainties in the CRCM-simulated North American climate. Climate Dynamics 30: 113–132.

33. Ugarte DG De La Torre, Ray DE (2000) Biomass and bioenergy applications of the POLYSYS modeling framework. Biomass Bioenergy 18: 291–308.

34. Ray DE, Moriak TF (1976) POLYSIM: A National Agricultural Policy Simulator. Agricultural Sector Models for the United States 28: 14–21.

35. Dicks MR, Campiche J, Ugarte DDLT, Hellwinckel C, Bryant HL, et al. (2009) Land use implications of expanding biofuel demand. J. of Agric. And App. Econ. 41(2): 435–453.

36. Khanal S, Anex RP, Anderson CJ, Herzmann DE, Jha MK (2013) Implications of biofuel policy-driven land cover change for rainfall erosivity and soil erosion in the United States. GCB Bioenergy: doi:10.1111/gcbb.12050.

37. Miguez F, Maughan M, Bolero GA, Long SL (2012) Modeling spatial and dynamic variation in growth, yield, and yield stability of the bioenergy crops Mixcanthus x giganteus and Panicum virgatum across the conterminous United States, Global Change Biology-Bioenergy doi:10.1111/j.1757-1707.2011.01150.x.

38. Mitchell R, Schmer MR (2012) Switchgrass harvest and storage. In A. Monti (ed.) Switchgrass: A valuable biomass crop for energy (Green Energy and Technology), pp. 113–127.

39. Chen F, Dudhia J (2001) Coupling an Advanced Land Surface-Hydrology Model with the Pen State-NCAR MM5 Modeling System. Part I: Model Implementation and Sensitivity. Monthly Weather Review1 29: 569–585.

40. Rosero E, Gulden LE, Yang Z, Goncalves L, Nui G, et al. (2011) Ensemble Evaluation of Hydrologically Enhanced Noah-LSM: Partitioning of the Water Balance in High-Resolution Simulations over the Little Washita River Experimental Watershed. J. Hydrometeor 12: 45–64.

41. Chiew FHS (2006) Estimation of rainfall elasticity of streamflow in Australia. Hydrological Sciences 51: 613–625.

42. Schaake JC (1990) From climate to flow. In: Climate change and US water resources, Waggoner, P.E., Eds.; J. Wiley and Sons, pp. 177–206.

43. Arora VK (2002) The use of the aridity index to assess climate change effect on annual runoff. Journal of Hydrology 265: 164–177.

44. Zhang L, Dawes W, Walker G (2001) Response of mean annual evapotranspiration to vegetation changes at catchment scale. Water Resources Research 37: 701–708.

45. Koster RD, Dirmeyer PA, Guo Z, Bonan G, Chan E, et al. (2004) Regions of strong coupling between soil moisture and precipitation. Science 305: 1138–1140.

46. Zhang L, Dawes W, Walker G (2001) Response of mean annual evapotranspiration to vegetation changes at catchment scale. Water Resources Research 37: 701–708.

47. Pike J (1964) The estimation of annual run-off from meteorological data in a tropical climate. Journal of Hydrology 2: 116–123.

48. Budyko MI (1963) Evaporation under natural conditions, Israel Program for Scientific Translations; [available from the Office of Technical Services, US Dept. of Commerce, Washington].

49. Turc L (1954) The water balance of soils. Relation between precipitation evaporation and flow. Ann Agron. 5: 491–569.

50. Ol'Dekop E (1911) On evaporation from the surface of river basins. Trans. Met. Obs. lurevskogo, Univ. Tartu, 4.

51. Schreiber P (1904) Über die Beziehungen zwischen dem Niederschlag und der Wasserführung der Flüsse in Mitteleuropa. Meteorologische Zeitschrift 21: 441–452.

52. Wu Y, Liu S (2012) Impacts of biofuels production alternatives on water quantity and quality in the Iowa River Basin. Biomass and Bioenergy 36: 182–191.

53. Braun M, Caya D, Frigon A, Slivitzky M (2011) Internal variability of Canadian RCM's hydrological variables at the basin scale in Quebec and Labrador. Journal of Hydrometeorology 13: 443–462.

54. Solomon S, Qin D, Manning M, Chen Z, Marquis M, et al. (2007) Climate Change 2007: The Physical Science Basis, IPCC Fourth Assessment Report, Miller, H.L., Eds.; Cambridge University Press, United Kingdom and New York, NY, USA, pp. 996.

55. Marland G, Pielke RA Sr, Apps M, Avissar R, Betts RA, et al. (2003) The climatic impacts of land surface change and carbon management, and the implications for climate-change mitigation policy. Climate Policy 3: 149–157.

56. Tu J (2009) Combined impact of climate and land use changes on streamflow and water quality in eastern Massachusetts, USA. Journal of Hydrology 379: 268–283.

57. Guo G, Hu Q, Jiang T (2007) Annual and seasonal streamflow responses to climate and land-cover changes in the Poyang Lake Basin China. Journal of Hydrology 355: 106–122.

58. Hu Q, Willson GD, Chen X, Akyuz A (2004) Effects of climate and landcover change on stream discharge in the Ozark highlands, USA. Environ. Model Assess 10: 9–19.

11

Comparative Landscape Genetics of Three Closely Related Sympatric Hesperid Butterflies with Diverging Ecological Traits

Jan O. Engler[1,2]*, **Niko Balkenhol**[2], **Katharina J. Filz**[3,4], **Jan C. Habel**[5], **Dennis Rödder**[1]

1 Zoological Research Museum Alexander Koenig, Bonn, Germany, **2** Department of Wildlife Sciences, University of Göttingen, Göttingen, Germany, **3** Department of Biogeography, Trier University, Trier, Germany, **4** Museum of Natural History Dortmund, Dortmund, Germany, **5** Department of Ecology and Ecosystemmanagement, Technical University Munich, Freising-Weihenstephan, Germany

Abstract

To understand how landscape characteristics affect gene flow in species with diverging ecological traits, it is important to analyze taxonomically related sympatric species in the same landscape using identical methods. Here, we present such a comparative landscape genetic study involving three closely related Hesperid butterflies of the genus *Thymelicus* that represent a gradient of diverging ecological traits. We analyzed landscape effects on their gene flow by deriving inter-population connectivity estimates based on different species distribution models (SDMs), which were calculated from multiple landscape parameters. We then used SDM output maps to calculate circuit-theoretic connectivity estimates and statistically compared these estimates to actual genetic differentiation in each species. We based our inferences on two different analytical methods and two metrics of genetic differentiation. Results indicate that land use patterns influence population connectivity in the least mobile specialist *T. acteon*. In contrast, populations of the highly mobile generalist *T. lineola* were panmictic, lacking any landscape related effect on genetic differentiation. In the species with ecological traits in between those of the congeners, *T. sylvestris*, climate has a strong impact on inter-population connectivity. However, the relative importance of different landscape factors for connectivity varies when using different metrics of genetic differentiation in this species. Our results show that closely related species representing a gradient of ecological traits also show genetic structures and landscape genetic relationships that gradually change from a geographical macro- to micro-scale. Thus, the type and magnitude of landscape effects on gene flow can differ strongly even among closely related species inhabiting the same landscape, and depend on their relative degree of specialization. In addition, the use of different genetic differentiation metrics makes it possible to detect recent changes in the relative importance of landscape factors affecting gene flow, which likely change as a result of contemporary habitat alterations.

Editor: Daniele Canestrelli, Tuscia University, Italy

Funding: JOE received financial support by the German Federal Environmental Foundation (DBU) fellowship programme. KJF was funded by the Friedrich-Ebert-Stiftung. The funders had no role in study design, data collection and analysis, decision to publish, or preparation of the manuscript.

Competing Interests: The authors have declared that no competing interests exist.

* Email: j.engler.zfmk@uni-bonn.de

Introduction

In the theory of island biogeography, McArthur & Wilson [1] predicted the evolution of biodiversity on islands based on two key factors: habitat size and isolation. Later, this island based model was adopted to explain population structure of organisms in mainland ecosystems consisting of habitat patches surrounded by a semi- or non-permeable matrix. This mainland transformation of the theory of island biogeography inspired the fundamental paradigm of the metapopulation concept [2–3] and also of the neutral theory in both macroecology and population genetics [4–5]. Ultimately, island biogeography theory revolutionizes our thinking on habitat fragmentation and conservation biology (summarized in [6]). Apart from habitat size and isolation, spatial biodiversity patterns are also influenced by additional factors such as habitat quality [7], intrinsic characteristics of species-specific dispersal behaviour [8–9] and ecological tolerance [10] of species. Importantly, population responses are highly species-specific, when the quality of the landscape matrix in between suitable habitat patches is reduced [11]. This would also have consequences for global biodiversity [12–13] and large scale conservation efforts [14].

Understanding the effects of the landscape matrix on realized dispersal and functional population connectivity is also a major focus of landscape genetics [15–17]. Incorporating spatial landscape information with population genetic data goes far beyond the classical analysis of isolation-by-distance (IBD) [18]. Species respond differently to the landscape, in terms of their dispersal, which ultimately affects the rates of gene flow among local populations [19–20]. While the classical isolation-by-distance approach introduced by Wright [18] accounts for the geographic (Euclidean) distance between sampled populations only, other approaches such as the recently proposed isolation-by-resistance (IBR) concept [21] accounts for these species-specific responses to different landscape components that impede or favor gene-flow across a given landscape matrix.

Many studies assess landscape effects on gene flow in only a single species. However, to understand how landscape effects on gene flow differ between species, and to take effective conservation actions, it is important to analyze multiple species in the same landscape using identical methods [22]. However, past studies comparing different species mostly focused on species that inhabited comparable habitats, but were taxonomically independent [19–20,23]. A different comparative approach is to analyze landscape genetic relationships in closely related taxa inhabiting the same landscape. Such a focus on taxonomically related sympatric species (i.e. congeneric species which have the same or overlapping geographic ranges, regardless of whether or not they co-occur at the same locality) allows the assessment of traits that gradually change between the congeners independently from confounding effects that may arise in relation to different evolutionary histories or environments, respectively [24–25]. Next to dispersal propensity, niche breadth (i.e. the degree of specialization on specific habitat traits) is a very important trait in this respect, as it is directly associated with the available habitat within a landscape.

Generalist species can be found in a broader variety of ecosystems, showing higher abundances and broader spatial distributions. In contrast, specialist species demanding certain habitat conditions are often geographically restricted to specific habitats and usually occur in lower local abundances [26]. Apart from ecological demands, connectivity among local populations is further influenced by the dispersal propensity of species. Typically, sedentary species are mostly characterized by rather limited individual exchange compared to species with strong dispersal behavior. These ecological and behavioral traits also affect the genetic structure of generalist versus specialist species [10,26–27]. Organisms with specific habitat demands and restricted dispersal behavior should generally be characterized by low gene flow resulting in strong genetic differentiation. In contrast, species with weaker habitat specificity and higher dispersal propensity should show increased levels of gene flow, leading to a lack of genetic differentiation. Importantly, landscape influences on gene flow and resulting genetic patterns could also differ between generalist and specialist species inhabiting the same landscape.

In this study, we present a comparative landscape genetic analysis involving three closely-related butterfly species, to assess the impact of landscape parameters (i.e. land use, topography and climatic conditions) on the genetic structure of sympatric species with different ecological traits. We re-analyzed a molecular dataset taken from a previous study [28], where landscape effects were previously ignored, involving three congeneric, but ecologically divergent skipper species of the genus *Thymelicus* (Hubner 1890). The three species include the generalist *T. lineola*, which occurs in high abundances and shows strong dispersal propensity; the specialist *T. acteon* which is sedentary and occurs restricted to specific habitats; and *T. sylvestris*, which lies in between these two extremes in terms of habitat specificity and dispersal abilities. Using these three species, we (i) investigate the impact of ecological traits on species-specific functional landscape connectivity and (ii) determine the overall relevance of landscape characteristics for connectivity in each species, as well as the individual importance of topography, climatic conditions and land-use parameters. We hypothesized that species-specific landscape effects on gene flow should follow the cline of specialization in the three Hesperid butterflies, with strongest landscape effects on genetic differentiation in the most specialized *T. acteon* and weakest landscape effects in the generalist *T. lineola*.

Material and Methods

Ethics statement

The research was conducted under permission, to collect butterflies and to work in several protected areas, by the local authorities of Saarbrücken (Germany, Saarland), Koblenz (Germany, Rhineland-Palatinate), Luxembourg, and Metz (Loraine, France). Imagoes of the respective species were stored in liquid nitrogen until genetic analysis.

Study area and species

Our study area is located in the south-west of Germany and includes adjacent parts of France and Luxembourg (Fig. 1, Table S1). The sampling sites covered an area of approximately 120 km in north-south direction and 100 km in east-west direction. The landscape is characterised by a mosaic of residential areas, agricultural land, meadows, forests and semi-natural calcareous grasslands. Especially grasslands, but also some meadows and forest skirts provide suitable habitats for the three selected *Thymelicus* species, acting as valuable retreats and stepping stones [29].

The three selected model species *T. sylvestris*, *T. lineola* and *T. acteon* are closely related to each other with *T. lineola* and *T. acteon* being most distant related and where *T. sylvestris* clusters to a monophylum with *T. acteon* (Material S1). The three species show different habitat demands, even if they are co-occurring at suitable grassland patches: *T. lineola* occupies a broad ecological niche [30] and exhibits strong dispersal behaviour [31]. This combination of wide occurrence and strong dispersal behaviour results in a widespread, spatially continuous distribution. In contrast, *T. acteon* demands specific habitat characteristics like xerothermic climatic conditions and consequently occurs only in highly restricted, geographically disjunct calcareous grasslands. The third, intermediate species, *T. sylvestris* stands in-between both extremes, showing a broad ecological tolerance [30], similar to the generalist *T. lineola*, but shows a rather restricted dispersal behaviour [31].

Molecular data and genetic cluster analysis

For our comparisons, we used a population genetic dataset based on 15 polymorphic allozymes published previously by [28] who did not account for landscape effects. Several studies have shown that the implications as drawn from allozymes and, where available, microsatellites loci were highly congruent in butterflies [32–34]. Here, the use of allozymes instead of other marker systems such as microsatellites has two advantages. 1) In Lepidopterans, locus-specific microsatellites are difficult to find and suitable polymorphic loci are consequently rare [35–38]. This is most likely due to almost identical flanking regions in the Lepidopteran microsatellite DNA [36,39]. However, specificity of these regions is a crucial prerequisite for successful primer annealing [39]. 2) From a landscape genetic perspective, the use of potentially adaptive marker systems might be beneficial in the detection of spatial genetic differentiation in contrast to neutral marker systems, because spatial signals in markers under selection would appear more rapidly [40].

The data set comprised in total 1,063 individuals (417 *T. sylvestris*, 380 *T. lineola*, 160 *T. acteon*) sampled at 12 locations which were distributed across the same study area. Sample sizes ranged from 17 to 44 individuals per species and location. *Thymelicus sylvestris* and *T. lineola* were sampled at identical locations, while *T. acteon* was not found at four of the sampled locations and the data set was supplemented by one additional location (Fig. 1). The 15 enzyme systems provide the following 18 loci: MDH (2 loci), G6PDH, ACON, MPI, AAT (2 loci), FUM, PGI, ME, HBDH, APK, PGM, 6PGDH, IDH (2 loci), GPDH and PEP$_{Phe-Pro}$. Most of these enzymes showed polymorphisms, except enzyme ME in *T. lineola* and GPDH in *T. sylvestris*. Details about the analytical procedure and the specific running conditions are given in [28]. We used the resulting dataset to

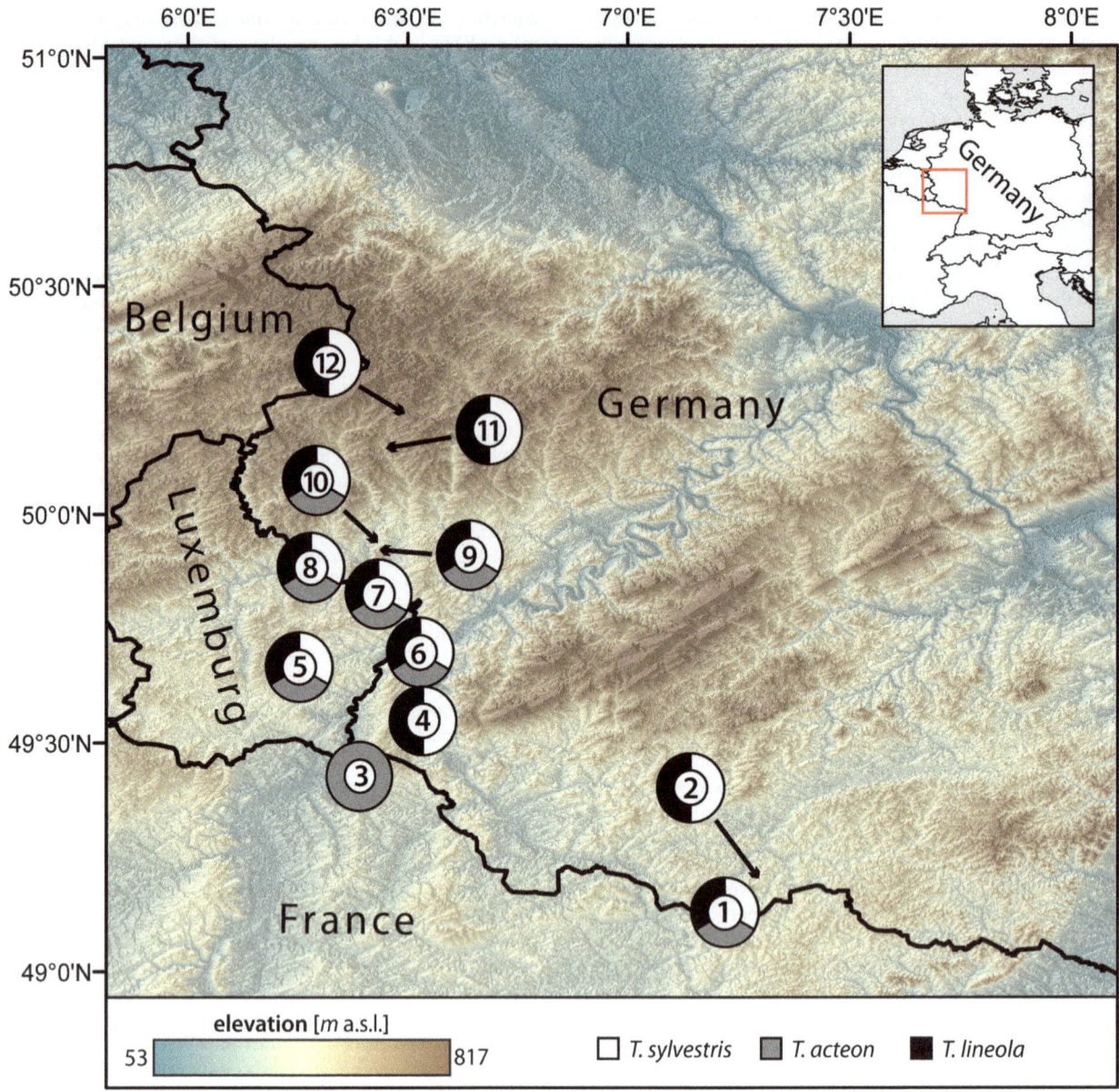

Figure 1. Locations of populations studied for all three *Thymelicus* species in southwestern Germany and adjoining areas in France and Luxemburg.

estimate pairwise F_{ST} and D_{est} for each species in programmes ARLEQUIN 3.1 [41] and SMOGD [42], respectively. The use of these two different measures of inter-population differentiation was recently recommended [43], because of the different underlying assumptions of either measure so that their combination might provide a more detailed impression into the underlying evolutionary processes of differentiation (see [43] and discussion in this study for further details). Tests for Hardy-Weinberg equilibrium and summary statistics for genetic diversity and differentiation were also calculated in ARLEQUIN 3.1.

Prior to inferring landscape effects on genetic differentiation, the number of genetic groups (K) as well as their spatial delineation was evaluated for each species separately using the genetic clustering method implemented in the software GENELAND [44]. This was necessary because (*i*) genetic differences can occur without any obvious landscape pattern (e.g. along secondary

contact zones after postglacial expansion from distinct refugia or through anthropogenic introductions from another source population), which in turn would lead to (*ii*) erroneous conclusions on isolation-by-distance IBD/isolation-by-resistance IBR analyses on spatially independent structured data. GENELAND assigns georeferenced individuals to genetics clusters (K) that maximize Hardy-Weinberg-and Linkage-Equilibrium. K was treated as unknown to allow GENELAND to vary K within a given range between 1 and the maximum number of populations depending on the species (i.e. 7 in *T. acteon* and 11 in both *T. sylvestris* and *T. lineola*). Markov Chains were run for 3,000,000 generations and sampled every 1000th generation, after an initial burn-in of 300 samples after thinning (10%). Markov Chains with these settings were run 10 times independently and the iteration with the highest log posterior probability was chosen for inferring the most likely K and individual assignments.

Modelling landscape effects on genetic differentiation

To test for landscape influences on genetic differentiation in each species, we modeled multiple species distribution models (SDM) incorporating topographic, bioclimatic and/or land use features. We then used resulting SDMs as resistance surfaces to derive inter-population connectivity estimates based on electrical circuit theory, and statistically compared these connectivity estimates to actual genetic differentiation. SDMs are increasingly applied for resistance surface parameterization in landscape genetic studies [23,45] even under longer evolutionary time scales [46,47], since they avoid the subjective parameterization of resistance surfaces which was criticized in the past [48].

Species records

To model SDMs for the three *Thymelicus* species in the study area, presence data were taken from personal observations of JCH, D. Louy and T. Schmitt (Germany) covering the years 2003–2012. Further presence data were added from high resolution records downloaded from the GBIF database (www.gbif.org). The final datasets comprised 67 records for *T. sylvestris*, 62 for *T. lineola* and 28 records for *T. acteon*. Given their specific habitat demands and the sampling effort that was performed across the study area for either species (Fig. 1), we are confident to have compiled a representative sample that covered the realized distribution of the species in our study area.

Environmental layers

For construction of the SDMs, we used freely available GIS based environmental layers. Bioclimatic data based on monthly averaged temperature and precipitation data with 30 arc seconds spatial resolution was obtained from the Worldclim Database (Vers. 1.4; www.worldclim.org; [49]). The comprehensive set of 19 bioclimatic variables are thought to be highly relevant for shaping species' Grinnellian (abiotic) niches [50]. In order to minimize the degree of inter-correlation among the variables (i.e. to keep pair-wise Pearson's $R^2 < 0.75$), we selected a subset of variables (bio3, 7, 8, 9, 10, 11, 12, 15, 18, see Tab. 3/Appendix S3 for definitions) which were assumed to be most relevant for the study species. Topography-related data were derived from the SRTM Shuttle mission in 90 meters resolution (available through USGS seamless server; Table S2). Based on the altitude layers, we calculated slope and aspect using ArcGIS 9.3 (ESRI Redlands, California, USA). Finally, CORINE land use related data was obtained from the European Environmental Agency (www.eea.europa.eu). We either used CORINE 2006 data to assess current habitat availability as well as CORINE 1990 data for assessing recent land use changes. All environmental layers were re-sampled to uniform grid resolution of 90 m.

Calculating the Potential Connectivity Model

We defined a set of hypotheses based on the available environmental data and generated five variable sets for comparing landscape effects on species-specific gene flow (therein called scenarios, Table S2). These scenarios represent various habitat characteristics (i.e. climate, topography and land use) that were found to be important for butterfly distributions at different spatial scales in previous studies [51–57]. Based on these variable sets and the respective species records, we computed species distribution models (SDMs) with the software Maxent 3.3.3e [58] to generate maps displaying habitat suitability for each species under a given scenario. As many other presence-pseudoabsence SDM algorithms, Maxent links environmental conditions at presence records of a given taxon to those environmental conditions available within a specific geographic area (background) to predict spatial patterns of environmental suitability. The SDM output represents the likelihood of species potential occurrence across a geographic area of interest (projection; for a detailed description see: [59]). We used Maxent instead of other available algorithms because it frequently outperforms other approaches [60–61], even if the number of presence locations is rather limited [62–63]. We ran Maxent with the default settings but used a bootstrap approach, which allows random selection of 70% of presence locations for model training and the remaining 30% for model testing. This procedure was repeated for 100 times and an averaged map of suitable habitats was generated across all repetitions. As output we selected the logistic format which ranges linearly from 0 (not suitable) to 1 (fully suitable). For model evaluation, the area under the receiver operating characteristic curve (AUC) was used [64]. In particular, the AUC as internally computed in Maxent is a measure for the ability of the model to distinguish the given presence records from the background data accounting for the proportion of the study area which is predicted to be suitable for the target species [58]. The AUC ranges between 0.5 (random prediction) to 1.0 (perfect discrimination between presence and pseudo-absence).

For the land use change scenario, we used land use data from CORINE 2006 as a categorical environmental layer - just as we had done for the land use scenario. However, we subsequently projected the model fit onto the CORINE 1990 layer to assess habitat change in terms of a stability surface. The stability surface is the average of both CORINE layers, with high values indicating suitable habitat patches that remain stable over the 16 years time span, whereas low values represent low habitat suitability, a strong habitat change in time, or both. This approach for calculating stability surfaces is commonly used to estimate land use change and habitat suitability across time (see [46–47] for examples).

The resulting SDMs were used as conductance surfaces (i.e. high values indicate good conductivity between two sites, whereas low values indicate poor conductivity [65]) in Circuitscape v.3.4.1 to calculate resistances to movement and gene flow among sampling locations [65]. Circuitscape is based on electrical circuit theory, which was recently adapted from electrical engineering for the assessment of landscape ecological questions [65]. In particular, Circuitscape defines nodes (grid cells) and associated unit resistors (the resistance value) that connecting two nodes and calculates resistance distances between focal locations based on a nodal analysis algorithm as described in [21]. As the habitat matrix had a very high extent (i.e. ~7.6 Mio. cells), we chose a four-neighbor-connection scheme in order to meet the available computational capacities. It has been previously shown that four and eight-neighbor-connection scheme lead to highly similar outcomes [66].

Comparing connectivity estimates with genetic data

Resulting resistance values among locations were statistically compared to estimates of genetic differentiation (i.e. F_{ST} and D_{est}) using linear regression models as well as multiple regressions on distance matrices (MRDM) [67] in R v.2.14.1 [68]. For the linear regression models, the Akaike Information Criterion corrected for small sample sizes (AICc) was used for model comparisons within each species [69]. Despite their sensitivity for non-independence in pair-wise comparisons, multi-model inference based on information theory has been frequently applied in landscape genetic analyses [19–20] as the error entering the comparison was assumed to be equal for each model, which did not affects model ranking and thus still allows for assessing the relative model performance. To ascertain results obtained with the AIC model selection, we also estimated significance of MRDM models using 1,000 permutations. For MRDMs, the *ecodist* package for R was used [70].

Results

Genetic structures

No significant deviation from Hardy-Weinberg equilibrium was detected for any population in the respective species. Genetic diversity was comparatively low in *T. lineola* (mean ± SE; $AR = 1.78 \pm 0.17$ $H_E = 9.6 \pm 2.1$, $H_O = 9.2 \pm 2.1$), while *T. acteon* showed highest genetic diversities ($AR = 1.88 \pm 0.18$, $H_E = 14.9 \pm 2.9$, $H_O = 12.5 \pm 2.6$). *Thymelicus sylvestris* showed an intermediate level of genetic diversity, as compared to its congeners ($AR = 1.80 \pm 0.10$, $H_E = 11.9 \pm 1.5$, $H_O = 11.0 \pm 1.4$). The genetic differentiation was low in *T. lineola* ($F_{ST} = 0.0081$; $D_{est} = 0.0012$; p = n.s.), while we detected highest genetic differentiation for *T. acteon* ($F_{ST} = 0.0718$; $D_{est} = 0.0143$; p<0.0001). Again, *Thymelicus sylvestris* showed an intermediate level of genetic differentiation, with a rather low among-population variance ($F_{ST} = 0.0179$; $D_{est} = 0.0039$; p<0.0001) (Table 1).

Genetic clustering results

The posterior density and log-likelihood levels of all GENELAND runs stabilized long before the end of the Markov Chains, indicating that convergence was reached (Figure S1). For each of the species, all 10 replicate MCMC runs converged on K = 1 panmictic cluster (Appendix S4), indicating no absolute barriers affecting IBD or IBR assumptions.

Species Distribution Models

AUC values derived from the SDMs ranged from 'poor' (AUC = 0.66, scenarios 'land use' and 'land use change' in *T. sylvestris*, Table 2) to 'good' (AUC = 0.86, scenario 'all' in *T. lineola*, Table 2) according to the classification scheme for model quality from [71] adapted from [64]. Variable contributions in multi-factorial SDMs (scenarios 'climate', 'topography' and 'all') differed between species (Table 3). For the topography scenario, *slope* contributed most to the SDM in all three species, followed by *aspect* and *altitude* (Table 3). In *T. acteon* a different set of variables had higher explanative power with respect to the climate scenario. Here, *precipitation of the warmest quarter* (bio18) was most important, followed by a set of temperature related variables (bio3, 7, 8, 9, 11; Table 3). In contrast, *Thymelicus lineola* and *T. sylvestris* had very similar variable contributions as a result of the highly similar distribution of occurrence records. In these species, the *mean temperature of the coldest quarter* followed by the *temperature annual range* contributed to more than half of the

total model (Table 3). Finally, considering the entire predictor set, a combination of *slope* and *land use* contributed most in all species, but where *T. lineola* and *T. sylvestris* had again more similar variable contributions rather than *T. acteon* (Table 3). In accordance, *T. lineola* and *T. sylvestris* showed similar potential distributions containing large continuous areas of high suitability, whereas *T. acteon* shows a highly patchy distribution with large unsuitable areas surrounding potential habitat patches (Fig. 2).

Landscape effects of genetic differentiation

Results obtained with the various SDM-based connectivity estimates differed strongly among the three model species (Table 2). The generalist species *T. lineola* showed neither IBD nor any form of IBR using F_{ST} (max ΔAICc = 0.86). Using D_{est}, the IBD scenario produced the best model (AICc = −806.68, $\omega = 0.48$) however with a weak relationship ($R^2 = 0.045$, p = 0.064). Furthermore, MRDM showed no landscape related signals for either estimate of genetic differentiation in *T. lineola*, suggesting that gene flow in this species is not affected by any spatial or landscape features at this scale. The most specialized species, *T. acteon* showed no significant IBD, but significant IBR for two scenarios (land use & land use change) with both F_{ST} and D_{est} under multi-model inference. These signals become also prominent using MRDM for inference, even though models were slightly insignificant at p = 0.05 (land use change F_{ST}: $R^2 = 0.232$, p = 0.051/D_{est}: $R^2 = 0.190$, p = 0.102). The combined results from AIC and MRDM suggest that land use and land use change both affect genetic differentiation among *T. acteon* populations. Genetic differentiation in *Thymelicus sylvestris* corresponded most strongly to the connectivity estimates derived from the SDM incorporating all variables (AICc = −271.89, $\omega = 0.67$) using F_{ST} and the information-theoretic approach. The climate related scenario was also within the most reliable models under AICc (ΔAICc = 1.65, $\omega = 0.29$). However, MRDM suggested that land use and land use change were also important for explaining genetic differentiation in this species. The opposite becomes obvious using D_{est} as differentiation metric. Here, the information theoretic approach reveals climate, land use and land use change as highly informative, with climate being most important (AIC = −723.08, $\omega = 0.45$). Surprisingly, the scenario covering the entire variable set contributed nearly no information (ΔAICc = 4.44, $\omega = 0.05$). In addition, MRDM highlighted only climate as significantly related to genetic differentiation. In summary, the combined results of different differentiation metrics and inference methods suggest that the climatic conditions across the study site deliver the most important and stable relationship for adjusting gene flow in the intermediate species, with additional effects of land use. Classical IBD received less support against IBR models

Table 1. Summary statistics for genetic diversity and differentiation for the three *Tymelicus* buttlerflies.

	T. lineola	T. acteon	T. sylvestris	source
AR	1.78±0.17	1.88±0.18	1.80±0.10	Louy *et al.* 2007
H_E	9.6±2.1	14.9±2.9	11.9±1.5	Louy *et al.* 2007
H_O	9.2±2.1	12.5±2.6	11.0±1.4	Louy *et al.* 2007
P_{tot}	52.0±9.7	66.0±9.1	42.9±7.9	Louy *et al.* 2007
P_{95}	36.4±9.4	49.3±13.4	32.3±4.2	Louy *et al.* 2007
F_{ST}	0.0081	0.0755	0.0179	Louy *et al.* 2007
D_{est}	0.0012	0.0143	0.0039	Habel *et al.* 2013

Table 2. Comparison of the genetic structure in three *Thymelicus* butterflies with different landscape parameter sets.

	SDM	Linear regression model					MRDM	
	AUC	AICc	ΔAICc	ω	R^2	p	R^2	p
Model F$_{ST}$								
T. lineola								
Fst~distance	-	−321.65		0.21	−0.003	0.359	0.016	0.569
Fst~topography	0.76	−321.44	0.21	0.19	−0.007	0.424	0.012	0.603
Fst~climate	0.81	−321.41	0.23	0.19	−0.007	0.431	0.012	0.602
Fst~all	0.86	−320.87	0.77	0.14	−0.017	0.741	0.002	0.828
Fst~landusechange	0.68	−320.84	0.81	0.14	−0.018	0.795	0.001	0.854
Fst~landuse	0.67	−320.78	0.86	0.14	−0.019	0.893	0.000	0.926
T. acteon								
Fst~landusechange	**0.69**	**−94.70**		**0.56**	**0.202**	**0.009**	**0.232**	0.051
Fst~landuse	**0.71**	**−93.90**	0.80	**0.37**	**0.179**	**0.014**	**0.209**	0.069
Fst~distance	-	−88.12	6.58	0.02	−0.009	0.393	0.028	0.433
Fst~climate	0.79	−87.87	6.83	0.02	−0.018	0.476	0.020	0.748
Fst~topography	0.79	−87.44	7.26	0.01	−0.034	0.737	0.004	0.821
Fst~all	0.84	−87.41	7.29	0.01	−0.035	0.771	0.003	0.772
T. sylvestris								
Fst~all	**0.85**	**−273.53**		**0.67**	**0.252**	**<0.0001**	**0.266**	**0.002**
Fst~climate	**0.78**	**−271.89**	1.65	**0.29**	**0.229**	**<0.0001**	**0.244**	**0.010**
Fst~land use	0.66	−266.28	7.26	0.02	0.147	0.002	0.162	**0.024**
Fst~land use change	0.66	−265.52	8.02	0.01	0.135	0.003	**0.151**	**0.035**
Fst~distance	-	−263.78	9.75	0.01	0.107	0.009	0.123	0.068
Fst~topography	0.78	−262.73	10.81	0.00	0.09	0.015	0.106	0.102
Model D$_{est}$								
T. lineola								
Dest~distance	-	**−806.68**		**0.48**	**0.045**	0.064	0.063	0.176
Dest~topography	0.76	−803.93	2.75	0.12	−0.004	0.373	0.015	0.559
Dest~climate	0.81	−804.10	2.58	0.13	−0.001	0.329	0.018	0.535
Dest~landusechange	0.68	−803.59	3.09	0.10	−0.010	0.493	0.009	0.647
Dest~all	0.86	−803.31	3.37	0.09	−0.015	0.652	0.004	0.780
Dest~landuse	0.67	−803.10	3.58	0.08	−0.019	0.951	0.000	0.968
T. action								
Dest~landusechange	**0.69**	**−274.21**		**0.45**	**0.159**	**0.021**	0.190	0.102
Dest~landuse	**0.71**	**−274.09**	0.12	**0.42**	**0.155**	**0.022**	0.186	0.090
Dest~climate	0.79	−269.81	4.40	0.05	0.015	0.244	0.052	0.608

Table 2. Cont.

Model F_{ST}	SDM AUC	Linear regression model AICc	ΔAICc	ω	R²	p	MRDM R²	p
Dest~distance	-	−268.92	5.29	0.03	−0.016	0.460	0.021	0.614
Dest~all	0.84	−268.41	5.79	0.02	−0.035	0.765	0.004	0.784
Dest~topography	0.79	−268.36	5.85	0.02	−0.037	0.845	0.001	0.893
T. sylvestris								
Dest~climate	0.78	**−723.08**		**0.45**	**0.099**	**0.011**	**0.115**	**0.049**
Dest~land use	0.66	−721.12	1.96	0.17	0.066	0.033	0.083	0.086
Dest~land use change	0.66	−721.18	1.89	0.18	0.067	0.032	0.084	0.085
Dest~distance	-	−720.50	2.57	0.12	0.055	0.046	0.073	0.118
Dest~all	0.85	−718.64	4.44	0.05	0.023	0.139	0.041	0.259
Dest~topography	0.78	−717.61	5.47	0.03	0.004	0.272	0.023	0.410

Genetic differentiation was inferred by F_{ST} (upper half) and D_{est} (lower half) respectively. SDM AUC values for each scenario (excepting classical IBD) showing the model quality are given as well as parameters for both, linear regression models and multiple regression based on distance matrices (MRDM). Bold values highlight models with highest support (ΔAICc<2 in combination with a significant R^2 in linear regression models; significant R^2 in MRDMs).

(Table 2, Figure S2) in all species. Interestingly, topography seems to play no role at all for any of the species.

Discussion

Studying taxonomically related species inhabiting the same environment makes it possible to infer how species-specific ecological traits affect population genetic structuring without confounding effects of different landscapes or phylogenetic history [24]. By conducting a comparative landscape genetic study involving ecologically diverging Hesperid butterflies, we found different impacts of landscape parameters on the genetic structure of the three study species.

The obtained results show strong genetic differentiation and high genetic diversities in the specialist species *T. acteon*, and low genetic differentiation with accompanying low genetic diversities in the generalist species *T. lineola* with *T. sylvestris* standing in-between the two congeners. The amount of genetic diversity is typical for butterflies in this region (reviewed in [26]). Our analyses indicate that climate has a strong impact on the connectivity of *T. sylvestris* but that other variables (such as land use) might have become more influential in the most recent times. Land use as well as changes in land use patterns (i.e. assessed over a 16yr period) influences the connectivity of *T. acteon* populations. In contrast, *T. lineola* populations were panmictic, lacking any landscape related effects on genetic differentiation at this spatial scale.

Diverging responses to identical landscape conditions

Our data illustrate that closely related species representing a gradient of ecological traits (i.e. from generalist to specialist/from highly mobile to rather philopatric) also show a gradient of changing genetic structures and even more interesting of changing landscape genetic associations (Fig. 3). This highlights that ecological traits determine the species-specific resistance of the landscape matrix, so that its effect on population connectivity can differ strongly among closely-related species inhabiting the same landscape.

The strong genetic differentiation in *T. acteon* is concordant with its patchy occurrence predicted in our SDMs (Fig. 2b), which were best explained by the land-use parameters derived from the CORINE dataset. Furthermore, land-use related scenarios were the only ones that host an IBR-related signal among all competing scenarios in this species (Table 2). Here, the two scenarios 'land-use' and 'land-use-change' fit equally well, irrespective of the genetic differentiation metric or statistical inference method used. Thus, the landscape genetic signal in this specialist species is highly consistent among different analyses, leading to high certainty of inferences.

The slight differences between these two scenarios might be stochastic. However, since there is also consistence about the ranking across all approaches (i.e. land use change steadily explains slightly more variance under each situation than land use), land-use-change might be even more important, when addressing land-use-change over an even larger time period than the 16 years used here. Unfortunately, there is no information available to assess past land-use-changes covering this large geographical extent further into the past. Keeping time-lags between fragmentation and genetic responses accompanying these fragmentations in mind (e.g. as reviewed in [72]) there is some evidence that 16 years are not adequate to detect genetic impacts of altered habitats in this time period in a species with an annual generation time. Changes over this period result just in slightly different resistance surfaces between the scenarios 'land-use' and 'land-use-change'. Nevertheless, *T. acteon* is becoming increas-

Table 3. Averaged variable contributions for the scenarios 'topography', 'climate' and 'all'.

Scenario	Variable	*T. acteon*	*T. lineola*	*T. sylvestris*
Topography	alt	7.1	12.1	10.6
	aspect	21.1	29.7	32.3
	slope	71.8	58.2	57.1
Climate	bio3 (isothermality)	12.0	9.7	8.7
	bio7 (temperature annual range)	10.3	23.8	23.5
	bio8 (mean temperature of wettest quarter)	12.0	3.4	4.0
	bio9 (mean temperature of driest quarter)	11.3	10.1	10.0
	bio10 (mean temperature of warmest quarter)	3.6	4.6	5.7
	bio11 (mean temperature of coldest quarter)	16.1	32.0	31.4
	bio12 (annual precipitation)	5.2	10.0	11.3
	bio15 (precipitation seasonality)	5.4	3.3	3.0
	bio18 (precipitation of warmest quarter)	24.3	3.1	2.4
all	land use	37.7	24.9	23.2
	alt	1.3	8.1	7.7
	aspect	9.3	12.4	14.4
	slope	31.2	24.0	29.7
	bio3 (isothermality)	2.8	3.3	3.3
	bio7 (temperature annual range)	2.3	10.0	8.0
	bio8 (mean temperature of wettest quarter)	4.2	1.9	2.0
	bio9 (mean temperature of driest quarter)	1.0	0.1	0.3
	bio10 (mean temperature of warmest quarter)	0.1	0.7	0.4
	bio11 (mean temperature of coldest quarter)	2.1	6.5	4.7
	bio12 (annual precipitation)	2.1	5.8	4.6
	bio15 (precipitation seasonality)	1.5	1.1	0.8
	bio18 (precipitation of warmest quarter)	4.5	1.2	0.8

Note that land use dependent scenarios are not shown herein as they contain one single variable.

ingly vulnerable in large parts of Europe [73] and has likely declined during the past 30 years within the study area due to habitat loss [29]. Thus, the slightly stronger signal of the land-use-change scenario in comparison to the land-use scenario might become even more prominent when extrapolating these changes further decades into the past, highlighting habitat loss as serve danger for this species.

The genetic diversities (such as heterozygosity or mean number of alleles) are highest in *T. acteon* compared to the other two species. This result is somewhat surprising, as the consequence of restricted gene flow and strong geographic restriction of local populations usually leads to rising genetic differentiation and declining genetic diversity, as frequently observed for species demanding specific habitat qualities and/or sedentary dispersal behaviour [10,74–76]. However, there are also examples where genetic diversities in rare species exceed those of their common congeners [77–79]. This contrasting pattern to neutral genetic theory might be a result from hybridization ([80], but see [78]) or because of time-lags that display the past genetic diversity, when connectivity between populations was much higher than today [79,81]. Indeed, genetic differentiation responds to habitat changes quicker than genetic diversity [82–83] so that the high genetic diversity observed for *T. aceton* may not yet reflect the negative consequences of on-going habitat alterations for this species.

In contrast to the specialist *T. acteon*, the generalist *T. lineola* represents opposing genetic features: the species shows a broad ecological amplitude and a much higher mobility [31]. This combination led to higher abundance pattern in combination with increased inter-population migration rates. These species traits lead to a rather panmictic genetic structure in our study area that appears to prevent landscape genetic relationships or IBD. This coherence between wide ecological amplitudes, high rates of individual exchanges (e.g. gene flow) and thus low genetic differentiation were frequently observed in other studies [84–85]. However, it needs to be considered that on a larger study extent, barriers such as oceans, large lakes, mountain ranges might become important for gene flow acting on a macro-scale [86–87]. The landscape matrix in our study area did not enable the assessment of such macro-scale effects, since the landscape matrix is rather continuous at this scale and large barriers are lacking, as indicated by the GENELAND results.

Finally, the species standing in-between these two extremes, *T. sylvestris*, has an abundance like *T. lineola* but shows a sedentary dispersal behavior comparable to that of *T. acteon* [31]. The reduced dispersal propensity of this species coupled with its wide occurrence makes the colonization of a habitat nearby much more likely than of far distant habitats. Consequently, we obtain IBD and IBR signals for many sets of variables in this species (Table 2). However, when combining the information from the different assessment methods (F_{ST} vs. D_{est}/multimodel inference vs.

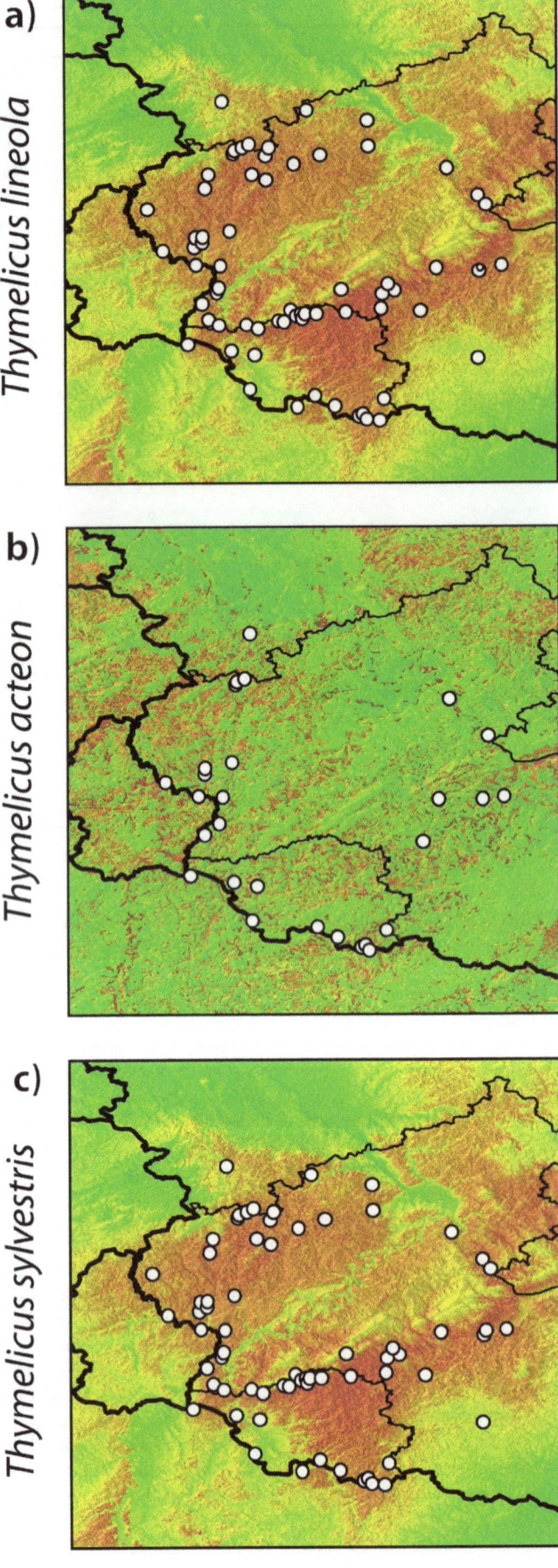

Figure 2. SDM output for *Thymelicus lineola* **(A)** *T. acteon* **(B) and** *T. sylvestris* **(C) respectively.** White circles on SDMs are presence locations used for modeling; Warmer colors (red) indicate higher suitability depending on the best model as presented in Table 2 (climate for *T. sylvestris*; land use change for *T. acteon*; note that *T. lineola* does not have a best model because of its panmictic state. Therefore, also climate is represented here).

MRDM), landscape resistance based on the climate scenario was most important, delivering a consistent strong signal across the different inference methods used (Table 2, see also below). This contrasts to the IBR of *T. acteon*, where climate plays no role at all. In contrast to land-use, climate acts on a meso-scale at our study area (i.e. masking larger areas of the study extent rather than small habitat patches). In *T. sylvestris* the climate related SDM revealed high resistances along river valleys as well as on the higher elevations of the low mountain ranges (Fig 2c). These potential barriers act at a much larger scale and extent compared to the small and patchy habitat islands enclosed by more or less unfavourable habitats in *T. acteon*. Consequently, the different landscape features contributing to the IBR signals in these two species highlight the importance of scale and shape of the connective elements (or their respective barriers) in the landscape matrix where methodological shortcomings can be excluded (Engler, unpublished). However, the obtained IBR models explain only up to 24% of the variance in our dataset. That in turn indicates that the remaining variance of our data can only be explained by additional factors such as ecological traits and habitat requirements. These can be even more relevant for butterfly species than habitat size and habitat isolation, e.g. as shown for the Heath butterfly *Coenonympha tullia* [7]. Nevertheless, the extent of the relationships in our IBD/IBR comparisons are in concert with other studies [88] indicating that gene flow can be interpreted

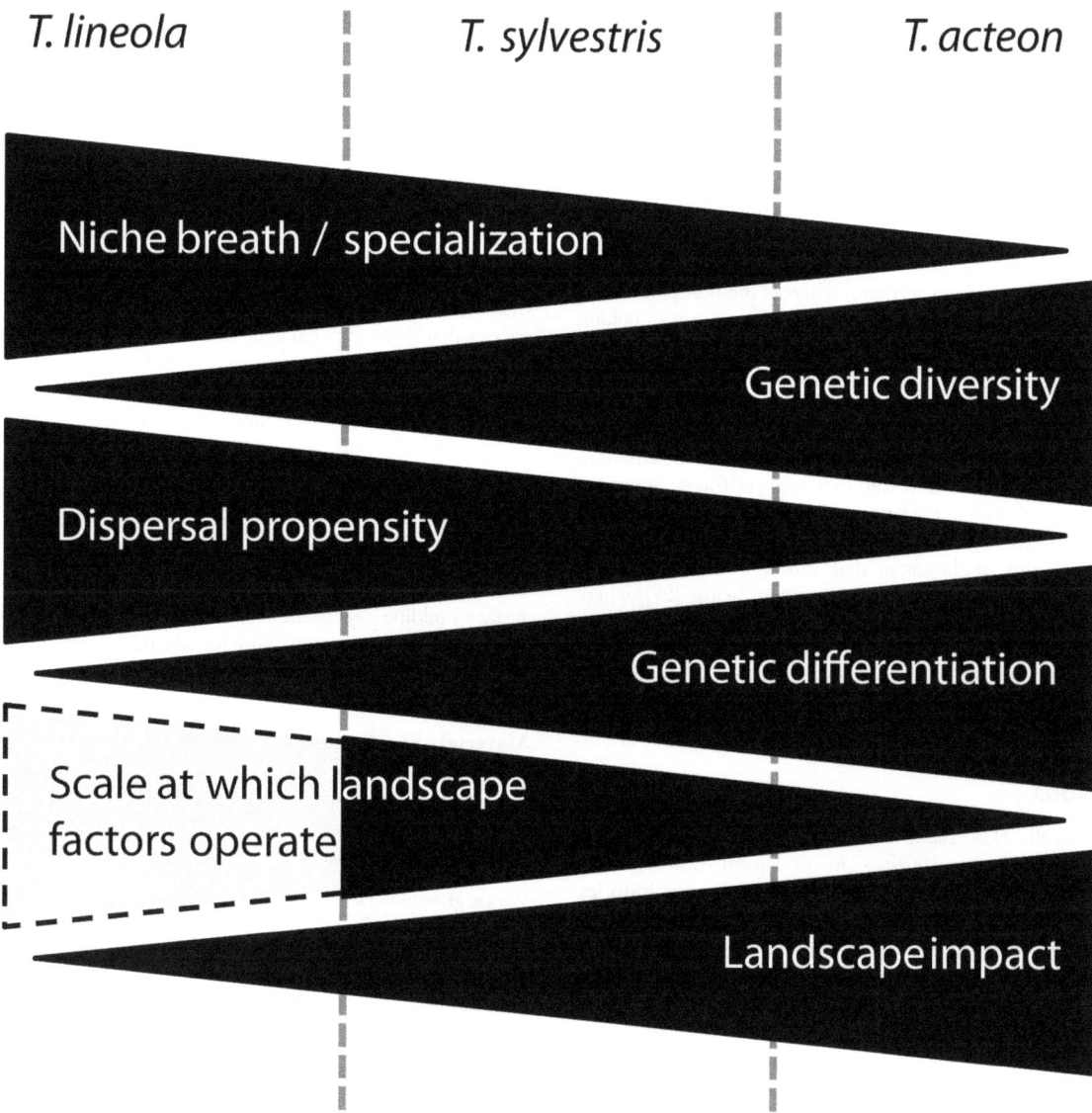

Figure 3. Schematic illustration about the gradual effects forcing on the three *Thymelicus* **species.** Hatched area highlights the hypothesized effect of landscape on gene flow in *T. lineola* on the macro-scale which was not testable in the study area.

as an important component out of a variety of mechanisms influencing population genetic structure.

Accounting for F_{ST} and D_{est} in landscape genetic studies

Interestingly, in the case of *Thymelicus sylvestris*, the prominent signal under F_{ST} arising from the SDM using all landscape variables becomes completely eliminated when using D_{est} as a differentiation metric. The fact that different metrics can lead to different conclusions is also evident in the ongoing debate about the utility of different genetic differentiation measures [89–94]. For example, traditional F_{ST}-like metrics are more sensitive to recent demographic changes (which depends e.g. on effective population size) than metrics which are independent of effective population size, such as D_{est} [43,90,93–94]. This makes F_{ST} more sensible to effects of gene flow or drift in comparison to D_{est}. Thus, from a landscape genetic perspective, using different types of differentiation metrics allows to test for the contribution of landscape effects in contemporary versus past times. If landscape composition change over time (and consequently the amount of gene flow mediated by the landscape), F_{ST} would respond much quicker to those changes while D_{est} remains rather stable over time. In the case of *Thymelicus sylvestris*, this means that D_{est} may highlight the landscape effect (here climate) of highest importance for gene flow in this species in former times, whereas F_{ST} highlights more recent landscape effects on genetic structure that involves also other landscape elements beside climate such as land use and topographical elements.

In contrast to the climate-only scenario, connectivity estimates involving all variables did not give highest importance to climatic factors. In particular, land use and slope contribute almost 54% of the total importance of this scenario, whereas the best performing variables from the climate scenario, bio11 (mean temp of coldest quarter) and bio7 (temp annual range) that contribute together 54.9%, contributing under the full model just 12.7% of the total importance. This might highlight the change of landscape factors important to gene flow in this species. As *T. sylvestris* is indeed common but not very mobile, anthropogenic land transformations of the past decades might now lead to a stronger fragmentation of populations which ultimately lead to changes in the contributions of landscape factors shaping gene flow as shown elsewhere [95]. Consequently, this might mean that this species is just at the tipping point of being of conservation concern (*sensu* [27]) where population trends swapping from stable to decreasing. Its congeners *T. acteon* and *T. lineola* showing both consistent results across the different metrics underpinning their stable state in terms of their abundance (insentient and widespread vs. sensible and endangered) and specialization (generalist vs. specialist).

Conclusions

Taxonomically close relatives serve as ideal model systems to study interspecific characteristics in ecological traits without confounding effects derived from different evolutionary histories. Yet, studies investigating the role of landscape on gene flow of closely related taxa inhabiting the same environment are still scarce. Our results reveal that even between sibling species, gene flow is affected by the landscape in very different ways. Thus, it is challenging to predict landscape genetic relationships in one species from a study involving another species, even if the two species are taxonomically closely related. Nevertheless, some generalizations are possible for specialist versus generalist species. In our study, the genetic structure of the generalist species with high dispersal propensities remained unaffected by the current landscape matrix, whereas specialist species were highly sensitive to fine scale habitat features. Changes of these features might therefore affect specialists more readily than generalist species with the negative consequences for their genetic setup. Species with an intermediate degree of specialization (here *T. sylvestris*) also interact with the landscape but at coarser scales in comparison to specialist species (here *T. acteon*). However, in light of global change such species might be on the highest risk due to negative genetic effects such as inbreeding depression, because changes in the habitat matrix can push former meta-population into isolated remnants [27]. This becomes also evident in *T. sylvestris* comparing the genetic structure under either F_{ST} or D_{est}. Further studies focusing on the degree of habitat specialization in addition to dispersal capabilities are needed, ideally conducted with closely related taxa in other areas. Such comparative studies will greatly expand our current understanding of landscape genetic relationships and ultimately lead to more effective conservation and management of biodiversity.

Supporting Information

Figure S1 Estimation of the number of panmictic clusters for each species. A) Convergence of the MCMC after thinning (see methods for details). Values prior to burn-in (indicated as red dashed line) were not considered as chain does not reached convergence. B) Frequency of the estimated number of populations along the chain after burn-in.

Figure S2 Scatterplots showing the differences of isolation by distance patterns with isolation by resistance patterns in the two species that show a spatial genetic structure (*Thymelicus sylvestris* is shown at the upper half, *T. acteon* at the lower half). Note that just the most prominent isolation by resistance pattern is shown (i.e. climate in *T. sylvestris* and land use change in *T. acteon*).

Table S1 Geographic coordinates of the sampling locations. ID numbers correspond to those stated in Fig. 1.

Table S2 Description of the landscape data used for resistance surface building depending on the scenario assumed. Note that the scenarios 'land use' and 'land use change' used the same data source. SDM refers to species distribution model.

Material S1 Evolutionary history of the three *Thymelicus* butterflies.

Acknowledgments

We are grateful to Deborah A. Dawson, Alain Frantz, Joseph Chipperfield, and Lisette P. Waits for valuable comments on an earlier draft of this article.

Author Contributions

Conceived and designed the experiments: JOE NB DR JCH. Performed the experiments: JCH JOE DR NB. Analyzed the data: JOE NB DR JCH KJF. Contributed reagents/materials/analysis tools: JCH JOE NB DR. Wrote the paper: JOE JCH NB DR KJF.

References

1. MacArthur R-H, Wilson E-O (1967) The theory of island biogeography. Princeton: Princeton University Press. 205 p.
2. Levins R (1969) Some demographic and genetic consequences of environmental heterogeneity for biological control. Bull Entomol Soc America 15: 237–240.
3. Hanski I (1999) Metapopulation ecology. Oxford: Oxford University Press. 319 p.
4. Hu XS, He F, Hubbel SP (2006) Neutral theory in macroecology and population genetics. Oikos 113: 548–556.
5. Lomolino MV, Brown JH (2009) The reticulating phylogeny of island biogeography theory. Quart Rev Biol 84: 357–390.
6. Laurance WF (2009) Beyond island biogeography theory: understanding habitat fragmentation in the real world. In: Losos JB, Ricklefs RE, editors. The theory of island biogeography revisited. Princeton: Princeton University Press. 476 p.
7. Dennis RLH, Eales H (1997) Patch occupancy in Coenonympha tullia (Müller 1764) (Lepidoptera: Satyrinae): Habitat quality matters as much as patch size and isolation. J Insect Conserv 1: 167–176.
8. Conradt L, Roper TJ, Thomas CD (2001) Dispersal behaviour of individuals in metapopulations of two British butterflies. Oikos 95: 416–424.
9. van Dyck H, Baguette M (2005) Dispersal behaviour in fragmented landscapes: routine or special movements? Basic App Ecol 6: 535–545.
10. Habel JC, Meyer M, Schmitt T (2009a) The genetic consequence of differing ecological demands of a generalist and a specialist butterfly species. Biodiv Conserv 18: 1895–1908.
11. Goodwin BJ (2003) Is landscape connectivity a dependent or independent variable? Landscape Ecol 18: 687–699.
12. Fahrig L (2003) Effects of habitat fragmentation on biodiversity. Ann Rev Ecol Evol Sys 34: 487–515.
13. Hof C, Levinsky I, Araújo MB, Rahbek C (2011) Rethinking species' ability to cope with rapid climate change. Glob Chan Biol 17: 2987–299.
14. Seiferling IS, Proulx R, Peres-Neto PR, Fahrig L, Messier C (2012) Measuring protected-area isolation and correlations on isolation with land-use intensity and protection status. Conserv Biol 26: 610–618.
15. Holderegger R, Wagner HH (2008) Landscape genetics. Bioscience 58: 199–208.
16. Storfer A, Murphy MA, Spear SF, Holderegger R, Waits LP (2010) Landscape genetics: where are we now? Mol Ecol 19: 3496–3514.
17. Manel S, Holderegger R (2013) Ten years of landscape genetics. Trends Ecol Evol 28: 614–621.
18. Wright S (1943) Isolation by distance. Genetics 28: 114–138.
19. Goldberg CS, Waits LP (2010) Comparative landscape genetics of two pond-breeding amphibian species in a highly modified agricultural landscape. Mol Ecol 19: 3650–3663.
20. Richardson JL (2012) Divergent landscape effects on population connectivity in two co-occurring amphibian species. Mol Ecol 21: 4437–4451.
21. McRae BH (2006) Isolation by resistance. Evolution 60: 1551–1561.
22. Schwenk WS, Donovan TM (2011) A multispecies framework for landscape conservation planning. Conserv Biol 25: 1010–1021.
23. Poelchau MF, Hamrick JL (2012) Differential effects of landscape-level environmental features on genetic structure in three codistributed tree species in Central America. Mol Ecol 21: 4970–4982.
24. Steele CA, Baumsteiger J, Storfer A (2009) Influence of life-history variation on the genetic structure of two sympatric salamander taxa. Mol Ecol 18: 1629–1639.
25. Dawson MN (2012) Parallel phylogeographic structure in ecologically similar sympatric sister taxa. Mol Ecol 21: 987–1004.
26. Habel JC, Rödder D, Lens L, Schmitt T (2013) The genetic signature of ecologically different grassland Lepidopterans. Biodiv Conserv 22: 2401–2411.
27. Habel JC, Schmitt T (2012) The burden of genetic diversity. Biol Conserv 147: 270–274.
28. Louy D, Habel JC, Schmitt T, Assmann T, Meyer M, et al. (2007) Strongly diverging population genetic patterns of three skipper species: the role of habitat fragmentation and dispersal ability. Conserv Gen 8: 671–681.
29. Wenzel M, Schmitt T, Weitzel M, Seitz A (2006) The serve decline of butterflies on western German calcareous grasslands during the last 30 years: a conservation problem. Biol Conserv 128: 542–552.
30. Asher J, Warren M, Fox R, Harding P, Jeffcoate G, et al. (2001) The millennium atlas of butterflies in Britain and Ireland. Oxford: Oxford University Press. 433 p.
31. Bink FA (1992) Ecologische Atlas van de Dagvlinders van Noordwest-Europa. Haarlem: Schuyt & Co. Uitgevers en Importeurs. 512 p.
32. Meglécz E, Pecsenye K, Varga Z, Solignac M (1998) Comparison of differentiation pattern at allozyme and microsatellite loci in Parnassius mnemosyne (Lepidoptera) populations. Hereditas 128: 95–103.
33. Habel JC, Zachos FE, Finger A, Meyer M, Louy D, et al. (2009b) Unprecedented long term genetic monomorphism in an endangered relict butterfly species. Conserv Gen 10: 1659–1665.
34. Habel JC, Rödder D, Schmitt T, Nève G (2011) Global warming will affect the genetic diversity and uniqueness of Lycaena helle populations. Glob Chan Biol 17: 194–205.
35. Meglécz E, Solignac M (1998) Microsatellite loci for Parnassius mnemosyne (Lepidoptera). Hereditas 128: 179–180.
36. Meglécz E, Petenian F, Danchin E, Coer D'Acier A, Rasplus J-Y, et al. (2004) High similarity between flanking regions of different microsatellites detected

within each of two species of Lepidoptera: Parnassius Apollo and Euphydrias aurinia. Mol Ecol 13: 1693–1700.
37. Habel JC, Meyer M, Schmitt T, Assmann T (2008) Polymorphic microsatellite loci in the endangered butterfly Lycaena helle (Lepidoptera: Lycaenidae). Eur J Enthomol 105: 361–362.
38. Finger A, Zachos FE, Schmitt T, Meyer M, Assmann T, et al. (2008) The genetic status of the Violet Copper Lycaena helle – a relict of the cold past in times of global warming. Ecography 32: 382–390.
39. Zang D-X (2004) Lepidopteran microsatellite DNA: redundant but promising. Trends Ecol Evol 19: 507–509.
40. Landguth E, Balkenhol N (2012) Relative sensitivity of neutral versus adaptive genetic data for assessing population differentiation. Conserv Gen 13: 1421–1426.
41. Excoffier L, Larval G, Schneider S (2005) Arlequin ver. 3.0: an integrated software package for population genetics data analysis. Evol Bioinform Onl 1: 47–50.
42. Crawford NG (2009) SMOGD: software for the measurement of genetic diversity. Mol Ecol Res 10: 556–557.
43. Leng L, Zhang D (2011) Measuring population differentiation using Gst or D? A simulation study with microsatellite DNA markers under a finite island model and nonequilibrium conditions. Mol Ecol 20: 2494–2509.
44. Guillot G, Santos F (2009) A computer program to simulate multilocus genotype data with spatially auto-correlated allele frequencies. Mol Ecol Res 9: 1112–1120.
45. Ortego J, Riordan EC, Gugger PF, Sork VL (2012) Influence of environmental heterogeneity on genetic diversity and structure in an endemic southern Californian oak. Mol Ecol 21: 3210–3223.
46. Bell RC, Parra JL, Tonione M, Hoskin CJ, Mackenzie JB, et al. (2010) Patterns of persistence and isolation indicate resilience to climate change in montane rainforest lizards. Mol Ecol 19: 2531–2544.
47. Devitt TJ, Cameron Devitt, SE, Hollingworth BD, McGuire JA, Moritz C (2013) Montane refugia predict population genetic structure in the Large-blotched Ensatina salamander. Mol Ecol 22: 1650–1665.
48. Spear SF, Balkenhol N, Fortin M-J, McRae BH, Scribner K (2010) Use of resistance surfaces for landscape genetic studies: considerations for parameterization and analysis. Mol Ecol 19: 3576–3591.
49. Hijmans RJ, Cameron SE, Parra JL, Jones PG, Jarvis A (2005) Very high resolution interpolated climate surfaces for global land areas. Int J Climat 25: 1965–1978.
50. Beaumont LJ, Hughes L, Poulsen M (2005) Predicting species distributions: use of climatic parameters in BIOCLIM and its impact on predictions of species' current and future distributions. Ecol Mod 186: 250–269.
51. Weiss SB, Murphy DD, White RR (1988). Sun, slope, and butterflies: topographic determinants of habitat quality for Euphydryas editha. Ecology 69: 1486–1496.
52. Warren MS, Hill JK, Thomas JA, Asher J, Fox R, et al. (2001) Rapid responses of British butterflies to opposing forces of climate and habitat change. Nature 414: 65–69.
53. Pe'er G, Saltz D, Thulke HH, Motro U (2004) Response to topography in a hilltopping butterfly and implications for modeling nonrandom dispersal. Anim Behav 68: 825–839.
54. Pe'er G, Heinz SK, Frank K (2006) Connectivity in heterogeneous landscapes: analyzing the effect of topography. Landscape Ecol 21: 47–61.
55. Pin Koh L (2007) Impacts of land use change on South-east Asian forest butterflies: a review. J App Ecol 44: 703–713.
56. Marini L, Fontana P, Kimek S, Battisti A, Gaston KJ (2009) Impact of farm size and topography on plant and insect diversity of manages grasslands in the Alps. Biol Conserv 142: 394–403.
57. Filz KJ, Engler JO, Stoffels J, Weitzel M, Schmitt T (2013) Missing the target? A critical view on butterfly conservation efforts on calcareous grasslands. Biodiv Conserv 22: 2223–2241.
58. Phillips SJ, Anderson RP, Schapire RE (2006) Maximum entropy modeling of species geographic distributions. Ecol Mod 190: 231–259.
59. Elith J, Phillips SJ, Hastie T, Dudík M, Chee YE, et al. (2011) A statistical explanation of MaxEnt for ecologists. Divers Distrib 17: 43–47.
60. Elith J, Graham CH, Anderson RP, Dudík M, Ferrier S, et al. (2006) Novel methods improve prediction of species' distribution from occurrence data. Ecography 29: 129–151.
61. Heikkinen RK, Luoto M, Araújo MB, Virkkala R, Thuiller W, et al. (2006) Methods and uncertainties in bioclimatic envelope modelling under climate change. Prog Phys Geogr 30: 751–777.
62. Hernandez PH, Graham CH, Master LL, Albert DL (2006) The effect of sample size and species characteristics on performance of different species distribution methods. Ecography 29: 773–785.
63. Wisz MS, Hijmans RJ, Li J, Peterson AT, Graham CH, Guisan et al. (2008) Effects of sample size on the performance of species distribution models. Divers Distrib 14: 763–773.
64. Swets JA (1988) Measuring the accuracy of diagnostic systems. Science 240: 1285–1293.

65. McRae BH, Dickson BG, Keitt TH, Shah VB (2008) Using circuit theory to model connectivity in ecology, evolution, and conservation. Ecology 89: 2712–2724.

66. McRae BH, Beier P (2007) Circuit theory predicts gene flow in plant and animal populations. PNAS 104: 19885–19890.

67. Lichstein JW (2007) Multiple regression on distance matrices: a multivariate spatial analysis tool. Plant Ecol 188: 117–131.

68. R development core team (2012) R: a language and environment for statistical computing. Available: http://www.R-project.org.

69. Burnham KP, Anderson DR (2002) Model selection and multimodel inference: a practical information-theoretic approach. New York: Springer. 488 p.

70. Goslee SC, Urban DL (2007) The *ecodist* package for dissimilarity-based analysis of ecological data. J Stat Softw 22: 1–19.

71. Araújo MB, Pearson RG, Thuiller W, Erhard M (2005) Validation of species-climate impact models under climate change. Glob Chan Biol 11: 1504–1513.

72. Keyghobadi N (2007) The genetic implications of habitat fragmentation for animals. Can J Zool 85: 1049–1064.

73. van Swaay CAM, Warren M (1999) Red data book of European butterflies (Rhopalocera). Strasbourg: Council of Europe Publishing. 269 p.

74. Kassen R (2002) The experimental evolution of specialists, generalists, and the maintenance of diversity. J Evol Biol 15: 173–190.

75. Packer L, Zayed A, Grixti C, Ruz L, Owen RE, et al. (2005) Conservation genetics of potentially endangered mutualisms: reduced levels of genetic variation in specialist versus generalist bees. Conserv Biol 19: 195–202.

76. Zachos FE, Althoff C, v Steynitz Y, Eckert I, Hartl GB (2007) Genetic analysis of an isolated red deer (*Cervus elaphus*) population showing signs of inbreeding depression. Eur J Wildl Res 53: 61–67.

77. Gitzendanner MA, Soltis PS (2000) Patterns of genetic variation in rare and widespread plant congeners. Am J Bot 87: 783–792.

78. Ellis JR, Pashley CH, Burke JM, McCauley DE (2006) High genetic diversity in a rare and endangered sunflower as compared to a common congener. Mol Ecol 15: 2345–2355.

79. Kadlec T, Vrba P, Kepka P, Schmitt T, Konvicka M (2010) Tracking the decline of the once-common butterfly: delayed oviposition, demography and population genetics in ther hermit *Chazara briseis*. Anim Conserv 13: 172–183.

80. Richards ZT, van Oppen MJH (2012) Rarity and genetic diversity in Indo-Pacific *Acropora* corals. Ecol Evol 2: 1867–1888.

81. Severns PM, Liston A, Wilson MV (2011) Habitat fragmentation, genetic diversity, and inbreeding depression in a threatened grassland legume: is genetic rescue necessary? Conserv Gen 12: 881–893.

82. Balkenhol N, Pardini R, Cornelius C, Fernandes F, Sommer S (2013) Landscape-level comparison of genetic diversity and differentiation in a small mammal inhabiting different fragmented landscapes of the Brazilian Atlantic Forest. Conserv Gen 14: 355–367.

83. Keyghobadi N, Roland J, Matter SF, Strobeck C (2005) Among- and within-patch components of genetic diversity respond at different rates to habitat fragmentation: an empirical demonstration. Proc R Soc London B 272: 553–560.

84. Brouat C, Chevallier H, Meusnier S, Noblecourt T, Rasplus J-Y (2004) Specialization and habitat: spatial and environmental effects on abundance and genetic diversity of forest generalist and specialist *Carabus* species. Mol Ecol 13: 1815–1826.

85. Habel JC, Schmitt T (2009) The genetic consequences of different dispersal behaviours in Lycaenid butterfly species. Bull Entomol Res 99: 513–523.

86. Lee-Yaw JA, Davidson A, McRae BH, Green DM (2009) Do landscape processes predict phylogeographic patterns in the wood frog? Mol Ecol 18: 1863–1874.

87. Kekkonen J, Seppä P, Hanski IK, Jensen H, Väisänen RA, et al. (2011) Low genetic differentiation in a sedentary bird: house sparrow population genetics in a contiguous landscape. Heredity 106: 183–190.

88. Groot AT, Classen A, Inglis O, Blanco CA, López Jr J, et al. (2011) Genetic differentiation across North America in the generalist moth *Heliothis virescens* and the specialist *H. subflexa*. Mol Ecol 20: 2676–2692.

89. Heller R, Siegismund HR (2009) Relationship between three measures of genetic differentiation G_{ST}, D_{EST} and G'_{ST}: how wrong have we been? Mol Ecol 18: 2080–2083.

90. Ryman N, Leimar O (2009) G_{ST} is still a useful measure of genetic differentiation – a comment on Jost's *D*. Mol Ecol 18: 2084–2087.

91. Jost L (2009) D vs. G_{ST}: Response to Heller and Siegismund (2009) and Ryman and Leimar (2009). Mol Ecol 18: 2088–2091.

92. Gerlach G, Jueterbock A, Kraemer P, Deppermann J, Harmand P (2010) Calculations of population differentiation based on G_{ST} and D: forget G_{ST} but not all of statistics! Mol Ecol 19: 3845–3852.

93. Meirmans PG, Hedrick PW (2011) Assessing population structure F_{ST} and related measures. Mol Ecol Res 11: 5–18.

94. Raeymaekers JAM, Lens L, Van den Broeck F, Van Dongen S, Volckaert FAM (2012) Quantifying population structure on short timescales. Mol Ecol 21: 3458–3473.

95. Pavlacky Jr DC, Goldizen AW, Prentis PJ, Nicholls JA, Lowe AJ (2009) A landscape genetics approach for quantifying the relative influence of historic and contemporary habitat heterogeneity on the genetic connectivity of a rainforest bird. Mol Ecol 18: 2945–2960.

12

Landscape Genetics for the Empirical Assessment of Resistance Surfaces: The European Pine Marten (*Martes martes*) as a Target-Species of a Regional Ecological Network

Aritz Ruiz-González[1,2,3]*, **Mikel Gurrutxaga**[2,4], **Samuel A. Cushman**[5], **María José Madeira**[1,2], **Ettore Randi**[3,6], **Benjamin J. Gómez-Moliner**[1,2]

1 Department of Zoology and Animal Cell Biology, University of the Basque Country, UPV/EHU, Vitoria-Gasteiz, Spain, **2** Systematics, Biogeography and Population Dynamics Research Group, Lascaray Research Center, University of the Basque Country, UPV/EHU, Vitoria-Gasteiz, Spain, **3** Conservation Genetics Laboratory, National Institute for Environmental Protection and Research, ISPRA, Ozzano dell'Emilia, Bologna, Italy, **4** Department of Geography, University of the Basque Country, UPV/EHU, Vitoria-Gasteiz, Spain, **5** U.S. Forest Service, Rocky Mountain Research Station, Flagstaff, AZ, United States of America, **6** Department 18/Section of Environmental Engineering, Aalborg University, Aalborg, Denmark

Abstract

Coherent ecological networks (EN) composed of core areas linked by ecological corridors are being developed worldwide with the goal of promoting landscape connectivity and biodiversity conservation. However, empirical assessment of the performance of EN designs is critical to evaluate the utility of these networks to mitigate effects of habitat loss and fragmentation. Landscape genetics provides a particularly valuable framework to address the question of functional connectivity by providing a direct means to investigate the effects of landscape structure on gene flow. The goals of this study are (1) to evaluate the landscape features that drive gene flow of an EN target species (European pine marten), and (2) evaluate the optimality of a regional EN design in providing connectivity for this species within the Basque Country (North Spain). Using partial Mantel tests in a reciprocal causal modeling framework we competed 59 alternative models, including isolation by distance and the regional EN. Our analysis indicated that the regional EN was among the most supported resistance models for the pine marten, but was not the best supported model. Gene flow of pine marten in northern Spain is facilitated by natural vegetation, and is resisted by anthropogenic landcover types and roads. Our results suggest that the regional EN design being implemented in the Basque Country will effectively facilitate gene flow of forest dwelling species at regional scale.

Editor: Jesus E. Maldonado, Smithsonian Conservation Biology Institute, United States of America

Funding: This study has been funded by the Basque Government through the Research group on "Systematics, Biogeography and Population Dynamics" (Ref. IT317-10; GIC10/76; IT575/13) and by the University of the Basque Country (UPV-EHU) and the Department of Environment, Territorial Planning, Agriculture and Fisheries (Basque Government) through IKT S.A under the University-Enterprise research program (Ref. UE07/02). Ruiz-González holds a Post doc fellowship awarded by the Dept. of Education Universities and Research of the Basque Government (Ref. DKR-2012-64). Several samples analysed in this study have been obtained in the framework of different carnivore surveys funded by regional or national administrations (Spanish Ministry of Environment, Regional Governments of Navarre and Aragon, Alava Provincial Council). The funders had no role in study design, data collection and analysis, decision to publish, or preparation of the manuscript.

Competing Interests: The authors have declared that no competing interests exist.

* Email: aritz.ruiz@ehu.es

Introduction

Long-term biodiversity conservation requires the preservation of ecological and evolutionary processes, such as gene flow, dispersal movements and population range shifts [1]. The ability of individuals to move across changing landscapes is crucial for maintaining regional populations [2,3]. The preservation of these processes requires, in turn, that landscape connectivity be preserved, especially when we take into account the synergetic effects of habitat fragmentation and climate change [1]. Landscape connectivity is defined as the degree to which landscape facilitates or impedes movement of organisms among resource patches [4]. Connectivity is species-specific and reflects the response of individuals to landscape features and the patterns of dispersal and gene flow that result from these individual responses [5].

Thus, landscape connectivity depends to a large extent on how the spatial configuration of habitat and land use interact with the movement ecology of particular species [6].

Ecological networks have been promoted as coherent systems composed of core areas linked by ecological corridors capable of facilitating the dispersal, migration and gene flow of wild species in landscapes and regions [7–9]. They are configured and managed with the objective of maintaining ecological functions and conserving biodiversity [7]. Although the development of ecological networks is based on the precautionary principle and on ecological theory [8], the absence of empirical evidence regarding their effectiveness and the difficulty in obtaining this evidence has been a focus of criticism about the extent to which they have in fact ensured landscape connectivity and increased biodiversity conservation [10,11].

In the design of ecological networks there is a need to predict regional ecological corridors and to quantify the degree of expected landscape connectivity between specific areas [3,9,11–13]. 'Least-cost modeling' is one commonly employed approach for designing ecological corridors [9,14], in which resistance values are assigned to distinct habitat or land use types and the least-cost paths (LCP) between specific locations are calculated using a geographical information system (GIS). How landscape influences effective distances between locations is calculated as the accumulated cost through the least cost paths [14,15]. However, for most organisms, setting the resistance values is a difficult process in which expert judgment and data available in the literature play an important role [16–19].

Accurate identification of the potential factors that drive gene flow in heterogenous landscapes and the scales at which they are acting is a foundation of reliable mapping of corridors [9,18]. Thus, reliable development of corridors must be based on a correct representation of the local resistance relative to the movement ecology of the organism of focus [9,18]. Landscape genetics, a research area that integrates landscape ecology, population genetics and spatial statistics, provides a valuable framework for testing the influence of landscape structure and composition on dispersal and gene flow [20,21]. It facilitates quantification of the resistance to gene flow a given landscape element poses [12,22]. Thus, one of the principal applications of landscape genetics in landscape planning and conservation biology is to empirically test and optimize resistance maps [23–26]. This facilitates the optimal design of ecological corridors [3,16,23], the detection of barriers to gene flow [27–29] and the identification of the landscape features which favour or impede dispersal [30–35].

Landscape genetics has shifted towards individual-based sampling and analysis, especially when organisms are continuously distributed [12,22]. However, sufficient sample collection for this purpose is a difficult task, especially in rare and elusive species in which sampling is a limiting factor [36]. In this context, non-invasive genetic sampling allows us to address studies of wildlife species without the need to capture or even observe them [37–40].

In 2005 a regional ecological network was established in the Basque Country (North Spain) by delimiting the ecological corridors linking forest protected areas [41]. A functional group of forest mammal species was selected to guide the development of a generic resistance map, which would, in turn, serve as a basis least-cost modeling of the network of ecological corridors linking these core areas. These mammals were considered suitable target species due to their sensitivity to recent fragmentation and homogenization dynamics in the regional landscape, such as road construction, urbanization and agrarian intensification [41,42]. The resistance map was parameterized through bibliographical review and expert opinion and was based on the assignment of different resistance levels to each land use [41]. The regional government of the Basque country incorporated that coherent ecological network as a reference for the environmental assessment of plans, programs and projects in 2005 [41]. In addition to its intrinsic internal relevance, the Basque country has been chosen for its crucial role in the regulation of biotic flows in south-western Europe [43]. This is because of its strategic location between two important biodiversity reservoirs in south-western Europe, the mountain chains of the Pyrenees and the Cantabrian Range [43–45]. Consequently, the preservation and restoration of connectivity in this transitional area between mountain ranges requires reliable knowledge about ecological responses of organisms to landscape composition and structure [45].

Among the set of functional forest mammals used in the design of the coherent regional ecological network, the European pine marten (*Martes martes*) is the most forest dependent species [46]. The pine marten is generally associated with forest habitats, mainly mature forests [46–48]. Deforestation and forest fragmentation limit the distribution and density of pine martens [48–50], which are believed to need a minimum woodland area to survive (ca. 2km²) [47] and tend to avoid treeless areas [48,51,52]. Their occurrence patterns are affected by forest patch size, percentage of woodland cover, food abundance, sex, age class and habitat fragmentation levels [47,51]. Given their strong associations with high forest structural complexity, the species is particularly sensitive to human influences on their habitats, including habitat loss and landscape-scale effects of habitat fragmentation [48,51,53]. Nonetheless, they have also been recently reported in fragmented landscapes characterized by isolated, small forest fragments within an agricultural landscape matrix [48,51,54], suggesting they are not as obligately interior-forest dependent as previously described [51]. However, in such landscapes, linear features, such as hedgerows and small woods, play a key role to connect adjacent forest patches [48,54,55].

Consequently, the pine marten is a species which is well suited to studies focused on the effects of forest fragmentation on genetic structure and gene flow [56]. However, whether habitat characteristics that predict marten occupancy act as barriers to dispersal, influencing gene flow and population genetic structure across the landscape, is largely unknown [56].

The main objective of this research is to evaluate a large suite of alternative resistance hypotheses for the pine marten and compare the most supported empirical model with the expert-derived landscape resistance model used to parameterize the corridor network for the Basque Country. Specifically, we aim to evaluate (1) different binary landscape resistance maps which cover a gradient from greater to lesser preference of the pine marten for forest environments in order to identify which land uses favour or impede genetic interchange in the study area; and secondly (2) whether or not the resistance map with which the regional ecological network was originally designed in the Basque Country was correctly parametrized to reflect European pine marten gene flow.

If there is no effect of landscape structure on dispersal and gene flow in martens, then we expected either: (a) panmixia, where there is no genetic pattern, or (b) isolation-by-distance, where genetic differences increase with geographic distance [57]. If landscape structure influences marten dispersal, then we expected (c) isolation by landscape resistance [30]. Given that the most consistent marten-habitat relation appears to be a general association with forest habitats, and avoidance of open, non-forested habitats [47,48,51,52], we expected that open and human altered landscapes would act as a barrier for martens, and hence that landscape structure would have an effect on gene flow. In addition, we hypothesize that the intervening landscape features between forest patches (i.e., matrix) could also play a key role to substantially affect pine marten dispersal, and consequently the connectivity between forest environments [48,55,58].

Methods

Study area and spatial data

The region of the Basque Country is located in the northern Iberian Peninsula (Fig. 1) within the Atlantic and Mediterranean biogeographical regions. It comprises an area of 7,235 km² and has an average human population density of 298 inhabitants per square kilometer. Forests cover 28%, forestry plantations 29%,

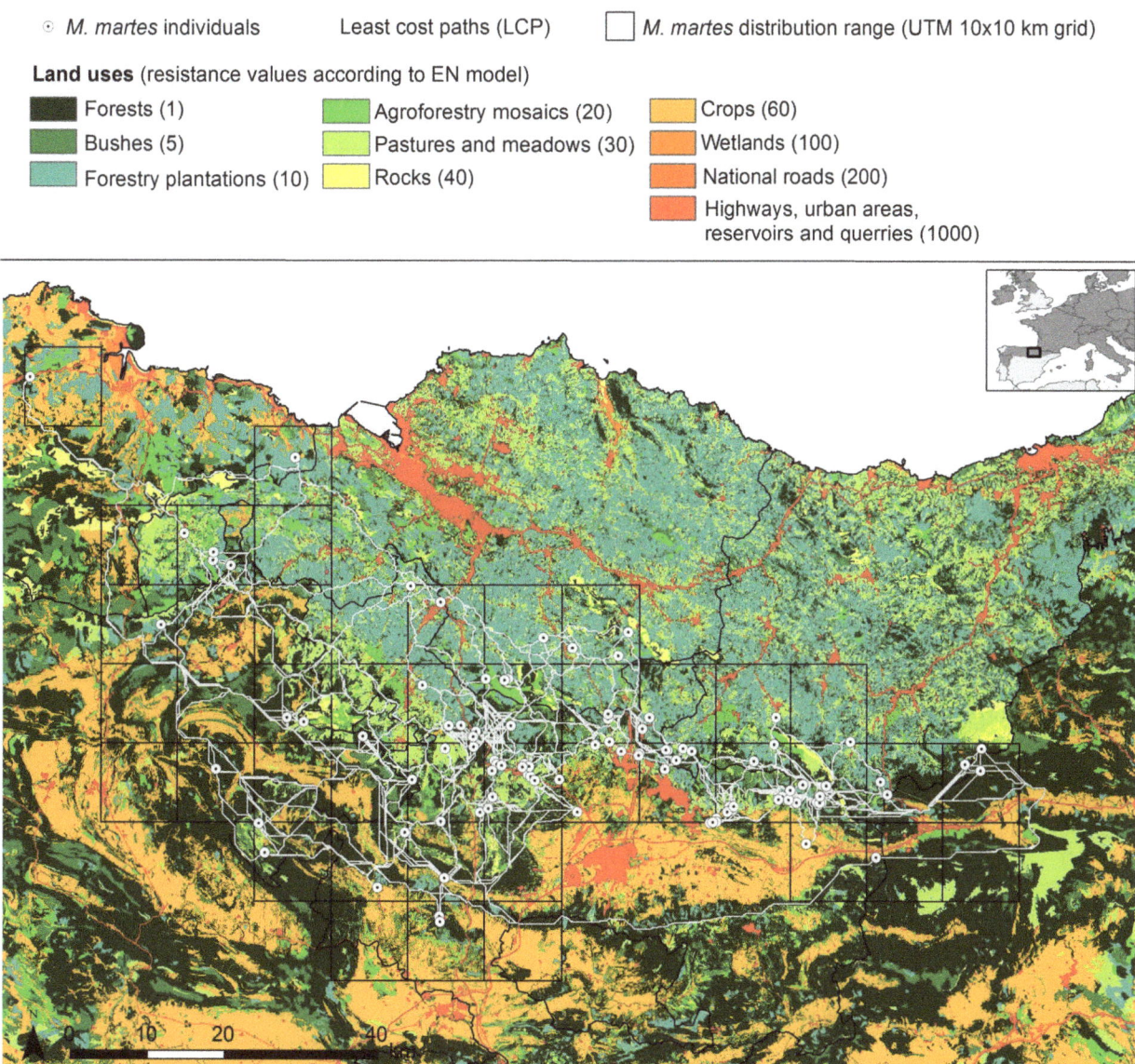

Figure 1. Ecological network resistance map (EN) and LCP analysis between European pine marten individuals in the study area.
Least cost paths (LCP) obtained between the 101 pine marten individuals in accordance with the EN resistance map, analogous to that used in the design of the corridors in the ecological network of the Basque Country (North Spain) [41]. Resistance values for each land use are indicated in brackets.

non-wooded mountains 24%, cultivated land 14%, and urban land and infrastructures 5.7% of the land area, respectively.

Land use information was obtained in vector format from the most recent forest map of Spain [59] and from national road network maps [60].

Non-invasive genetic sampling and species identification

We used non-invasive scat sampling to collect genetic samples from the *Martes* sp. (*Martes martes* and *Martes foina*) in the study area between 2004 and 2010. Thus, no specific permissions were required for faecal sampling purposes, as the sampling was carried out without needing to intervene directly in the species in focus. We conducted a multi-stage sampling scheme, in which samples from a pilot study were used to assess the appropriateness of the sampling with respect to the research questions. Thus, two scat-based surveys were conducted between 2004 and 2010 across the

sympatric range of both species in the study area. The first survey, conducted in 2004–2005, was used to initially estimate the distribution range of the two sympatric species of the genus *Martes* in the study area and isolate genetic samples of the focal species (*M. martes*). The second, conducted in 2006–2010, was used to refine species distribution information and to obtain a higher number of *M. martes* samples for microsatellite genotyping after a genetic species identification process [52] Aiming to homogenously cover the wide study area and obtain the highest number of different individuals we prioritized our sampling to faecal samples that were separated a minimum of 1km apart (i.e. potentially avoiding re-samplings of the same individual). We also prioritized fresh scat samples to increase genotyping success [61]. Additionally, fresh tissue specimens from road-killed pine martens were included in the data base, when possible. Tissue specimens were collected by authorized veterinarian personnel of the Wildlife

Rehabilitation Centre of Martioda (Alava Regional Council. Department of Environment. Biodiversity section), in line with the laws and ethical protocols governing wildlife management (Law 42/2007) and were submitted to Department of Zoology and Animal Cell Biology (UPV/EHU) for further DNA analyses. No animals were sacrificed for the only purposes of this study. Therefore, a formal approval by an Institutional Animal Care and Use Committee was not necessary. Universal Transversal Mercator (UTM) coordinates were recorded for all the samples collected using a global positioning system (Garmin eTtrex) [61]. The faecal samples were stored in autoclaved tubes containing ethanol 96% and frozen at $-20°C$ until processed [52]. DNA was isolated from tissues and scat using the Qiagen DNeasy Tissue DNA (Qiagen, Hombrechtikon, Switzerland) and DNA Stool MiniKit (Qiagen, Hombrechtikon, Switzerland) according to the manufacturer's instructions, respectively. As pine marten faeces cannot be distinguished from those of the sympatric stone marten (*M. foina*), which is widespread in the study area, and can also be easily confused with those of other carnivores [62], a molecular technique was applied for the identification of faecal samples. Species identification was accomplished by a polymerase chain reaction – restriction fragment length polymorphism (PCR-RFLP) method, providing for an effective genetic identification of sympatric marten species following the method described in Ruiz-González et al. [52].

Microsatellite analyses and individual identification

Identification of individual pine martens used nuclear DNA following methods in Ruiz-González et al. [61]. All the faecal samples identified by the PCR-RFLP method [52] as pine marten were genotyped at 15 variable microsatellite loci (Table S1) using a multiplex protocol specifically designed for degraded faecal DNA analysis [61] and following a modified multitube-approach [63]. The multitube-approach of 4 independent replicates followed by a stringent criteria to construct consensus genotype (i.e. accepting heterozygotes if the two alleles were seen at least in two replicates and homozygotes if a single allele was seen at least in three replicates) is a commonly used approach in non invasive genetic studies leading to a low probability of retaining a false homozygote or false allele error (e.g. [64–66]). Briefly, DNA quality was initially screened by PCR-amplifying each DNA sample four times at four loci (Multiplex 1: MP0188; MP0059; Gg-7; Ma-1), since the results obtained for this four loci are indicative of the genotyping success for the full panel of 15 microsatellites [61].

Only samples showing> 50% positive PCRs were further amplified four times at the remaining 11 loci. Samples with ambiguous results after four amplifications per locus or with < 50% successful amplifications across loci were removed from further analysis as they were not considered reliable genotypes. Multiplex PCR products were run on an ABI (Foster City, CA) 3130XL automated sequencer (Applied Biosystems), with the internal size standard GS500 LIZ (Applied Biosystems). Fragment analyses were conducted using the ABI software Genemapper 4.0.

RELIOTYPE software [67] was used to assess genotype reliability obtained by 4 independent replicates. Samples that were not reliably typed at all loci after 4 replicates (at score threshold $R = 0.95$) were discarded from the analysis. GIMLET software v 1.3.4 [68] was used to calculate the probabilities of identity (PID and PID-sibs) so as to quantify the efficacy in discriminating the fifteen loci in combination. Consensus genotypes from four replicates were reconstructed using GIMLET, accepting heterozygotes if the two alleles were seen at least in two replicates and homozygotes if a single allele was seen at least in three replicates (e.g. [64–66]). GIMLET was also used to estimate

genotyping errors: allelic dropout (ADO) and false alleles (FA) [63,69].

The raw microsatellite data and geographic coordinates of the 101 pine marten individuals are included in Table S2.

Genetic diversity and pairwise individual genetic distances

We summarized genetic variation through the number of alleles per locus (A), expected (HE) and observed (HO) heterozygosities using GENETIX v 4.05.2 [70]. Estimates of pairwise linkage disequilibria for each pair of loci and deviation from Hardy Weinberg equilibrium (HWE) genotypic proportions at each locus were tested using the exact test implemented in GenePop version 4.0 [71]. Statistical significance was evaluated by running a Markov Chain Monte Carlo (MCMC) consisting of 10,000 batches of 10,000 iterations each, with the first 10,000 iterations discarded before sampling [72]. Significance levels were adjusted with sequential Bonferroni correction in order to correct for the effect of multiple tests [73], (i.e. $\alpha = 0.05/\text{number markers}$). MICRO-CHECKER software [74] was used to check for potential scoring errors and the presence of null alleles. The Rousset's a_r inter-individual genetic distance [75] was computed using the program SPAGeDI [76] since this parameter of relatedness does not rely on a reference population [77] and has been successfully applied to infer the effect of landscape on genetic structure of continuously distributed vertebrates [32,78–80].

Construction of landscape resistance models

We produced different resistance maps representing 59 different hypotheses about the resistance of different land use types using ARCGIS version 9.3 [81], with a raster cell size set to 50 m (Table 1; File S1). As suggested by Anderson et al. [82] the sampling grain selected (i.e. 50×50 m) is adequate to infer landscape effects on gene flow as is smaller than the average home-range size of the study species (i.e.> $0.5 Km^2$, [47]). In addition, this resolution allows representation of small landscape patches, but also those smaller elements in the landscape that will be crucial for the resulting effective distances, including linear elements such roads and highways [14].

1) Isolation by distance: Our first hypothesis and null model was a test of isolation by distance across a uniform resistance landscape [57,83]. In this model we assumed movement could occur with equal facility in any direction, with all raster cell values equal in resistance (i.e. resistance value 1).

2) Binary Landscape resistance maps: Our second set of hypotheses propose that some land uses promote genetic connectivity for forest dependant species, such as the pine marten, that specialize in such habitats [47,48], while others resist gene flow. Thus, different binary resistance maps were developed to evaluate the specific land uses which were favourable and unfavourable to the dispersal movements of the martens (i.e. habitat *vs* non-habitat-model). As pine martens are believed to need a minimum woodland area to survive (ca. $2 km^2$) [47] and tend to avoid treeless areas [48,49,51], we expected to find a positive effect of closed-canopy forest habitats and negative effect of open areas and human transformed landscapes on gene flow. Thus, the different binary resistance maps created (Land_A, Land_B, Land_C, Land_D, Land_E, Land_F and Land_G) covered a gradient from greater to lesser preference of the focal species for forest environments, ranging from strictly forest land (Land_A) up to and including open spaces (Land_G) (Table 1, Fig. 2). Therefore, we classified land use data as habitat *vs.* non-habitat and parameterized the models according to a range of plausible resistance values. As there is not a general rule for the

Table 1. Resistance values corresponding to the resistance maps taken evaluated.

Land uses	IBD	Binary landscape resistance maps														Ecological Network	
		A_x	B_x	C_x	D_x	E_x	F_x	G_x	Ab_x	Bb_x	Cb_x	Db_x	Eb_x	Fb_x	Gb_x	EN	ENnb
		Land_A_x to Land_G_x							Land_A_x to Land_G_x								
Forests	1	1	1	1	1	1	1	1	1	1	1	1	1	1	1	1	1
Forestry plantations	1	X	1	1	1	1	1	1	X	1	1	1	1	1	1	10	10
Scrubland	1	X	X	1	1	1	1	1	X	X	1	1	1	1	1	5	5
Agroforestry mosaics	1	X	X	X	1	1	1	1	X	X	X	1	1	1	1	20	20
Pastures and meadows	1	X	X	X	X	1	1	1	X	X	X	X	1	1	1	30	30
Rocks	1	X	X	X	X	X	1	1	X	X	X	X	X	1	1	40	40
Crops	1	X	X	X	X	X	X	1	X	X	X	X	X	X	1	60	60
Wetlands	1	X	X	X	X	X	X	X	100	100	100	100	100	100	100	100	100
National roads	1	X	X	X	X	X	X	X	200	200	200	200	200	200	200	200	50
Highways, urban areas, reservoirs and quarries	1	X	X	X	X	X	X	X	1000	1000	1000	1000	1000	1000	1000	1000	50

Binary landscape resitance maps: 1) *Land_A-Land_G:* Binary resistance maps, on a gradient from greater to lesser preference of the focal species in relation to forest environment; 2) *Land_Ab-Land_Gb:* Maps with letter "b" include the barrier effect of national roads, highways, urban areas, reservoirs and quarries. All the models were evaluated for 4 different resistance values (X = 5, 25, 50, 100) (e.g. Land_A_5 correspond to Land_A model with resistance value of 5). *Ecological Network resistance map:* 1) *EN:* a resistance map analogous to that used in the design of the ecological network of the Basque country; *ENnb:* a variant of the latter which diminishes the barrier effect of national roads, highways, urban areas, water reservoirs and quarries.

Table 2. Results of causal modeling of landscape resistance on genetic distance in European pine marten according to Mantel and partial mantel tests.

Binary Landscape Resistance Models (Land_A to Land_G)

Model/Resistance Values	1) Simple mantel G*L R	G*L p	Rank	2) Partial mantel test G*L|Dis R	G*L|Dis p	Rank	3) Partial mantel G*Dis|L R	G*Dis|L p	CMS?
Land_A *Forest*									
5	0.221	0.0001	49	0.057	0.202	54	−0.007	0.877	N
25	0.219	0.0001	53	0.066	0.215	51	0.047	0.317	N
50	0.217	0.0001	56	0.074	0.170	49	0.066	0.156	N
100	0.211	0.0001	58	0.080	0.138	46	0.089	0.053	N
Mean (±SE)	**0.2169 (±0.0044)**			**0.0692 (±0.0099)**					
Land_B *Forest + Forestry plantations)*									
5	0.216	0.0001	57	0.027	0.463	58	0.011	0.764	N
25	0.224	0.0001	47	0.068	0.134	50	0.009	0.838	N
50	0.227	0.0001	46	0.082	0.085	45	0.024	0.612	N
100	0.232	0.0001	40	0.099	0.051	41	0.040	0.380	N
Mean (±SE)	**0.2247 (±0.0069)**			**0.0689 (±0.0309)**					
Land_C *Forest + forestry plantations + Scrublands*									
5	0.218	0.0001	54	0.041	0.128	56	−0.006	0.863	N
25	0.228	0.0001	45	0.080	0.043	47	−0.009	0.840	Y
50	0.232	0.0001	41	0.091	0.033	43	0.008	0.868	Y
100	0.239	0.0001	33	0.111	0.018	37	0.021	0.648	Y
Mean (±SE)	**0.2291 (±0.0088)**			**0.0806 (±0.0292)**					
Land_D *Forest + forestry plantations + scrublands + agroforestry mosaics*									
5	0.220	0.0001	52	0.052	0.161	55	−0.019	0.608	N
25	0.229	0.0001	44	0.083	0.041	44	−0.012	0.801	Y
50	0.235	0.0001	37	0.100	0.038	40	−0.001	0.981	Y
100	0.242	0.0001	31	0.116	0.026	33	0.017	0.721	Y
Mean (±SE)	**0.2313 (±0.0095)**			**0.0877 (±0.0272)**					
Land_E *Forest + forestry plantations + scrublands + agroforestry mosaics + pastures*									
5	0.220	0.0001	51	0.062	0.049	53	−0.035	0.265	Y

Binary Landscape Resistance Models (Land_Ab to Land_Gb)

Model/Resistance Values	1) Simple mantel test G*L R	G*L p	Rank	2) Partial mantel test G*L|Dis R	G*L|Dis p	Rank	3) Partial mantel G*Dis|L R	G*Dis|L p	CMS?
Land_Ab									
5	0.243	0.0001	28	0.126	0.0005	26	−0.045	0.256	Y
25	0.243	0.0001	29	0.117	0.0138	32	0.008	0.861	Y
50	0.238	0.0001	34	0.112	0.0279	36	0.038	0.421	Y
100	0.229	0.0001	42	0.107	0.0445	38	0.067	0.159	Y
Mean (±SE)	**0.2381 (±0.0063)**			**0.1152 (±0.0078)**					
Land_Bb									
5	0.242	0.0001	30	0.125	0.0001	27	−0.050	0.161	Y
25	0.245	0.0001	23	0.124	0.0022	28	−0.028	0.511	Y
50	0.245	0.0001	25	0.121	0.0083	29	−0.004	0.938	Y
100	0.241	0.0001	32	0.117	0.0195	31	0.028	0.539	Y
Mean (±SE)	**0.2432 (±0.0018)**			**0.1220 (±0.0037)**					
Land_Cb									
5	0.245	0.0001	24	0.135	0.0002	21	−0.059	0.105	Y
25	0.250	0.0001	14	0.140	0.0005	18	−0.045	0.305	Y
50	0.251	0.0001	11	0.137	0.0022	20	−0.024	0.587	Y
100	0.248	0.0001	17	0.129	0.0085	25	0.009	0.859	Y
Mean (±SE)	**0.2487 (±0.0028)**			**0.1349 (±0.0045)**					
Land_Db									
5	0.246	0.0001	22	0.137	0.0001	19	−0.062	0.090	Y
25	0.251	0.0001	12	0.142	0.0006	15	−0.047	0.280	Y
50	0.253	0.0001	8	0.140	0.0024	17	−0.027	0.564	Y
100	0.251	0.0001	13	0.134	0.0064	22	0.005	0.915	Y
Mean (±SE)	**0.2501 (±0.0031)**			**0.1382 (±0.0036)**					
Land_Eb									
5	0.248	0.0001	19	0.149	0.0001	13	−0.078	0.016	Y

Table 2. Cont.

Model/ Resistance Values	1) Simple mantel		Rank	2) Partial mantel test		Rank	3) Partial mantel		CMS?	1) Simple mantel test		Rank	2) Partial mantel test		Rank	3) Partial mantel		CMS?
	G*L	G*L		G*L/Dis	G*L/Dis		G*Dis/L	G*Dis|L		G*L	G*L		G*L/Dis	G*L/Dis		G*Dis/L	G*Dis/L	
	R	p		R	p		R	p		R	p		R	p		R	p	
25	0.234	0.0001	39	0.113	0.009	35	−0.061	0.132	Y	0.255	0.0001	7	0.167	0.0001	6	−0.089	0.015	Y
50	0.243	0.0001	27	0.133	0.004	24	−0.064	0.142	Y	0.259	0.0001	**4**	0.171	0.0001	**2**	−0.086	0.032	Y
100	0.253	0.0001	10	0.150	0.002	11	−0.061	0.176	Y	0.261	0.0001	**2**	0.170	0.0002	**4**	−0.074	0.097	Y
Mean (±SE)	**0.2373 (±0.0138)**			**0.1146 (±0.0380)**						**0.2556 (±0.0060)**			**0.1638 (±0.0103)**					
Land_F *Forest + forestry plantations + scrublands + agroforestry mosaics + pastures + rocky areas*										**Land_Fb**								
5	0.220	0.0001	50	0.062	0.044	52	−0.035	0.266	Y	0.248	0.0001	20	0.149	0.0001	12	−0.078	0.016	Y
25	0.234	0.0001	38	0.114	0.008	34	−0.062	0.134	Y	0.255	0.0001	6	0.167	0.0001	**5**	−0.089	0.014	Y
50	0.243	0.0001	26	0.134	0.004	23	−0.064	0.140	Y	0.259	0.0001	**3**	0.171	0.0001	**1**	−0.086	0.028	Y
100	0.253	0.0001	9	0.150	0.003	10	−0.061	0.184	Y	0.262	0.0001	**1**	0.170	0.0002	**3**	−0.074	0.088	Y
Mean (±SE)	**0.2374 (±0.0138)**			**0.1150 (±0.0381)**						**0.2557 (±0.0060)**			**0.1640 (±0.0104)**					
Land__G *Forest + forestry plantations + scrublands + agroforestry mosaics + pastures + rocky areas + croplands*										**Land_Gb**								
5	0.217	0.0001	55	0.041	0.027	57	−0.018	0.351	Y	0.249	0.0001	16	0.157	0.0001	7	−0.089	0.007	Y
25	0.223	0.0001	48	0.074	0.002	48	−0.039	0.125	Y	0.249	0.0001	15	0.157	0.0001	8	−0.087	0.009	Y
50	0.229	0.0001	43	0.096	0.000	42	−0.051	0.077	Y	0.248	0.0001	18	0.152	0.0001	9	−0.081	0.017	Y
100	0.237	0.0001	36	0.120	0.000	30	−0.063	0.043	Y	0.246	0.0001	21	0.142	0.0001	16	−0.071	0.036	Y
Mean (±SE)	**0.2263 (±0.0082)**			**0.0827 (±0.0338)**						**0.2481 (±0.0014)**			**0.1519 (±0.0069)**					
ENnb										**EN**								
	0.237	0.0001	35	0.103	0.032	39	−0.003	0.947	Y	0.256	0.000	5	0.146	0.0018	14	−0.030	0.519	Y

Model definitions according to Table 1. There are 3 Mantel tests comprising causal modeling: (1) G*L—simple Mantel test between the candidate model and genetic distance; (2) G*L|Dis—partial Mantel test between the candidate model and genetic distance, partialling out Euclidean distance; (3) G*D|L—partial Mantel test between the Euclidean and genetic distance, partialling out the candidate model. For a candidate model to be supported tests (1) and (2) must be significant, while test (3) must be negative or non-significant. Mantel tests meeting each criterion are italicized. Ranking of each model according to Mantel and partial Mantel r values is included. CMS? Indicates if the model is supported within the causal modeling framework (Y) or not (N).

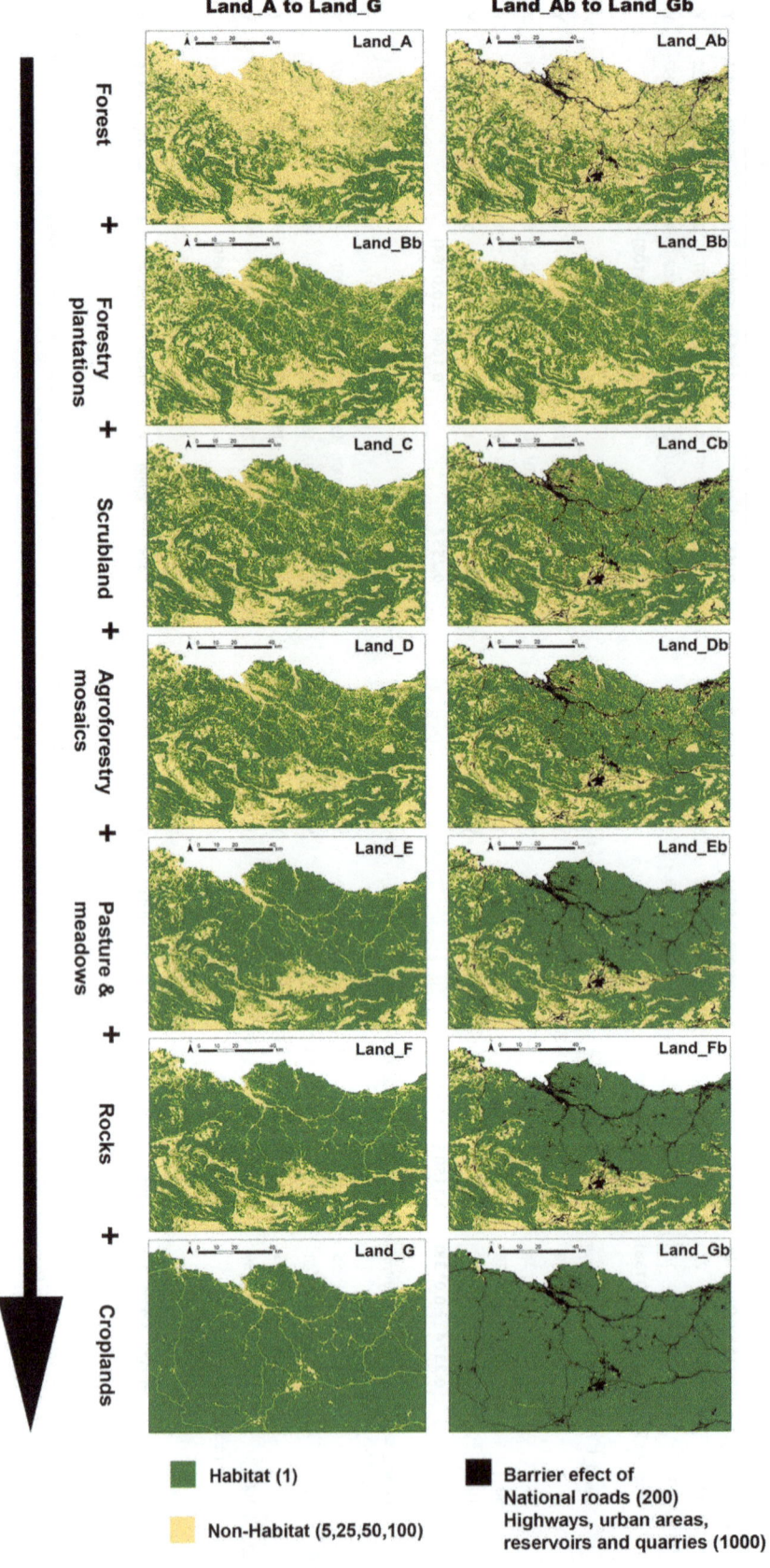

Figure 2. Binary landscape resistance maps on a gradient from greater to lesser preference of the focal species in relation to forest environment. Binary landscape resistance maps, on a gradient from greater to lesser preference of the focal species in relation to forest environment (Land_A to Land_G). Green-coloured cells represent "Habitat" (resistance value 1) and yellow-coloured cells "Non-Habitat" (resistance

values evaluated, 5, 25, 50,100). Models Land_Ab to Land_Gb additionally include black-coloured cells representing the barrier effect of national roads (resistance value 200), highways, urban areas, reservoirs and quarries (resistance value 1000).

assignment of resistance values to non-habitat, we explored 4 different resistance values (5, 25, 50, 100) of non-habitat relative to habitat to verify if it could affect the detectability of landscape genetic relationships, as it has been previously shown in both empirical [24] and simulation [84] studies. In this way, preferential land uses for dispersal were assigned a value of 1 (i.e. Habitat), while non-favourable to dispersal habitat (i.e. non-habitat) sites were assigned a value of 5, 25, 50 or 100, depending on the scenario.

For each of the resistance maps described above, we tested the effect of potential anthropogenic barriers, including national roads (resistance 200), highways, urban areas, reservoirs and quarries (resistance 1000) (Land_Ab to Land_Gb resistance maps; Table 1), using the same resistance values of the ecological network resistance map (see below), for comparative purposes [41]. As each resistance map (Land_A to Land_G; Land_Ab to Land_Gb) was explored for 4 different resistance values, we evaluated 56 different binary resistance maps (i.e. without barrier effect: Land_A_x to Land_G_x; With barrier effect: Land_Ab_x to Land_Gb_x; where X corresponds to the 4 evaluated resistances values per model: 5, 25, 50, and 100) (Table 1).

3) Ecological network resistance map: We evaluated the resistance map previously utilized in the design of ecological corridors linking forest Natura 2000 areas of the Basque Country [41]. The resistance map was based on the assignment of different resistance levels to each land use and parameterized through bibliographical review and expert opinion [41]. The resistance surface outlined in [41] was updated with the new available spatial data regarding land uses in the study area [59,60] (Table 1). Raster breaks in linear barrier elements were avoided by the reinforcement of the size of national roads and highways [14]. Sections of highways which run through viaducts or tunnels were assigned the resistance value corresponding to the land use of the surrounding area. Additionally, a second resistance map was used, with a view to testing the effect of noticeably decreasing the resistance value attributed to the potential barrier effects of national roads, highways, urban areas, reservoirs and quarries (ENnb map) (Table 1).

The raw ascii file of the EN is included in File S1. Following the resistance values outlined in Table 1, all the 59 resistance maps evaluated can be produced from the raw ascii EN resistance map (File S1).

The effective and Euclidean distances between each pair of individuals were calculated with PATHMATRIX 1.1 [15]. Pairwise effective distances between individuals were calculated as the accumulated cost through the least cost paths (LCP) throughout each resistance surface [14,15] (Fig. 1).

We proposed 59 alternative landscape models: 1) 56 binary landscape resistance maps; 2) two complex resistance maps based on the resistance surfaces used to develop the regional ecological network (EN and ENnb) (Table 1), and 3) the null model of Isolation by Distance.

Relationship between genetic and geographical distances within a reciprocal causal modeling framework

Mantel correlations between genetic distance and alternative resistance hypotheses. The pairwise genetic distances matrix (Rousset's a_r) was correlated with different matrices of geographical and (cost) distances encompassing a total number of 5151 pairwise comparisons, including: i) Euclidean distance, to determine whether the patterns of differentiation follow an isolation by distance pattern (null hypothesis) and ii) the effective distances calculated for each of the 58 resistance maps, to infer landscape structure effects on gene flow. The correlation between distance matrixes was calculated by means of the Mantel test [85] and partial Mantel tests [86] as implemented in the ECODIST package [86] in R version 2.7 (R Development Core Team 2008) with 10,000 permutations. Given the potential sensitivity of Mantel tests to non-linear relationships between genetic and cost-distances [87], we compared results between two sets of analyses, one log transforming the effective and Euclidean distances, and one using the original untransformed cost-distance matrices.

Factorial hypothesis cube randomization: Evaluation of the unimodality of support across landscape models. When hypotheses are constructed across a quantitative range of values for a parameter, it is possible to evaluate the degree to which the analysis indicates a unimodal peak of support for a global best model [30]. The degree of unimodality of model support in a factorial hypothesis cube is one measure of the reliability of model results [13]. This is done by computing the differences in support (in our case partial Mantel r values) among all neighbouring cells (i.e. different models) in the hypothesis cube and comparing the sum of those differences to the distribution of the sum of differences from a large number of randomizations of the hypothesis cube (e.g. [13]). We evaluated the unimodality of support across the 56 binary resistance hypotheses (i.e. without barrier effect: Land_Ax to Land_Gx; With barrier effect: Land_Abx to Land_Gbx; where X corresponds to the 4 evaluated resistances values per model: 5, 25, 50, and 100) for the transformed and untransformed analyses using the randomization procedure introduced by Cushman et al. [13], in which the order of hypotheses in the hypothesis cube is randomized a large number of times and each time the difference in partial Mantel r (partialling out distance) is calculated between neighboring hypotheses in the cube. The sum of squared neighbor distances from the actual hypothesis cube (Actual Sum Differences, ASD) is then compared to the distribution of squared neighbor distances in the randomized hypothesis cubes (Mean Sum Randomized Differences, MSRD). We conducted this analysis with 1,000,000 randomizations of the hypothesis cube for both the untransformed and transformed analysis. If no randomizations produce a sum of squared neighbor distances as small as observed, it is strong evidence that the analysis has shown a strong peak of support (i.e. unimodal support).

Original causal modeling. In addition to the reciprocal causal modeling approach [84] (see below), we conducted the original causal modeling [30] as a comparative framework, in which the 58 alternative landscape resistance models are tested against the null model of isolation by distance (IBD) as described in Cushman et al. [30]. There were 3 sets of diagnostic Mantel and partial Mantel tests to complete the causal modeling. These included: (i) simple Mantel tests between genetic distance and landscape resistances; (ii) partial Mantel tests between genetic distance and landscape cost distances, partialling out the effects of Euclidean distance; (iii) partial Mantel tests between genetic distances and Euclidean distance, partialling out the effects of landscape resistance. To infer an effect of a landscape resistance scenario on dispersal, we expected (i) and (ii) to be significant, and we expected (iii) to be negative or non-significant if that scenario

a)

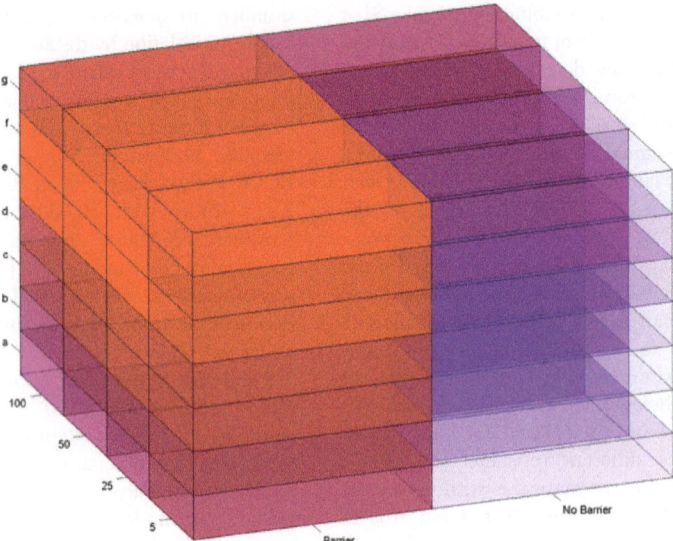

b)

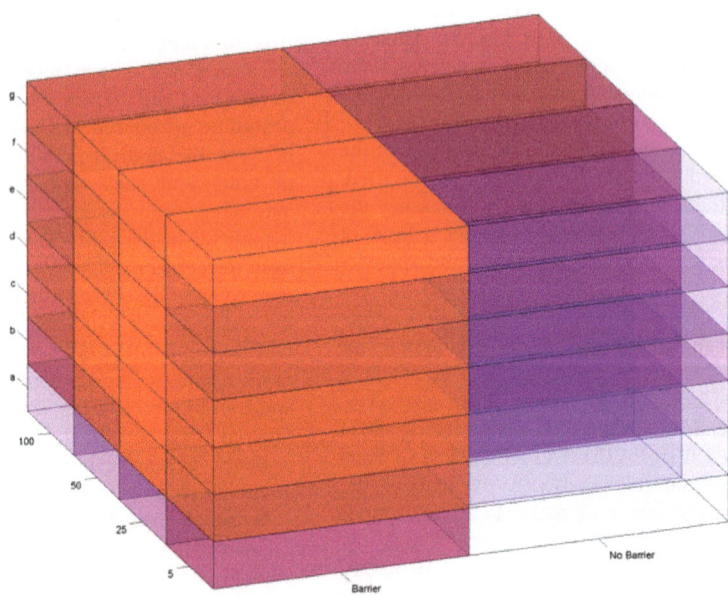

Figure 3. Factorial hypothesis cube randomization. Visualization of the 56 binary landscape-resistance hypotheses after the effects of geographical distance are partialed out on the a) log transformed and b) untransformed cost distances. The cubes each represent one of the 56 binary landscape-resistance models. The cubes are colored in a gradient from blue to red, with red being the most supported models based on the partial Mantel r value. The Mantel r values corresponding to each cube are found in Table 2 and Table S3 for the log transformed and the untransformed matrices, respectively.

'correctly' explained population connectivity in our study population [30].

Reciprocal causal modeling. We used (partial) Mantel tests in a reciprocal causal modeling framework [84] to analyse the influence of landscape structure on gene flow and to determine the extent to which possible landscape resistance models (i.e. resistance maps) explained the spatial pattern of genetic distance between individuals. Cushman and Languth [88] found that the inherent

Table 3. Factorial randomization of the hypothesis cube.

	Untransformed	Transformed
Rank	1	1
Actual Sum Differences (ASD)	19.6325	17.81
Mean Sum Randomized Differences (MSRD)	26.4738	24.17
SD error from MSRD	6.59E+04	6.97E+04

high correlation among alternative resistance models results in a high risk of spurious correlations using simple Mantel tests. Several refinements, including causal modeling [30], have been developed to reduce the risk of affirming spurious correlations and to assist model selection. However, Cushman et al. [84] showed these still suffer from elevated Type I error rates. Consequently, Cushman et al. [84] proposed "reciprocal causal modeling" which they showed greatly lessens Type I error rates in landscape genetic analysis [89]. In reciprocal causal modeling, each alternative resistance hypothesis is tested against all others with partial Mantel tests. A matrix of relative support is calculated by taking the difference between a) the partial Mantel r of each candidate model partialling out each alternative model, and b) the partial Mantel test of the alternative model partialling out the candidate model [84]. A fully supported hypothesis will have positive values of this difference with all alternative models, and no alternative models will have positive values compared to the supported model.

Results

Non-invasive sample collection and species identification

Out of 733 faecal samples collected from the entire study area, 141 were discarded because they were not fresh or because they presumably belong to the same individual (samples separated by < 1km). 494 out of 592 analyzed samples were classified as *Martes* sp. (*M. martes* and *M. foina*) based on genetic species identification results. Thus, unequivocal species identification was possible in 83.45% of the samples. We effectively identified 232 faecal samples as stone marten and 262 as pine marten. Additionally, we obtained 57 tissue samples from road-killed pine martens.

Out of 262 faecal samples identified as pine marten, 108 were not included to the microsatellite genotyping procedure. These samples correspond to the sampling period from 2004–2005, which was used for a first distribution assessment of sympatric martens in the study area and were not potentially fresh enough for microsatellite analysis. Thus, 213 pine marten samples (154 faecal samples and 59 tissue samples) were used for microsatellite genotyping.

Individual identification, genotype checking and genetic diversity

The first quality-screening test, based on 4 replicates of four loci, was not passed by 73 non-invasive samples (47.40%), which were immediately discarded. The remaining 81 samples (52.59%) were amplified at the other 11 loci. After multiple-tubes genotyping, 27 samples from this sub-set (17.53% from the total analyzed samples) were then discarded because they showed <50% PCR success, or because of high failure rates. Full multilocus microsatellite genotypes were obtained for the remaining 54 samples (66.67% from the samples that passed the screening and 35.06% from the total samples analyzed) all showing reliability score> 0.95 [67].

The observed average error rates across loci were: ADO = 0.188 and FA = 0.017. PID analysis showed that the set of 15 loci would produce an identical genotype with a probability of 1.69×10^{-10}, and with a probability of 4.45×10^{-5} for a full-sib, suggesting no "shadow effect" (i.e. all the genotypes identify distinct individuals; [90], and that matching genotypes were recaptures of the same individual).

After a regrouping procedure, we identified 42 individual genotypes from faecal samples. All of the 59 tissue samples were correctly genotyped at 15 loci and all provided new individuals. In total we identified 113 genotypes that corresponded with 101 different individuals. The number of times each individual was detected in the survey varied from 1 to 3, with a total number of 12 re-samplings. Complete genetic profiles and the geographic coordinates for the 101 pine marten individuals are included in Table S2.

The average observed (HO) and expected (HE) heterozygosity values were 0.53 and 0.58, respectively (Table S1). All 15 loci were variable with total numbers of alleles ranging between 3 and 8 per locus. The overall pine marten dataset showed a significant deficit of heterozygotes as compared to Hardy-Weinberg expectations (p <0.001). Despite the broad scale of sampling, the majority of loci were in Hardy-Weinberg proportions (13 out of 15). Only loci Mp0188 and Lut-435 were out of Hardy-Weinberg proportions (Table S1). These results suggest signs of a Wahlund effect, due to the existence of an isolation by distance (Euclidean or effective) pattern in the study area. Linkage disequilibrium was not apparent for any pair of loci after performing Bonferroni corrections.

Correlation between genetic and effective distances

Factorial hypothesis cube randomization. We evaluated the unimodality of support across the 56 binary resistance maps for the log transformed and untransformed data to determine which form of the data should be used for subsequent analyses. After the effects of distance are partialled out, ranking the models by partial Mantel r value provides a means to determine which hypotheses have the greatest support and to identify the most related model to the genetic structure (Table 2, Fig. 3). According to the results outlined in Figure 3, there is a more coherent, unimodal pattern of support in the transformed analysis than the untransformed analysis. Additionally, factorial randomization of the hypothesis cube, in both the transformed and untransformed analyses, no instance of 1,000,000 randomizations produced a sum of squared differences between neighboring hypotheses (MSRD) as small as the actual sum of squared differences (ASD) in partial Mantel r values (Table 3), indicating very high unimodality in both forms of analysis. However, the transformed analysis had higher total support for optimal unimodal support of the best hypothesis as indicated by the larger number of standard errors of MSRD between neighboring hypotheses across the 1,000,000 randomizations (Table 3). Accordingly, all subsequent analyses are restricted to the log transformed resistance distances. As indicated

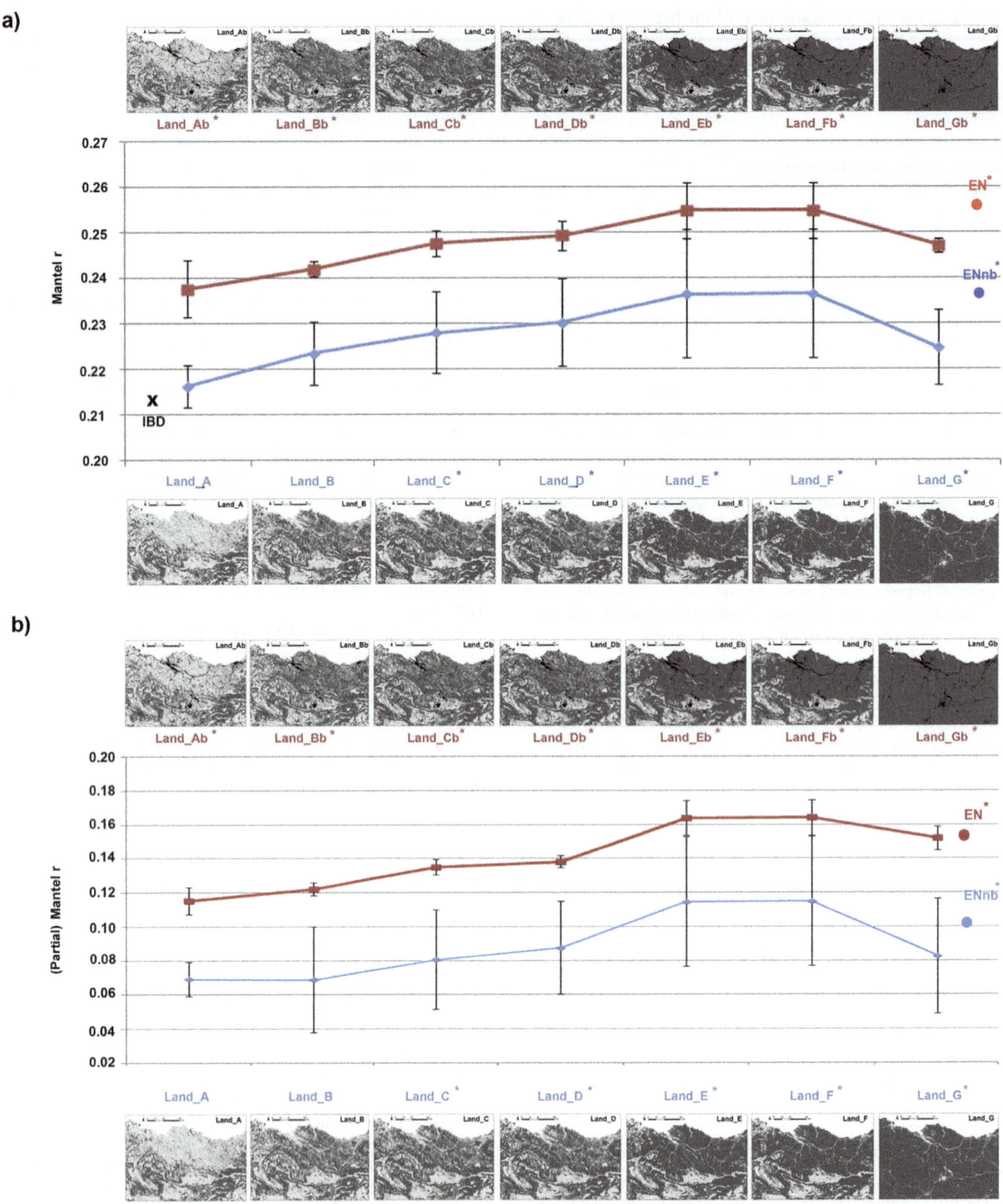

Figure 4. Mantel r results for the different landscape resistance maps evaluated (log transformed). a) Pearson correlation coefficients (Mantel r) between genetic distance and effective distance (log transformed) and **b)** Pearson correlation coefficients (Partial Mantel r) between genetic distance and effective distance (log transformed) after factoring out the effect of the Euclidean distance in the different landscape resistance maps examined. Models marked with an asterisk correspond to the models supported within the causal modeling framework [30].

by the hypothesis cube (Fig. 3), the different resistance values evaluated (5, 25, 50, 100) slightly modified the (partial) Mantel correlation results obtained for each model for both the log transformed (Table 2; Fig. S1 and Fig. S2) and the untransformed distances (Table S3; Fig. S3 and Fig. S4), but overall a consistent pattern was obtained.

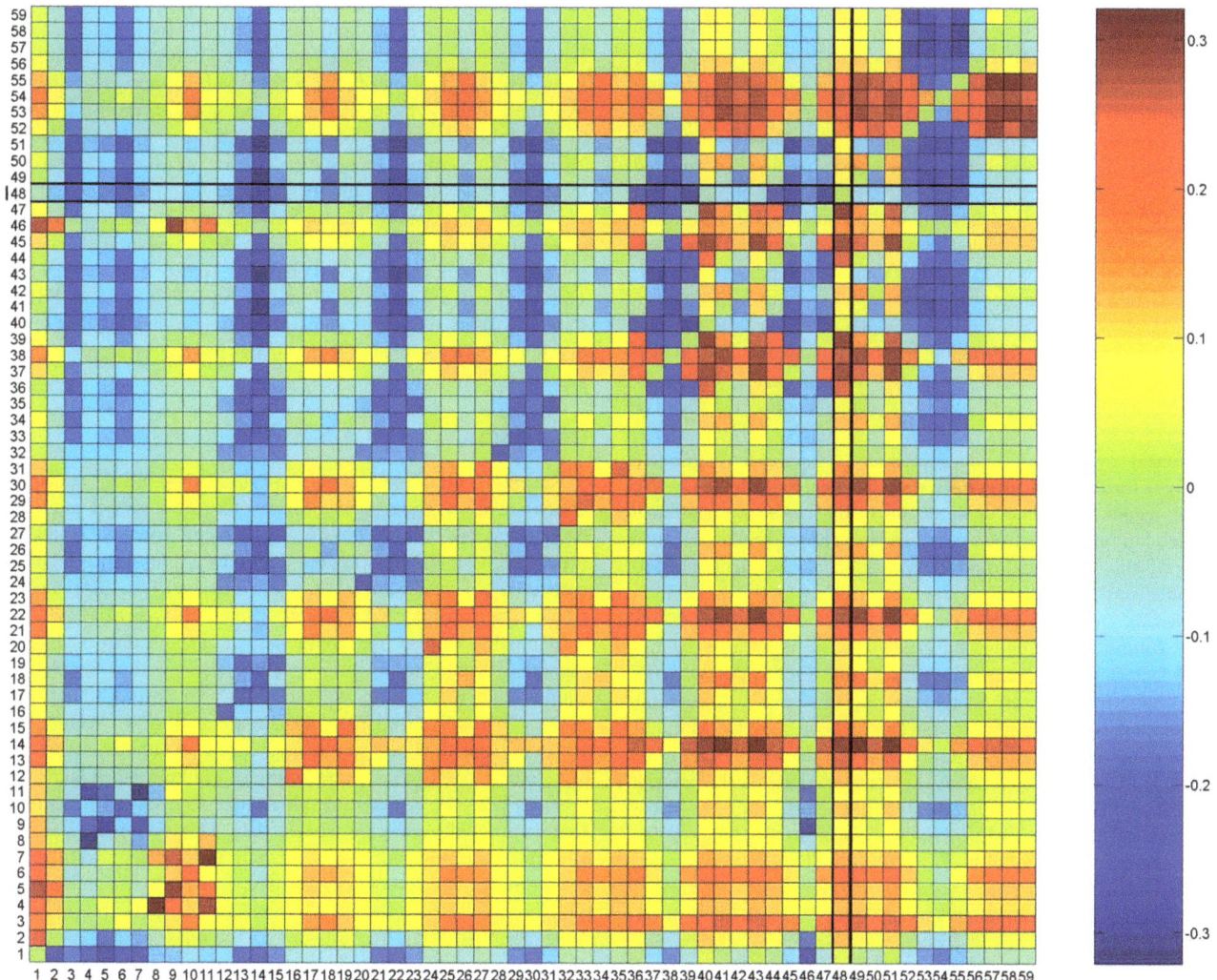

Figure 5. Reciprocal causal modeling results. Results of reciprocal causal modeling on the log transformed cost distances. A single resistance model (Model 48-Land_Fb100) is supported in analysis of the transformed cost distances. Columns indicate focal models, and rows indicate alternative models. The color gradient from blue to red indicates support for the focal model independent of the alternative model (e.g. focal model | alternative model – alternative model | focal model is positive). A fully supported model would have all positive values in the vertical dimension (e.g. that model is supported independently of all other models), and all negative values in the horizontal dimension (no other model is supported independently of the focal model). Model number and associated resistance map: 1 - EN, 2 - ENnb, 3 - Geo_dist, 4 - Land_A100, 5 - Land_A25. 6- Land_A5. 7 - Land_A50, 8 - Land_Ab100, 9 - Land_Ab25, 10 - Land_Ab5. 11 - Land_Ab50. 12 - Land_B100, 13 - Land_B25, 14 - Land_B5, 15 - Land_B50, 16 - Land_Bb100, 17 - Land_Bb25, 18 - Land_Bb5, 19 - Land_Bb50, 20 - Land_C100, 21 - Land_C25, 22- Land_C5, 23 - Land_C50, 24- Land_Cb100, 25 - Land_Cb25, 26 - Land_Cb5, 27 - Land_Cb50, 28 - Land_D100, 29 - Land_D25, 30 - Land_D5, 31 - Land_D50, 32 - Land_Db100, 33 - Land_Db25, 34 - Land_Db5, 35 - Land_Db50, 36 - Land_E100, 37 - Land_E25, 38 - Land_E5, 39 - Land_E50, 40 - Land_Eb100, 41 - Land_Eb25, 42 - Land_Eb5, 43 - Land_Eb50, 44 - Land_F100, 45 - Land_F25, 46 - Land_F5, 47 - Land_F50, 48 - Land_Fb100, 49 - Land_Fb25, 50 - Land_Fb5, 51 - Land_Fb50, 52 - Land_G100, 53 - Land_G25, 54 - Land_G5, 55 - Land_G50, 56 - Land_Gb100, 57 - Land_Gb25, 58 - Land_Gb5, 59 - Land_Gb50.

Simple Mantel correlations between genetic differentiation and alternative landscape models. *1) Isolation by distance:* A significant positive correlation was obtained between the genetic distances and Euclidean distances (r = 0.214; p <0.0001), bearing clear witness to the existence of a pattern of isolation by distance (IBD) (Table 2, Fig. 3). However, when the models were ranked based on Mantel r, all the landscape models performed better than the null model (Table 2, Fig. 4).

2) Binary resistance maps: All of the simple Mantel tests were significant when analyzed in log-transformed form (Table 2). The correlation between genetic and effective distance gradually increased on including, in addition to natural forest (Land_A), forestry plantations (Land_B), scrublands (Land_C), agroforestry

mosaics (Land_D), and pastures and meadows (Land_E) as environments favouring dispersal (Table 2, Fig. 4). This correlation did not change on including rocky areas as dispersal environments (Land_F), while it decreased on including cultivated land (Land_G). The same pattern was obtained with models which specifically increase the cost value of the main barrier features (national roads, highways, urban areas, reservoirs and quarries; Land_Ab to Land_Gb resistance maps), but with an increase of Mantel r values with respect to Land_A to Land_G models (Table 2, Fig. 4), indicating that including barrier effects due to linear features improves the resistance model. The correlation reached its maximum value on including barrier effects in resistance maps Land_E and Land_F (i.e. resistance maps

Land_Eb and Land_Fb; r = 0.256±0.0060; p <0.0001) (Table 2, Fig. 4). The different resistance values (5, 25, 50, 100) slightly modified the correlation results obtained for each model (see Figure S2 and S4 for further details), with the highest correlations for Land_Fb100 (Table 2, Table S3; Fig. 4).

3) Ecological network resistance map: The effective distances calculated on the basis of the EN map were positively correlated with genetic distances and explained a slightly higher proportion of the observed genetic variance than the Euclidean distances (EN, r = 0.256; p <0.0001; Table 2, Fig. 4). The degree of correlation when using the ENnb map was less than that obtained with EN, though still greater than that obtained using Euclidean distance (r = 0.237; p <0.0001; Table 2, Fig. 4). However, the original model used in the design of the ecological network (EN), which included a higher barrier effect for national roads, highways, urban areas, reservoirs and quarries was better supported than the alternative model (ENnb).

Partial Mantel correlations between genetic differentiation and alternative resistance hypotheses. We found significant effects of nearly all of the landscape resistance models (46 out of 52), as the relationship between genetic distance and effective distance was always significant when Euclidean distance was factored out of the relationship (p <0.05) (Table 2; Table S3).

1) Binary landscape resistance maps: The correlation values after factoring out the effect of Euclidean distance showed the same pattern of increase of that obtained by means of a simple Mantel test, with Land_Fb50 showing the highest partial Mantel r correlation (Table 2; Figure 4 and Fig. S2). However, Land_A$_x$, Land_B$_x$, Land_C5 and Land_D5 were not significant when the Euclidean distance was partialled out. Factorial support cubes indicate a clear unimodal peak of support in models Land_Fb50, Land_Eb50, Land_Fb100, Land_Eb100 (Figure 3) that are ranked from first to fourth according to partial r values. Similarly, these models have the highest simple Mantel r values of all of the evaluated models (Table 2; Figure 4). These results suggest that there is a strong peak of support for Land_Fb50 with a clear similarity with Land_Eb50, Land_Fb100 and Land_Eb100. Thus, the best supported models were associated with minimum resistance to movement on forest, forestry plantations, scrublands, agroforestry mosaics and pastures habitats and clear support for the barrier effect.

2) Ecological network resistance map: Both EN and ENnb models appeared better supported than the null model of IBD as this latter retained a significant positive relationship with a$_r$-based genetic distance after factoring out the effects of Euclidean distance, but less supported than the top resistance models (Land_Eb50, Land_Fb100, Land_Eb100; Table 2).

Original and reciprocal causal modeling. Using the original form of causal modeling approach [30] we found that all binary resistance hypotheses except Land_Ax, Land_Bx, Land_C5 and Land_D5 were supported (Table 2).

Using the novel Cushman et al. [84] method of "reciprocal causal modeling", only one resistance model out of the 59 candidate models was fully supported. The single supported model is model number 48, Land_Fb100. The reciprocal causal modeling method shows that the indexes of relative support of this model [i.e. calculated by taking the difference between 1) the partial Mantel r value of the candidate model partialling out each alternative model and 2) the partial Mantel r of the alternative model partialling out the candidate model (reciprocal partial Mantel test)] are all positive (Fig. 5). In addition to model Land_Fb100, several others are nearly perfectly supported. Land_Eb100 is supported independently of all but model

Land_Fb100. Similarly, Land_Fb50 is supported independently of all except Land_Fb100 and Land_Eb100. Model Land_Eb50 is supported independently of all but Land_Fb100, Land_Fb50, and Land_Eb100. This verifies the peak of support seen in the hypothesis cube, with highest support for Land_Fb100, followed by Land_Eb100, Land_Fb50, and Land_Eb50 [i.e. models that included in addition to natural forest, forestry plantations, scrublands, agroforestry mosaics, pastures and meadows (Land Eb) and rocky areas (Land_Fb) as environments favouring dispersal and importance effect of roads as potential barriers to gene flow]. Importantly, the Ecological Network model EN was supported independently of all other models except Land_Fb100, Land_Eb100, Land_Fb50 and Land_Eb50, indicating that it is a highly effective surrogate for landscape resistance to pine marten gene flow.

Discussion

Recent studies suggested that individual-based landscape genetic analysis using partial Mantel tests in a causal modeling framework have high power to correctly identify landscape resistance as a driving process and reject spurious correlations with isolation by distance [84,88]. Cushman et al. [84] showed that the reciprocal causal modeling method we employed here substantially reduces the frequency of Type I errors. In this regards, Castillo et al. [91] recently showed that simulations support the reciprocal causal modeling with partial Mantel tests approach as an effective means to identifying the relationship between gene flow and landscape variables. Although there has been recent controversy over the use of Mantel tests in landscape genetics [92–94], a preferable alternative has yet to be identified that does not also suffer drawbacks [84]. There is no one-size-fits-all approach, and the most appropriate methodology will depend on the research question and landscape under investigation [92]. We use here the more robust modeling framework, proposed by Cushman et al. [84], that is based on the relative support of each candidate model and includes a reciprocal causal modeling step in the model optimization process. Moreover, we used the original causal modeling approach [30], factorial hypothesis cube randomization and ranking by simple Mantel test values as a comparative framework to further explore the performance of these several approaches.

Our results clearly indicate that a standard isolation-by-distance model is not sufficient to explain the observed genetic pattern, and including landscape variables through different resistance maps significantly improves the prediction of the target species gene flow. One model was supported in the transformed analysis using reciprocal causal modeling. This model, Land_Fb100, indicates that pine marten gene flow in northern Spain is facilitated by forests, forestry plantations, scrubland, agroforestry mosaics and pastures and meadows, and that crops have roughly 100 times higher resistance than optimal habitat. Further, this uniquely supported model indicates that anthropogenic barriers, such as national roads, highways, urban areas, reservoirs and quarries and wetlands likely pose much greater resistance to marten gene flow. This suggests that the population connectivity of pine martens in the study area may be vulnerable to habitat loss and fragmentation processes, due to the presence of anthropogenic barriers as has been previously suggested in other forest-dependant species [25,32,95].

The reciprocal causal modeling approach [84] clearly improved discrimination among the competing models, from only one fully supported model (i.e. Land_Fb100) versus more than 80% of models supported equivocally by the original form of causal modeling [30]. Indeed, recent studies showed that the novel reciprocal causal modeling approach is a strong improvement over other methods [91,96]. Even thought reciprocal causal modeling has a greater discrimination to detect the top model, the Cushman et al. [30] method (i.e. causal modeling + model rank + hypothesis cube) reached nearly identical conclusions. We found that ranking models based on simple or partial Mantel r gave consistent support. Likewise, our evaluation of unimodality of support in the hypothesis cube (e.g. [13,24,30]) verified the same peak of support. Reciprocal causal modeling resolved the Type I error problem that we saw in our results from the original form of causal modeling (evaluating support relative to isolation by distance), and was consistent with the peak of support seen in the factorial hypothesis cube. Thus, both forms of causal modeling are complementary and provided independent support to the obtained results, as did evaluating unimodality of support in the hypothesis cube.

In landscape genetics a large number of pairwise genetic relatedness measures have been applied to infer effects of landscape structure on gene flow [12,97], with Rousset's a_r or â [75] being one of the most widely used measures (e.g. [79,80,98,99]). Watts et al. [100] proposed a differentiation statistic (ê) that seems to improve Rousset's â performance for populations with large neighborhood-size ($D\sigma2$) values (i.e. weak IBD pattern). However, differences in statistical performance usually highly depend on the data set used and the sampling scheme as well as how well the data set meets the underlying model assumptions [77]. Thus, and even if the results among different genetic distance measures have generally agreed [24,29,30,101] further studies based on empirical analyses of genetic patterns and simulation modeling are needed to properly evaluate the potential effect of different genetic distance estimator on disentangling landscape effects on gene flow.

Influence of land uses on gene flow: insights into pine marten ecology

The most consistent marten-habitat relation appears to be a general association with forest habitats, and avoidance of open, non-forested habitats [48,51,52]. Thus, the marten's unwillingness to cross open habitats may restrict the species' ability to disperse and colonise new forested areas [51,55]. Ruiz-González et al. [52] found that pine marten occurrence in the study area is highly dependent on the presence of forest and consequently sensitive to forest fragmentation as has been previously suggested in other studies across Europe [49,51]. Nevertheless, the presence of forest habitats is not the only factor which explains pine marten gene flow in the study area, indicating that the habitat selection and gene flow of pine martens may be driven by different factors [17,25,31].This may be because gene flow is driven by mating and dispersal events and habitat selection reflects the behaviour of individual organisms to maximize fitness within home ranges (e.g. [102]).

Our results suggest that it is not only forest masses which serve as favourable environments for dispersal. Scrubland, agroforestry mosaics and grassland habitats also potentially favour dispersal, since the correlation increases as, step by step, these environments are included as predictor variables of pine marten gene flow [Land_B(b) <Land_C(b) <Land_D(b) <Land_E(b)<Land_F(b)]. Original causal modeling identified the same pattern and suggested that Land_Eb and Land_Fb for resistance values of 50 and 100 are the most supported models. Likewise, novel reciprocal causal modeling highlighted similar results, identifying Land_Fb100 as the uniquely supported model, due to its greater discriminatory power [84].

These results are in consonance with recent ecological studies of European pine martens, based on radio tracking, which provide new data substantially differing from traditional descriptions in the scientific literature as strictly forest dependant species [48,51,55]. These studies show that martens are not exclusively confined to extensive forest patches but that they also use other patches including scrubland and agroforestry mosaics [48,51,54,55]. Indeed, the inclusion of scrub habitat in marten home ranges is likely to be related to its role in the connectivity of forest habitats [48,55]. In the same way, the improvement in correlation obtained by including pastures and meadows indicates that the species does not always avoid crossing these open spaces areas when there is forest habitat in the immediate vicinity as has been previously suggested by radiotracking data [48]. This is precisely the case in the area under study, where pastures and meadows are typically found in the immediate vicinity of forest. However, the inclusion of homogeneous croplands reduces the correlation between genetic distance and effective distance, suggesting that zones with intensive agriculture potentially impede species dispersal. This could be due to the scarcity of natural vegetation in these zones and the distance separating them from forest in the study area [41].

Additionally, models that increase the barrier effect of major roads and urban areas leads to a substantial improvement in the correlation between genetic distances and cost distances. The correlation with models Land_Ab-Land_Fb was greater than that obtained with models Land_A-Land_F, and the uniquely supported model in reciprocal causal modeling included the barrier effect (Land_Fb100). This suggests that the potential barrier effect of these land uses could have a synergic effect within a fragmented landscape, decreasing the gene flow due to road avoidance behaviour and/or road mortality [103–106].

Similar landscape genetics studies have also been conducted on other forest dependant *Martes* sp., providing contrasting results regarding landscape effects on gene flow [25,32,107–110]. Similar to the results found in this study, Broquet et al. [32], found that American marten (*Martes americana*) dispersal in Ontario is impeded by the loss and fragmentation of suitable habitat. Wasserman et al. [25] showed that gene flow in the Northern Idaho American marten population is driven by a gradient function of elevation, which was a proxy for snowpack, with marten avoiding lower elevations and dispersing in mid to high elevation montane forests. In contrast, Koen et al. [110] found that marten dispersal across Ontario can best be described as neighbour-mating with no directional bias caused by forest-management induced landscape structure, resulting in a pattern of isolation by distance, suggesting that Ontario landscape is well connected with respect to suitable marten habitat. These contrasting landscape hypotheses governing gene flow could be explained by the different limiting factors that could be acting in each of landscape under study [22].

Even though no previous individual-based landscape genetics data was available for the pine marten, Mergey et al. [58] found that genetic diversity is not associated with habitat fragmentation metrics in France, in spite of the existence of a high degree of forest fragmentation in the studied marten populations. However, this result does not demonstrate that the pine marten gene flow is not affected by forest fragmentation processes. Thus, a more detailed individual-based landscape genetics analysis (Larroque et al. Unpublished data), could provide better insights into the landscape processes governing gene flow and an interesting comparative framework with Spanish pine marten populations.

Empirical evaluation of ecological network resistance maps through landscape genetics

Maps of ecological corridors are commonly used in land use planning, but unfortunately are more often the product of expert opinion rather than empirical data [9,10]. Thus, using landscape genetic analysis, we could partially solve this limitation by studying the gene flow of a target species with regards to the resistance maps used to design the ecological networks [11,12]. Here, the parameterization found in the resistance map which was used to design the regional corridors linking forest protected areas of the Basque Country (north Spain) [41] was adequate to explain pine marten gene flow, with one of the highest partial Mantel r value (r = 0.145) of all the evaluated models. Based on reciprocal causal modeling only Land_Fb100, Land_Eb100, Land_Fb50 and Land_Eb50 were supported independently of EN. Even though Land_Fb100 better explains pine marten gene flow, the high mantel correlation value between the cost distances for Land_Fb100 and EN (Mantel r = 0.9362 p<0.001) suggests EN is a good proxy. This indicates that the EN model used to develop regional connectivity networks among protected areas [41] likely performs very well as a surrogate for landscape resistance for pine marten.

Thus, the resistance map with which the regional ecological network was originally designed in the Basque Country (EN), appears to have high congruence with one of its official target species at regional scale [41]. This is a welcome finding, given that most past evaluations of expert-derived resistance values found that they performed poorly in comparison to empirically optimized models [18,24]. Given the importance of pine marten as a bio-indicator of species associated with natural vegetation [41], our results emphasize the importance of incorporating regional corridors into land use planning and management to preserve landscape connectivity for forest dwelling species.

The influence of resistance values and logarithmic transformations to detect landscape genetic relationships

Since our ability to detect the effects of landscape structure on genetic differentiation depends on both the landscape features used and the relative costs of each feature, different resistance values could provide different results [17,19,83]. Previous studies have found that the degree of contrast in resistance to gene flow in habitat as compared to non-habitat could affect whether or not a given landscape configuration will significantly affect genetic differentiation [22]. We found an increasing reliability of predictions as resistance contrast increased, with several models only supported by 25, 50 and 100 resistance values and unimodal peak of support for 50 and 100 resistance values. The hypothesis that was uniquely supported by reciprocal causal modeling indicated that non-habitat was 100 times more resistant than habitat, and anthropogenic barriers may impart additional resistance as high as 1000 times that of optimal habitat.

Some landscape genetic studies have found that the untransformed geographic distances perform better than logarithmic transformation (e.g. [24,30]), while most previously studies used transformed distances without any evaluation. However, the relationship between cost distances and genetic distances is highly dependant on the study area and the focal species. For example, at small extents the relationship between cost and genetic distances is nearly linear and the untrasformed correlations may fit the data better [24,30]. However, when the study area is large in extent relative to the dispersal ability of the species, as in the present study, the relationship between cost distance and genetic distance will be nonlinear and the logarithm transform will improve fit. Thus, taking into account the potential bias due to an incorrect use of transformations, we propose that future landscape genetics should evaluate the unimodality of support among the hypotheses as a means to determine the degree to which the transformation improves the analysis.

Conclusions

This paper presents a comprehensive individual-based landscape genetic analysis of the European pine marten, and the first formal use of landscape genetics to evaluate the effectiveness of regional ecological networks. We compared results from several methods of model selection and found that ranking based on Mantel r or partial Mantel r, the unimodality of support in the hypothesis cube, causal modeling and reciprocal causal modeling all identified the same best models of landscape resistance for European pine marten in northern Spain. Reciprocal causal modeling appeared to provide the strongest differentiation among hypotheses and enabled the identification of a single, independently supported model. Gene flow of European pine marten is facilitated by natural land cover, such as forest, scrublands and pastures and meadows, and is resisted by anthropogenic land uses and linear barriers such as major roads. We confirm that the resistance map used to develop the regional ecological network in the Basque Country is a close surrogate to the empirically optimized resistance model for marten.

Supporting Information

Figure S1 Mantel Correlation values for a) Land_A-Land_G and b) Land_Ab-Land_Gb models for the 4 different cost values evaluated on the log transformed cost distances.

Figure S2 Partial Mantel Correlation values for a) Land_A-Land_G and b) Land_Ab-Land_G models for the 4 different cost values evaluated on the log transformed cost distances.

Figure S3 Mantel Correlation values for a) Land_A-Land_G and b) Land_Ab-Land_G models for the 4 different cost values evaluated on the untransformed cost distances.

Figure S4 Partial Mantel Correlation values for a) Land_A-Land_G and b) Land_Ab-Land_G models for the 4 different cost values evaluated on the untransformed cost distances.

Table S1 Summary of the genetic variability. Genetic variability of the 15 microsatellite loci multiplexed used in this study. Number of alleles (N_A), Observed (H_O) and expected (H_E) heterozygosities for each locus and for whole data set.. Loci marked with an asterisk deviated from Hardy-Weinberg proportions.

Table S2 Sample locations and microsatellite data. Complete genetic profiles (15 microsatellite loci) and the geographic coordinates for the 101 pine marten individuals

Table S3 Results of causal modeling of landscape resistance on genetic distance in European pine marten according to Mantel and partial mantel tests for the untransformed distances.

File S1 Ecological network resistance map. Raw '.asc' file of the EN resistance map from which all the resistance maps evaluated can be produced following the resistance values outlined in Table 1.

Acknowledgments

The authors wish to thank the following persons and institutions for supplying some of the tissue and faecal samples used in this study: Patricia Lizarraga, Ricardo Gutierrez and Laura Elorza (CRF Martioda—Alava Provincial Council); Txema Fernandez and Javier Sesma (IKT S:A); the technical staff and rangers of the Natural Parks of Alava (Juan Carlos Ortíz, Ricardo Ortíz, María Elena García, Kepa García, Sonia Benítez, Elisabeth Cabanillas, Iker Ayala, Arantza Puente, Jesús Gómez, Iñaki Martínez, Lidia Lacha, Jokin Sáez de Camara); Rangers from Alava (Asier Martínez, Tomás Landa, Javier Pinedo, Mario Corral) and Bizkaia (Igor Aginako, Julio Ruiz Guijarro, Ignacio Martínez, Eneko Díaz, Juan Manuel Pérez de Ana, David Vado, Ander Eguia) Provincial Councils; Joseba Carreras (Alava Provincial Council); Jokin Larumbe (Navarre Government); Dr. Fermín Urra (GAVRN); Asun Gómez (TRAGSATEC); Dr. Madis Podra (Asoc. Visón Europeo); Nerea Ruiz de Azua (EKOS, S.L.); Gorka Belamendia (CEA); Felipe Canales and Miguel Ángel Campos (Consultora de Recursos Naturales, S.L); Enrique Arberas; Oskar Berdión; Javier López de Luzuriaga; Dr. Iñigo Zuberogoitia (ICARUS); Dr. Marta Barral (NEIKER); Rebeca Pérez; Haizea Aguirre; Pablo Pérez; Amaia García de Albéniz; Alvaro Osés; and Gerardo Domínguez. Especial thanks to Maria Vergara, Oihana Razkin, Luis Javier Chueca and Peio Lozano (UPV/EHU), the staff of IKT S.A and colleagues of the Conservation genetics laboratory (ISPRA) for their valuable help at different stages of the study.

Author Contributions

Conceived and designed the experiments: ARG MG SC ER BJGM. Performed the experiments: ARG MG MJM. Analyzed the data: ARG MG SC. Contributed reagents/materials/analysis tools: ER BJGM. Wrote the paper: ARG MG SC. Revised the manuscript: ARG MG SC MJM ER BJGM.

References

1. Opdam P, Wascher D (2004) Climate change meets habitat fragmentation: linking landscape and biogeographical scale levels in research and conservation. Biological Conservation 117: 285–297.

2. Fahrig L (2007) Non-optimal animal movement in human-altered landscapes. Functional Ecology 21: 1003–1015.

3. Cushman SA, McKelvey KS, Schwartz MK (2009) Use of Empirically Derived Source-Destination Models to Map Regional Conservation Corridors. Conservation Biology 23: 368–376.

4. Taylor PD, Fahrig L, Henein K, Merriam G (1993) Connectivity is a vital element of landscape structure. Oikos 68: 571–573.

5. Brooks CP (2003) A scalar analysis of landscape connectivity. Oikos 102: 433–439.

6. Cushman SA, Landguth EL (2012) Multi-taxa population connectivity in the Northern Rocky Mountains. Ecological Modelling 231: 101–112.

7. Bennett G, Wit P (2001) The development and application of ecological networks: a review of proposals, plans and programmes. Amsterdam: AIDEnvironment.

8. Jongman R, Pungetti G (2004) Ecological networks and greenways: concept, design, implementation. Cambridge: Cambridge University Press. pp. 368

9. Cushman S, McRae B, Adriansen F, Beier P, Shirley M, et al. (2013) Biological corridors and connectivity. In: MacDonald D, editor. Conservation in Theory and Practice. New York: Wiley. pp. 32.

10. Boitani L, Falcucci A, Maiorano L, Rondinini C (2007) Ecological networks as conceptual frameworks or operational tools in conservation. Conservation Biology 21: 1414–1422.

11. Luque S, Saura S, Fortin M-J (2012) Landscape connectivity analysis for conservation: insights from combining new methods with ecological and genetic data. Landscape Ecology 27: 153–157.

12. Segelbacher G, Cushman SA, Epperson BK, Fortin MJ, Francois O, et al. (2010) Applications of landscape genetics in conservation biology: concepts and challenges. Conservation Genetics 11: 375–385.

13. Cushman SA, Chase M, Griffin C (2010) Mapping landscape resistance to identify corridors and barriers for elephant movement in Southern Africa. In: Cushman SA, Huettmann F, editors. Spatial Complexity, Informatics and Wildlife Conservation. Tokyo: Springer. pp. 349–367.

14. Adriaensen F, Chardon JP, De Blust G, Swinnen E, Villalba S, et al. (2003) The application of 'least-cost' modelling as a functional landscape model. Landscape and Urban Planning 64: 233–247.

15. Ray N (2005) PATHMATRIX: a geographical information system tool to compute effective distances among samples. Molecular Ecology Notes 5: 177–180.

16. Beier P, Majka DR, Spencer WD (2008) Forks in the road: Choices in procedures for designing wildland linkages. Conservation Biology 22: 836–851.

17. Spear SF, Balkenhol N, Fortin MJ, McRae BH, Scribner K (2010) Use of resistance surfaces for landscape genetic studies: considerations for parameterization and analysis. Molecular Ecology 19: 3576–3591.

18. Zeller KA, McGarigal K, Whiteley AR (2012) Estimating landscape resistance to movement: a review. Landscape Ecology 27: 777–797.

19. Rayfield B, Fortin M-J, Fall A (2010) The sensitivity of least-cost habitat graphs to relative cost surface values. Landscape Ecology 25: 519–532.

20. Manel S, Schwartz MK, Luikart G, Taberlet P (2003) Landscape genetics: combining landscape ecology and population genetics. Trends in Ecology & Evolution 18: 189–197.

21. Manel S, Holderegger R (2013) Ten years of landscape genetics. Trends in Ecology & Evolution 28: 614–621.

22. Cushman SA, Shirk AJ, Landguth EL (2013) Landscape genetics and limiting factors. Conservation Genetics 14: 263–274.

23. Epps CW, Wehausen JD, Bleich VC, Torres SG, Brashares JS (2007) Optimizing dispersal and corridor models using landscape genetics. Journal of Applied Ecology 44: 714–724.

24. Shirk AJ, Wallin DO, Cushman SA, Rice CG, Warheit KI (2010) Inferring landscape effects on gene flow: a new model selection framework. Molecular Ecology 19: 3603–3619.

25. Wasserman TN, Cushman SA, Schwartz MK, Wallin DO (2010) Spatial scaling and multi-model inference in landscape genetics: Martes americana in northern Idaho. Landscape Ecology 25: 1601–1612.

26. Bull RAS, Cushman SA, Mace R, Chilton T, Kendall KC, et al. (2011) Why replication is important in landscape genetics: American black bear in the Rocky Mountains. Molecular Ecology 20: 1092–1107.

27. Zalewski A, Piertney SB, Zalewska H, Lambin X (2009) Landscape barriers reduce gene flow in an invasive carnivore: geographical and local genetic structure of American mink in Scotland. Molecular Ecology 18: 1601–1615.

28. Perez-Espona S, McLeod JE, Franks NR (2012) Landscape genetics of a top neotropical predator. Molecular Ecology 21: 5969–5985.

29. Blair C, Weigel DE, Balazik M, Keeley ATH, Walker FM, et al. (2012) A simulation-based evaluation of methods for inferring linear barriers to gene flow. Molecular Ecology Resources 12: 822–833.

30. Cushman SA, McKelvey KS, Hayden J, Schwartz MK (2006) Gene flow in complex landscapes: Testing multiple hypotheses with causal modeling. American Naturalist 168: 486–499.

31. Centeno-Cuadros A, Roman J, Delibes M, Antonio Godoy J (2011) Prisoners in Their Habitat? Generalist Dispersal by Habitat Specialists: A Case Study in Southern Water Vole (Arvicola sapidus). Plos One 6.

32. Broquet T, Ray N, Petit E, Fryxell JM, Burel F (2006) Genetic isolation by distance and landscape connectivity in the American marten (Martes americana). Landscape Ecology 21: 877–889.

33. Wang YH, Yang KC, Bridgman CL, Lin LK (2008) Habitat suitability modelling to correlate gene flow with landscape connectivity. Landscape Ecology 23: 989–1000.

34. Schwartz MK, Copeland JP, Anderson NJ, Squires JR, Inman RM, et al. (2009) Wolverine gene flow across a narrow climatic niche. Ecology 90: 3222–3232.

35. Blair C, Jimenez Arcos VH, Mendez de la Cruz FR, Murphy RW (2013) Landscape Genetics of Leaf-Toed Geckos in the Tropical Dry Forest of Northern Mexico. Plos One 8.

36. Beja-Pereira A, Oliveira R, Alves PC, Schwartz MK, Luikart G (2009) Advancing ecological understandings through technological transformations in noninvasive genetics. Molecular Ecology Resources 9: 1279–1301.

37. Taberlet P, Waits LP, Luikart G (1999) Noninvasive genetic sampling: look before you leap. Trends in Ecology & Evolution 14: 323–327.

38. Piggott MP, Taylor AC (2003) Remote collection of animal DNA and its applications in conservation management and understanding the population biology of rare and cryptic species. Wildlife Research 30: 1–13.

39. Waits LP, Paetkau D (2005) Noninvasive genetic sampling tools for wildlife biologists: A review of applications and recommendations for accurate data collection. Journal of Wildlife Management 69: 1419–1433.

40. Schwartz MK, Monfort SL (2008) Genetic and Endocrine Tools for Carnivore Surveys. In: Long R, MacKay P, Ray J, Zielinski W, editors. Noninvasive survey methods for North American carnivores. Washington D.C.: Island Press. pp. 228–250.

41. Gurrutxaga M, Lozano PJ, del Barrio G (2010) GIS-based approach for incorporating the connectivity of ecological networks into regional planning. Journal for Nature Conservation 18: 318–326.

42. Jongman RHG (2002) Homogenisation and fragmentation of the European landscape: ecological consequences and solutions. Landscape and Urban Planning 58: 211–221.

43. Jongman R, Bouwma I, Van Doorn A (2006) Indicative map of the Pan–European ecological network in Western Europe. Wageningen: Alterra. 104 p.

44. Gurrutxaga M, Rubio L, Saura S (2011) Key connectors in protected forest area networks and the impact of highways: A transnational case study from the Cantabrian Range to the Western Alps (SW Europe). Landscape and Urban Planning 101: 310–320.

45. Mallarach J, Rafa M, Sargatal J (2010) Cantabrian Mountains-Pyrénées-Massif Central-Western Alps great mountain corridor. In: Worboys GL, Francis WL, Lockwood M, editors. Connectivity conservation management A global guide London: Earthscan. pp. 269–279.

46. Proulx G, Aubry K, Birks J, Buskirk S, Fortin C, et al. (2004) World distribution and status of the genus Martes in 2000. In: Harrison D, Fuller A, Proulx G, editors. Martens and fishers (Martes) in human- altered environments: an international perspective. New York: Springer-Verlag. pp. 77–98.

47. Zalewski A, Jedrzejewski W (2006) Spatial organisation and dynamics of the pine marten Martes martes population in Bialowieza Forest (E Poland) compared with other European woodlands. Ecography 29: 31–43.

48. Pereboom V, Mergey M, Villerette N, Helder R, Gerard JF, et al. (2008) Movement patterns, habitat selection, and corridor use of a typical woodland-dweller species, the European pine marten (Martes martes), in fragmented landscape. Canadian Journal of Zoology-Revue Canadienne De Zoologie 86: 983–991.

49. Brainerd SM, Rolstad J (2002) Habitat selection by Eurasian pine martens Martes martes in managed forests of southern boreal Scandinavia. Wildlife Biology 8: 289–297.

50. Kurki S, Nikula A, Helle P, Linden H (1998) Abundances of red fox and pine marten in relation to the composition of boreal forest landscapes. Journal of Animal Ecology 67: 874–886.

51. Mergey M, Helder R, Roeder J-J (2011) Effect of forest fragmentation on space-use patterns in the European pine marten (Martes martes). Journal of Mammalogy 92: 328–335.

52. Ruiz-Gonzalez A, Rubines J, Berdion O, Gomez-Moliner BJ (2008) A non-invasive genetic method to identify the sympatric mustelids pine marten (Martes martes) and stone marten (Martes foina): preliminary distribution survey on the northern Iberian Peninsula. European Journal of Wildlife Research 54: 253–261.

53. Bright PW (2000) Lessons from lean beasts: conservation biology of the mustelids. Mammal Review 30: 217–226.

54. Balestrieri A, Remonti L, Ruiz-Gonzalez A, Gomez-Moliner BJ, Vergara M, et al. (2010) Range expansion of the pine marten (Martes martes) in an agricultural landscape matrix (NW Italy). Mammalian Biology 75: 412–419.

55. Caryl FM, Quine CP, Park KJ (2012) Martens in the matrix: the importance of nonforested habitats for forest carnivores in fragmented landscapes. Journal of Mammalogy 93.

56. Schwartz M, Ruiz-González A, Pertoldi C, Masuda R (2012) Martes conservation genetics: assessing within species movements, units to conserve and connectivity cross ecological and evolutionary time. In: Aubry K, Zielinski W, Raphael M, Proulx G, Buskirk S, editors. Biology and conservation of marten, sables, and fisher: a new synthesis. New York: Cornell University Press. pp. 398–428.

57. Wright S (1943) Isolation by distance. Genetics 28: 114–138.

58. Mergey M, Larroque J, Ruette S, Vandel J-M, Helder R, et al. (2012) Linking habitat characteristics with genetic diversity of the European pine marten (Martes martes) in France. European Journal of Wildlife Research 58: 909–922.

59. Environment SMft (2006) Mapa forestal de España 1:50.000. Madrid: Ministerio de Medio Ambiente.

60. Institut SNG (2008) Base cartográfica numérica BCN200. Ministerio de Fomento.

61. Ruiz-Gonzalez A, Jose Madeira M, Randi E, Urra F, Gomez-Moliner BJ (2013) Non-invasive genetic sampling of sympatric marten species (Martes martes and Martes foina): assessing species and individual identification success rates on faecal DNA genotyping. European Journal of Wildlife Research 59: 371–386.

62. Davison A, Birks JDS, Brookes RC, Braithwaite TC, Messenger JE (2002) On the origin of faeces: morphological versus molecular methods for surveying rare carnivores from their scats. Journal of Zoology 257: 141–143.

63. Taberlet P, Griffin S, Goossens B, Questiau S, Manceau V, et al. (1996) Reliable genotyping of samples with very low DNA quantities using PCR. Nucleic Acids Research 24: 3189–3194.

64. Frantz AC, Pope LC, Carpenter PJ, Roper TJ, Wilson GJ, et al. (2003) Reliable microsatellite genotyping of the Eurasian badger (Meles meles) using faecal DNA. Molecular Ecology 12: 1649–1661.

65. Stenglein JL, Waits LP, Ausband DE, Zager P, Mack CM (2010) Efficient, Noninvasive Genetic Sampling for Monitoring Reintroduced Wolves. Journal of Wildlife Management 74: 1050–1058.

66. Brzeski KE, Gunther MS, Black JM (2013) Evaluating River Otter Demography Using Noninvasive Genetic Methods. Journal of Wildlife Management 77: 1523–1531.

67. Miller CR, Joyce P, Waits LP (2002) Assessing allelic dropout and genotype reliability using maximum likelihood. Genetics 160: 357–366.

68. Valiere N (2002) GIMLET: a computer program for analysing genetic individual identification data. Molecular Ecology Notes 2: 377–379.

69. Pompanon F, Bonin A, Bellemain E, Taberlet P (2005) Genotyping errors: Causes, consequences and solutions. Nature Reviews Genetics 6: 847–859.

70. Belkhir K, Borsa P, Chikhi L, Raufaste N, Bonhomme F (2004) Genetix 4.02, Logiciel sous windows pour la génétique des populations. Université de Montpellier II, Montpellier, France.

71. Raymond M, Rousset F (1995) Genepop (Version-1.2) - Population-Genetics Software for Exact Tests and Ecumenicism. Journal of Heredity 86: 248–249.

72. Guo SW, Thompson EA (1992) Performing the Exact Test of Hardy-Weinberg Proportion for Multiple Alleles. Biometrics 48: 361–372.

73. Rice WR (1989) Analyzing Tables of Statistical Tests. Evolution 43: 223–225.

74. Van Oosterhout C, Hutchinson WF, Wills DPM, Shipley P (2004) MICRO-CHECKER: software for identifying and correcting genotyping errors in microsatellite data. Molecular Ecology Notes 4: 535–538.

75. Rousset F (2000) Genetic differentiation between individuals. Journal of Evolutionary Biology 13: 58–62.

76. Hardy OJ, Vekemans X (2002) SPAGEDi: a versatile computer program to analyse spatial genetic structure at the individual or population levels. Molecular Ecology Notes 2: 618–620.

77. Vekemans X, Hardy OJ (2004) New insights from fine-scale spatial genetic structure analyses in plant populations. Molecular Ecology 13: 921–935.

78. Coulon A, Cosson JF, Angibault JM, Cargnelutti B, Galan M, et al. (2004) Landscape connectivity influences gene flow in a roe deer population inhabiting a fragmented landscape: an individual-based approach. Molecular Ecology 13: 2841–2850.

79. Blair ME, Melnick DJ (2012) Scale-Dependent Effects of a Heterogeneous Landscape on Genetic Differentiation in the Central American Squirrel Monkey (Saimiri oerstedii). Plos One 7.

80. Dudaniec RY, Rhodes JR, Wilmer JW, Lyons M, Lee KE, et al. (2013) Using multilevel models to identify drivers of landscape-genetic structure among management areas. Molecular Ecology 22: 3752–3765.

81. ESRI (2009) ArcMap version 9.3. Redlands, CA, USA.: Environmental Systems Research Institute.

82. Anderson CD, Epperson BK, Fortin MJ, Holderegger R, James PMA, et al. (2010) Considering spatial and temporal scale in landscape-genetic studies of gene flow. Molecular Ecology 19: 3565–3575.

83. Jenkins DG, Carey M, Czerniewska J, Fletcher J, Hether T, et al. (2010) A meta-analysis of isolation by distance: relic or reference standard for landscape genetics? Ecography 33: 315–320.

84. Cushman SA, Wasserman TN, Landguth EL, Shirk AJ (2013) Re-Evaluating Causal Modeling with Mantel Tests in Landscape Genetics. Diversity 5: 51–72.

85. Mantel N (1967) Detection of disease clustering and a generalized regression approach. Cancer Research 27: 209–220.

86. Smouse PE, Long JC, Sokal RR (1986) Multiple-regression and correlation extensions of the mantel test of matrix correspondence. Systematic Zoology 35: 627–632.

87. Rousset F (1997) Genetic differentiation and estimation of gene flow from F-statistics under isolation by distance. Genetics 145: 1219–1228.

88. Cushman SA, Landguth EL (2010) Spurious correlations and inference in landscape genetics. Molecular Ecology 19: 3592–3602.

89. Castillo JA, Epps CW, Davis AR, Cushman SA (2014) Landscape effects on gene flow for a climate-sensitive montane species, the American pika. Molecular Ecology 23: 843–856.

90. Mills LS, Citta JJ, Lair KP, Schwartz MK, Tallmon DA (2000) Estimating animal abundance using noninvasive DNA sampling: Promise and pitfalls. Ecological Applications 10: 283–294.

91. Castillo JA, Epps CW, Davis AR, Cushman S (2014) Landscape effects on gene flow for a climate-sensitive montane species, the American pika. Molecular Ecology 23: 843–856.

92. Balkenhol N, Gugerli F, Cushman SA, Waits LP, Coulon A, et al. (2009) Identifying future research needs in landscape genetics: where to from here? Landscape Ecology 24: 455–463.

93. Guillot G, Rousset F (2013) Dismantling the Mantel tests. Methods in Ecology and Evolution 4: 336–344.

94. Graves TA, Beier P, Royle JA (2013) Current approaches using genetic distances produce poor estimates of landscape resistance to interindividual dispersal. Molecular Ecology 22: 3888–3903.

95. Coulon A, Guillot G, Cosson JF, Angibault JMA, Aulagnier S, et al. (2006) Genetic structure is influenced by landscape features: empirical evidence from a roe deer population. Molecular Ecology 15: 1669–1679.

96. Cushman SA, Max TL, Whitham TG, Allan GJ (2014) River network connectivity and climate gradients drive genetic differentiation in a riparian foundation tree. Ecological Applications 24:1000–1014.

97. Guillot G, Leblois R, Coulon A, Frantz AC (2009) Statistical methods in spatial genetics. Molecular Ecology 18: 4734–4756.

98. Latch EK, Boarman WI, Walde A, Fleischer RC (2011) Fine-Scale Analysis Reveals Cryptic Landscape Genetic Structure in Desert Tortoises. Plos One 6.

99. Quemere E, Crouau-Roy B, Rabarivola C, Louis EE, Chikhi L (2010) Landscape genetics of an endangered lemur (*Propithecus tattersalli*) within its entire fragmented range. Molecular Ecology 19: 1606–1621.

100. Watts PC, Rousset F, Saccheri IJ, Leblois R, Kemp SJ, et al. (2007) Compatible genetic and ecological estimates of dispersal rates in insect (*Coenagrion mercuriale*: Odonata: Zygoptera) populations: analysis of 'neighbourhood size' using a more precise estimator. Molecular Ecology 16: 737–751.

101. Row JR, Blouin-Demers G, Lougheed SC (2010) Habitat distribution influences dispersal and fine-scale genetic population structure of eastern foxsnakes (*Mintonius gloydi*) across a fragmented landscape. Molecular Ecology 19: 5157–5171.

102. Cushman SA, Lewis JS (2010) Movement behavior explains genetic differentiation in American black bears. Landscape Ecology 25: 1613–1625.

103. Balkenhol N, Waits LP (2009) Molecular road ecology: exploring the potential of genetics for investigating transportation impacts on wildlife. Molecular Ecology 18: 4151–4164.

104. Cushman SA, Compton BW, McGarigal K (2010) Habitat fragmentation effects depend on complex interactions between population size and dispersal ability: Modeling influences of roads, agriculture and residential development across a range of life-history characteristics. In: Cushman SA, Huettman F, editors. Spatial complexity, informatics and wildlife conservation. Tokyo: Springer. pp. 369–387.

105. Jackson ND, Fahrig L (2011) Relative effects of road mortality and decreased connectivity on population genetic diversity. Biological Conservation 144: 3143–3148.

106. Cushman SA, Lewis JS, Landguth EL (2013) Evaluating the intersection of a regional wildlife connectivity network with highways. Movement Ecology 1: 1–12.

107. Cushman SA, Raphael MG, Ruggiero LF, Shirk AS, Wasserman TN, et al. (2011) Limiting factors and landscape connectivity: the American marten in the Rocky Mountains. Landscape Ecology 26: 1137–1149.

108. Wasserman TN, Cushman SA, Littell JS, Shirk AJ, Landguth EL (2013) Population connectivity and genetic diversity of American marten (*Martes americana*) in the United States northern Rocky Mountains in a climate change context. Conservation Genetics 14: 529–541.

109. Garroway CJ, Bowman J, Wilson PJ (2011) Using a genetic network to parameterize a landscape resistance surface for fishers, *Martes pennanti*. Molecular Ecology 20: 3978–3988.

110. Koen EL, Bowman J, Garroway CJ, Mills SC, Wilson PJ (2012) Landscape resistance and American marten gene flow. Landscape Ecology 27: 29–43.

13

The Impact of Selective-Logging and Forest Clearance for Oil Palm on Fungal Communities in Borneo

Dorsaf Kerfahi[1,2]**, Binu M. Tripathi**[1]**, Junghoon Lee**[2]**, David P. Edwards**[3]*, **Jonathan M. Adams**[1]*

1 Department of Biological Sciences, Seoul National University, Seoul, Republic of Korea, **2** School of Chemical and Biological Engineering, Interdisciplinary Program of Bioengineering, Seoul National University, Seoul, Republic of Korea, **3** Department of Animal and Plant Sciences, University of Sheffield, Sheffield, United Kingdom

Abstract

Tropical forests are being rapidly altered by logging, and cleared for agriculture. Understanding the effects of these land use changes on soil fungi, which play vital roles in the soil ecosystem functioning and services, is a major conservation frontier. Using 454-pyrosequencing of the ITS1 region of extracted soil DNA, we compared communities of soil fungi between unlogged, once-logged, and twice-logged rainforest, and areas cleared for oil palm, in Sabah, Malaysia. Overall fungal community composition differed significantly between forest and oil palm plantation. The OTU richness and Chao 1 were higher in forest, compared to oil palm plantation. As a proportion of total reads, Basidiomycota were more abundant in forest soil, compared to oil palm plantation soil. The turnover of fungal OTUs across space, true β-diversity, was also higher in forest than oil palm plantation. Ectomycorrhizal (EcM) fungal abundance was significantly different between land uses, with highest relative abundance (out of total fungal reads) observed in unlogged forest soil, lower abundance in logged forest, and lowest in oil palm. In their entirety, these results indicate a pervasive effect of conversion to oil palm on fungal community structure. Such wholesale changes in fungal communities might impact the long-term sustainability of oil palm agriculture. Logging also has more subtle long term effects, on relative abundance of EcM fungi, which might affect tree recruitment and nutrient cycling. However, in general the logged forest retains most of the diversity and community composition of unlogged forest.

Editor: Andrew Hector, University of Oxford, United Kingdom

Funding: This project was funded by a National Research Foundation grant (NRF-2013-031400), Ministry of Education, Science and Technology, South Korea. The funders had no role in study design, data collection and analysis, decision to publish, or preparation of the manuscript.

* Email: david.edwards@sheffield.ac.uk (DPE); foundinkualalumpur@yahoo.com (JMA)

Introduction

Tropical forests are one of the world's most important reservoirs for biodiversity [1]. They contain an exceptional concentration of the world's species, but are being reduced in area faster than any other ecosystem [2]. Roughly half of the world's natural extent of tropical forest has been logged or converted to different land uses [3]. Some 403 million hectares of tropical rainforests have been included in timber estates and slated for selective logging [4], and between 2000 and 2010, approximately 13 million hectares of forest within the tropics were cleared for agricultural activities [5,6], including oil palm plantations [7].

Anthropogenic disturbances in tropical forests are causing a dramatic decline in global biodiversity, and in associated biological processes that maintain the productivity and sustainability of ecosystems [8]. Several studies have shown that selective logging does not drastically impact the overall species richness and diversity of tropical forest [9–12], however, it has been shown that the impact of selective logging could be anything between fairly mild and severe depending on the intensity of logging [13] with changes in the composition of species, as forest-interior specialists decline and edge-tolerant, gap specialists increase in abundance [14,15]. In contrast, the conversion of both primary and logged

forest to agricultural land uses has been to shown to have a far greater negative impact on biodiversity than does logging. Conversion to agriculture results in a major reduction in biodiversity, again across a host of animal and plant taxa [16]. The conversion of primary and logged forest to agricultural plantations also results in a substantial decrease in the functional diversity of tropical ecosystems, with implications for the provision of ecosystem functions, whereas logging has lesser impacts on these metrics [17,18]. As well as affecting plants and larger animals, land use change also affects the soil biota. Land use change affects soil pH, carbon and nutrient content [19,20], causing shifts in soil microbial communities [21–23].

Fungi constitute one of the most diverse and dominant groups of organisms in soil, and they play important ecological roles in the ecosystem as decomposers, pathogens and plant mutualists [24,25]. Understanding the structure and diversity of soil fungal communities is fundamental to the understanding of their function in the ecosystem and their impact on plant communities [26]. However, while minimal work has been done in the tropics to assess the effect of land use changes on soil bacterial communities [23,27], until recently relatively little was known about the impacts of tropical land use change on soil fungal communities. Various studies have suggested that forest clearance to tree plantations or agricultural crops shifts soil fungal communities, linked to strong changes in soil properties [22,28]. However, previous studies of forest clearance to other forms of agriculture on forest fungal communities have mostly been limited to techniques that give relatively low taxonomic resolution (i.e. T-RFLP and PCR-DGGE). Nevertheless, a recent study by McGuire et al. [29], which used high throughput sequencing to analyze soil fungal communities in Southeast Asian tropical forests in west Malaysia, showed that conversion of primary forest to oil palm plantations alters fungal community composition and function, whereas primary and logged forests were more similar in composition and nutrient cycling potential. However, there is a need for further studies to understand the impacts of logging cycles on soil fungal communities and of the conversion of logged forest to agriculture, since logged forests now dominate the tropics [3] and are much more likely to be converted to agriculture than primary forests [2,6].

In this study, we also focused on the rainforests of the Sundaland region, of Southeast Asia, but some 1,800 km away in east Malaysia (Borneo). Across the Sundaland region, the primary forest has been subject to differing degrees of logging intensity. Much of the region's forest (about 50%) has never been logged, while many areas have been subject to one or two logging cycles [30]. Also, oil palm is one of the most rapidly expanding crops in this region. This provides an opportunity to study the effect of different intensities of logging on the soil fungal community and also to evaluate if conversion of forest to oil palm plantation has a stronger impact on soil fungi than logging, as is the case for numerous macroscopic taxa. Our objective here was to understand whether land use change has an impact on the structure and diversity of fungal communities in the Yayasan Sabah (YS) logging concession in Malaysian, Borneo. We compared the fungal communities in forests with different logging histories (unlogged, once-logged and twice-logged), and oil palm plantations. We examined whether there are differences in α and â-diversity, as well as community composition. These results may provide important information for soil management policies, and estimation of ecological impact of land use change in this region.

Materials and Methods

Study area

The study area is located within the Yayasan Sabah (YS) logging concession and contiguous oil palm plantation areas, in Sabah, Malaysian Borneo (4°58′ N, 117°48′ E). The forests in this area are naturally dominated by valuable timber tree species belonging to the family Dipterocarpaceae [31]. Due to logging for the wood industry and clearance for palm oil plantations, the area of forest in Borneo - as elsewhere in the tropics - has been dramatically reduced in recent decades [32].

Fieldwork was conducted in the Ulu Segama-Malua Forest Reserve (US-MFR) and adjacent oil palm estates in Sabah, Borneo. Some areas of forest were logged between 1970 and 1990, and some of these were then re-logged between 2000 and 2007. During the first logging rotation, approximately 113 m^3 per hectare (range 73 m^3 to 166 m^3) of commercially valuable trees > 0.6 m diameter were extracted. During the second logging rotation, an additional 31 m^3 per hectare (range 15 m^3 to 72 m^3) of timber were removed [10,31]. Selectively logged forest in the US-MFR is adjacent to the 45,200 ha Danum Valley Conservation Area (DVCA) and Palum Tambun Watershed Reserve, containing large areas of unlogged forest [10,16]. Oil palm plantations are situated to the north and south of the US-

MFR, with mature palms of 20 to 30 years old, planted at a density of 100 trees per hectare.

Soil sampling and DNA extraction

From September to October 2012, twenty-four transects each of 200 m in length were located across four different land uses: unlogged (primary) forests, once-logged and twice-logged forests, and oil palm plantations, with six transects per habitat. Within each habitat, distances between transects ranged from 500 m to 65 km, whereas across habitats, distances between transects ranged from 1 km to 67 km. From each transect, at 50 m intervals, approximately 50 g from the top 5 cm of soil (excluding the leaf litter layer) was taken in a sterile plastic bag using a trowel, giving five samples of soil per transect. The trowels were thoroughly cleaned with ethanol between successive transect sampling. All soil samples were then sieved (2 mm) in laboratory to homogenize the sample and stored at –20°C until DNA extraction [10]. Twenty-four soil DNA extractions (one for each transect) were performed using 0.3 g of soil, with the Power Soil DNA extraction kit (MO BIO Laboratories, Carlsbad, CA, USA) following the directions described by the manufacturer.

PCR amplification and pyrosequencing

Fungal DNA was amplified using ITS primers targeting the internal transcribed spacer (ITS) region 1 and 2. Forward primers comprised the 454 Fusion Primer A-adaptor, a specific multiplex identifier (MID) barcode, and the ITS1F primer (5′-CTTGGTCATTTAGAGGAAGTAA-3′) [33], while the reverse primer was composed of the B-adapter and ITS4 primer (5′-TCCTCCGCTTATTGATATGC-3′) [34]. Polymerase chain reactions (PCR) were performed in 50 µl reactions using the following temperature program: 95°C for 10 min s; 30 cycles of 95°C for 30 s, 55°C for 30 s, 72°C for 30 s; and 72°C for 7 min. The PCR products were purified using the QIAquick PCR purification kit (Qiagen) and quantified using PicoGreen (Invitrogen) spectrofluorometrically (TBS 380, Turner Biosystems, Inc. Sunnyvale, CA, USA). 50 ng of purified PCR product for each sample were combined in a single tube and sent to Macrogen Inc. (Seoul, Korea) for sequencing using 454/Roche GS FLX Titanium Instrument (Roche, NJ, USA).

Sequence processing

Initial quality filtering and denoising were performed following the 454 SOP in the mothur pipeline [35]. The ITS1 region was verified and extracted using the ITS1 extractor for fungal ITS sequences [36]. Putative chimeric sequences were detected and removed via the Chimera Uchime algorithm contained within mother [37]. Operational taxonomic units (OTUs) were assigned using the QIIME implementation of UCLUST [38], with a threshold of 97% pairwise identity. OTUs were classified taxonomically using the classify command in mothur at 80% Naïve Bayesian bootstrap cutoff with 1000 iterations against the UNITE database [39]. Ectomycorrhizal (EcM) fungi were determined by matching taxonomy assignments with established EcM lineages as determined by recent phylogenetic and stable isotope data [40]. The 454 sequence run has been deposited in the NCBI Sequence Read Archive under accession number SRP041467.

Statistical analysis

To correct for differences in number of reads, which can bias diversity estimates, all samples were rarified to 3,347 reads per sample. To test for effects of land use types on the OTU richness

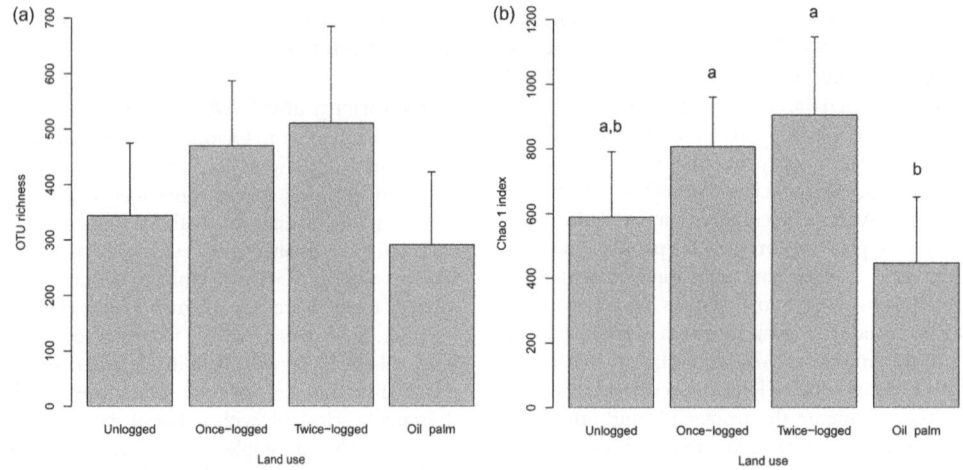

Figure 1. Diversity indices of the fungal community across different land uses in Sabah, Malaysian Borneo. (a) OTU richness and (b) Chao1 index. Pairwise comparisons are shown; different letters denote significant differences between groups at P<0.05.

and diversity indices, we used a linear model (LM) for normal data or generalized linear model (GLM) for non-normal data, considering land use as the major factor. We used the same procedure to test whether relative abundance of the most abundant phyla differed among different land use types. We also assessed the effect of land use on the relative abundance at the order and genus levels within those phyla that showed significant differences due to land use. Post-hoc Tukey tests were used for pairwise comparisons. When neither a linear nor a generalized linear model fitted the data, we used a Kruskall-Wallis test to assess the effect of land use on the relative abundance of fungal taxa, with the Bonferroni correction to assess pairwise comparisons.

To test whether species composition results may have been influenced by pseudoreplication within study sites, we used a Mantel test (Mantel Nonparametric Test Calculator 2.0) [41] to compare transect matrices of fungal compositional to geographic

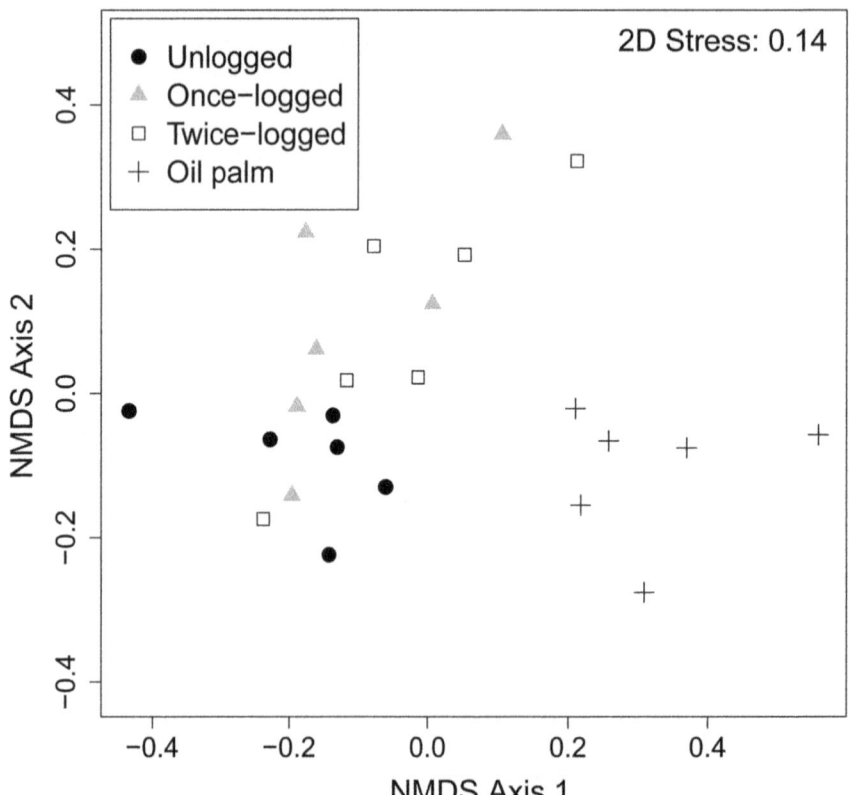

Figure 2. Non-metric multidimensional scaling (NMDS) ordination showing clustering of fungal communities among different land uses in Sabah, Malaysian Borneo.

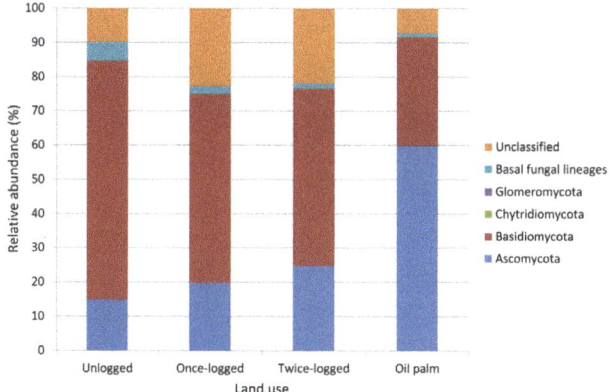

Figure 3. Relative abundance of dominant fungal phyla among different land uses in Sabah, Malaysian Borneo.

distance between pairs of transects within a site and between pairs of transects across the entire dataset [42,43].

The OTU-based community similarity was calculated by using the Bray-Curtis index [44]. Non-metric multidimensional scaling (NMDS), to visualize the change in species composition across the land use types, was conducted in Primer-E software (Version 6, Plymouth, UK). Then, we tested the difference among different land use types using an analysis of similarity (ANOSIM).

We measured β-diversity amongst land use types following Anderson et al. [45], which is defined as the variation in community structure without defining a particular gradient or direction. Therefore, we estimated true β-diversity following Whittaker [46] in [47] for every land use type. In addition, we calculated β-diversity as the average distance from each site to the group centroid [45]. The betadisper function in R was used to test if β-diversity shows any difference between land use types. True β-diversity (i.e. $S/\bar{\alpha}$) for each pair of samples within each of the four land uses was estimated by the following equation:

$$\frac{S}{\bar{\alpha}} = \frac{a+b+c}{(2a+b+c)/2}$$

Where S is the total number of OTUs in two samples, $\bar{\alpha}$ is the average number of OTUs for both samples, a is shared OTUs between both samples, b is OTUs found only in sample 1 and c are OTUs found only in sample 2. To compare true β-diversity among land uses, we used a linear model using land use as factor, and sample as random factor to control for pseudoreplication, as every sample is used in more than one comparison within each land use. Post-hoc Tukey tests were used for pairwise comparisons among different land uses.

Results

A total of 114,744 quality sequences were obtained from the 24 soil samples, with coverage ranging from 3,347–6,456 sequences. After rarifying to 3,347 reads per sample, we obtained a total of 80,328 sequences, and of these around 84% sequences were classified up to phylum level with a total of 5,327 OTUs (defined at ≥97% sequence similarity level). Fungal OUT richness (i.e. number of OTUs) was marginally significantly different across land use types ($F_{3,24} = 2.98$, P = 0.05; Fig. 1a), with lowest levels of OTU richness observed in oil palm plantations compared to primary and logged forests. Predicted OTU richness calculated using the Chao1 estimator was significantly higher in logged forests than oil palm plantations (F3,24 = 4.74, P = 0.01; Fig. 1b), whereas Shannon index did not show any variation among land uses (F3,24 = 1.93, P = 0.15).

The NMDS plots of pairwise Bray–Curtis dissimilarities showed that fungal communities were clustered significantly across land use types (ANOSIM: R = 0.51, P<0.001; Fig. 2). Mantel tests showed no effect of distance on the composition of fungal communities across different land use types in Borneo (all P> 0.07). The majority of fungal sequences recovered in our study belonged to the Basidiomycota and Ascomycota, with relative abundances of 52% and 29%, respectively (Fig. 3). The basal

Table 1. Comparison of relative abundance of the dominant fungal orders within the phyla Ascomycota and Basidiomycota among land uses[a].

Taxa	F or χ^{2b}	df	P	Pairwise comparisons[c]
Ascomycota				
Helotiales	3.67	3, 24	0.02	Once-logged > oil palm
Hypocreales	5.54	3, 24	0.006	Once-logged/twice-logged/unlogged < oil palm
Pleosporales	6.42	3, 24	0.003	Once-logged/unlogged < oil palm
Basidiomycota				
Agaricales	10.9	3, 24	0.0001	Once-logged/twice-logged/unlogged > oil palm
Russulales	11.3	3, 24	0.0001	Unlogged > Once-logged/twice-logged/oil palm
Sebacinales	11.1*	3	0.01	Unlogged/twice-logged > oil palm
Sporidiobolales	15.1*	3	0.001	Once-logged/twice-logged/unlogged < oil palm
Thelephorales	10.0*	3	0.01	Unlogged/twice-logged > oil palm
Trichosporonales	13.3	3, 24	0.0001	Once-logged/twice-logged/unlogged > oil palm

[a]Only orders for which significant differences were found are shown.
[b]Effect of land use on relative abundance evaluated by linear or generalized linear model or by the Kruskal-Wallis test (*).
[c]Pairwise comparisons by *post hoc* Tukey test for linear/generalized linear models or P values Bonferroni-corrected for Kruskal-Wallis. Differences were considered significant at a P value of <0.05.

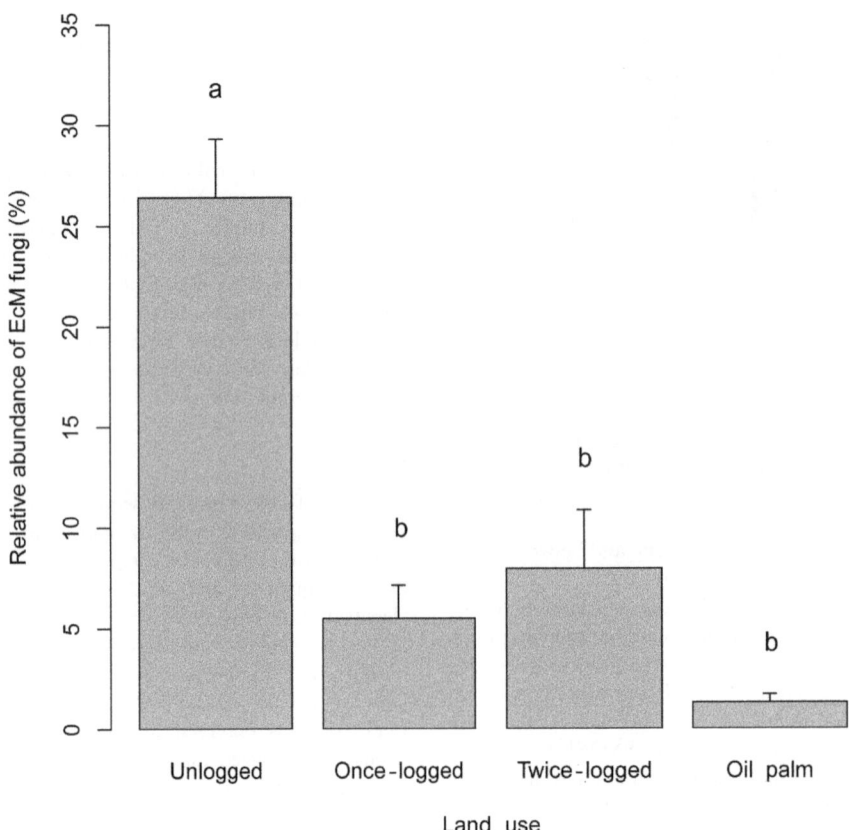

Figure 4. Relative abundance (means ± SD) of ectomycorrhizal (EcM) fungal sequences among different land uses in Sabah, Malaysian Borneo. Pairwise comparisons are shown; different letters denote significant differences between groups at P<0.05.

fungal lineages represented 2%, followed by Glomeromycota and Chytridiomycota with less than 1%, and 15% of the detected sequences were unclassified (Fig. 3). We found a significant change in the relative abundance of the two most dominant phyla: Basidiomycota ($F_{3,24} = 4.23$, $P = 0.01$) and Ascomycota ($F_{3,24} = 4.81$, $P = 0.01$) along different land uses. The abundance of Basidiomycota was greater in the unlogged forest, intermediate in the logged forest, and least in the oil palm. Conversely, the oil palm plantations had greater abundance of Ascomycota compared to forest soils (Fig. 3). At the order level, there were significant differences in the relative abundances of the most dominant orders (Table 1). The results revealed that the relative abundance of *Agaricales, Russulales, Thelephorales, Trichosporonales, Sebacinales* and *Helotiales* were significantly higher in the forest than oil palm plantations (P<0.05; Table 1). However, *Hypocreales, Sporidiobolales* and *Pleosporales*, were more abundant in oil palm plantations than unlogged and logged forests (P<0.05; Table 1).

A total of 11,421 sequences belonged to known groups of ectomycorrhizal (EcM) fungi, with the EcM fungi representing around 10% (11,421 sequences) of the total detected fungal sequences, with 180 OTUs. The relative abundance of EcM sequences was significantly different across land use types, with highest and lowest relative abundances observed in primary forest (mean relative abundance = 26%) and oil palm plantations (mean relative abundance = 0.5%), respectively ($\chi^2 = 18.04$, P<0.001; Fig. 4). From our soil samples, we identified 14 genera belonging to EcM fungi with *Russula* as the most dominant genus (65% of total EcM sequences), followed by *Tomentella, Sebacina* and *Lactarius*. The relative abundance of these four dominant EcM

genera combined were significantly higher in forest soils compared to oil palm plantations (P<0.05). For *Russula* alone, we also found a difference in abundance between unlogged (greatest abundance), once-logged (less abundant), and twice-logged forest (lowest abundance of *Russula*) (P<0.05 in each case).

The β-diversity, measured as the average distance of all samples to the centroid in each land use type, did not show significant difference among land uses ($F_{3,24} = 2.77$, P = 0.06). However, there was a significant effect of forest conversion to oil palm plantations on fungal true β-diversity (i.e. S/α; $F_{3,24} = 3.85$, P = 0.01), with oil palm having lowest true β-diversity compared to unlogged and logged forests (Fig. 5). Logging did not produce any significant change in true β-diversity.

Discussion

Our results showed that fungal OTU richness and Chao1 index differed among land uses. Oil palm plantations had lower OTU richness, which can be attributed to the effects of anthropogenic intervention and forest conversion. This contrasts with the findings on bacterial communities of Lee-Cruz et al. [27] in the same study site, where they found that OTU richness and diversity indices did not differ among land uses, and that α-diversity was similar in forests and oil palm plantations.

We found that the structure of fungal communities differed most fundamentally between forests and oil palm plantations. This finding mirrors that of McGuire et al. [29], who found that the community of soil fungi collected across three different land uses in Malaysia differed between oil palm plantations and forests. Changes in fungal community composition in logged forest and

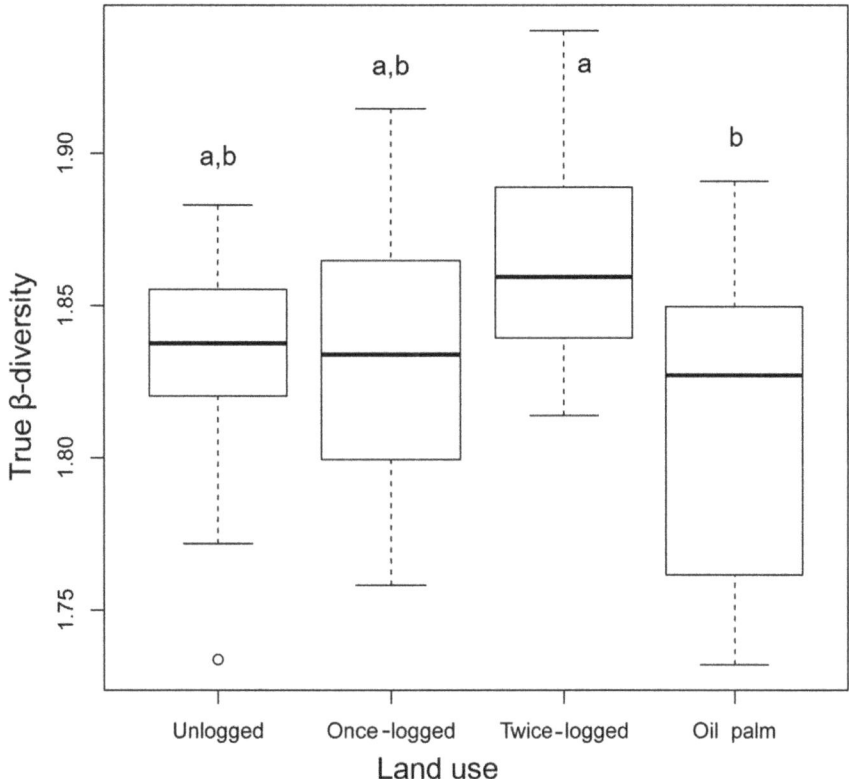

Figure 5. Fungal community true β-diversity (i.e. $S/\bar{\alpha}$) among the four land uses in Sabah, Malaysian Borneo. Boxes show the lower quartile, the median and the upper quartile. Pairwise comparisons are shown; different letters denote significant differences between groups at P< 0.05.

former forest areas could directly impact the functioning of soil communities and their ability to provide key ecosystem services, such as decomposition and nutrient recycling [22]. Given that fertilizer prices are predicted to rise dramatically in the coming decades [48], these results suggest that improvements in agricultural methods by establishing diversified farms could be necessary for sustaining vital soil biodiversity and ecosystem-service values.

The abundance pattern of fungal taxa detected here is similar to the soils of the Neotropics and elsewhere, where Basidiomycota and Ascomycota are also the most prevalent groups [49–53]. The most abundant orders of Basidomycota in the forest areas we sampled were *Russulales*, *Agaricales*, *Sporidiobolales*, *Telephorales*, *Trichosporonales* and *Sebacinales*. They all became less abundant in oil palm plantations compared to primary forest. These orders are generally reported to be the most varied and abundant groups of fungi in forests around the world [40,54,55], they have also been characterized as being lignicolous, saprobic or mycorrhizal, associated with litter decomposition, and they are known to degrade plant-derived cellulose [56]. In our forest plots, the ectomycorrhizal orders *Russulales* and *Telephorales* were the most common. This pattern has also been found in forests from the Neotropics and African tropics [57–59].

The lower abundance of Basidiomycota in oil palm soils may be due to the lack of large quantities of coarse woody debris, often derived from roots or branches of forest trees, in oil palm plantations. However, it may be worth investigating whether the change in abundance of Basidiomycota - and change in their overall community structure - with conversion of forest to oil palm may have implications for long term nutrient processing in the oil palm soils. Since Basidiomycota as a group are often able to break

down relatively recalcitrant substrates and changes in their abundance and community composition might impede nutrient recycling in oil palm soils. The converse increase in Ascomycota, may be seen in terms of the relatively lignin-poor and nutrient-rich character of most of the organic matter reaching the soil from roots and leaves of oil palm, and from the herbaceous weedy layer that grows under the palms.

We found that both a history of logging, and forest conversion to oil palm plantation, resulted in shifts in EcM fungal communities. The most drastic effects were with forest conversion to oil palm. However, logging history also had detectable effects EcM fungi relative abundance. Ectomycorrhizas are one of the most important widespread types of mycorrhiza in forests of the cool temperate and boreal latitudes [60], and they also form an important group and often the dominant ecologically and economically important minority of Dipterocarpaceae family trees in tropical Asia [61–65].

The lower abundance of EcM fungi in logged forests might be due to a thinner, more incomplete root mat following past logging disturbance. The much lower abundance of EcM fungi in oil palm plantation soils could be partly due to the much lower abundance of potential host roots (e.g. Dipterocarpaceae which are absent from the palm plantations). Such fungi have generally been found to recover slowly from disturbance even when potential host plants are present [66]. Among the detected genera in our study, we found that *Russula*, *Sebacina*, *Lactarius* and *Tomentella* showed significant impact of land use change. *Russula* was relatively the most abundant EcM genus in our samples: this genus has been found to be common on roots of dipterocarp forests, tropical and southern hemisphere angiosperm forests [62,67,68].

This study also investigated the impacts of land use change on the turnover of fungal communities across space (β-diversity), and results suggest that there is a spatial homogenization of fungal communities in oil palm agriculture compared to forest: fungal β-diversity in oil palm plantations was lower than forest. Habitat conversion to agriculture also reduces β-diversity of soil bacteria in Amazonia [69]. In contrast, across the same land use system and sites in Borneo as we studied here, there was actually an increase in β-diversity of bacteria with both logging and conversion to oil palm [27].

There are some caveats that accompany our findings. We worked in only one biogeographic region and on only one form of agriculture. It is thus important to replicate this work in other logging systems and in other key expanding crops, including soya, sugar cane, and cacao. We also did not explicitly demonstrate the impacts of changing fungal composition on soil ecosystem functions and services: this is a major knowledge gap, with critical importance to the development of sustainable logging and, in particular, agricultural systems in the tropics.

In conclusion, the conversion of both primary and logged forest to oil palm drives a change in the overall fungal community, including EcM fungi abundance, and an associated decrease in total fungal community beta-diversity. This finding invites further studies that investigate the long-term implications of such changes for agricultural sustainability. There was a more subtle long-term impact of logging on fungal communities. Most measurable features of the unlogged forest fungal community remained unchanged after logging. However, there were significant changes in EcM fungal abundance due to logging, which could have a pervasive impact since EcM fungi are thought to play a key role in tree growth and community structure. Despite this, the lack of drastic changes in the overall forest fungal community structure following logging strengthens the view that logged forest is not necessarily an irretrievably damaged and drastically altered system, and that protecting it from conversion to oil palm may still have considerable conservation benefits.

Acknowledgments

We thank Yayasan Sabah, Danum Valley Management Committee, Sabah Parks, the state Secretary, Sabah Chief Minister's Departments, and the Sabah Biodiversity Council for permission to conduct research. We also, thank the Royal Society's Southeast Asian Rainforest Research Program (SEARRP) and the Borneo Rainforest Lodge for logistical support and site access.

Author Contributions

Conceived and designed the experiments: DPE JMA. Performed the experiments: DK BMT DPE. Analyzed the data: DK BMT DPE JL. Contributed reagents/materials/analysis tools: DK BMT DPE JMA JL. Contributed to the writing of the manuscript: DK BMT DPE JMA.

References

1. Whitmore TC (1990) An introduction to tropical rain forests. Oxford: Clarendon Press. 226 pp.

2. Asner GP, Rudel TK, Aide TM, Defries R, Emerson R (2009) A contemporary assessment of change in humid tropical forests. Conserv Biol 23: 1386–1395.

3. Wright SJ (2005) Tropical forests in a changing environment. Trends Ecol Evol 20: 553–560.

4. Blaser J (2011) Status of tropical forest management 2011: International Tropical Timber Organization.

5. Food and Agriculture Organization of the United Nations (2010) Global forest resources assessment 2010 : main report. Rome: FAO. 340 p.

6. Hansen MC, Stehman SV, Potapov PV, Loveland TR, Townshend JR, et al. (2008) Humid tropical forest clearing from 2000 to 2005 quantified by using multitemporal and multiresolution remotely sensed data. Proc Natl Acad Sci U S A 105: 9439–9444.

7. Morris RJ (2010) Anthropogenic impacts on tropical forest biodiversity: a network structure and ecosystem functioning perspective. Philos Trans R Soc Lond B Biol Sci 365: 3709–3718.

8. Achard F, Eva HD, Stibig H-J, Mayaux P, Gallego J, et al. (2002) Determination of Deforestation Rates of the World's Humid Tropical Forests. Science 297: 999–1002.

9. Gibson L, Lee TM, Koh LP, Brook BW, Gardner TA, et al. (2011) Primary forests are irreplaceable for sustaining tropical biodiversity. Nature 478: 378–381.

10. Edwards DP, Larsen TH, Docherty TD, Ansell FA, Hsu WW, et al. (2011) Degraded lands worth protecting: the biological importance of Southeast Asia's repeatedly logged forests. Proc Biol Sci 278: 82–90.

11. Berry NJ, Phillips OL, Lewis SL, Hill JK, Edwards DP, et al. (2010) The high value of logged tropical forests: lessons from northern Borneo. Biodivers Conserv 19: 985–997.

12. Woodcock P, Edwards DP, Newton RJ, Vun Khen C, Bottrell SH, et al. (2013) Impacts of intensive logging on the trophic organisation of ant communities in a biodiversity hotspot. PLoS One 8: e60756.

13. Burivalova Z, Şekercioğlu ÇH, Koh LP (2014) Thresholds of Logging Intensity to Maintain Tropical Forest Biodiversity. Curr Biol 24: 1–6.

14. Hamer K, Hill J, Benedick S, Mustaffa N, Sherratt T, et al. (2003) Ecology of butterflies in natural and selectively logged forests of northern Borneo: the importance of habitat heterogeneity. J Appl Ecol 40: 150–162.

15. Cleary DF, Boyle TJ, Setyawati T, Anggraeni CD, Loon EEV, et al. (2007) Bird species and traits associated with logged and unlogged forest in Borneo. Ecol Appl 17: 1184–1197.

16. Edwards DP, Hodgson JA, Hamer KC, Mitchell SL, Ahmad AH, et al. (2010) Wildlife-friendly oil palm plantations fail to protect biodiversity effectively. Conserv Lett 3: 236–242.

17. Edwards F, Edwards D, Larsen T, Hsu W, Benedick S, et al. (2013) Does logging and forest conversion to oil palm agriculture alter functional diversity in a biodiversity hotspot? Anim Conserv 17: 163–173.

18. Baraloto C, Hérault B, Paine C, Massot H, Blanc L, et al. (2012) Contrasting taxonomic and functional responses of a tropical tree community to selective logging. J Appl Ecol 49: 861–870.

19. McGrath DA, Smith CK, Gholz HL, Oliveira FdA (2001) Effects of Land-Use Change on Soil Nutrient Dynamics in Amazônia. Ecosystems 4: 625–645.

20. Murty D, Kirschbaum MUF, McMurtrie RE, McGilvray H (2002) Does conversion of forest to agricultural land change soil carbon and nitrogen? a review of the literature. Global Change Biol 8: 105–123.

21. Cornejo FH, Varela A, Wright SJ (1994) Tropical Forest Litter Decomposition under Seasonal Drought: Nutrient Release, Fungi and Bacteria. Oikos 70: 183–190.

22. Lauber CL, Strickland MS, Bradford MA, Fierer N (2008) The influence of soil properties on the structure of bacterial and fungal communities across land-use types. Soil Biol Biochem 40: 2407–2415.

23. Tripathi B, Kim M, Singh D, Lee-Cruz L, Lai-Hoe A, et al. (2012) Tropical Soil Bacterial Communities in Malaysia: pH Dominates in the Equatorial Tropics Too. Microb Ecol 64: 474–484.

24. Orgiazzi A, Lumini E, Nilsson RH, Girlanda M, Vizzini A, et al. (2012) Unravelling soil fungal communities from different Mediterranean land-use backgrounds. PloS one 7: e34847.

25. Wu TH, Chellemi DO, Martin KJ, Graham JH, Rosskop EN (2007) Discriminating the effects of agricultural land management practices on soil fungal communities. Soil Biol Biochem 39: 1139–1155.

26. Martin F, Cullen D, Hibbett D, Pisabarro A, Spatafora JW, et al. (2011) Sequencing the fungal tree of life. New Phytol 190: 818–821.

27. Lee-Cruz L, Edwards DP, Tripathi BM, Adams JM (2013) Impact of Logging and Forest Conversion to Oil Palm Plantations on Soil Bacterial Communities in Borneo. Appl Environ Microbiol 79: 7290–7297.

28. Lupatini M, Jacques RJS, Antoniolli ZI, Suleiman AKA, Fulthorpe RR, et al. (2013) Land-use change and soil type are drivers of fungal and archaeal communities in the Pampa biome. World J Microbiol Biotechnol 29: 223–233.

29. McGuire KL, D'Angelo H, Brearley FQ, Gedallovich SM, Babar N, et al. (2014) Responses of soil fungi to logging and oil palm agriculture in Southeast Asian tropical forests. Microbial Ecol.

30. Wilcove DS, Giam X, Edwards DP, Fisher B, Koh LP (2013) Navjot's nightmare revisited: logging, agriculture, and biodiversity in Southeast Asia. Trends Ecol Evol 28: 531–540.

31. Fisher B, Edwards DP, Giam XL, Wilcove DS (2011) The high costs of conserving Southeast Asia's lowland rainforests. Front Ecol Environ 9: 329–334.

32. Gibbs HK, Ruesch AS, Achard F, Clayton MK, Holmgren P, et al. (2010) Tropical forests were the primary sources of new agricultural land in the 1980s and 1990s. Proc Natl Acad Sci U S A 107: 16732–16737.

33. Gardes M, Bruns TD (1993) ITS primers with enhanced specificity for basidiomycetes-application to the identification of mycorrhizae and rusts. Mol Ecol 2: 113–118.

34. White TJ, Bruns T, Lee S, Taylor J (1990) Amplification and direct sequencing of fungal ribosomal RNA genes for phylogenetics. PCR protocols: a guide to methods and applications 18: 315–322.

35. Schloss PD, Gevers D, Westcott SL (2011) Reducing the effects of PCR amplification and sequencing artifacts on 16S rRNA-based studies. PLoS One 6: e27310.

36. Nilsson RH, Veldre V, Hartmann M, Unterseher M, Amend A, et al. (2010) An open source software package for automated extraction of ITS1 and ITS2 from fungal ITS sequences for use in high-throughput community assays and molecular ecology. Fungal Ecol 3: 284–287.

37. Edgar RC, Haas BJ, Clemente JC, Quince C, Knight R (2011) UCHIME improves sensitivity and speed of chimera detection. Bioinformatics 27: 2194–2200.

38. Edgar RC (2010) Search and clustering orders of magnitude faster than BLAST. Bioinformatics 26: 2460–2461.

39. Abarenkov K, Nilsson RH, Larsson KH, Alexander IJ, Eberhardt U, et al. (2010) The UNITE database for molecular identification of fungi - recent updates and future perspectives. New Phytol 186: 281–285.

40. Tedersoo L, May T, Smith M (2010) Ectomycorrhizal lifestyle in fungi: global diversity, distribution, and evolution of phylogenetic lineages. Mycorrhiza 20: 217–263.

41. Liedloff A (1999) Mantel nonparametric test calculator. Version 2.0. School of Natural Resource Sciences, Queensland University of Technology, Australia.

42. Ghazoul J (2002) Impact of logging on the richness and diversity of forest butterflies in a tropical dry forest in Thailand. Biodiver Conserv 11: 521–541.

43. Ramage BS, Sheil D, Salim HM, Fletcher C, MUSTAFA NZA, et al. (2013) Pseudoreplication in tropical forests and the resulting effects on biodiversity conservation. Conserv Biol 27: 364–372.

44. Magurran AE (2004) Measuring biological diversity. Maldan, MA: Blackwell Pub. viii, 256 p.

45. Anderson MJ, Crist TO, Chase JM, Vellend M, Inouye BD, et al. (2011) Navigating the multiple meanings of beta diversity: a roadmap for the practicing ecologist. Ecol Lett 14: 19–28.

46. Whittaker RH (1960) Vegetation of the Siskiyou mountains, Oregon and California. Ecol Monogr 30: 279–338.

47. Koleff P, Gaston KJ, Lennon JJ (2003) Measuring beta diversity for presence-absence data. J Anim Ecol 72: 367–382.

48. Piesse J, Thirtle C (2009) Three bubbles and a panic: An explanatory review of recent food commodity price events. Food Policy 34: 119–129.

49. O'Brien HE, Parrent JL, Jackson JA, Moncalvo JM, Vilgalys R (2005) Fungal community analysis by large-scale sequencing of environmental samples. Appl Environ Microbiol 71: 5544–5550.

50. Wubet T, Christ S, Schöning I, Boch S, Gawlich M, et al. (2012) Differences in soil fungal communities between European beech (Fagus sylvatica L.) dominated forests are related to soil and understory vegetation. PloS One 7: e47500.

51. Gomes NC, Fagbola O, Costa R, Rumjanek NG, Buchner A, et al. (2003) Dynamics of fungal communities in bulk and maize rhizosphere soil in the tropics. Appl Environ Microbiol 69: 3758–3766.

52. Oros-Sichler M, Gomes NC, Neuber G, Smalla K (2006) A new semi-nested PCR protocol to amplify large 18S rRNA gene fragments for PCR-DGGE analysis of soil fungal communities. J Microbiol Methods 65: 63–75.

53. Bridge PD, Newsham KK (2009) Soil fungal community composition at Mars Oasis, a southern maritime Antarctic site, assessed by PCR amplification and cloning. Fungal Ecol 2: 66–74.

54. Geml J, Laursen GA, Timling I, McFarland JM, Booth MG, et al. (2009) Molecular phylogenetic biodiversity assessment of arctic and boreal ectomycorrhizal Lactarius Pers. (Russulales; Basidiomycota) in Alaska, based on soil and sporocarp DNA. Mol Ecol 18: 2213–2227.

55. Geml J, Laursen GA, Herriott IC, McFarland JM, Booth MG, et al. (2010) Phylogenetic and ecological analyses of soil and sporocarp DNA sequences reveal high diversity and strong habitat partitioning in the boreal ectomycorrhizal genus Russula (Russulales; Basidiomycota). New Phytol 187: 494–507.

56. Kuramae EE, Hillekens RHE, de Hollander M, van der Heijden MGA, van den Berg M, et al. (2013) Structural and functional variation in soil fungal communities associated with litter bags containing maize leaf. Fems Microbiol Ecol 84: 519–531.

57. Ba AM, Duponnois R, Moyersoen B, Diedhiou AG (2012) Ectomycorrhizal symbiosis of tropical African trees. Mycorrhiza 22: 1–29.

58. Smith ME, Henkel TW, Aime MC, Fremier AK, Vilgalys R (2011) Ectomycorrhizal fungal diversity and community structure on three co-occurring leguminous canopy tree species in a Neotropical rainforest. New Phytol 192: 699–712.

59. Tedersoo L, Sadam A, Zambrano M, Valencia R, Bahram M (2010) Low diversity and high host preference of ectomycorrhizal fungi in Western Amazonia, a neotropical biodiversity hotspot. Isme Journal 4: 465–471.

60. Smith SE, Read DJ (2010) Mycorrhizal symbiosis: Academic press.

61. Brearley FQ (2012) Ectomycorrhizal Associations of the Dipterocarpaceae. Biotropica 44: 637–648.

62. Peay KG, Kennedy PG, Davies SJ, Tan S, Bruns TD (2010) Potential link between plant and fungal distributions in a dipterocarp rainforest: community and phylogenetic structure of tropical ectomycorrhizal fungi across a plant and soil ecotone. New Phytol 185: 529–542.

63. Moyersoen B, Becker P, Alexander I (2001) Are ectomycorrhizas more abundant than arbuscular mycorrhizas in tropical heath forests? New Phytol 150: 591–599.

64. Haug I, Weiß M, Homeier J, Oberwinkler F, Kottke I (2005) Russulaceae and Thelephoraceae form ectomycorrhizas with members of the Nyctaginaceae (Caryophyllales) in the tropical mountain rain forest of southern Ecuador. New Phytol 165: 923–936.

65. Natarajan K, Senthilarasu G, Kumaresan V, Riviere T (2005) Diversity in ectomycorrhizal fungi of a dipterocarp forest in Western Ghats. Curr Sci 88: 1893–1895.

66. Peay KG, Schubert MG, Nguyen NH, Bruns TD (2012) Measuring ectomycorrhizal fungal dispersal: macroecological patterns driven by microscopic propagules. Mol Ecol 21: 4122–4136.

67. Tedersoo L, Jairus T, Horton BM, Abarenkov K, Suvi T, et al. (2008) Strong host preference of ectomycorrhizal fungi in a Tasmanian wet sclerophyll forest as revealed by DNA barcoding and taxon-specific primers. New Phytol 180: 479–490.

68. Riviere T, Diedhiou AG, Diabate M, Senthilarasu G, Natarajan K, et al. (2007) Genetic diversity of ectomycorrhizal Basidiomycetes from African and Indian tropical rain forests. Mycorrhiza 17: 415–428.

69. Rodrigues JL, Pellizari VH, Mueller R, Baek K, Jesus Eda C, et al. (2013) Conversion of the Amazon rainforest to agriculture results in biotic homogenization of soil bacterial communities. Proc Natl Acad Sci U S A 110: 988–993.

14

Resource-Mediated Indirect Effects of Grassland Management on Arthropod Diversity

Nadja K. Simons[1]*, **Martin M. Gossner**[1], **Thomas M. Lewinsohn**[2], **Steffen Boch**[3], **Markus Lange**[4], **Jörg Müller**[5], **Esther Pašalić**[1], **Stephanie A. Socher**[3], **Manfred Türke**[1], **Markus Fischer**[3,6], **Wolfgang W. Weisser**[1]

1 Terrestrial Ecology Research Group, Department of Ecology and Ecosystem Management, School of Life Sciences Weihenstephan, Technische Universität München, Freising, Germany, 2 Department of Animal Biology, Institute of Biology, University of Campinas, Campinas, Sao Paulo, Brazil, 3 Institute of Plant Sciences, University of Bern, Bern, Switzerland, 4 Max-Planck-Institute for Biogeochemistry, Jena, Germany, 5 Institute of Biochemistry and Biology, University of Potsdam, Potsdam, Germany, 6 Biodiversity and Climate Research Centre, Senckenberg Gesellschaft für Naturforschung, Frankfurt/Main, Germany

Abstract

Intensive land use is a driving force for biodiversity decline in many ecosystems. In semi-natural grasslands, land-use activities such as mowing, grazing and fertilization affect the diversity of plants and arthropods, but the combined effects of different drivers and the chain of effects are largely unknown. In this study we used structural equation modelling to analyse how the arthropod communities in managed grasslands respond to land use and whether these responses are mediated through changes in resource diversity or resource quantity (biomass). Plants were considered resources for herbivores which themselves were considered resources for predators. Plant and arthropod (herbivores and predators) communities were sampled on 141 meadows, pastures and mown pastures within three regions in Germany in 2008 and 2009. Increasing land-use intensity generally increased plant biomass and decreased plant diversity, mainly through increasing fertilization. Herbivore diversity decreased together with plant diversity but showed no response to changes in plant biomass. Hence, land-use effects on herbivore diversity were mediated through resource diversity rather than quantity. Land-use effects on predator diversity were mediated by both herbivore diversity (resource diversity) and herbivore quantity (herbivore biomass), but indirect effects through resource quantity were stronger. Our findings highlight the importance of assessing both direct and indirect effects of land-use intensity and mode on different trophic levels. In addition to the overall effects, there were subtle differences between the different regions, pointing to the importance of regional land-use specificities. Our study underlines the commonly observed strong effect of grassland land use on biodiversity. It also highlights that mechanistic approaches help us to understand how different land-use modes affect biodiversity.

Editor: Christian Rixen, WSL Institute for Snow and Avalanche Research SLF, Switzerland

Funding: The work was funded by the DFG Priority Program 1374 "Infrastructure-Biodiversity-Exploratories" (DFG-WE 3081/21-1.). www.dfg.de/spp/en. Additional funds for exchange visits to analyze data were provided by Unicamp (PRP/Faepex) for TML and MMG and by DAAD (Project TUMBRA) for NKH. www.toek.wzw.tum.de/index.php?id=117. This work was supported by the German Research Foundation (DFG) and the Technische Universität München within the funding programme Open Access Publishing. The funders had no role in study design, data collection and analysis, decision to publish, or preparation of the manuscript.

Competing Interests: The authors have declared that no competing interests exist.

* Email: nadja.simons@tum.de

Introduction

Negative effects of intensive grassland land use on biodiversity have been found for many taxa including plants [1], herbivorous and carnivorous arthropods [2–4], pollinating insects [5] and birds [6]. Despite the growing consensus that intensive land use has generally negative effects on biodiversity [7], the particular mechanisms that lead to a decrease in biodiversity are often not fully understood [8], because land use consists of various modes that can have opposing or additive effects on biodiversity. In semi-natural grasslands important land-use modes are mowing, grazing and fertilization. Several observational and experimental studies found decreasing diversities of plants and arthropods with increasing frequency of mowing events e.g. [4,9–11], with increasing fertilization intensity e.g. [12,13] or with increasing

grazing intensity [14–16]. Whereas effects of land-use modes on plants are often direct, e.g. when mowing hinders seed set of late-flowering plants, effects on higher trophic levels such as insect herbivores or carnivores may be either direct or mediated by changes in the plant community.

Several hypotheses have been proposed to describe the effects of differently diverse plant communities on the diversity and abundance of herbivores and predators. The 'Resource Heterogeneity Hypothesis' (RHH) predicts that more diverse resources provide more niches for a greater number of specialized species at higher trophic levels [17–20], i.e. herbivore diversity is promoted by increased plant diversity and predator diversity increases in response to herbivore diversity – a positive trophic cascade. In contrast to the RHH, the 'More Individuals Hypothesis' (MIH) proposes that diversity of consumers increases when resource quantity increases, i.e. plant biomass for herbivores and herbivore biomass for predators [21]. According to the MIH, this positive effect of resource abundance (or biomass) on consumer diversity is mediated by an increase in total consumer abundance. Borer et al. [22] studied plant diversity effects on arthropod diversity in experimental plant communities and found that more diverse plant communities hosted more diverse herbivore communities, but that this effect was mediated by higher overall plant and herbivore biomass. While the RHH assumes a direct link between resource and consumer diversity, the MIH assumes an indirect link through resource and consumer abundance, i.e. resource abundance increases consumer abundance which in turn increases consumer diversity. Thus, arthropod biomass and diversity may be differently affected by increasing land-use intensity depending on the mechanistic relationships between land use and the herbivore and carnivore communities.

We combined detailed information on grassland land-use modes (fertilization, grazing and mowing) with biomass and species richness data of plants and arthropods to analyse effects of land-use intensity on the arthropod community. Although experimental short-term manipulation of land use can elucidate immediate effects of the land-use modes (such as mowing), it is not clear whether these mechanisms operate similarly when different land-use practices are combined in agricultural grasslands and under long term conditions. This can only be assessed by studying grasslands which have been used as meadows or pastures continuously for several years or decades. In our study, we build on the work of Socher et al. [1] who tested for direct and indirect effects of grassland land use on plant diversity and biomass. We studied effects of land use on the arthropod community including both direct and indirect (via the plant community) chains of effects. The study system includes grasslands which have been managed for at least 20 years and comprise a wide range of land-use intensities. By including three different regions in Germany and by sampling in two consecutive years, we were able to consider the generality of observed patterns.

Based on the two hypotheses mentioned above and previous knowledge on mechanisms of grassland land use, we defined two alternative models. In the first model ('Resource Heterogeneity Model') we tested whether land-use intensity affects herbivore and predator diversity via changes in the diversity of their respective resources (RHH). According to the RHH we expected positive effects of plant diversity on herbivore diversity and of herbivore diversity on predator diversity (Figure 1 A). In the second model ('Resource Abundance Model'), we tested whether effects of land-use intensity affect arthropods through changes in resource quantity (MIH) and added herbivore and predator biomass to the model (Figure 1 B). According to the MIH, we expected plant biomass to have a positive effect on herbivore diversity through herbivore biomass and positive effects of herbivore biomass on predator diversity, respectively.

Within both models (Figure 1 A & B), we expected mowing to decrease plant diversity (by a loss of disturbance-sensitive species) and to decrease plant biomass (through mechanical disturbances during the growing period). Grazing was expected to increase plant diversity (increasing number of niches for plants or preventing competitive exclusion) and decrease plant biomass (recurrent disturbance of plant growth). Another possible plant species response to grazing or mowing is overcompensating for tissue loss leading to the opposite expectation (i.e. increase of plant biomass following grazing or mowing). However, overcompensation has usually been demonstrated for particular species, not at the community level e.g. [23,24], where a decrease of plant biomass is more likely. Fertilization was expected to decrease plant

diversity (dominance of fast-growing species) and increase plant biomass (increased nutrient input). Based on the findings of Socher et al. [1] we expected plant biomass to be negatively correlated with plant diversity.

In our study we asked the following questions: 1) How do land-use modes, singly or in combination, affect arthropod diversity? 2) How are effects of land-use intensity on herbivore diversity mediated by the responses of plants to land use (do they follow the 'Resource Heterogeneity Hypothesis' or the 'More Individuals Hypothesis')? 3) Are the responses of predators to land use governed by the same mechanisms as responses of herbivores?

Materials and Methods

Study sites and land use

The Biodiversity Exploratory research program (www.biodiversity-exploratories.de) was established in 2006 within three regions in Germany: (1) Schorfheide-Chorin (SCH) in north-east Germany (3–140 m a.s.l., 53°02′ N 13°83′ E, annual mean precipitation 500–600 mm, mean temperature 8–8.5°C). (2) Hainich-Dün (HAI) in central Germany (285–550 m a.s.l., 51°20′ N 10°41′ E, 500–800 mm, 6.5–8°C). (3) Swabian Alb (ALB) in south-west Germany (460–860 m a.s.l., 48°43′ N 9°37′ E, 700–1000 mm, 6–7°C). Within each region, 50 plots of 50 m × 50 m size serve as basis for surveys of biodiversity or ecosystem processes. Those plots were chosen from a total of 500 candidate plots in each region on which initial vegetation and land-use surveys were conducted. The 50 plots per region cover the whole regional gradient of land-use modes and intensity [25].

Land-use modes on the studied grasslands include mowing with different frequency (meadows), grazing by different kinds of livestock (cattle and sheep pastures) or both mowing and grazing (mown pastures). Grasslands are either unfertilized or fertilized with different amounts of fertilizer. During the study years all grassland plots continued to be managed by farmers in the same way as the surrounding grasslands. Grassland land use was assessed each year (since 2006) by standardized interviews with farmers and land-owners. From these interviews, we derived information on fertilization, grazing and mowing: Fertilization intensity was calculated as the total amount of nitrogen applied per hectare and year, in the form of organic or chemical fertilizer. For grazing, information on livestock units and the duration of grazing periods were combined as a measure of grazing intensity. Mowing intensity was calculated as the number of mowing events per year.

For our analyses we used land-use information from the two years prior to sampling and the sampling year. Mowing intensity included mowing events in the sampling year only up to the sampling event (e.g. for samples taken in June, mowing events later than June in the sampling year were not included). As mowing not only has long-term but also short-term effects on arthropods (e.g. Humbert et al. [26] showed that mowing with machinery leads to high mortality in Orthoptera), we included the number of days between the arthropod sampling day and the last mowing event prior to the respective sampling as an additional land-use variable ('Time after mowing'). As two arthropod samplings were conducted per year (see below), we used the mean number of days for each year. For a detailed description of land-use intensity calculations and an overview of land-use information see Appendix S1 as well as Table S1 & Table S2.

Plant and arthropod sampling and plot selection

Plant diversity and plant biomass on the plots were assessed between mid-May and mid-June in 2008 & 2009 using vegetation surveys and aboveground biomass harvests following the methods

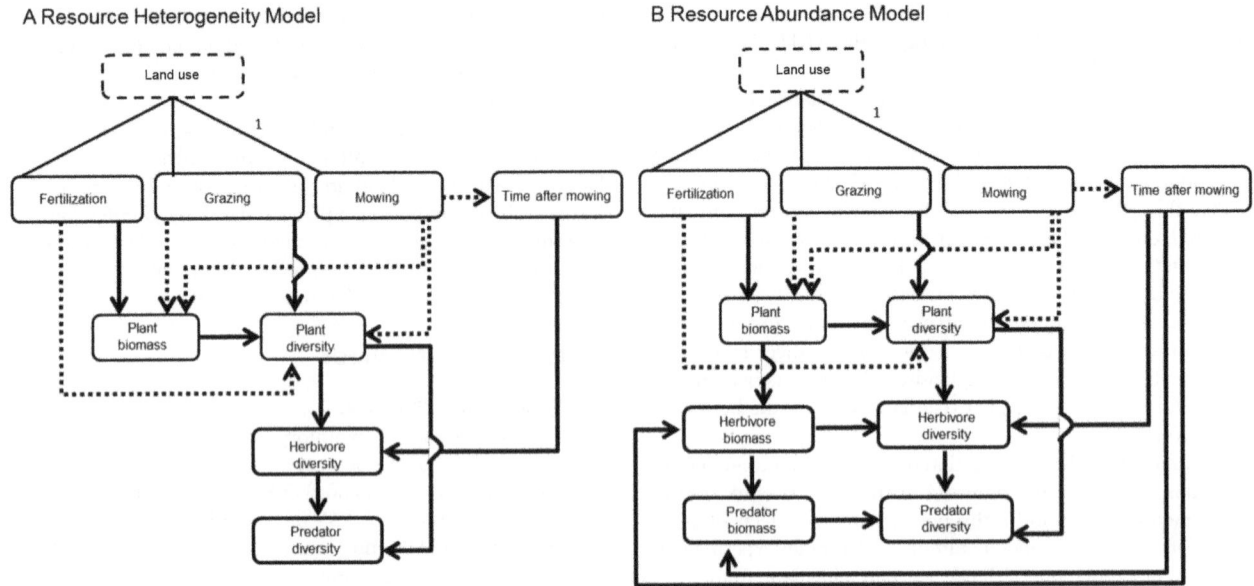

Figure 1. Concepts for structural equation models. 'Land use' is a latent (i.e. unmeasured) variable influencing the management components. The path coefficient for the correlation between 'Land use' and cutting frequency was a priori set to 1. The arrows indicate expected causal effects between measured variables; solid arrows represent expected positive effects and dotted arrows expected negative effects. For more detailed information on model definition and variable calculation see the Method section and Appendix S1.

described by Socher et al. [1]: On a 4 m × 4 m subplot in each plot all vascular plant species were identified in the field following the nomenclature of Wisskirchen & Häupler [27], categorized into functional groups (grasses, herbs, legumes) and their cover (%) was estimated. As vegetation was generally relatively uniform across the plot, we consider our subplot sample as representative for the 50 m × 50 m plot. Plant biomass was clipped at a height of 2 to 3 cm above ground and dried for 48 hours at 80°C before weighing. Occasionally occurring shrubs and litter were excluded before drying. In 2008, biomass samples were taken in two 25 cm × 50 cm subplots next to the vegetation survey plot. In 2009, biomass was harvested on eight subplots of 50 cm × 50 cm size adjacent to one side of the vegetation survey plot to better reflect small-scale variability. In both years, total plant biomass [g/m²] was recorded as the mean from the subplots. In mown and grazed plots, the subplots were fenced until biomass harvest to obtain productivity estimates unaffected by current year grazing or mowing. For the analyses we used plant species richness per plot as measure for plant diversity.

We sampled arthropods by sweep-netting with a total of 60 double-sweeps (one double-sweep is defined as moving the net from the left to the right and back perpendicular to the walking direction) along three plot borders (150 m in total) in June and August 2008 & 2009 during periods of dry weather conditions. The two sampling months were chosen because they cover the main activity period of arthropods and were found to represent the variation in diversity among plots equally well as sampling in five months (tested with a subset of plots in 2009, data not shown). We chose sweep-netting over suction sampling because of logistic difficulties with suction sampling on our large number of plots and because suction sampling was found to underrepresent Heteroptera (Gossner et al., unpublished data from our plots and [28]).

Sampling was conducted within seven days or fewer per region and within two weeks in all regions. Arthropods were preserved in 70% ethanol, sorted by taxonomic group and identified to species level by taxonomic experts. We focused on Araneae, Hemiptera

(Cicadomorpha, Fulgoromorpha and Heteroptera), Coleoptera and Orthoptera because of their numerical dominance in the studied grasslands. Only adult individuals were considered for the analysis, because identification of juveniles is often difficult. Samples from both months within one year were pooled for analysis. Because sampling effort was standardized there was no need to adjust for differences in the number of sampled individuals. Accordingly, we used species richness as diversity measure.

Due to external influences, such as aggressive cows preventing us from entering several plots, the number of plots with a complete dataset (plant and arthropod data) differed somewhat between years: in 2008 we considered 124 plots and in 2009 we considered 141 plots (see Table 1 for the number of plots per region). Of those plots, 117 plots were considered in both years.

Ethics Statement

Field work permits were issued by the responsible state environmental offices of Baden-Württemberg, Thüringen and Brandenburg (according to § 72 BbgNatSchG, i.e. nature conservation law of Brandenburg). Out of the 150 initially sampled plots, 17 are part of grassland areas that are protected under the Habitats Directive (FFH), 58 are protected as LSG (landscape conservation site), 17 are protected as NSG (nature conservation site) and 5 are situated within a former military training area. Except the 5 plots which are situated in the former military training area and 6 plots that are situated on state property, all other grasslands are privately owned by the farmers. One individual of *Phytoecia cylindrica* (Coleoptera, Linnaeus, 1758), which is protected under BArtSchV (Federal Protection of Species Order) was sampled in 2008.

Classification of trophic guilds and arthropod biomass assessment

All identified arthropod species were assigned to one of three trophic guilds (herbivores, predators and decomposers) based on

Table 1. Mean ± standard error for the sampled variables within each of the three regions.

	Vascular plants		Herbivores			Predators		
	No. species	Biomass (g)	No. species	Abundance	Biomass (g)	No. species	Abundance	Biomass (g)
2008								
Swabian Alb N=43	33.3±10.3	598.1±302.9	33.3±10.6	628.6±570.9	15.3±14.3	5.4±3.5	10.6±8.3	0.4±0.4
Hainich-Dün N=38	25.9±9.6	592.6±228.8	31.4±12.3	307.4±205.0	9.6±8.6	5.6±2.8	15.8±10.0	0.5±0.5
Schorfheide-Chorin N=43	16.0±3.7	752.0±281.7	26.7±7.7	281.1±145.0	8.8±4.9	5.1±2.7	13.2±9.5	0.3±0.3
F-value$_{2,121}$	46.72	4.56	4.64	11.89	5.09	0.29	3.25	0.91
P	<0.001	0.012	0.011	<0.001	0.008	0.752	0.042	0.404
Means over all regions	25.0±2.2	648.0±11.0	30.0±1.9	410.0±19.6	11.2±3.1	5.0±1.3	13.0±2.6	0.4±0.02
2009								
Swabian Alb N=47	32.1±11.2	272.1±123.1	20.5±5.6	180.9±144.5	6.7±4.8	3.8±2.0	4.6±2.8	0.3±0.3
Hainich-Dün N=48	33.6±12.1	299.7±128.7	16.6±7.1	73.0±93.4	2.8±4.4	4.3±2.5	8.5±7.2	0.2±0.2
Schorfheide-Chorin N=46	19.8±4.2	331.2±130.5	22.1±8.0	163.2±105.1	4.7±4.0	5.2±2.9	9.8±7.2	0.3±0.6
F-value$_{2,140}$	27.25	2.49	7.648	11.76	9.25	4.1	9.0	0.5
P	<0.001	0.086	<0.001	<0.001	<0.001	0.019	<0.001	0.61
Means over all regions	29.0±2.2	301.0±7	20.0±1.6	139.0±10.6	4.7±2.2	4.0±1.2	8.0±2.3	0.3±0.7

N = number of plots sampled in the respective year. Arthropod biomass was estimated from the species' mean length (see text). Means were tested for differences between regions with ANOVA. Degrees of freedom are indicated with the F-value$_{df\ Region,\ df\ Residuals}$. Abundances are expresses as number of adult individuals whereas biomass is given in grams. Species lists of herbivores and predators can be found in Appendix S5 and Appendix S6, original datasets are available from the Dryad Digital Repository.

their known main food resource as adults (Appendix S2). Because decomposers made up less than 2% of individuals in 90% of our samples, they were not included in the analysis. Arthropod biomass per plot was estimated by applying the general power function developed by Rogers et al. [29] to each sampled species:

$$biomass[mg] = 0.305 * L^{2.62}$$

Where L is the mean length (in mm) of a species, derived from the same sources as given for the identification of feeding guilds (Appendix S2). The contribution of a species to total biomass was then calculated as: species-specific biomass × abundance of the species. Biomass for each plot was calculated for herbivores and predators separately by summing over all species. We used biomass rather than abundances as measure for resource abundance to be consistent with plant assessments.

Statistical analysis

To compare means of the response variables among regions, we used ANOVA (aov) with 'Region' as explanatory variable in R [30].

For structural equation modelling, there are, in principle three main approaches [31]: 1) Strictly confirmatory, where a single model is defined and tested against the data. 2) Alternative modelling, where similar models are tested and compared by model selection for best fit. 3) Model generating, where one tentative initial model is simplified until the model fit cannot be improved. We used a combination of approaches 1 and 3 by defining two models based on the 'Resource Heterogeneity' (RHH) and 'More Individuals' (MIH) hypotheses (confirmatory approach, Figure 1) but using step-wise deletion of paths within these models when the model structure did not fit the data (model generating). The second step was included to assess whether the discrepancies between model and data structure were due to erroneous assumptions about interactions among variables (in this case model fit would be improved by step-wise deletion) or due to missing (i.e. not measured) variables (in that case model fit would not be improved).

Grassland land use was included in the models by the variables fertilization, grazing and mowing as well as 'Time after mowing' (as described in the first section of Materials and Methods). In addition, we added a latent (i.e. unmeasured) variable 'Land use' to describe the combined effect of land use. This latent variable accounts for correlations between the different land-use modes. As latent variables have no underlying data which the model algorithm can use to calculate its variance, either the latent's variance or one regression weight between the latent and one of the connected variables has to be fixed to 1. We chose to fix the regression weight between the latent variable and mowing frequency because mowing is a good descriptor of land-use intensity and it is correlated with both fertilization and grazing intensity [32]. The two other regression weights (between 'Land use' and grazing or fertilization) and the latent's variance can then be estimated by the model algorithm. To test for a non-linear effect of grazing intensity, we tried a quadratic transformation of grazing intensity, but because it did not improve model fit we kept the linear relationship. As the correlations between the land-use modes varied between regions, the model sometimes did not converge; in these cases, we included correlations between mowing and fertilization and between mowing and grazing instead of the latent variable. Except for the additional variables and pathways that were added in the 'Resource Abundance Model', all pathways

were expected to be identical in both models (Fig. 1 A & B), as motivated in the Introduction.

We used the software package 'sem' in R for structural equation modelling [30] which fits models using Maximum Likelihood based on observed and expected covariance matrices. The model fit is estimated as the overall model p-value which indicates if the two covariance matrices are significantly different from each other ($p<0.05$, bad model fit) or not ($p>0.05$, good model fit [31], p.128f). A step-wise selection procedure was applied by hand using the corrected Akaike's Information Criterion (AICc) for models without appropriate fit (model p-value <0.05). If a second measure of model quality, the Goodness-of-fit (GoF) index, was above 0.75 for the final model, models without adequate fit were still used for comparison among regions.

To calculate the total effect of 'Land use' on the plant and arthropod variables, the standardized path coefficients within each possible pathway between 'Land use' and the respective variable were multiplied and all resulting products were summed (e.g. coefficients from 'Land use' to fertilization and from fertilization to plant biomass were multiplied and added to the products of pathways including grazing and mowing). Total effects among plant and arthropod variables were calculated accordingly. Data was transformed as necessary (further details in Appendix S3).

Results

Over all plots we recorded 271 vascular plant species in 2008 and 281 vascular plant species in 2009; 54,660 herbivore individuals from 392 species were sampled in 2008 and 19,507 herbivores from 335 species in 2009; 1,823 predator individuals from 162 species were sampled in 2008 and 1,075 predators from 154 species in 2009. The average plant biomass per plot was 648 g/m^2 in 2008 and 301 g/m^2 in 2009, mean estimated herbivore biomass per plot was 11.2 g and 4.7 g (2008 and 2009 respectively) and mean estimated predator biomass per plot was 0.4 g and 0.3 g (including 1 and 5 plots with no predators in 2008 and 2009 respectively). The lower plant and herbivore biomass in 2009 compared to 2008 can be attributed to a shorter time-span between the onset of the vegetation period and sampling (as calculated for 'Time after mowing', see Appendix S1). Increasing plant biomass was correlated with a decrease in plant species diversity (see below) and with an increase in relative cover of grass species (compared to herbaceous species) ($R^2_{adj} = 0.269$ in 2008 and $R^2_{adj} = 0.081$ in 2009, $P<0.001$ in both years and all regions; Spearman's correlation).

In 2008 and 2009, plant and herbivore species richness as well as herbivore abundance and herbivore biomass differed significantly between regions and were highest in ALB (Table 1). Plant biomass also differed significantly between regions in 2008 and was highest in SCH while in 2009 it was similar among regions (Table 1). Predator species richness differed between regions only in 2009 (being highest in SCH) and was very similar among regions in 2008. Predator abundances differed between regions both in 2008 and 2009 and were lowest in ALB. Predator biomass did not differ between regions in both years (Table 1).

Resource Heterogeneity Model

The Resource Heterogeneity Model fitted the 2008 and 2009 data well for each region (Figure 2). Except for the omission of the latent variable 'Land use' in three cases, no further path-selection procedure was applied. Despite this omission, correlations among the land-use modes were consistent between years and regions, where fertilization was positively correlated with 'Land use' (or

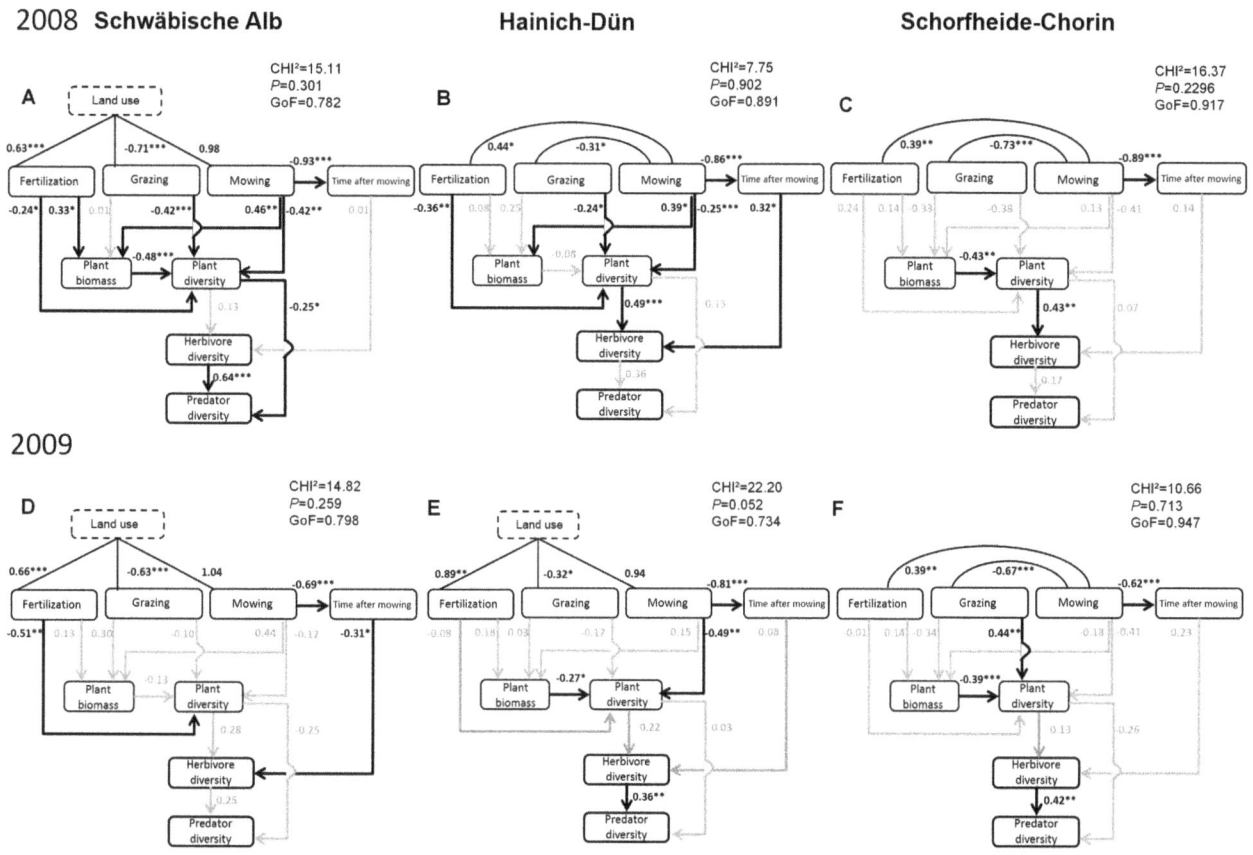

Figure 2. Standardized regression weights and significance levels from the Resource Heterogeneity Model. Models are shown for 2008 (A, B, C) and 2009 (D, E, F) for the Swabian Alb (A, D), Hainich-Dün (B, E) and Schorfheide-Chorin (C, F). Black solid lines and numbers indicate significant paths, grey arrows and numbers indicate non-significant paths. Correlation graphs for plant and arthropod measures can be found in Appendix S4. GoF = Goodness of fit. Significance levels: P<0.05: */P<0.01: **/P<0.001: ***. Total effects in Tables 2 & 3 were calculated using the standardized regression weights, e.g. plant diversity effects on predator diversity in Swabian Alb 2008 were calculated as 0.13 (coefficient from plant diversity to herbivore diversity)*0.64 (herbivore diversity to predator diversity) *−0.25 (plant diversity to predator diversity) = −0.17.

mowing) and grazing was negatively correlated with 'Land use' (or mowing).

Fertilization generally had the predicted negative effect on plant diversity and a positive effect on plant biomass, although effects were significant only in ALB (for diversity in 2008 & 2009 and for biomass only in 2008) and HAI (for diversity in 2008). We found positive but not significant effects of grazing intensity on plant biomass (Figure 2) but grazing intensity consistently reduced plant diversity (except for a significant increase in SCH in 2009). Mowing frequency consistently increased plant biomass and decreased plant diversity (although only significant in ALB and HAI and mostly in 2008). Differences in effect direction were found in SCH (positive effect of fertilization on plant diversity, and negative effects of grazing and mowing on plant biomass) but all of those effects were not significant.

As expected, plant biomass was always negatively related to plant diversity even though the paths for HAI 2008 and ALB 2009 were not significant. Effects of plant diversity on herbivore diversity were positive in all regions and years, as predicted by the RHH (although only significant in HAI and SCH). Similarly, we found a positive effect of herbivore diversity on predator diversity in all regions and years (here it was significant in 3 of 6 cases).

When summing the direct and indirect effects of 'Land use' on the target variables, we found total effects to be positive for plant

biomass and negative for plant diversity (Table 2). The only exception was SCH, due to the strong influence of grazing (negative on plant biomass in 2008 and positive on plant diversity in 2009). The total effect of 'Land use' on herbivore diversity was negative in all regions and both years, whereas total effects on predator diversity were generally very weak (<0.1). The total effects of plant diversity on herbivore diversity were all positive as they are identical to the direct effects in this model. Total effects of plant diversity on predator diversity were, however, generally weak. Overall, 'Land use' decreased plant diversity directly and indirectly through an increase in plant biomass, which led to a decrease in herbivore diversity and a following decrease in predator diversity.

Resource Abundance Model

The Resource Abundance Model fitted the 2008 data adequately well for HAI and SCH, but not for ALB, even after applying path-selection procedures. For 2009, we found adequate model fits for all three regions after applying path-selection to HAI (Figure 3). Compared with the Resource Heterogeneity Model, the relationships among land-use modes remained the same, except that the latent variable could be kept in the model for HAI in 2008 (because the model converged in the first run). The direct effects of the land-use modes on plant diversity and biomass changed slightly in magnitude but not in direction compared with

Table 2. Standardized total effects derived from the Resource Heterogeneity Model for both years.

		Land use (mode) effects on				Plant diversity effects on	
		Plant biomass	Plant diversity	Herbivore diversity	Predator diversity	Herbivore diversity	Predator diversity
Swabian Alb	2008	0.65	−0.59	−0.09	0.09	0.13	−0.17
	2009	0.35	−0.44	0.10	0.03	0.28	0.06
Hainich-Dün							
Fertilization	2008	0.08	−0.37	−0.18	−0.06	0.49	0.32
Grazing	2008	0.25	−0.26	−0.13	−0.05		
Mowing	2008	0.39	−0.56	−0.54	−0.19		
	2009	0.28	−0.56	−0.19	−0.08	0.22	0.11
Schorfheide-Chorin							
Fertilization	2008	0.41	0.18	0.10	0.01	0.43	0.15
Grazing	2008	−0.33	−0.24	−0.10	0.0		
Mowing	2008	0.17	0.01	−0.01	0.0		
Fertilization	2009	0.14	−0.06	0.0	0.02	0.13	−0.20
Grazing	2009	−0.03	0.46	0.06	0.02		
Mowing	2009	−0.19	−0.10	−0.16	−0.09		

Total effects were calculated by multiplying the standardized path coefficients on the single pathways between two variables and summing up those values for all possible pathways. Standardized total effects can range between −1 and 1. Effects are shown from the first row on the second row. For a description on how total effects were calculated see last paragraph in Material and Methods and refer to legend of Figure 2 for an example. Original datasets are available from the Dryad Digital Repository.

the first model and between years. The change in magnitude is due to the fact that effects were standardized (i.e. values are relative to each other) which is influenced by the change in number of variables included in the model. As in the first model, plant biomass was negatively related to plant diversity (although not significant in HAI 2008) and plant diversity had a positive effect on herbivore diversity (although only significant in ALB & HAI 2008 and SCH 2009; Figure 3).

We did not find a significant direct effect of plant biomass on herbivore biomass as predicted by the MIH in any region or year. As the relationship between plant and herbivore biomass may be weakened if there are shifts in the relative abundances of differently sized herbivore species, we also tested the Resource Abundance Model using arthropod abundances instead of biomass but again did not find any effect of plant biomass on herbivore abundances, neither in 2008 or 2009 (Figure S1: Standardized regression weights and significance levels from the resource abundance model including arthropod abundances). Herbivore biomass had a significant positive effect on herbivore diversity in all regions and both years. Herbivore biomass also had a positive effect on predator biomass, significant in three out of six cases (ALB 2008, 2009, SCH 2009). Predator biomass itself had a significant positive effect on predator diversity in all regions and both years.

Total effects of 'Land use' on plant and arthropod diversity as well as on plant biomass were identical in direction and similar in

effect strength to the Resource Heterogeneity Model for both years, including the reverse effects in SCH due to grazing. Total effects of plant diversity on herbivore diversity were positive, but total effects of plant biomass on herbivore diversity were absent or very weak. Total effects of herbivore biomass on predator diversity were positive (Table 3).

In summary, 'Land use' increased plant biomass, but this effect was not leading to an increase in herbivore biomass. Nevertheless, increasing herbivore biomass generally increased predator biomass which in turn increased predator diversity.

Discussion

We tested whether effects of land-use intensity on the diversity of arthropod herbivores and predators are mediated by changes in the diversity or biomass of their respective resources, i.e. plants and herbivores. Although the strength of effects varied between regions and years, we found negative effects of land-use intensity on the diversity of both arthropod groups which were mediated by different pathways. For herbivores, the results were consistent with the 'Resource Heterogeneity Hypothesis', as we found significant effects of plant diversity on herbivore diversity with no evidence for effects mediated by plant biomass. For predators, results were more consistent with the 'More Individuals Hypothesis', i.e. predator diversity and biomass were more strongly affected

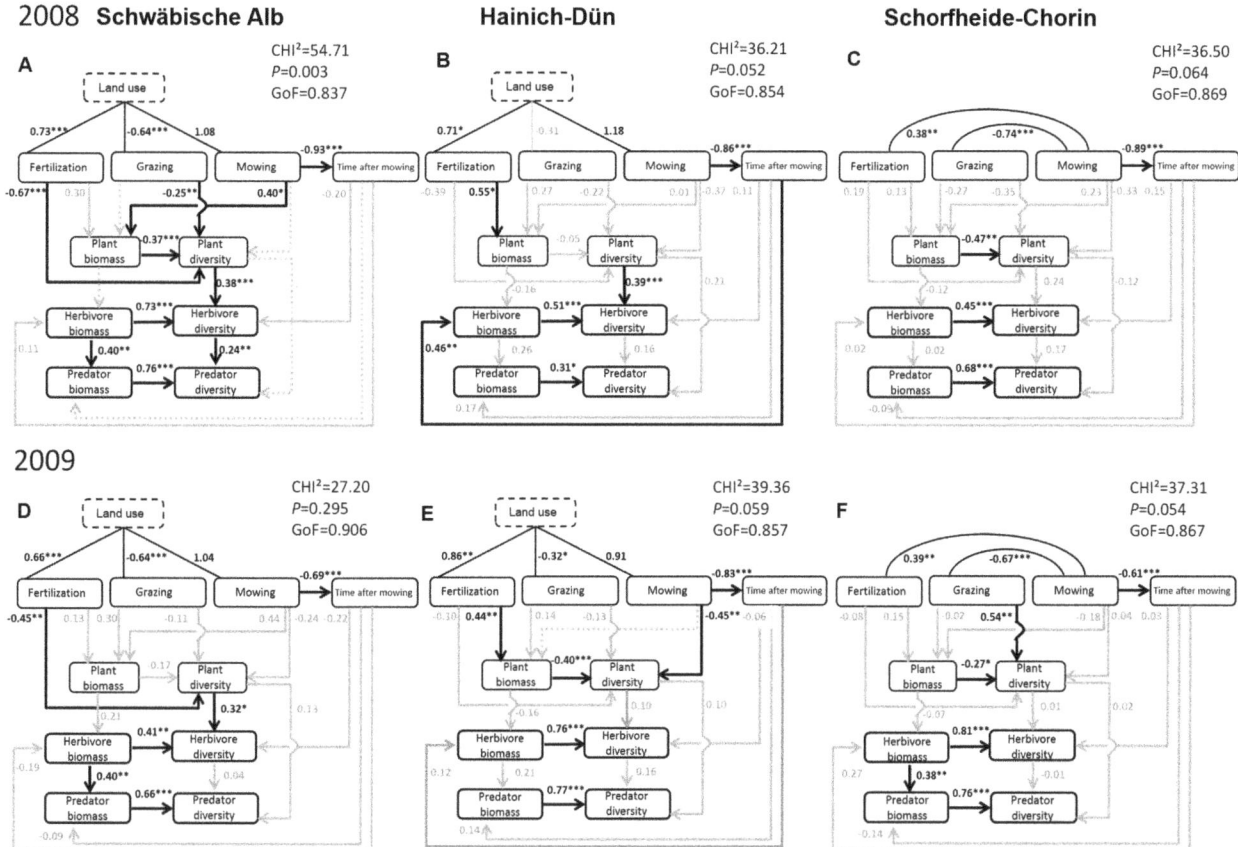

Figure 3. Standardized regression weights and significance levels from the Resource Abundance Model. Models are shown for 2008 (A, B, C) and 2009 (D, E, F) for the Swabian Alb (A, D), Hainich-Dün (B, E) and Schorfheide-Chorin (C, F). Black solid lines and numbers indicate significant paths, grey arrows and numbers indicate non-significant paths. Grey dotted paths were excluded during the step-wise selection procedure. Correlation graphs for plant and arthropod measures can be found in Appendix S4. GoF = Goodness of fit. Significance levels: $P<0.05$: */$P<0.01$: **/ $P<0.001$: ***.

Table 3. Standardized total effects derived from the Resource Abundance Model for both years.

		Land use (mode) effects on				Plant diversity effects on		Plant biomass effects on	Herbivore biomass effects on
		Plant biomass	Plant diversity	Herbivore diversity	Predator diversity	Herbivore diversity	Predator diversity	Herbivore diversity	Predator diversity
Swabian Alb	2008	0.65	−0.56	−0.09	−0.06	0.38	0.09	−0.14	0.48
	2009	0.05	−0.48	0.07	0.02	0.32	0.15	0.04	0.28
Hainich-Dün	2008	0.84	−0.71	−0.70	−0.36	0.39	0.27	−0.10	0.16
	2009	−0.37	−0.43	−0.53	−0.38	0.77	0.52	0.03	0.66
Schorfheide-Chorin	2008					0.24	−0.08	0.17	0.09
Fertilization	2008	−0.28	0.32	0.09	−0.02				
Grazing	2008	0.23	−0.46	−0.12	0.03				
Mowing	2008	0.13	−0.39	−0.25	0.06				
	2009					0.01	0.03	−0.06	0.28
Fertilization	2009	−0.02	−0.08	0.0	0.0				
Grazing	2009	−0.18	0.59	0.01	0.02				
Mowing	2009	0.15	0.0	0.01	0.02				

Total effects were calculated by multiplying the standardized path coefficients on the single pathways between two variables and summing up those values for all possible pathways. Standardized total effects can range between −1 and 1. Effects are shown from the first row on second row. For an example on how effects were calculated see legend of Figure 2.

indirectly by changes in herbivore biomass than by direct effects of herbivore diversity.

Total land-use effects on arthropods and differences between regions

The negative total effect of land-use intensity on herbivore and predator diversity is consistent with previous reports on the consequences of grassland land use on arthropods [2–4], but the effects of the individual land-use modes (especially grazing) differed between regions and years and were only partly consistent with expectations. For instance, moderate grazing was found to have a positive effect on arthropod diversity in several studies [33] but we found only weak total effects of grazing intensity on arthropod diversity (<0.2). A wider range of grazing intensities in our study system compared to other studies might explain the absence of a positive effect in our case; e.g. Dennis et al. [34] found higher abundance and diversity of arthropods under moderate grazing intensity of sheep compared to high grazing intensity. The different types of livestock in our study system likely changed effects of grazing as well because herbivore diversity in 2008 was significantly lower on cattle-grazed plots compared to sheep-grazed plots and mixed grazing by cattle and horses had a significant positive effect on predator diversity compared with cattle or sheep grazing (Figure S2: Effects of livestock type on plant and insect species richness in 2008). Differences in the grazing gradients between regions and changes in grazing practices between years can also explain the change from a negative effect of grazing to a positive effect of grazing on plant diversity in SCH.

The different correlations among the land-use modes in the different regions are, in fact, a striking result of our study. Whereas all modes were significantly correlated with the latent variable 'Land use' in ALB, fertilization and grazing were sometimes (HAI) or always (SCH) independent of each other in the other regions. The discrepancies were probably caused by the range of land-use options realized in the different regions. In ALB, where most of the grasslands are managed by small farming enterprises or farming families, we find extensively grazed, unfertilized plots, e.g. sheep pastures on nutrient-poor hillsides, as well as intensively grazed, fertilized plots. Thus, grazing and fertilization are closely linked in this region. In the other two regions, plots with low grazing intensity sometimes receive low fertilizer input (for instance, in organic farming practices) but in other cases they are highly fertilized mown pastures that are only grazed at the end of the plant growth period. This weakened the correlation between grazing and fertilization.

Effects of land-use intensity on plant biomass and diversity

Socher et al. [1] extensively discussed the effects of fertilization, grazing and mowing on plant biomass and diversity using the same plots as the present study, thus here we only summarize the main points: Increasing fertilization intensity generally decreased plant diversity and increased biomass as found in many preceding studies. Negative effects of high mowing frequency on plant diversity support previous findings from different types of grasslands and various regions [35]. The negative effect of grazing on plant diversity appears to contrast the general finding that grazing increases plant diversity [36]. However, most previous studies compared grazed with ungrazed sites and thus found that grazing increased plant diversity via increased sward heterogeneity [11,37]. In our case, grazing ranged from no grazing to very high grazing intensities, which could result in a non-linear or hump-shaped effect of grazing on plant diversity. As including a non-

linear effect of grazing did not improve our model (see Method section) it seems likely that the negative effect of very high grazing intensities is exceeding the positive effect moderate grazing has compared to non-grazed sites.

Effects of changes in the plant community on herbivores

Our results showed clear evidence that land-use effects on herbivore diversity are mediated by plant diversity as predicted by the 'Resource Heterogeneity Hypothesis'. The fact that we did not find effects on herbivore diversity mediated by plant biomass is in contrast to findings from experimental plant communities [22], which showed that arthropod diversity was only indirectly affected by plant diversity through increased plant biomass. This indirect effect on arthropod diversity was additionally mediated by arthropod biomass (measured as biovolume) and therefore followed the 'More Individuals Hypothesis'[22]. The differences between results from our study in managed grasslands and results from experimental plant communities may be due to the absence of correlations between the proportion of particular plant functional groups and plant diversity in biodiversity experiment, as both variables are manipulated similarly. In contrast, the grasslands in our study system showed an increasing cover proportion of grasses with decreasing plant diversity. As found by Haddad et al. [38] the presence of grass species (which are productive but have low nutritional quality for herbivores) led to a decrease in total insect abundance by 25% even though total plant biomass increased. Only when grass species were absent and all plants were of higher nutritional quality, insect abundance was best explained by plant biomass. Hence, the higher cover of grass species on grasslands with low plant diversity in our study system led to a higher plant biomass but at the same time could not sustain higher herbivore biomass possibly because the overall nutritional value for the arthropods did not increase together with plant biomass.

Effects of changes in the herbivore community on predators

We found direct and indirect effects between herbivore and predator diversity, indicating mechanisms in accordance with both the 'Resource Heterogeneity' and the 'More Individuals Hypothesis'. One example would be the ALB 2008 where both direct and indirect effects were significant and strong. This indicates a complementary role of both mechanisms which might be the result of different predator groups reacting to either one of the mechanisms. In our study, the total effect of herbivore biomass on predator diversity was stronger in five out of six cases than the direct effect of herbivore diversity (compare Figure 3 and Table 3). This agrees with results from a plant diversity experiment, where effects of plants on herbivores were consistent with the 'Resource Heterogeneity Hypothesis' but effects of herbivores on predators were more in agreement with the 'More Individuals Hypothesis' [12]. Further research is needed to understand how predator diversity is affected by land use, as herbivore biomass was not affected by any of the included factors. This is relevant for sustainable land use of grasslands in agricultural dominated landscapes, because high predator abundance and diversity enhances biocontrol potential e.g. [39].

Where to go from here

We tested the effect of land-use intensity on important grassland herb-dwelling arthropods over a wide geographic range. To achieve standardized sampling in the vegetation layer across a large number of plots, sweep-net sampling was used which is a suitable method to representatively sample important herbivores (e.g. Heteroptera) as well as predators (e.g. Araneae) among arthropods (e.g. [40–42]). Nevertheless, it is a less well-performing method to sample other functional groups such as pollinators (butterflies and bees) [40] or ground-dwelling species [43]. Additional methods (such as suction sampling, pitfall traps or pan traps) might therefore be advised if a study's focus is not only on herb-dwelling species. Disentangling the differences which might apply to different functional groups within herbivores and predators (e.g. sucking vs. chewing herbivores or predators with different hunting strategies) will further increase our understanding of land-use effects. One promising approach was recently proposed by Lavorel et al. [44] who included producer and consumer traits in structural equation models to understand how land use affects ecosystem services through changes in the trait composition of the groups which provide the services.

Conclusions

Our results emphasize the importance of studying indirect effects of land-use intensity on the arthropod community, as they showed that herbivores and predators respond to changes of different aspects of their resources. We confirm that herbivore diversity is responding positively to higher plant diversity in grasslands, whereas herbivore biomass matters more than diversity for how predator diversity is affected by land use. By including different regions we showed on the one hand that the negative effects of high fertilization intensity and high mowing frequency on arthropod diversity are consistent over large scales; but on the other hand the variability of land-use traditions clearly indicates that findings cannot be easily extended to a wider geographical context. Our results thus not only emphasize the importance of land use for biodiversity changes, but also the need for more differentiated approaches to disentangle how different land-use modes have different effects on biodiversity, and how chains of effects differ for different aspects of biodiversity.

Supporting Information

Figure S1 Standardized regression weights and significance levels from the resource abundance model including arthropod abundances. Models are shown after step-wise deletion of non-significant paths. Black solid lines and numbers indicate significant paths; grey arrows indicate non-significant paths. Grey, dotted paths were excluded during the step-wise selection procedure. Significance level: p<0.05: */p< 0.01: **/p<0.001: ***.

Figure S2 Effects of livestock type on plant and insect species richness in 2008. Means per plot and standard errors are shown. Horizontal lines indicate significant differences based on Tukey's HSD test. Significance levels: p<0.05: */p<0.01: **/ p<0.001: ***.

Table S1 Mean and range of land-use activities in the three regions and during the years considered for the analysis with samplings from 2008.

Table S2 Mean and range of land-use activities in the three regions and during the years considered for the analysis with samplings from 2009.

Appendix S1 Assessment and calculation of land-use information.

Appendix S2 Classification of trophic guilds.

Appendix S3 Structural equation model setup and path-selection procedure.

Appendix S4 Bivariate correlations between model variables.

Appendix S5 List of arthropod species sampled in 2008

Appendix S6 List of arthropod species sampled in 2009.

Acknowledgments

Special thanks go to Leonardo Ré Jorge, who gave valuable comments during the discussion on the model structure. We thank Luis Sikora for his help with sweep-netting the grasslands and Roland Achtziger, Theo Blick, Boris Büche, Michael-Andreas Fritze, Günter Köhler, Frank Köhler, Franz Schmolke and Thomas Wagner for identifying the arthropods and providing data on trophic guilds and body size. A database on Orthoptera was kindly provided by Frank Dziock. We thank the managers of the three exploratories, Swen Renner, Sonja Gockel, Kerstin Wiesner and Martin Gorke for their work in maintaining the plot and project infrastructure; Andreas Hemp and Uta Schumacher for insuring successful field work in the Schorfheide; Simone Pfeiffer and Christiane Fischer giving support through the central office, Michael Owonibi for managing the central data base and Eduard Linsenmair, Dominik Hessenmöller, Jens Nieschulze, Ingo Schöning, François Buscot, Ernst-Detlef Schulze and the late Elisabeth Kalko for their role in setting up the Biodiversity Exploratories project.

Author Contributions

Conceived and designed the experiments: MF MMG WWW. Performed the experiments: SB ML JM EP SAS MT. Analyzed the data: NKS MMG TML. Contributed to the writing of the manuscript: NKS MMG TML SB ML JM EP SAS MT MF WWW.

References

1. Socher SA, Prati D, Boch S, Müller J, Klaus VH, et al. (2012) Direct and productivity-mediated indirect effects of fertilization, mowing and grazing on grassland species richness. Journal of Ecology 100: 1391–1399.
2. Bell JR, Wheater CP, Cullen WR (2001) The implications of grassland and heathland management for the conservation of spider communities: a review. Journal of Zoology 255: 377–387.
3. Di Giulio M, Edwards PJ, Meister E (2001) Enhancing insect diversity in agricultural grasslands: the roles of management and landscape structure. Journal of Applied Ecology 38: 310–319.
4. Nickel H, Hildebrandt J (2003) Auchenorrhyncha communities as indicators of disturbance in grasslands (Insecta, Hemiptera) - a case study from the Elbe flood plains (northern Germany). Agriculture Ecosystems & Environment 98: 183–199.
5. Weiner CN, Werner M, Linsenmair KE, Bluthgen N (2011) Land use intensity in grasslands: Changes in biodiversity, species composition and specialisation in flower visitor networks. Basic and Applied Ecology 12: 292–299.
6. Chamberlain DE, Fuller RJ, Bunce RGH, Duckworth JC, Shrubb M (2000) Changes in the abundance of farmland birds in relation to the timing of agricultural intensification in England and Wales. Journal of Applied Ecology 37: 771–788.
7. Allan E, Bossdorf O, Dormann CF, Prati D, Gossner MM, et al. (2014) Interannual variation in land-use intensity enhances grassland multidiversity. Proceedings of the National Academy of Sciences of the United States of America 111: 308–313.

8. Littlewood NA, Stewart AJA, Woodcock BA (2012) Science into practice - how can fundamental science contribute to better management of grasslands for invertebrates? Insect Conservation and Diversity 5: 1–8.

9. Marini L, Fontana P, Scotton M, Klimek S (2008) Vascular plant and Orthoptera diversity in relation to grassland management and landscape composition in the European Alps. Journal of Applied Ecology 45: 361–370.

10. Morris MG, Lakhani KH (1979) Responses of grassland invertebrates to management by cutting. 1. Species-diversity of Hemiptera. Journal of Applied Ecology 16: 77–98.

11. Woodcock B, Potts S, Tscheulin T, Pilgrim E, Ramsey A, et al. (2009) Responses of invertebrate trophic level, feeding guild and body size to the management of improved grassland field margins. Journal of Applied Ecology 46: 920–929.

12. Haddad NM, Crutsinger GM, Gross K, Haarstad J, Knops JMH, et al. (2009) Plant species loss decreases arthropod diversity and shifts trophic structure. Ecology Letters 12: 1029–1039.

13. van den Berg LJL, Vergeer P, Rich TCG, Smart SM, Guest D, et al. (2011) Direct and indirect effects of nitrogen deposition on species composition change in calcareous grasslands. Global Change Biology 17: 1871–1883.

14. Ryder C, Moran J, Mc Donnell R, Gormally M (2005) Conservation implications of grazing practices on the plant and dipteran communities of a turlough in Co. Mayo, Ireland. Biodiversity and Conservation 14: 187–204.

15. Sjodin NE, Bengtsson J, Ekbom B (2008) The influence of grazing intensity and landscape composition on the diversity and abundance of flower-visiting insects. Journal of Applied Ecology 45: 763–772.

16. Watkinson AR, Ormerod SJ (2001) Grasslands, grazing and biodiversity: editors' introduction. Journal of Applied Ecology 38: 233–237.

17. Hutchinson GE (1959) Homage to Santa Rosalia or Why are there so many kinds of animals? American Naturalist 93: 145–159.

18. Lewinsohn TM, Roslin T (2008) Four ways towards tropical herbivore megadiversity. Ecology Letters 11: 398–416.

19. Southwood TRE, Brown VK, Reader PM (1979) The relationships of plant and insect diversities in succession. Biological Journal of the Linnean Society 12: 327–348.

20. Strong DR, Jr., Lawton JH, Southwood TRE (1984) Insects on Plants: Community Patterns and Mechanisms. Cambridge, MA: Harvard University Press.

21. Srivastava DS, Lawton JH (1998) Why more productive sites have more species: An experimental test of theory using tree-hole communities. American Naturalist 152: 510–529.

22. Borer ET, Seabloom EW, Tilman D, Novotny V (2012) Plant diversity controls arthropod biomass and temporal stability. Ecology Letters 15: 1457–1464.

23. Andreasen C, Hansen CH, Moller C, Kjaer-Pedersen NK (2002) Regrowth of weed species after cutting. Weed Technology 16: 873–879.

24. Becklin KM, Kirkpatrick HE (2006) Compensation through rosette formation: the response of scarlet gilia (Ipomopsis aggregata: Polemoniaceae) to mammalian herbivory. Canadian Journal of Botany-Revue Canadienne De Botanique 84: 1298–1303.

25. Fischer M, Bossdorf O, Gockel S, Hansel F, Hemp A, et al. (2010) Implementing large-scale and long-term functional biodiversity research: The Biodiversity Exploratories. Basic and Applied Ecology 11: 473–485.

26. Humbert JY, Ghazoul J, Walter T (2009) Meadow harvesting techniques and their impacts on field fauna. Agriculture Ecosystems & Environment 130: 1–8.

27. Wisskirchen R, Haeupler H (1998) Standardliste der Farn- und Blütenpflanzen Deutschlands. Stuttgart (Hohenheim): Eugen Ulmer.

28. Brook A, Woodcock B, Sinka M, Vanbergen A (2008) Experimental verification of suction sampler capture efficiency in grasslands of differing vegetation height and structure. Journal of Applied Ecology 45: 1357–1363.

29. Rogers LE, Hinds WT, Buschbom RL (1976) A general weight versus length relationshop for insects. Annals of the Entomological Society of America 69: 387–389.

30. R Core Team (2013) R: A language and environment for statistical computing. 3.0.2 ed. Vienna, Austria: R Foundation for Statistical Computing.

31. Grace JB (2006) Structural Equation Modeling and Natural Systems. Cambridge, UK: Cambridge University Press.

32. Blüthgen N, Dormann CF, Prati D, Klaus VH, Kleinebecker T, et al. (2012) A quantitative index of land-use intensity in grasslands: Integrating mowing, grazing and fertilization. Basic and Applied Ecology 13: 207–220.

33. Scohier A, Dumont B (2012) How do sheep affect plant communities and arthropod populations in temperate grasslands? Animal 6: 1129–1138.

34. Dennis P, Skartveit J, McCracken DI, Pakeman RJ, Beaton K, et al. (2008) The effects of livestock grazing on foliar arthropods associated with bird diet in upland grasslands of Scotland. Journal of Applied Ecology 45: 279–287.

35. Hopkins A, Wilkins RJ (2006) Temperate grassland: key developments in the last century and future perspectives. Journal of Agricultural Science 144: 503–523.

36. Marion B, Bonis A, Bouzille JB (2010) How much does grazing-induced heterogeneity impact plant diversity in wet grasslands? Ecoscience 17: 229–239.

37. Rook AJ, Dumont B, Isselstein J, Osoro K, WallisDeVries MF, et al. (2004) Matching type of livestock to desired biodiversity outcomes in pastures - a review. Biological Conservation 119: 137–150.

38. Haddad NM, Tilman D, Haarstad J, Ritchie M, Knops JMH (2001) Contrasting effects of plant richness and composition on insect communities: A field experiment. American Naturalist 158: 17–35.

39. Geiger F, Bengtsson J, Berendse F, Weisser WW, Emmerson M, et al. (2010) Persistent negative effects of pesticides on biodiversity and biological control potential on European farmland. Basic and Applied Ecology 11: 97–105.

40. Buffington ML, Redak RA (1998) A comparison of vacuum sampling versus sweep-netting for arthropod biodiversity measurements in California coastal sage scrub. Journal of Insect Conservation 2: 99–106.

41. Doxon E, Davis C, Fuhlendorf S (2011) Comparison of two methods for sampling invertebrates: vacuum and sweep-net sampling. Journal of Field Ornithology 82: 60–67.

42. Spafford R, Lortie C (2013) Sweeping beauty: is grassland arthropod community composition effectively estimated by sweep netting? Ecology and Evolution 3: 3347–3358.

43. Standen V (2000) The adequacy of collecting techniques for estimating species richness of grassland invertebrates. Journal of Applied Ecology 37: 884–893.

44. Lavorel S, Storkey J, Bardgett R, Bello F, Berg M, et al. (2013) A novel framework for linking functional diversity of plants with other trophic levels for the quantification of ecosystem services. Journal of Vegetation Science 24: 942–948.

15

Spatial and Temporal Variations of Ecosystem Service Values in Relation to Land Use Pattern in the Loess Plateau of China at Town Scale

Xuan Fang[1], Guoan Tang[1]*, Bicheng Li[2], Ruiming Han[3]

1 Key Laboratory of Virtual Geographic Environment, Ministry of Education, School of Geography Science, Nanjing Normal University, Nanjing, China, **2** Research Center of Soil and Water Conservation and Ecological Environment, Chinese Academy of Sciences, Yangling, Shaanxi, China, **3** School of Geography Science, Nanjing Normal University, Nanjing, China

Abstract

Understanding the relationship between land use change and ecosystem service values (ESVs) is the key for improving ecosystem health and sustainability. This study estimated the spatial and temporal variations of ESVs at town scale in relation to land use change in the Loess Plateau which is characterized by its environmental vulnerability, then analyzed and discussed the relationship between ESVs and land use pattern. The result showed that ESVs increased with land use change from 1982 to 2008. The total ESVs increased by 16.17% from US$ 6.315 million at 1982 to US$ 7.336 million at 2002 before the start of the Grain to Green project, while increased significantly thereafter by 67.61% to US$ 11.275 million at 2008 along with the project progressed. Areas with high ESVs appeared mainly in the center and the east where largely distributing orchard and forestland, while those with low ESVs occurred mainly in the north and the south where largely distributing cropland. Correlation and regression analysis showed that land use pattern was significantly positively related with ESVs. The proportion of forestland had a positive effect on ESVs, however, that of cropland had a negative effect. Diversification, fragmentation and interspersion of landscape positively affected ESVs, while land use intensity showed a negative effect. It is concluded that continuing the Grain to Green project and encouraging diversified agriculture benefit to improve the ecosystem service.

Editor: Ricardo Bomfim Machado, University of Brasilia, Brazil

Funding: This study was sponsored by the Jiangsu Planned Projects for Postdoctoral Research Funds (No. 1401033C), the National Natural Science Foundation of China (No. 41401441), and the Priority Academic Program Development of Jiangsu Higher Education Institutions (PAPD) (No. 164320H101). The funders had no role in study design, data collection and analysis, decision to publish, or preparation of the manuscript.

Competing Interests: The authors have declared that no competing interests exist.

* Email: tangguoan@njnu.edu.cn

Introduction

Ecosystem contributes to human welfare by providing goods and services directly and indirectly [1–2]. With widely spreading of environmental problems, ecosystem service received increasing attention. Many studies showed human factors, such as urban sprawl [3,4,5], socioeconomic changes [6], agricultural policies [7,8], could affect natural or artificial ecosystems. Land use, an original and foundational human activity and represents the most substantial human alteration to systems on the planet of earth for long-term study [9], plays an important role in providing ecosystem services, including biodiversity, water filtration, retention of soil, etc. [10] Inappropriate land use may lead to significant degradation of local and regional ecological services [11]. Moreover, there were studies showed that ecosystem service trade-offs could successful apply to land use planning [12,13]. Understanding the relationship between ecosystem services and land use change is essential for maintaining a healthy ecosystem and getting sustainable services.

The growing body of literatures focused on how ecosystem service changes in response to land use change of different regions [14,15,16,17,18]. However, these studied focus on the impact of land use type on ecosystem service, while the spatial pattern of land that reflects ecological processed and functions [19] get less attention. Monitoring the characteristic of landscape patterns including area, shape, diversity, etc., is helpful to deeply understand the relationship between ecosystem service and land use change and then to provide complete references for land use planning.

The Loess Plateau is the area suffered from the most severe soil erosion in the world, and it is also a major agricultural production region in China [20]. Long-term poor land use has resulted in vegetation destruction and accelerated soil erosion [21]. To control soil erosion and restore the ecosystem, the Grain for Green project converting slope cropland to grassland or forestland was implemented in 1999 by the Chinese Government [22]. The land use on the plateau under the project has changed significantly. Studying the ecosystem service in relation to land use change before and after the Grain to Green project was crucial for ecosystem protection and agricultural sustainability for the area. Researchers have analyzed ecosystem service at different scales within the Loess Plateau [17,18,23]. However, town is a basic administrative area in China. Exploring the characteristic of ecosystem services change at town scale is of practical significance to provide operable land use planning.

Ecosystem service values (ESVs) is monetary assessment of ecosystem services. This paper examined the characteristics of ESVs at Hechuan town, a typical town in the hilly and gully region of the Loess Plateau. The objectives of this study were: 1) to analyze the changes in land use pattern from 1982 to 2008; 2) to access the spatial and temporal variation in ESVs in response to land use during this period; 3) to quantitively analysis the relationship between ESVs and land use pattern; and 4) to discuss how land use management is favorable for ecosystem service supply and the ecological and economic sustainable development.

Data and Methods

2.1 Ethics statement

No specific permits were required for the described studies, and the work did not involve any endangered or protected species.

2.2 Study area

The study area, Hechuan town (106°18′43″~106°32′16″E, 35°54′59″~36°06′05″N), is located in Guyuan city of the Ningxia Hui Autonomous Region of northwest China (Fig. 1), consisting 12 villages with 16,524 people. The reasons that Hechuan Town was chosen as the study area were, on the one hand, Hechuan town has the typical characteristics of Loess Plateau including the

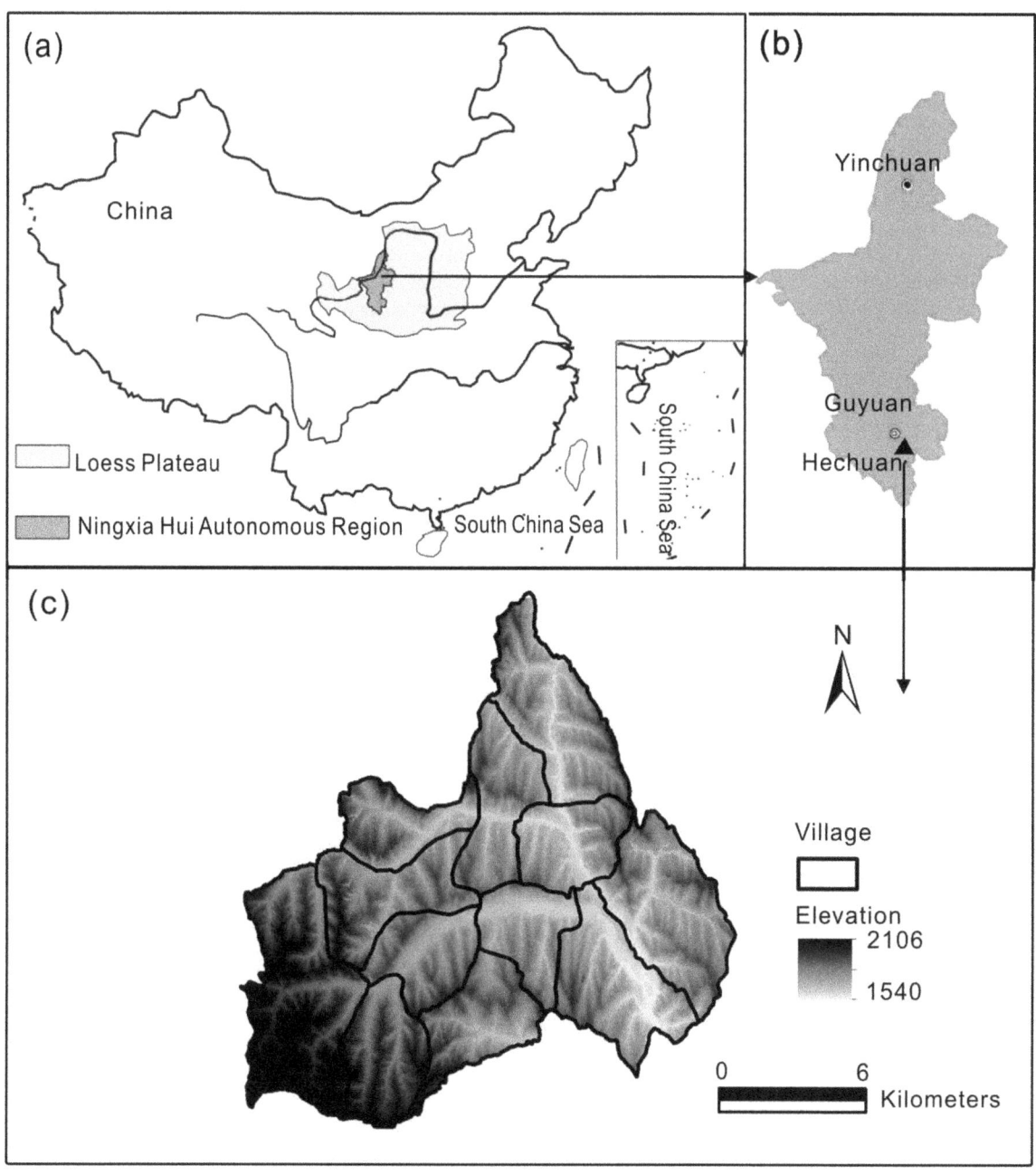

Figure 1. Location of the study area. Ningxia Province and the Loess Plateau, China (a), the location of Hechuan Town in the Loess area of Ningxia Province (b) and, the village boundary and the digital elevation model (DEM) map of Hechuan Town (c).

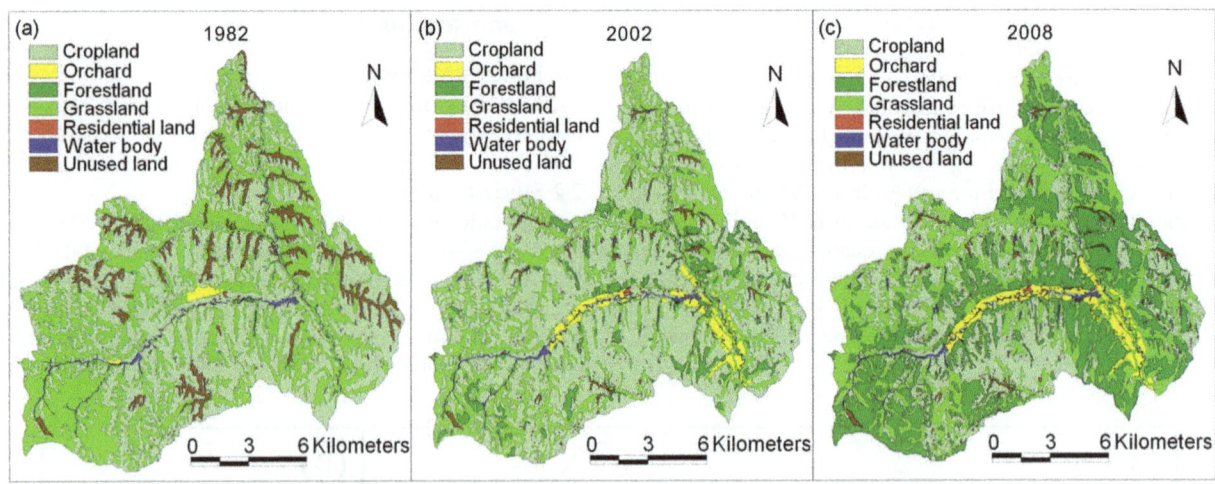

Figure 2. Land use maps of Hechuan town in 1982 (a), 2002 (b) and 2008 (c).

terrain of hill and gull, the fragile ecosystem and the backward economy; on the other hand, there was a long term ecological observation and experiment station in the study area, which facilitated the survey of land use and ecosystems. This town has an altitude ranging from 1540 to 2106 m, covering an area of 215.58 km^2. There exist the topographic differences in the town. The central area with river terrace stretches smoothly with a low elevation. The terrain in the northern area is fragmented while that of southern area is relatively simple. Hechuan town has a semi-arid continental temperate climate with the average annual temperature of 6.9°C and precipitation of 419 mm (1982–2002). Most of the annual precipitation is concentrated between June to September in the form of heavy storms that can cause severe soil erosion. The soil is composed of loessial soil and Dark loessial soils, which is erodible due to its weak cohesion and high infiltrability.

The ecosystem in Hechuan town is fragile with serious soil erosion and frequent natural disasters. Human disturbances of excessive land use, such as deforestation, overgrazing and over-reclamation further destructed the native natural grassland. Therefore, this area has long been in a vicious circle, endless cultivation and poverty. Since the early 1980s, a variety of comprehensive investigation of soil erosion was practiced by Chinese Academy of Sciences. Shanghuang watershed, located in the east of Hechuan town, was taken as a key test area. The

ecological restoration covering the whole town was started from implementing the Grain for Green project after 2002 (launched in 1999 by China government). Since then, abandoned cropland, shrubland (*Caragana korshinskii*, *Hippophae rhamnoides*) and artificial grassland (*Medicago sativa*) was generated, which made a significant change on landscape pattern and ecosystem components providing a variety of ecosystem services. Meanwhile, farming and grazing, the traditional way of living, had to be changed, and raising livestock, orchards, and migrant working diversified their incomes.

2.3 Data acquisition and preprocessing

Land use data was the key data for evaluating landscape pattern and ecosystem service. The land use data of 1982 was obtained by digitizing the land use patches from the 1:10,000 scale topographic maps of 1982, in which the information of land use types and its boundary are clearly shown. The 10 m resolution of remote sensing image could be considered to be corresponding with the scale of 1:50000 [24,25]. The land use data of 1982 acquired from 1:10000 topographic maps was therefore generalized to be at 1:50000 scale [26]. The land use data of 2002 and 2008 were respectively extracted from the 10 m resolution multispectral Spot-5 image of 2002 and 2008 by updating the land use patches of

Table 1. Equivalent weight factor of ecosystem service values (ESVs) per hectare of terrestrial ecosystem in China [30].

	Cropland	Forestland	Grass land	Water body	Barren land
Gas regulation	0.72	4.32	1.5	0.51	0.06
Climate regulation	0.97	4.07	1.56	2.06	0.13
Water supply	0.77	4.09	1.52	18.77	0.07
Soil formation and retention	1.47	4.02	2.24	0.41	0.17
Waste treatment	1.39	1.72	1.32	14.85	0.26
Biodiversity protection	1.02	4.51	1.87	3.43	0.40
Food production	1.00	0.33	0.43	0.53	0.02
Raw material	0.39	2.98	0.36	0.35	0.04
Recreation and culture	0.17	2.08	0.87	4.44	0.24
Total	7.90	28.12	11.67	45.35	1.39

Table 2. The ecosystem service values (ESVs) per hectare of different land use types in Hechuan town (US$·ha-1·yr-1).

	Cropland	Orchard	Forestland	Grass land	Water body	Unused land
Gas regulation	22.570	91.222	135.422	47.022	15.987	1.881
Climate regulation	30.407	88.244	127.585	48.902	64.576	4.075
Water supply	24.138	87.930	128.212	47.649	588.397	2.194
Soil formation and retention	46.081	98.118	126.018	70.219	12.853	5.329
Waste treatment	43.573	47.649	53.918	41.379	465.514	8.150
Biodiversity protection	31.975	99.999	141.378	58.620	107.523	12.539
Food production	31.348	11.912	10.345	13.480	16.614	0.627
Raw material	12.226	52.351	93.416	11.285	10.972	1.254
Recreation and culture	5.329	46.238	65.203	27.272	139.184	7.523
Total	247.647	623.663	881.498	365.828	1421.619	43.573

1982 one by one in visual interpretation method. The interpretation sign was established by understanding the Spot image characteristics and carrying out field surveys in order to further determine the relationship between the true ground and the image. The kappa accuracy index [27] was used to assess the accuracy of the interpretation. The stratified random sampling method was used to generate the reference points on the classified image for the accuracy test. These reference points were located in the field with a GPS with 5-m precision for ground truth. The total kappa indexes are all higher than 0.85, which are higher than the minimum acceptable (0.7) [28]. Considering the characteristic of the land use in study area and the interpretation level of the data and to facilitate the calculation of ESVs, the land use was classified into seven types: cropland, orchard, forestland, grassland, residential area, water area, and unused land (Fig. 2).

To acquire accurate area data of the land use for ESVs estimation and facilitate analyzing the spatial distribution of ESVs, the topographic maps and Spot images were transformed to the same projection and coordinate system (the Albers-Conical-Equal-Area projection system and Krasovsky 1940 coordinate system) before the extraction of land use data, and all acquired land use data were transformed to Arc-grid formats with the same grid size (10 m×10 m). The above data processing was completed using ERDAS and ArcGIS software.

2.4 Analysis on land use pattern

The transfer matrix analysis of land use was produced to understand how land use changed. Landscape metrics analysis was used for spatial pattern analysis of land use. Landscape metrics has been adopted widely; meanwhile, its abilities to indicate ecological process gained increasing attention [29,30,31]. Conceptual flaws in landscape pattern analysis, limitations inherent in landscape metrics and the improper use of pattern analysis may lead to the misuse of landscape metrics [32]. For better explanations and predictions of ecological phenomena from ecological pattern, the landscape metrics in this study was therefore selected by two steps. Firstly, the diversity, the fragmentation and the dominance of landscape were all considered, and then 34 metrics was selected, by understanding the knowledge of the landscape pattern and the ecological services indication of landscape metrics [33,34] and referring to the previous studies on landscape pattern [4,31,35,36]. Secondly, a correlation analysis for the 34 metrics was employed to ensure the low redundancy among landscape metrics. If the coefficient between two metrics was significant at 0.05 level, only one metric of them could be eventually selected.

Landscape-level metrics providing general landscape information and class-level metrics providing more specific information about variations at the local level and spatial patterns of land use classes [37] were used to monitor the characteristics of landscape pattern. The selected landscape-level metrics were patch density (PD), area-weighted mean shape index (SHAPE_AM), Interspersion and Justaposition Index (IJI), and Shannon's diversity index (SHDI). The selected class-level metrics were PD, the percentage of landscape (PLAND), SHAPE_AM and IJI. PD and SHAPE_AM could show the fragmentation of landscape. SHDI and PLAND reflect the dominance of some land use type and the diversity of landscape, respectively. IJI reflects whether the patches or classes are contiguous. Landscape metrics analysis was conducted with above metrics by FRAGSTATS 3.3, in which the eight-neighbor rule was used to derive the patch number. Besides these metrics, the land use intensity index (LUII) was also used to describe the landscape pattern. It was calculated by the following equation [31]:

$$LUII = \sum_{i=1}^{n} A_i \times C_i \qquad (1)$$

where $LUII$ is the land use intensity index, A_i is the percentage of for a give land use type i, and C_i is the coefficient value of intensity for a give land use type i, that is assigned 4 for build-ups, 3 for farmland and 2 for forest, orchard, grassland and water bodies, and 1 for unused land.

2.5 Estimation of ESVs

Costanza et al.'s model of ESVs estimation was adopted in this study [1,2]. The model classified ecosystem service into 17 types of service functions and estimated the ESVs by placing an economic value on different biomes [34]. For the defects of this model, such as overestimating the agriculture ESVs and underestimating the wetland ESVs, Xie et al. proposed refined coefficients for ESVs assessment both solving the above problem and making it apply to China [33,34]. Based on this model, the total ESVs in the study area was calculated using the following formulas:

$$ESV_k = \sum_f A_k VC_{kf} \qquad (2)$$

Table 3. Land use transition matrix from 1982 to 2002 and from 2002 to 2008 (%).

1982	2002								
	Cropland	Orchard	Forestland	Grassland	Residential land	Water body	Unused land	Total	Loss
Cropland	44.42	2.33	1.18	2.50	0.35	0.04	0.01	50.83	6.41
Orchard	0.17	0.05	0.26	0.09	0.00	0.00	0.00	0.57	0.53
Forestland	0.10	0.01	0.15	0.02	0.00	0.00	0.00	0.28	0.13
Grass land	12.63	0.39	4.74	22.15	0.01	0.08	0.00	40.01	17.86
Residential land	0.05	0.01	0.00	0.01	0.14	0.00	0.00	0.22	0.07
Water body	0.01	0.00	0.01	0.00	0.00	0.80	0.00	0.81	0.02
Unused land	1.39	0.00	0.16	3.98	0.00	0.07	1.66	7.27	5.61
Total	58.76	2.78	6.51	28.77	0.51	0.99	1.68	100.00	
Gain	14.35	2.73	6.36	6.62	0.37	0.19	0.01		

2002	2008								
	Cropland	Orchard	Forestland	Grass land	Residential land	Water body	Unused land	Total	Loss
Cropland	27.09	1.00	20.75	9.84	0.07	0.00	0.00	58.76	31.68
Orchard	0.00	2.75	0.01	0.00	0.01	0.00	0.00	2.78	0.03
Forestland	0.05	0.04	5.84	0.57	0.01	0.01	0.00	6.51	0.67
Grass land	0.12	0.00	7.17	21.42	0.02	0.04	0.00	28.77	7.35
Residential land	0.00	0.00	0.00	0.00	0.51	0.00	0.00	0.51	0.00
Water body	0.02	0.02	0.00	0.01	0.00	0.94	0.00	0.99	0.05
Unused land	0.00	0.00	0.27	0.08	0.00	0.00	1.33	1.68	0.35
Total	27.27	3.82	34.05	31.91	0.62	1.00	1.33	100.00	
Gain	0.18	1.07	28.21	10.50	0.11	0.05	0.00		

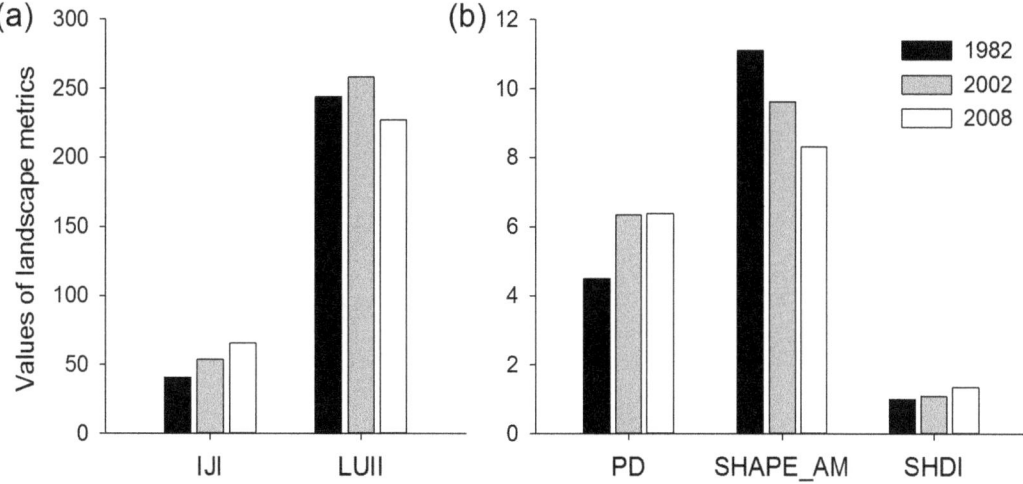

Figure 3. Landscape metrics at the landscape level in Hechuan Town in 1982, 2002 and 2008. IJI: Interspersion and Justaposition Index; LUII: land use intensity index; PD: patch density; SHAPE_AM: area-weighted mean shape index; SHDI: Shannon's diversity index.

$$ESV_f = \sum_k A_k VC_{kf} \qquad (3)$$

$$ESV = \sum_k \sum_f A_k VC_{kf} \qquad (4)$$

where ESV_k, ESV_f, ESV are the ESVs of land use type k, the ESVs of ecosystem service function type f, and the total ESVs respectively. A_k is the area (ha) for land use types. VC_{kf} is the value coefficient (US\$·ha-1·yr-1) for land use type k and ecosystem service function type f, which is the key for ESVs estimating. Xie et al.'s model was used to determine VC_{kf}, which can be expressed as follows:

$$VC_{kf} = R_{kf} \times V_f \qquad (5)$$

where R_{kf} is the equivalent weight factor of ecosystem service, V_f is food production values of agriculture land per area per year.

The equivalent weight factor was presented for customizing Chinese terrestrial ecosystem based on Costanza et al.'s model by surveying 500 Chinese ecologists (Table 1) [34]. It is the ratio of the ESVs to the economic value of average natural food production provided by agricultural land per hectare per year. The factors of land use types in our study were basically assigned based on the nearest ecosystems in Xie et al.'s model. However, minor adjustments were made. The equivalent weight factor of orchard which was not put forward clearly in Xie et al.'s model was determined by the mean of grassland and forestland by referring some researches [5,18]. The factor of unused land equates to that of barren land, and that of residential land was determined to zero.

The value of food production service of agriculture land per area per year was considered to be 1/7 of the actual price of food production in Xie et al.'s model. With the average actual food production of cropland in Hechuan town from 1982 to 2008 of 901.77 kg/ha which was get from *Statistic yearbook of the Yuanzhou District, Guyuan City, Ningxia Hui Autonomous Region* and the average grain price of US\$ 0.243 per kilogram (i.e. an equivalent of RMB Yuan 1.69 according to the average exchange

rate of 2008) in 2008, the value of food production service of cropland per area per year was calculated to be US\$ 31.348 (i.e. an equivalent of RMB Yuan 217.713 according to the average exchange rate of 2008). ESVs of one unit area of each land use types were then assigned as shown in Table 2.

After the ESVs were calculated by above processing, a sensitivity analysis was conducted to test the land use type's representative for ecosystem types and the certainty of the coefficients value for ecosystem service. A coefficient of sensitivity (CS) was used to indicate the degree of sensitivity of ESVs to a coefficients value, calculated by the following formula [5]:

$$CS = \left| \frac{(ESV_j - ESV_i)/ESV_i}{(VC_{jk} - VC_{ik})/VC_{ik}} \right| \qquad (6)$$

where ESV_j an ESV_i are the total ESVs of the initial status j and the adjusted status i, and VC_{jk} and VC_{ik} are the initial and adjusted coefficients. A 50% adjustment in the coefficients was made in the study. The greater the CS responded to the adjustment, the more critical is the use of an accurate coefficient [38]. A CS lower than 1 indicates the ESVs is inelastic to the coefficient and the estimation of ESVS is reliable. Otherwise, a CS greater than 1 indicates the estimation of ESVs is sensitive to the coefficient.

2.6 Correlation and regression analysis

The data of ESVs and landscape metrics was used to analysis the relationship between ecosystem service and land use pattern change. Because the spatial variation of landscape pattern exist among 12 villages in Hechuan town, the land use data of the three years (1982, 2002 and 2008) for the 12 villages can be considered as representing different landscape pattern on a time-for-space perspective [39]. Therefore, there were totally 36 sample data. Correlation and regression was employed for the relationship analysis, in which Multiple stepwise regression was specifically chosen considering the multicollinearity among landscape metrics. The dependents were the nine categories and total ESVs, while the corresponding independents were the landscape-level and class-level landscape metrics.

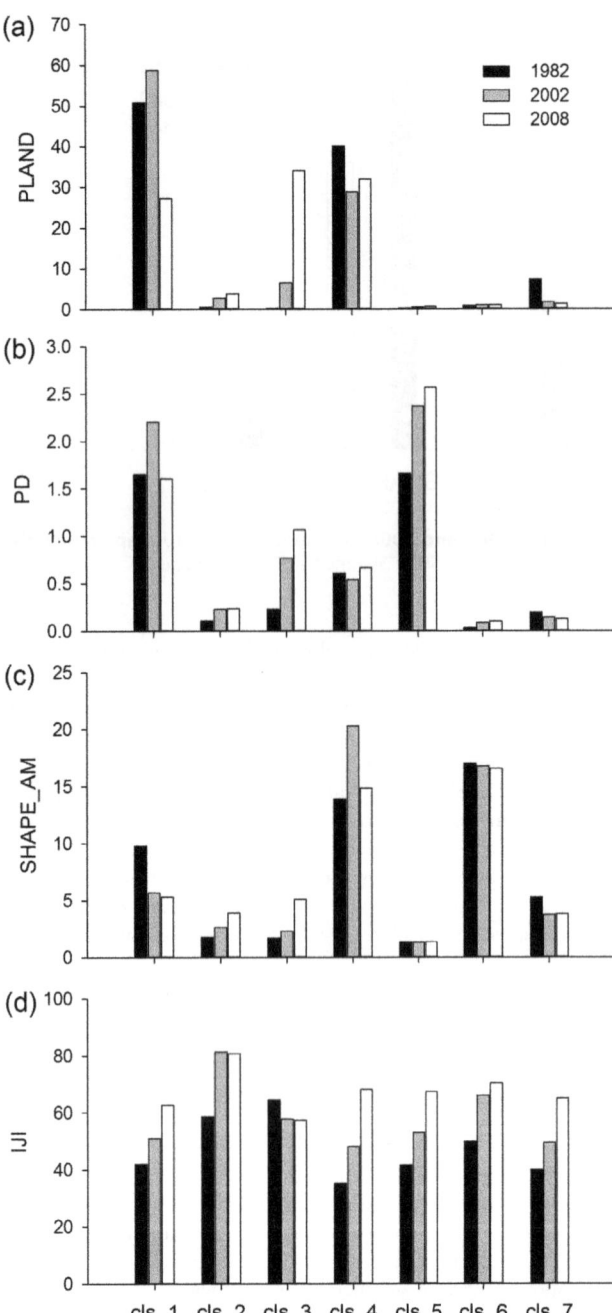

Figure 4. Landscape metrics at the class-level in Hechuan Town in 1982, 2002 and 2008. cls_1, cls_2, cls_3, cls_4, cls_5, cls_6, and cls_7 represent cropland, orchard, forestland, grassland, residential land, water body and unused land. PLAND: the percentage of landscape; PD: patch density; SHAPE_AM: area-weighted mean shape index; IJI: Interspersion and Justaposition Index.

Results

3.1 Changes of land use pattern

Table 3 showed the land use transition matrix. From 1982 to 2002, cropland as the dominant land use type increased from 50.83% to 58.76%. Grassland was the land use type with the largest change in area, decreasing from 40.01% to 28.77%. Orchard increased by 6.24% of total area, indicating the economic driving force of fruit trees on land use change. Forestland

Table 4. The change of ecosystem service values (ESVs) in Hechuan Town from 1982 to 2008.

		Cropland	Orchard	Forestland	Grass land	Water body	Unused land	Total
ESVs (10^6 US$ yr^-1)	1982	2.714	0.077	0.051	3.155	0.249	0.068	6.315
	2002	3.137	0.374	1.237	2.269	0.304	0.016	7.336
	2008	1.456	0.514	6.470	2.517	0.305	0.013	11.275
Change of ESVs (10^6 US$ yr^-1)	1982–2002	0.423	0.297	1.186	−0.887	0.054	−0.053	1.021
	2002–2008	−1.681	0.140	5.234	0.248	0.002	−0.003	3.939
	1982–2008	−1.258	0.437	6.419	−0.638	0.056	−0.056	4.960
Change of ESVS (%)	1982–2002	2.248	55.387	333.431	−4.045	3.145	−11.081	2.328
	2002–2008	−7.716	5.404	60.934	1.574	0.072	−2.992	7.731
	1982–2008	−6.674	81.578	1805.410	−2.913	3.232	−11.771	11.309
Average annual Change (%yr^-1)	1982–2002	0.112	2.769	16.672	−0.202	0.157	−0.554	0.117
	2002–2008	−1.286	0.901	10.155	0.262	0.012	−0.498	1.289
	1982–2008	−0.256	3.137	69.439	−0.112	0.124	−0.452	0.435

Table 5. Values of different ecosystem service functions in 1982, 2002, and 2008.

	1982			2002			2008		
	ESVs (10⁶ US\$·yr⁻¹)	%	Rank	ESVs (10⁶ US\$·yr⁻¹)	%	Rank	ESVs (10⁶ US\$·yr⁻¹)	%	Rank
Gas regulation	0.678	10.73	6	0.826	11.26	6	1.529	13.56	5
Climate regulation	0.791	12.53	5	0.936	12.75	5	1.540	13.65	4
Water supply	0.800	12.67	4	0.960	13.09	4	1.610	14.28	3
Soil formation and retention	1.141	18.06	1	1.260	17.17	1	1.764	15.65	1
Waste treatment	0.938	14.85	2	1.015	13.84	3	1.078	9.56	6
Biodiversity protection	0.915	14.49	3	1.054	14.37	2	1.738	15.42	2
Food production	0.466	7.38	7	0.506	6.90	7	0.367	3.25	9
Raw material	0.247	3.91	9	0.390	5.32	8	0.881	7.81	7
Recreation and culture	0.339	5.37	8	0.388	5.29	9	0.768	6.81	8
Total	6.315	100.00		7.336	100.00		11.275	100.00	

increased from 0.57% to 2.78%, reflecting that ecological restoration began to gain attention. From 2002 to 2008, cropland and forestland changed significantly, decreasing from 58.76% to 27.27% and increasing from 6.51% to 34.05% respectively. Land use structure was transferred from cropland dominated (58.76%) to cultivated land (27.27%), forestland (34.05%) and grassland (31.91%) relatively balanced distributed.

The most notable change of land use from 1982 to 2002 was the conversion from grassland to cropland and forestland with 12.63% and 4.74% of the total area respectively. The conversions from cropland (2.50%) and unused land (3.98%) to grassland were not adequate to compensate for the grass loss. From 2002 to 2008, the notable changes of land use were cropland to forestland, cropland to grassland, and grassland to forestland, with the rates of 20.75%, 9.84%, and 7.17% respectively. It was found that the conversion among land use types was more outstanding and concentrated than that before 2002, reflecting that the Grain for Green project as an ecological policy had great influence on land use change.

The results of landscape-level metric analysis were exhibited in Fig. 3. The significant increased PD from 1982 to 2002 reflected the landscape fragmentation. It was relative to the increase of patches on the land use types with intense human disturbance, such as cropland, residential land and artificial reservoir. Oppositely, the slight change of PD from 2002 to 2008 reflected that human disturbance became stable. The change of human disturbance was also demonstrated by the change of LUII which increased before 2002 and decreased after 2002. SHAPE_AM decreased in the study period, showing the landscape became more regular in shape. The increase of IJI suggested that the landscape became more contiguous and the ecological connectivity among land use types increased. SHDI increase obviously from 2002 to 2008, which related to that the land use structure became even.

Fig. 4 showed the change of class-level metrics. The PLAND of land use types indicated that cropland, forestland, and grassland had significantly influence on land use pattern. PD in orchard, forestland, and residential land increased obviously, attributing to the increasing area of these land use types and the fragmental terrain. SHAPE_AM showed that cropland and unused land became more regular in shape, while orchard and forestland more complicated. IJI increased generally in land use types. Orchard was the most contiguous with high IJI, which was relative to its concentrated distribution across the river terrace.

3.2 ESVs from 1982 to 2008

The ESVs of each land use type and the total ESVs was shown in Table 4. The total ESVs of Hechuan town was US\$ 6.315, US\$ 7.336 and US\$ 11.275 million in 1982, 2002 and 2008, respectively. From 1982 to 2002, the decline of ESVs caused by the decrease of grassland was offset by the increase of forestland, orchard and cropland, resulting that the total ESVs increased by US\$ 1.021 million. From 2002 to 2008, the total ESVs increased by US\$ 3.939 million, mainly due to the increase of forestland. The average annual change rate of total ESVs before and after 2002 was quite different, that is 0.81% and 8.95% respectively. It indicated the Grain to Green project implemented since 2002 had a significant effect on the ecosystem service. It was also shown from the value of ESVs produced by forestland occupying 57.39% of the total ESVs. Overall, the total ESVs increased US\$ 4.960 million during the study period, mainly due to the increase of ESVs by the increase of forestland and orchard far beyond the decrease of ESVs by the decrease of cropland and grassland. It was essentially because of the higher coefficient value of forestland and orchard than that of cropland and grassland.

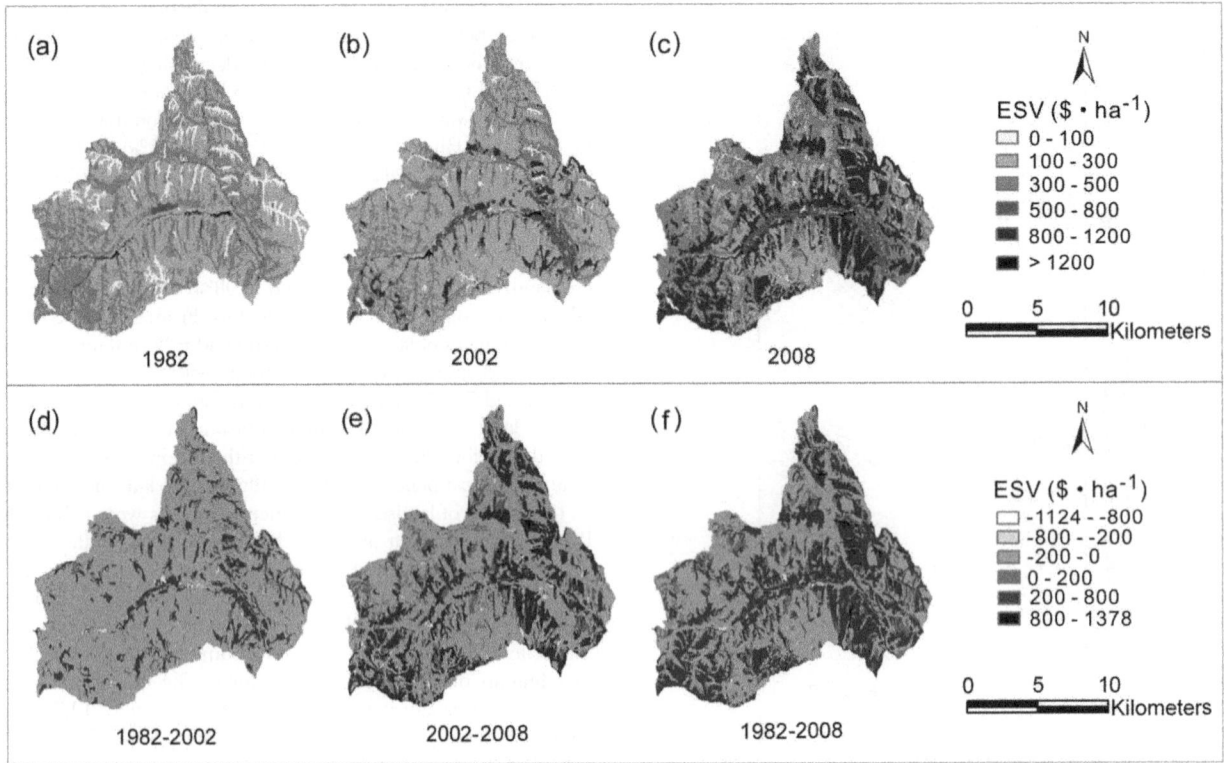

Figure 5. Spatial and temporal distribution of ecosystem service values (ESVs) in Hechuan Town from 1982 to 2008. The spatial distribution of ESVs in 1982 (a), 2002 (b) and 2008 (c), and the spatial-temporal changes of ESVs between time intervals from 1982 to 2002 (d), 2002 to 2008 (e) and 1982 to 2008 (f).

The ESVs of each ecosystem function type was shown in Table 5. Expect for food production, the values of ecosystem service functions increased especially after 2002. The decrease of food production was due to the great decline of cropland in the Grain to Green project. The ESVs proportion of each ecosystem function type to the total ESVs represented the contribution of each ecosystem function to the total ESVs. It was found that the functions of soil formation and retention, waste treatment, and food production were decline during 1982 to 2008, while other functions were improved. The rank of the contribution by each ecosystem service function was also estimated. It was basically stable except for relatively obvious decline in the rank of waste treatment and food production. In 2008, the rank order for each ecosystem service was as follows from high to low, soil formation and retention, biodiversity protection, water supply, climate regulation, gas regulation, waste treatment, raw material, recreation and culture, and food production. Soil formation and retention was the highest during the study period.

3.3 Spatial distribution of ESVs

Maps of ESVs in different periods (Fig. 5) showed the spatial distribution of ESVs of unit area in Hechuan town, directly reflecting the difference of ESVs among land use types. In 1982, the ESVs>4000 mostly appeared in the center of the town where river and river terrace located. It was because water body and orchard which intensely distributed in river terrace for its high water demand both had high ESVs. Therefore, due to the orchard increasing intensely and the forest increasing scatteredly, the increase of ESVs also mainly happened across the river terrace in 2002. Since 2008, the ESVs>4000 spread widely with the increase of forestland transformed from cropland. The lowest ESVs mostly

occurred in the gully where unused land was distributed in 1982. With vegetation recovery in the gully, the low ESVs happened from gully to terraced hillside where cropland with low ESVs was distributed in 2008. Fig. 5d–f showed the temporal change of ESVs spatial distribution. The change characteristic of 2002 to 2008 was adjacent to that during the total study period, reflecting that the change of ESVs mainly occurred after 2002, just after the Grain to Green project.

3.4 Relationship between ESVs and land use pattern

From the above analysis on the change of land use and ESVs in quantity and spatial distribution, we could infer there was some relationship between land use change and ecosystem service. To quantitively understand the relationship, the correlation analysis and regression analysis between ESVs and landscape pattern metrics was conducted.

Table 6 showed there existed significant correlations between ESVs and many landscape metrics (p<0.01), which explained that landscape pattern affected ESVs significantly. For example, the correlation coefficients between total ESVs and landscape metrics showed that there existed significantly positive relationship between SHDI (0.433), PLAND_3 (0.677), SHAPE_AM_3 (0.744), IJI_4 (0.513) and ESVs, and negative relationship between LUII (−0.634), PLAND_1 (−0.752) and ESVs. It reflected that the diversity and intensity of land use had important effects on total ESVs. It also reflected that cropland, forestland and grassland were the land use types which had significant effects on total ESVs. On quantity,the less the cropland and the more the forestland, the higher the total ESVs were. As to the landscape shape, the more regular the cropland and the more complex the forestland, the higher the total ESVs were. The higher the IJI of grassland, the

Table 6. Correlation coefficients between ecosystem service values (ESVs) and landscape pattern metrics.

	TESVs	ESVs_1	ESVs_2	ESVs_3	ESVs_4	ESVs_5	ESVs_6	ESVs_7	ESVs_8	ESVs_9
PD	0.035	0.497*	0.509*	0.539*	0.477*	0.516*	0.499*	−0.221	0.547	0.478*
SHAPE_AM	0.326	−0.216	−0.220	−0.188	−0.177	−0.088	−0.205	0.026	−0.290	−0.166
IJI	0.292	0.624*	0.635*	0.639*	0.597*	0.534*	0.621*	−0.293	0.687	0.586*
SHDI	0.433*	0.763*	0.764*	0.766*	0.741*	0.507*	0.765*	−0.636*	0.775*	0.765*
LUII	−0.634*	−0.681*	−0.658*	−0.618*	−0.675*	−0.113	−0.684*	0.977*	−0.599*	−0.734*
PLAND_1	−0.752*	−0.810*	−0.795*	−0.772*	−0.811*	−0.334	−0.815*	0.952*	−0.742*	−0.853*
PD_1	0.045	−0.055	−0.063	−0.056	−0.054	−0.113	−0.051	−0.134	−0.091	−0.022
SHAPE_AM_1	−0.476*	−0.369	−0.358	−0.330	−0.368	−0.063	−0.369	0.495	−0.328	−0.390
IJI_1	0.189	0.542*	0.552*	0.530*	0.510*	0.405	0.533*	−0.199	0.619*	0.485*
PLAND_2	0.323	0.527*	0.541*	0.590*	0.520*	0.595*	0.534*	−0.246	0.558	0.525*
PD_2	0.420	0.457*	0.471*	0.515*	0.449	0.529*	0.463*	−0.193	0.491	0.452
SHAPE_AM_2	0.159	0.450	0.468*	0.541*	0.439	0.629*	0.460*	−0.149	0.493	0.452
IJI_2	0.207	0.392	0.409	0.483*	0.376	0.583*	0.402	−0.115	0.438	0.396
PLAND_3	0.677*	0.984*	0.983*	0.941*	0.975	0.558*	0.980*	−0.770*	0.988*	0.961*
PD_3	0.276	0.631*	0.637*	0.629	0.625	0.477*	0.629*	−0.383	0.653	0.606*
SHAPE_AM_3	0.744*	0.828*	0.827*	0.780*	0.820	0.449*	0.821*	−0.623*	0.836*	0.799*
IJI_3	0.040	0.231	0.241	0.291	0.208	0.356	0.236	−0.069	0.276	0.231
PLAND_4	0.224	−0.192	−0.212	−0.203	−0.159	−0.298	−0.181	−0.264	−0.311	−0.110
PD_4	−0.294	0.199	0.201	0.190	0.160	0.101	0.192	−0.086	0.257	0.171
SHAPE_AM_4	0.455	−0.061	−0.065	−0.049	−0.028	−0.029	−0.053	−0.086	−0.125	−0.022
IJI_4	0.513*	0.717*	0.719*	0.705*	0.697*	0.457*	0.715*	−0.539*	0.739*	0.701*
PLAND_5	−0.035	0.290	0.313	0.381	0.279	0.578*	0.297	0.081	0.357	0.271
PD_5	−0.047	0.244	0.269	0.322	0.244	0.548*	0.248	0.188	0.314	0.208
SHAPE_AM_5	0.118	0.081	0.082	0.053	0.072	−0.009	0.075	−0.004	0.105	0.055
IJI_5	0.307	0.461*	0.470*	0.525*	0.446*	0.512*	0.471*	−0.312	0.482	0.477*
PLAND_6	0.047	0.139	0.167	0.360	0.148	0.852	0.170	0.064	0.153	0.207
PD_6	−0.160	0.122	0.137	0.198	0.105	0.371	0.128	0.080	0.172	0.118
SHAPE_AM_6	0.378	0.088	0.086	0.140	0.080	0.141	0.099	−0.218	0.067	0.137
IJI_6	0.020	0.201	0.217	0.276	0.197	0.438	0.208	0.028	0.239	0.197
PLAND_7	−0.385	−0.447	−0.474*	−0.525*	−0.507*	−0.769*	−0.456*	−0.041	−0.442	−0.422
PD_7	−0.270	−0.105	−0.129	−0.182	−0.154	−0.494*	−0.114	−0.245	−0.108	−0.089
SHAPE_AM_7	−0.236	−0.313	−0.329	−0.434	−0.348	−0.650*	−0.333	0.150	−0.279	−0.358
IJI_7	0.210	0.416	0.408	0.287	0.402	−0.083	0.394	−0.254	0.443	0.341

TESVs: the total ecosystem service values (ESVs); ESVs_1: the ESVs of gas regulation; ESVs_2 climate regulation; ESVs_3: the ESVs of water supply; ESVs_4: the ESVs of soil formation and retention; ESVs_5: the ESVs of waste treatment, ESVs_6: the ESVs of biodiversity protection; ESVs_7 the ESVs of food production; ESVs_8: the ESVs of raw material; ESVs_9: the ESVs of recreation and culture.
PD: patch density; SHAPE_AM: area-weighted mean shape index; IJI: Interspersion and Juxtaposition Index; SHDI: Shannon's diversity index; LUII: land use intensity index; PLAND: percentage of landscape. The 1, 2, 3, 4, 5, 6, 7 after the above landscape metrics respects different landscape, that is cropland, orchard, forestland, grassland, residential land, water body and unused land, respectively.
*significant at 0.01 level.

Table 7. Regression analysis between ecosystem service values (ESVs) and landscape patterns (n = 36).

Dependent	Standardized coefficients regression	R²	Sig.
Gas regulation	0.878×PLAND_3+0.166×PLAND_2-0.099×PLAND_1-0.068×IJI_1	0.990	*
Climate regulation	0.790×PLAND_3-0.197×PLAND_7-0.190×LUII+0.081×PLAND_2	0.998	*
Water supply	0.665×PLAND_3-0.317×PLAND_7-0.254×LUII+0.106×PLAND_2	0.955	*
Soil formation and retention	0.684×PLAND_3-0.301×PLAND_7-0.284×LUII+0.066×PLAND_2	0.998	*
Waste treatment	0.672×PLAND_6+0.365×PLAND_3-0.352×PLAND_7+0.051×PLAND_2+0.049×PLAND_5	0.993	*
Biodiversity protection	0.059×SHDI +0.861×PLAND_3-0.033×SHAPE_AM_3+0.133×PLAND_1	0.967	*
Food production	0.742×LUII-0.173×PLAND_3-0.052×SHDI+0.106×PLAND_1	0.991	*
Raw material	0.964×PLAND_3+0.091×SHDI+0.068×LUII	0.981	*
Recreation and culture	−0.747×PLAND_1+0.380×IJI_1	0.853	*
Total	-0.588×PLAND_1+0.569× SHAPE_AM_3-0.303×SHDI	0.709	*

*significant at 0.01 level.
PLAND_1: the percentage of cropland; PLAND_2: the percentage of orchard; PLAND_3: the percentage of forestland; PLAND_5: the percentage of residential land; PLAND_6: the percentage of water body; PLAND_7: the percentage of unused land; SHAPE_AM_3: the area-weighted mean shape index of forestland; IJI_1: the Interspersion and Justaposition Index of cropland; LUII: land use intensity index; SHDI: Shannon's diversity index.

higher the total ESVs were. This indicated that the connectivity of grassland was important for ecosystem service.

Correlation also occurred between ESVs of all the functions and landscape metrics (Table 6). However, the relationships between ESVs of different functions and landscape pattern were different. For example, the correlation between food production and landscape pattern was almost opposite from that between other ecosystem functions and landscape pattern. For example, PLAND_1 had a positive effect on food production; SHDI, PLAND_3, SHAPE_3, and IJI_4 had a negative effect on food production. It could infer that there were contradictions between food production and other ecosystem functions.

As shown in Table 7, the result of regression analysis further explained that the ESVs was correlated significantly with landscape pattern. The total ESVs could be predicted by PLAND on cropland, SHAPE_AM on grassland, and SHDI. ESVs of all kinds of ecosystem functions also could be explained by landscape metrics. These regression equations indicated that landscape-level metrics (such as SHDI and LUII) and class_level metrics (such as PLAND of forestland, orchard, and cropland, unused land, SHAPE of forestland, IJI of cropland) acted as predictors for categories of ecosystem services. Specifically, the proportion of forest (PLAND_3) accounted for almost all of the categories of ecosystem services.

Discussion

4.1 Reliability of ESVs

This study estimated ESVs by multiplying the area for each land use types by the corresponding value coefficients. As discussed in the previous researches, estimations using this method was coarse

with high variation and uncertainty for the following reasons, limitations on the economic evaluation [1], problems of double counting and scales [40,41,42], the complex, dynamic and nonlinear ecosystems [43], the imperfect matches of land use categories as proxies [38] and the accuracy of the ecosystem value coefficients [5]. This study also existed such uncertainty on ESVs estimation. For example, the value coefficient of orchard, determined by the average of forest and grassland, was an approximate estimation and need a further exploration. However, the estimation of temporal variation on ESVs was considered to be more reliable than that of cross-sectional analysis [5]. In addition, the sensitivity analysis of the estimated ESVs with 50% adjustment in the value coefficients was conducted. The result showed that the sensitivity coefficients of all land use categories were lower than 1 (Table 8), which suggested that despite of the above limitations, the estimated ESVs are reliable and useful for subsequent study.

4.2 Relationship between ESVs and landscape pattern

It is usually assumed that land use can affect the ecosystem service. Moreover, a few studies showed that there was a correlation between landscape pattern and ESVs [41,44]. This study signified this statement at town scale on the Loess Plateau. Land use configuration, land use intensity, landscape diversity, fragmentation and connectivity all affected ecosystem service.

The correlation analysis between ESVs and PLAND implied land use structure had significant impact on ecosystem service. Especially, the increase of forestland and the decrease of cropland played an important part in improving the ESVs in the past twenty years. It is closely related to the Grain to Green project comprehensively started in study area since 2002. In the project,

Table 8. The coefficient of sensitivity (CS) resulting from adjustment of ecosystem valuation coefficients.

	Cropland	Orchard	Forestland	Grass land	Water body	Unused land
1982	−0.430	−0.012	−0.008	−0.500	−0.039	−0.011
2002	−0.428	−0.052	−0.169	−0.309	−0.041	−0.002
2008	−0.129	−0.048	−0.574	−0.223	−0.027	−0.001

measures for optimizing land use structure were implemented, including restoring slope cropland into forest and grassland, banning grazing, transforming slopes into terraces, and building reservoirs, etc. Forestland and grassland increased by 423.19% (27.54% of the study area) and 10.93% (3.15% of the study area), and cropland decreased by 53.59%(31.49% of the study area) (Table 3). The increase of ESVs due to the increase of forestland occupied 46.28% of the total ESVs in 2008 (Table 4). The result of the correlation analysis between ESVs and PLAND reflected that vegetation recovery could strongly enhance ecosystem service, and it was coincident with many other studies on the Loess Plateau [17,18,23,45]. LUII, which also related to the proportion of land use types, implied the intensity of human activities. This study showed land use intensity had a negative effect on ecosystem service with negative correlation coefficients (−0.634) (Table 6). It was coincident with some studies on ESVs change under urbanization [5,31]. These studies showed that urbanization which means the increase of land use intensity led to considerable declines in ESVs.

Landscape diversity always presents high positive relevance with biodiversity [46]. Our results were coincident to previous statements given the positive relationships between SHDI and biodiversity conservation. However, there were studies reporting the negative relationships between them, in which the increase of SHDI was the result of rapid urban sprawl [31]. In our study, the increase of SHDI was because land use structure became more balanced, which was the result of the increase of forestland. In addition, landscape diversity could also promote agricultural production [47]. Our study disagreed with this statement, and showed that food production was weakened with landscape diversification. It was because that the increase of SHDI was the result of a larger number of conversion from cropland to forestland. Therefore, the relationship between landscape diversity and biodiversity conservation as well as food production should not be treat as the same but be understood considering the driving force of SHDI change.

Fragmentation could lead to declining habitat quality, lower wildlife survival, and limited movement of soil microorganisms [48], and subsequently cause the decrease of ecosystem service [30]. Our study disagreed with this statement. For example, PD of the total landscape, PD_Forest, PD_orchard and shape_ Forest revealed significantly positive impacts on most categories of ESVs (Table 6–7). The increase of PD and the decrease of connectivity of landscape were usually simultaneous, which is disagreed in our study (Fig. 3 and Fig. 4). The landscape became more contiguous as IJI shown. Table 6 showed the IJI had significantly positive impacts on ESVs. Especially, the increase of IJI of grassland promoted the total ESVs and all categories of ESVs. This maybe because the connectivity of landscape has contribution to habitat corridors [49] and forest production [50].

Based on the relationship between ESVs and landscape pattern, we could improve the ecosystem service by the adjustment of land use policy. On the one hand, continuing to implement the Grain to Green project is helpful for improving ESVS, because it could increase the vegetation coverage, decline the intensity of land use, and make cropland become regular by canceling the slope cropland. On the other hand, diversified agriculture gathering planing fruit trees, planting crops and breeding, which could promote the diversification of land use, should be encouraged to increase both ESVs and farmer's incomes.

Conclusion

ESVs at town scale in the Loess Plateau were estimated in Hechuan town of Ningxia Hui Autonomous Region from 1982 to 2008. It was concluded that ESVs varied with land use change. ESVs in 1982, 2002, and 2008 were US$ 6.315, US$ 7.336 and US$ 11.275 million respectively. Among all the land use types, forestland, grassland and cropland had important contribution (> 90%) on ESVs. The total ESVs increased slowly by 16.17% due to the decrease of grassland from 1982 to 2002, while the total ESVS increased significantly by 67.61% due to the increase of forestland from 2002 to 2008. Areas with high services level were mainly located in the center due to orchard and east due to forestland, while areas with low services level mainly located in the north and south sides due to cropland.

Land use pattern had a significant effect on ecosystem service in our study by analyzing and discussing the relationship between landscape pattern and ESVs. The proportion of forestland had a positive effect on ecosystem service while that of cropland had a negative effect on ESVs. The diversity and interspersion of landscape both had a positive effect on ESVs. Land use intensity which reflects the intensity of human activities had a negative effect on ESVs. Fragmentation had positive effect on ESVs, which was disagreed with the previous studies because the fragmentation in study area was related to the increased patch of such land use types as forestland, water body, orchard.

Based on the results of this study, it was conclude that land use pattern was important for ecosystem service. Therefore, we could improve the ecosystem service by the adjustment of land use policy. Continuing the Grain to Green project is reasonable and significant because it could increase the vegetation coverage and decline land use intensity. Diversified agriculture collecting planing fruit trees, growing food and breeding should be encouraged, because it could not only promote ecosystem service by increasing landscape diversification but also improve people's incomes.

Author Contributions

Conceived and designed the experiments: XF. Analyzed the data: XF. Contributed reagents/materials/analysis tools: GAT BCL. Contributed to the writing of the manuscript: XF GAT RMH.

References

1. Costanza R, Arge DR, Groot DR, Farber S, Grasso M, et al. (1997) The value of the world's ecosystem services and natural capital. Nature 387: 253−260.
2. Costanza R, Cumberland J, Daly H, Goodland R, Norgaard R (1997) An Introduction to ecological economics. Delray Beach Fla USA: St Lucie Press.
3. Kreuter UP, Harris HG, Matlock MD, Lacey RE (2001) Change in ESVs in the San Antonio area, Texas. Ecological Economics 39: 333−346.
4. Ronald CE, Yuji M (2013) Landscape pattern and ESV changes: Implications for environmental sustainability planning for the rapidly urbanizing summer capital of the Philippines. Landscape and Urban Planning 116: 60−72.
5. Li TH, Li WK, Qian ZH (2010) Variations in ESV in response to land use changes in Shenzhen. Ecological Economics 69: 1427−1435.
6. Cai YB, Zhang H, Pan WB, Chen YH, Wang XR (2013) Land use pattern, socio-economic development, and assessment of their impacts on ESV: study on

natural wetlands distribution area (NWDA) in Fuzhou city, southeastern China. Environ Monit Assess 185: 5111−5123.
7. Zaehle S, Bondeau A, Carter RT, Cramer W, Erhard M, et al. (2007) Projected changes in terrestrial carbon storage in europe under climate and land-use change, 1990–2100. Ecosystems 10: 380−401.
8. Eliska L, Jana F, Edward N, David V (2013) Past and future impacts of land use and climate change on agricultural ecosystem services in the Czech Republic. Land Use Policy, 33: 183−194.
9. Vitousek PM, Mooney HA, Lubchenco J, Melillo JM (1997) Human domination of earth's ecosystems. Science 277: 494−499.
10. Nasiri F, Huang GH (2007) Ecological viability assessment: A fuzzy multi-pleattribute analysis with respect to three classes of ordering techniques. Ecol Inform 2: 128−137.

11. Collin ML, Melloul AJ (2001) Combined land-use and environmental factors for sustainable groundwater management. Urban Water 3: 229–237.

12. Schmidta JP, Mooreb R, Alber M (2014) Integrating ecosystem services and local government finances into land use planning: A case study from coastal Georgia. Landscape and Urban Planning 122: 56–67.

13. Ernesto FV, Federico CF (2006) Land-use options for Del Plata Basin in South America: Tradeoffs analysis based on ecosystem service provision. Ecological Economics 57: 140–151.

14. Christine F, Susanne F, Anke W, Lars K, Franz M (2013) Assessment of the effects of forestland use strategies on the provision of ecosystem services at regional scale. Journal of Environmental Management 127: 96–116.

15. Ignacio P, Berta M, Pedro Z, David GDA, Carlos M (2014) Deliberative mapping of ecosystem services within and around Donana National Park (SW Spain) in relation to land use change. Reg Environ Change14: 237–251.

16. Mendoza-Gonzalez G, Martinez ML, Lithgow D, Perez-Maqueo O, Simonin P (2012) Land use change and its effects on the value of ecosystem services along the coast of the Gulf of Mexico. Ecological Economics 82: 23–32.

17. Su CH, Fu BJ (2013) Evolution of ecosystem services in the Chinese Loess Plateau under climatic and land use changes. Global and Planetary Change 101: 119–128.

18. Si J, Nasiri FZ, Han P, Li TH (2014) Variation in ESVs in response to land use changes in Zhifanggou watershed of Loess plateau: a comparative study. Environmental Systems Research 3: 2.

19. Turner MG, Gardner RH, O'Neill RV (2001) Landscape Ecology in theory and practice. New York: Springer-Verlag.

20. Ritsema CJ (2003) Introduction: soil erosion and participatory land use planning on the Loess Plateau in China. Catena 54: 1–5.

21. Fu BJ, Wang YF, Lu YH, He CS, Chen LD, et al. (2009) The effects of land-use combinations on soil erosion: a case study in the Loess Plateau of China. Progress in Physical Geography 33: 793–804.

22. Fu BJ, Chen DX, Qiu Y, Wang J, Meng QH (2002) Land Use Structure and Ecological Processes in the LoessHilly Area, China. Beijing: Commercial Press. (in Chinese).

23. Jing L, Zhiyuan R (2011) Variations in ESV in Response to Land use Changes in the Loess Plateau in Northern Shaanxi Province, China. Int. J. Environ. Res 5: 109–118.

24. Zhang TB, Tang JX, Liu DZ (2006) Feasibility of Satellite Remote Sensing Image About Spatial Resolution. Journal of Earth Sciences and Environment 28: 79–83.

25. Chu YF, Li ES, Lu J, Zhang KK (2007) The Adaptability Analysis to the Satellite Image Spatial Resolution and Mapping Scale. Hydrographic Surveying and Charting 27: 47–50.

26. Li Q, Liu C, Xi CY, Liu ML (2002) Cartographic Generalization of Digital Land Use Current Situation Map. Bulletin of Surveying and Mapping 9: 59–63.

27. Congalton RG (1991) A review of assessing the accuracy of classifications of remotely sensed data. Remote Sensing of Environment 37: 35–46.

28. Wang Y, Gao JX, Wang JS, Qiu J (2014) Value Assessment of Ecosystem Services in Nature Reserves in Ningxia, China: A Response to Ecological Restoration. PloS One 9: e89174. doi:10.1371/journal.Pone.0089174.

29. Ribeiro SC, Lovett A (2009) Associations between forest characteristics and socio-economic development: a case study from Portugal. Journal of Environmental Management 90: 2873–2881.

30. Su S, Jiang Z, Zhang Q, Zhang Y (2011) Transformation of agricultural landscapes under rapid urbanization: a threat to sustainability in Hang-Jia-Hu region, China. Applied Geography 31: 439–449.

31. Su SL, Xiao R, Jiang ZL, Zhang Y (2012) Characterizing landscape pattern and ESV changes for urbanization impacts at an eco-regional scale. Applied Geography, 34: 295–305.

32. Li H, Wu J (2004) Use and misuse of landscape indices. Landscape Ecology 19: 389–399.

33. Xie GD, Lu CX, Xiao Y, Zheng D (2003) The Economic Evaluation of Grassland Ecosystem Services in Qinghai Tibet Plateau, Journal of Mountain Science 21: 50–55. (in Chinese).

34. Xie GD, LU CX, Leng YF, Zheng D, Li SC (2008) Ecological assets valuation of Tibetan Plateau. Journal of Natural Resources 18: 190–196. (in Chinese).

35. Liu DL, Li BC, Liu Xianzhao Z, Warrington DN (2011) Monitoring land use change at a small watershed scale on the Loess Plateau, China: applications of landscape metrics, remote sensing and GIS. Environmental Earth Sciences 64: 2229–2239.

36. Pan WKY, Walsh SJ, Bilsborrow RE, Frizzelle BG, Erlien CM, et al. (2004) Farm-level models of spatial patterns of land use and land cover dynamics in the Ecuadorian Amazon. Agriculture, Ecosystems and Environment 101: 117–134.

37. de Groot RS, Wilson MA, Boumans RMJ (2002) A typology for the classification, description and valuation of ecosystem functions, goods and services. Ecological Economics 41: 393–408.

38. Kreuter UP, Harris HG, Matlock MD, Lacey RE (2001) Change in ESVs in the San Antonio area, Texas. Ecological Economics 39: 333–346.

39. Wu J, Jenerette GD, Buyantuyev A, Redman CL (2011) Quantifying spatiotemporal patterns of urbanization: the case of the two fastest growing metropolitan regions in the United States. Ecological Complexity 8: 1–8.

40. Turner RK, Paavola J, Coopera P, Farber S, Jessamya V, et al. (2003) Valuing nature: lessons learned and future research directions. Ecological Economics 46: 493–510.

41. Hein L, Koppen VK, de Groot RS, van Ierland EC (2006) Spatial scales, stakeholders and the valuation of ecosystem services. Ecological Economics 57: 209–228.

42. Konarska KM, Sutton PC, Castellon M (2002) Evaluating scale dependence of ecosystem service valuation: a comparison of NOAA-AVHRR and Landsat TM datasets. Ecological Economics 41: 491–507.

43. Limburg KE, O' Neill RV, Costanza R, Farber S (2002) Complex systems and valuation. Ecological Economics 41: 409–420.

44. Zhang MY, Wang KL, Liu HY, Zhang CH (2011) Responses of Spatial-temporal Variation of Karst Ecosystem Service Values to Landscape Pattern in Northwest of Guangxi, China. Chin. Geogra. Sci. 21: 446–453.

45. Deng L, Shangguan ZP, Li R (2012) Effects of the grain-for-green program on soil erosion in China. International Journal of Sediment Research 27: 120–127.

46. Nagendra H (2002) Opposite trends in response for the Shannon and Simpson indices of landscape diversity. Applied Geography, 22: 175–186.

47. Shrestha RP, Schmidt-Vogt D, Gnanavelrajah N (2010) Relating plant diversity to biomass and soil erosion in a cultivated landscape of the eastern seaboard region of Thailand. Applied Geography 30: 606–617.

48. Sherrouse BC, Clement JM, Semmens DJ (2011) A GIS application for assessing, mapping, and quantifying the social values of ecosystem services. Applied Geography 31: 748–760.

49. Li M, Zhu Z, Vogelmann JE, Xu D, Wen W, et al. (2011) Characterizing fragmentation of the collective forests in southern China from multitemporal Landsat imagery: a case study from Kecheng district of Zhejiang province.Applied Geography 31: 1026–1035.

50. Long JA, Nelson TA, Wulder MA (2010) Characterizing forest fragmentation: distinguishing change in composition from configuration. Applied Geography 30: 426–435.

16

Impact of Land-Use Intensity and Productivity on Bryophyte Diversity in Agricultural Grasslands

Jörg Müller[1]*, **Valentin H. Klaus**[2], **Till Kleinebecker**[2], **Daniel Prati**[3], **Norbert Hölzel**[2], **Markus Fischer**[1,3]

1 University of Potsdam, Institute of Biochemistry and Biology, Potsdam, Germany, **2** University of Münster, Institute of Landscape Ecology, Münster, Germany, **3** University of Bern, Institute of Plant Sciences, Bern, Switzerland

Abstract

While bryophytes greatly contribute to plant diversity of semi-natural grasslands, little is known about the relationships between land-use intensity, productivity, and bryophyte diversity in these habitats. We recorded vascular plant and bryophyte vegetation in 85 agricultural used grasslands in two regions in northern and central Germany and gathered information on land-use intensity. To assess grassland productivity, we harvested aboveground vascular plant biomass and analyzed nutrient concentrations of N, P, K, Ca and Mg. Further we calculated mean Ellenberg indicator values of vascular plant vegetation. We tested for effects of land-use intensity and productivity on total bryophyte species richness and on the species richness of acrocarpous (small & erect) and pleurocarpous (creeping, including liverworts) growth forms separately. Bryophyte species were found in almost all studied grasslands, but species richness differed considerably between study regions in northern Germany (2.8 species per 16 m^2) and central Germany (6.4 species per 16 m^2) due environmental differences as well as land-use history. Increased fertilizer application, coinciding with high mowing frequency, reduced bryophyte species richness significantly. Accordingly, productivity estimates such as plant biomass and nitrogen concentration were strongly negatively related to bryophyte species richness, although productivity decreased only pleurocarpous species. Ellenberg indicator values for nutrients proved to be useful indicators of species richness and productivity. In conclusion, bryophyte composition was strongly dependent on productivity, with smaller bryophytes that were likely negatively affected by greater competition for light. Intensive land-use, however, can also indirectly decrease bryophyte species richness by promoting grassland productivity. Thus, increasing productivity is likely to cause a loss of bryophyte species and a decrease in species diversity.

Editor: Francesco de Bello, Institute of Botany, Czech Academy of Sciences, Czech Republic

Funding: The DFG (German Research Foundation; http://www.dfg.de/en/index.jsp) funded the study in the framework of the Biodiversity Exploratories SSP 1374 (FI 1246/6-1). The funders had no role in study design, data collection and analysis, decision to publish, or preparation of the manuscript.

Competing Interests: The authors have declared that no competing interests exist.

* E-mail: muellerj@uni-potsdam.de

Introduction

Conservation of biodiversity is one of major ecological challenges nowadays [1]. In Central Europe, semi-natural grasslands are hotspots of biodiversity for both plants and animals [2–4]. However, these ecosystems severely declined in quantity and quality due to land-use change and intensification over the last few decades [5].

In semi-natural grassland ecosystems, investigations on relationships between land use and plant species diversity have only seldom considered bryophyte diversity [6]. Little is known about cryptogams such as bryophytes, although they are typical elements of grassland communities and conduce to fundamental ecosystem functions and processes such as carbon fixation [7–8] and the regulation of soil humidity and water retention capacity [9]. In addition, bryophytes can significantly promote or hamper germination and seedling establishment of vascular plants [10–12]. Furthermore, many bryophyte species are sensitive to environmental changes such as enrichment of nutrients, pollutants, or changes in humidity, and can thus serve as suitable ecological indicators for specific environmental conditions [13–16]. Similarly to vascular plants, bryophyte species can be affected by land use, either directly by mechanical impacts such as grazing and mowing,

by toxic impacts of high nitrogen applications [17] or indirectly through increased productivity leading to asymmetric light competition with tall-growing plant species [4]. However, despite their relevance only few ecological studies in grasslands included bryophytes. These investigations were usually restricted to specific habitats such as fens [18–20], mountain grasslands [21–22], and dry calcareous grasslands [4,23–27] or relied on artificial field experiments [15]. Bryophyte vegetation of rather common ecosystems like permanent agricultural grasslands was rarely studied or exhibited a very restricted species spectrum [13,15].

We investigated the diversity of bryophytes in 85 agricultural grasslands in two different regions in Germany. On these grasslands we assessed relationships between bryophytes, land-use intensity, and grassland productivity. We included aboveground vascular plant biomass, nutrient concentrations therein and mean Ellenberg indicator values for vascular plants to assess environmental and productivity-related impacts on bryophyte diversity [13,28,29]. We further distinguished between acrocarpous (erect and usually small, with sporophytes on the top of the branches) and pleurocarpous species (creeping, with sporophytes on lateral branches) growth forms to test whether they respond differently to land use and productivity as they represent different ecological gilds. Pleurocarpous species occupy usually larger patches and are more persistent than acrocarpous species. Many acrocarpous species are fast colonizing or ruderal bryophytes and can rapidly increase after on soil disturbances.

Specifically, we addressed the following questions: i) Which major gradients of land-use intensity (fertilizer application, mowing and grazing intensity) affect occurrence and composition of bryophyte vegetation in agricultural grasslands? and ii) Which measures of productivity – biomass production, biomass nutrient concentrations, or Ellenberg indicator values – are most strongly related to bryophyte species richness? And third, as the most important question we ask: iii) How is the relationship between bryophyte vegetation and productivity in agricultural grasslands?

Materials and Methods

Study Area and Land-use Intensity

The study involved 85 grassland plots of 50×50 m size within two regions belonging to the setup of the long-term and interdisciplinary project of the Biodiversity Exploratories [30]: i) the UNESCO Biosphere Reserve Schorfheide-Chorin in the Northeast and ii) Hainich-Dün consisting of the National Park "Hainich" with surroundings in Central Germany. In Hainich-Dün all plots were on calcareous mineral soils with large clay content (Cambisols and Stagnosols), whereas in Schorfheide-Chorin in addition to sandy mineral soils (Cambisols, Luvisols, and Gleysols) on the glacial moraines, about half of the plots occurred on drained organic fen soils (Histolols), which are frequently flooded in winter and early spring. All grasslands are of seminatural origin and regularly used as meadows, pastures or mown pastures. Most common grass species were *Poa pratensis*, *Dactylis glomerata* and *Bromus hordeaceus*. Among the herbs *Taraxacum officinale*, *Cerastium holosteoides* and *Trifolium repens* belong to the most frequent species. Plots were selected in a randomly stratified manner from a larger pool of 500 study plots per region to represent a wide gradient of land use typical for central European agricultural grasslands. Information on land use was inferred from standardized interviews with farmers containing detailed information on management practices of the last three years 2006–2008. For each plot we calculated mean intensities of grazing (number of livestock units × days grazing × ha^{-1}), fertilizer application (kg N × year^{-1} × ha^{-1}) and mowing (number of cuts × ha^{-1} × year^{-1}) as given in Blüthgen et al. (2012).

Vegetation Survey and Biomass Analyses

Following the nomenclature of Koperski et al. [31] we recorded bryophyte species richness and estimated total bryophyte cover in a 4×4 m subplot in all 85 plots (Hainich-Dün: $n = 43$; Schorfheide-Chorin: $n = 42$) in April 2009, the season when bryophytes are most easily recognized. Bryophyte species richness was further separated into acrocarpous and pleurocarpous species (including liverworts) according to Hill et al. [14].

From mid-May to mid-June, vascular plants were recorded at the same plots. Based on these vegetation relevés, we calculated mean Ellenberg indicator values for nutrients, reaction, light, and moisture per plot without cover weighting [13]. As estimation variable for productivity, aboveground biomass of vascular plants was harvested after recording on each plot simultaneously by cutting the vegetation 2 cm above ground on 1 m^2 as mixed samples of four randomly placed quadrates of 0.25 m^2. Occasionally occurring shrubs and litter were excluded from the biomass sampling. Temporary fences prevented our plots from mowing and grazing before sampling took place. Biomass samples were dried for 48 h at 80°C, weighed, and ground to pass a 0.5-mm screen. Total nitrogen (N) concentrations were determined with an elemental auto analyzer (NA 1500, Carlo Erba, Milan, Italy). For the analyses of phosphorus (P), potassium (K), calcium (Ca) and magnesium (Mg) samples were digested in a microwave oven system (MLS Start, Milestone, Bergamo, Italy) with concentrated nitric acid (65%) and hydrogen peroxide (30%) and analyzed by ICP-OES (Vista-PRO Axial, Varian, Palo Alto, USA). All analyses were run in duplicates and repeated if results differed by more than 10%.

Statistical Analyses

Canonical correspondence analysis (CCA) based on presence-absence data was used to assess patterns and gradients in the composition of bryophyte vegetation and to relate those to significant environmental factors (total inertia 2.74). Ordination was carried out with 83 relevés including all bryophyte species that occurred more than once in the dataset (30 species) and additional down-weighting of rare species. Furthermore, we used bi-plot scaling to optimize ordination for species data. All involved environmental variables (Ellenberg indicator values from vascular plants, vascular plant standing biomass, nutrient concentrations in biomass, nutrient ratios in biomass and the land-use intensity index) were standardized prior analysis (z-transformation) and forward selection with Monte Carlo permutation test (499 runs) was performed to assess the significance of environmental factors ($p < 0.05$). Further, inflation factors of significant variables were checked for co-linearity. Ordination was performed using Canocoo 4.5 [32]. The entire list of bryophyte species used for CCA ordination is given in Table S1.

Afterwards, we calculated Spearman rank correlations between axes scores and environmental variables.

Because of partly correlated proxy variables for land-use intensity and productivity (biomass production and nutrient content therein and Ellenberg indicator values), we calculated three separate multiple regression analyses as linear models to test for their effects on bryophyte diversity. Prior to linear model calculations, species richness data were square-root transformed to achieve normal distribution. All statistical tests were carried out with JMP (JMP 5.1, SAS institute, Cary, North Carolina, USA).

Results

Bryophyte Species Richness and Cover

In total, we recorded 44 different bryophyte taxa on 84 of the 85 grassland plots (Table 1) constituting between 0 and 43% of total plant species richness on plots (mean: $16.6 \pm 0.95\%$). The most frequently occurring species were the pleurocarpous *Brachythecium rutabulum* (91% of all plots), *Eurhynchium hians* (57%), *Amblystegium serpens* (22%), and the acrocarpous *Phascum cuspidatum* (53%), *Barbula unguiculata* (28%), *Bryum rubens* (16%), and - only on drained fen soils - *Physcomitrium pyriforme* (16%). Generally, pleurocarpous species were more frequent than acrocarpous species (Fig. 1). In average we recorded 4.59 ± 0.34 bryophyte species per 16 m plot but the species richness differed significantly among regions and were more than two times higher numbers in Hainich-Dün compared with Schorfheide-Chorin (Tables 1 and 2). This applied also for pleurocarpous and acrocarpous species separately. The total bryophyte species richness was positively correlated to species richness of vascular plants (Fig. 2). Mean bryophyte cover was relatively low in the studied grasslands ($11.5 \pm 1.6\%$) and varied likewise among plots and study regions (Tables 1 and 2).

The CCA-ordination underlined significant differences in bryophyte composition between but also within study regions by widely separating some grassland plots along the first axis (15.3% explained variance) but also along the second axis (7.0% explained

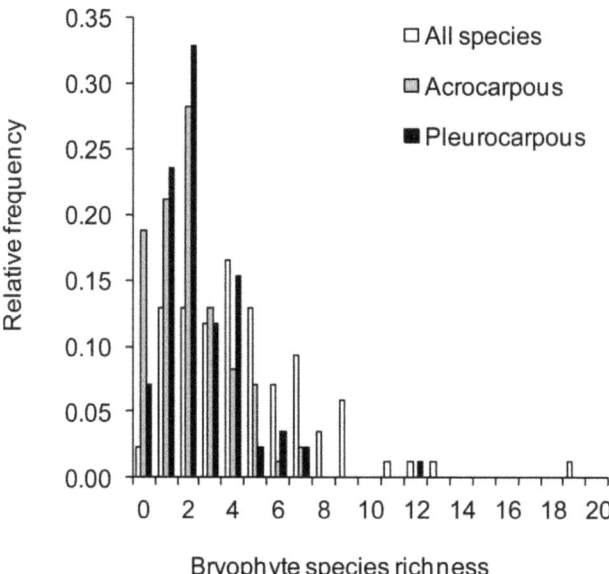

Figure 1. Frequency distribution of bryophyte species. Relative frequency distribution of the species richness of all, pleurocarpous, and acrocarpous bryophyte species per 16 m^2 plots in grasslands ($n = 85$).

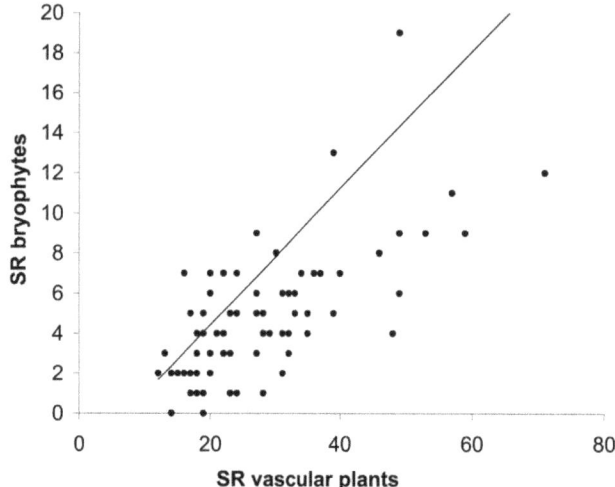

Figure 2. Relationship between species richness of bryophytes and vascular plants. Relationship between species richness of bryophytes and vascular plants ($n = 85$; $y = 0.17x - 0.15$; $R^2 = 0.47$; $F = 74.1$; $P < 0.0001$).

variance; Figure 3, Table S1). Due to strongly declining information content on the third and fourth axis (3.0 and 0.7%) we excluded them from further interpretation.

The first axis represents a gradient in especially moisture conditions (Ellenberg M indicator values, $\lambda = 0.34$, $F = 11.32$) but also in nutrient supply as revealed by Ellenberg N indicator values ($\lambda = 0.06$, $F = 2.40$) and N ($\lambda = 0.10$, $F = 3.57$) and K concentrations ($\lambda = 0.14$, $F = 5.28$) in vascular plant biomass (Table 3). These environmental factors widely separate some of the Schorfheide-Chorin plots from the rest of the plots. Only two bryophyte species, *Physcomitrium pyriforme* and *Eurhynchium speciosum*, had their main occurrence in these distinct plots in Schorfheide-Chorin, which are all situated on either Gleysols or Histosols, in contrast to all other species which either occur in both regions or mainly in plots in Hainich-Dün exhibiting less strong soil moisture (Fig. 3). The second ordination axis is characterized by soil reaction and arranges the plots assumingly along a gradient of soil pH.

Land-use Effects

Bryophyte vegetation was negatively affected by certain land-use measures (Table 2). While total and pleurocarpous species richness were negatively related to cutting frequency, pleurocarpous species richness was reduced by high levels of fertilizer application. Additionally, acrocarpous species richness exhibited a significant negative relationship to fertilizer application but only in the Hainich-Dün region. In contrast, grazing intensity revealed no significant effects on bryophyte species richness. Bryophyte cover was unrelated to all land-use measures under study (Table 2). In summary, bryophyte vegetation was significantly influenced by land-use intensity, but less strongly compared with regional differences.

Productivity Effects

Productivity affected total and pleurocarpous, but only to some extent acrocarpous bryophyte species richness (Table 2). Relationships between different measures of productivity and species

richness varied significantly among variables under consideration. Both pleurocarpous and total species richness were significantly negatively related to biomass production (Fig. 4). Only Ca concentrations in vascular plant biomass were related to total bryophyte species richness, while for pleurocarpous species richness N and Mg concentrations were also significant (Table 2; Fig. 4). Acrocarpous species were neither affected by biomass production nor by any of the measured nutrient concentrations in biomass (Table 2; Fig. 4). Similarly, bryophyte cover was not related to biomass production, but exhibited a negative relation to N concentrations in biomass (Table 2). Potassium concentrations were not significantly related to bryophyte species richness in regressions, but in ordination analysis (Table 3). Therefore, we explored relationships between K and total species richness in each region separately. We found a significant negative relationship in Hainich-Dün ($r_S = -0.38$; $p < 0.05$) and a positive in Schorfheide-Chorin ($r_S = 0.37$; $p < 0.05$). As further implied by ordination analysis (Fig. 2), mean Ellenberg indicator values for nutrients and moisture were strongly negatively associated with total, pleurocarpous and acrocarpous species richness (Fig. 4). Thus, Ellenberg indicator values explained much more variation in total species richness than more direct measures of productivity, such as biomass and nutrient concentrations.

Discussion

Bryophyte Vegetation in Agricultural Grasslands

Bryophyte species were present in the vegetation of nearly all investigated grasslands, demonstrating that bryophyte vegetation can significantly contribute to plant diversity in agricultural grasslands. However, strong differences in species richness, composition and cover between study regions were observed, underlining the necessity of regional replications in ecological studies before the generalization of findings is reliable [33].

Species richness in Schorfheide-Chorin, where half of the plots are situated on drained fen soils, was significantly lower compared to the second study region Hainich-Dün. High soil moisture turned out to significantly reduce bryophyte species richness in agricultural grasslands (Fig. 4). This decrease especially in drained fen grasslands directly depends on the detrimental impact of strong

Table 1. Summary about bryophyte, land-use intensity, productivity and indicator value data.

	Hainich-Dün				Schorfheide-Chorin			
	Min	Max	Mean	±SE	Min	Max	Mean	±SE
Total species richness	2	19	6.4	0.5[a]	0	9	2.8	0.3[b]
Acrocarpous bryophytes	0	7	2.8	0.3[a]	0	5	1.3	0.2[b]
Pleurocarpous bryophytes	1	12	3.5	0.3[a]	0	4	1.4	0.1[b]
Cover of bryophytes [%]	3	60	21.5	2.2[a]	0	7	1.3	0.3[b]
Vascular plant species	16	71	33.8	1.8[a]	12	31	19.9	0.6[b]
Fertilization intensity [kg N*ha^{-1}*year^{-1}]	0	140	24.9	5.2	0	163	21.8	6.1
Mowing intensity [Cuts*year^{-1}]	0	3.00	0.7	0.1	0	3	1.0	0.1
Grazing intensity [GVE*days^{-1}ha*year^{-1}]	0	1201	125	22	0	1362	110	27
Biomass [g/m^2]	95.7	599.5	295.8	19.3	109.3	699.3	340.3	20.8
Ca [g*kg^{-1}]	3.1	11.6	7.2	0.3	2.2	20.7	6.9	0.5
K [g*kg^{-1}]	15.4	37.1	25.1	0.7[a]	4.5	30.4	16.2	1.0[b]
Mg [g*kg^{-1}]	0.8	2.5	1.4	0.1[b]	0.7	5.6	2.1	0.2[a]
P [g*kg^{-1}]	1.3	3.7	2.6	0.1	1.6	3.9	2.6	0.1
N [g*kg^{-1}]	12.4	24.9	17.3	0.5[b]	8.6	35.2	19.9	1.0[a]
Ellenberg nutrient values	4.03	6.79	5.62	0.10[b]	4.82	7.38	6.11	0.08[a]
Ellenberg moisture values	4.29	5.69	4.99	0.05[b]	4.58	6.78	5.58	0.09[a]
Ellenberg light values	6.69	7.26	6.95	0.02[a]	6.60	7.23	6.86	0.02[b]
Ellenberg reaction values	6.00	7.20	6.69	0.04[a]	5.25	7.00	6.38	0.07[b]

Mean, minimum, maximum and SE of bryophyte diversity, vascular plant species, land-use measures, aboveground vascular plant biomass, nutrient content of biomass (Ca = calcium; K = potassium; Mg = magnesium; P = phosphorus; N = nitrogen), and mean Ellenberg indicator values for vascular plants for the two regions. Letters ([a], [b]) indicate significant group differences between the regions.

seasonal fluctuations of the water table, which suppress both species typical for wet and species typical for dry habitats [34]. Flooding events during winter are typical for the lowlands in this region. These suppress the colonization of all xeric and most mesic bryophyte species, while soil desiccation during summer due to drainage prevented the establishment of species-rich bryophyte communities that are rather typical of calcareous fens [18,35]. Under such conditions, vascular plants can acquire water from deeper levels due to their deep-reaching root system, outcompeting almost all small growing bryophytes on organic soils [36]. Additionally sandy soils on mineral sites cause faster desiccation and more imbalanced moisture dynamics on topsoil than loamy soils resulting in unsuitable conditions for bryophytes especially in precipitation poor regions like Schorfheide-Chorin. These moisture effects led to huge differences in total, acrocarpous as well as pleurocarpous bryophyte species richness between study regions. In addition to this we detected further effects of Ellenberg indicator values for reaction on bryophyte composition and species richness, illustrating the importance of soil conditions such as pH value for bryophyte vegetation. Furthermore, other factors on the regional scale such as land-use history may have also affected bryophyte species richness in this region, as reported for vascular plants [33]. Considering the low abundance of bryophytes in Schorfheide-Chorin, it became obvious that under these conditions bryophytes are prevented from fulfilling their ecological function to, for example, store nutrients during winter and as protective layer against surface wash off [37].

Land-use Intensity

In contrast to studies on vascular plant diversity in grasslands, land use and in particular grazing intensity were only weakly related to all kinds of bryophyte species richness and cover [13,38,39]. However, fertilizer application and, closely correlated with this [40], mowing intensity significantly decreased bryophyte species richness assumingly due to intolerance of high nitrogen loads of most bryophyte species [14]. Fertilizer application stimulates growth of tall grasses and herbs, which results in enhanced light competition on the ground causing the exclusion of bryophytes [37]. Nevertheless, in our study, land-use intensity was by far less closely related to the composition of bryophyte vegetation in grasslands than expected. Possibly, effects of moisture and nutrient conditions due to peat mineralization and subsequent internal fertilization processes in drained fen soils might have partially overruled effects of fertilizer application, grazing and mowing intensity.

Productivity Measures

Enhanced productivity had a strong detrimental influence on bryophyte species richness, clearly exceeding direct effects of land-use measures mentioned above (Table 2). Previous field studies and experiments stressed that increased nutrient levels fostering biomass production of vascular plants can be considerably harmful to bryophytes [36,41–45]. However, to our knowledge, we confirmed this relationship for the first time across a broad range of permanent agricultural grasslands.

Although vascular plant biomass and N and Mg concentrations therein were negatively related to bryophyte species richness, mean Ellenberg indicator values for nutrients were most closely related to bryophyte species richness. Compared with point measurements, mean Ellenberg values are particularly suitable for the integration of variation in environmental factors in time and space [12,46] and they are strongly related to biomass production

Table 2. Summaries of multiple regression analyses.

Model A

Source	df	Acrocarpous	Pleurocarpous	All species	Cover
Region	1	23.32***	56.18***	52.18***	273.59***
Fertilizer application	1	0.01	15.27**	2.94	0.59
Mowing intensity	1	5.44*	2.75	6.00*	0.10
Grazing intensity	1	0.36	0.64	0.01	1.89
Region × Fertilization	1	4.27*	3.27	0.55	0.01
Region × Mowing intensity	1	0.74	0.14	0.30	0.10
Region × Grazing intensity	1	0.31	0.43	0.74	3.40
Residual mean square	77	0.40	0.21	0.34	0.38
R adj.		0.25	0.46	0.40	0.77

Model B

Source	df	Acrocarpous	Pleurocarpous	All species	Cover
Region	1	20.74***	60.02***	56.41***	293.1***
Biomass	1	0.32	12.48**	6.04*	3.36
Ca	1	0.15	5.97*	4.02*	0.84
K	1	0.50	1.74	0.75	0.37
Mg	1	1.32	4.98*	4.53	2.37
P	1	0.72	1.64	1.11	0.05
N	1	0.8	6.29*	3.78	4.99*
Residual mean square	76	0.45	0.19	0.31	0.35
R adj.		0.16	0.51	0.45	0.78

Model C

Source	df	Acrocarpous	Pleurocarpous	All species	Cover
Region	1	25.26***	86.99***	86.36***	288.59***
Ellenberg nutrient values	1	9.33**	69.0***	58.72***	3.49
Ellenberg moisture values	1	4.28*	5.12*	6.03*	2.95
Ellenberg light values	1	1.14	0.83	0.03	2.19
Ellenberg reaction values	1	1.7	0.12	1.12	0.02
Residual mean square	79	0.37	0.14	0.21	0.36
R adj.		0.3	0.65	0.64	0.78

Summaries of multiple regressions of species richness of acrocarpous, pleurocarpous and all bryophytes as well as bryophyte cover on intensities of land-use procedures (Model A); aboveground biomass and their nutrient content (Model B), and mean Ellenberg indicator values for vascular plants (Model C) in the two regions.
Levels of significance:
*** = p<0.0001;
** = 0.0001<p<0.01,
* = 0.01<p<0.05.
Further details concerning single variables are given in Table 1.

[29]. Further, biomass estimation method might have lead to non-reliable results due to non-consideration of increasing re-growth ability of grasses after cutting or grazing in summer and variability due to different weather conditions among the two regions. Contradictory relationships between bryophyte species richness and K concentrations in biomass may result from different soil types among study regions. Contrary to grasslands on clay-rich mineral soils, grasslands on drained fen soils are often characterized by K deficiency due to high losses of K via leaching or removing of hay in intensive mowing regimes [7]. Along with enhanced N availability due to aerobic mineralization of peat, K deficiency causes higher Mg uptake by plants and therefore higher Mg concentrations in biomass [48]. Essentially, the enrichment of nitrogen in these soils favors highly competitive grasses and sedges, which effectively suppress bryophyte vegetation [33,49]. Thus, K is positively associated with bryophyte species richness in Schorfheide-Chorin due to higher numbers of species in grasslands on mineral soils compared to those on drained fen soils. Meanwhile, highly productive grassland vegetation on clay-rich soils accumulates K when fertilized with this nutrient [15]. This has most likely led to a negative relationship between K concentrations in biomass and bryophyte species richness on all

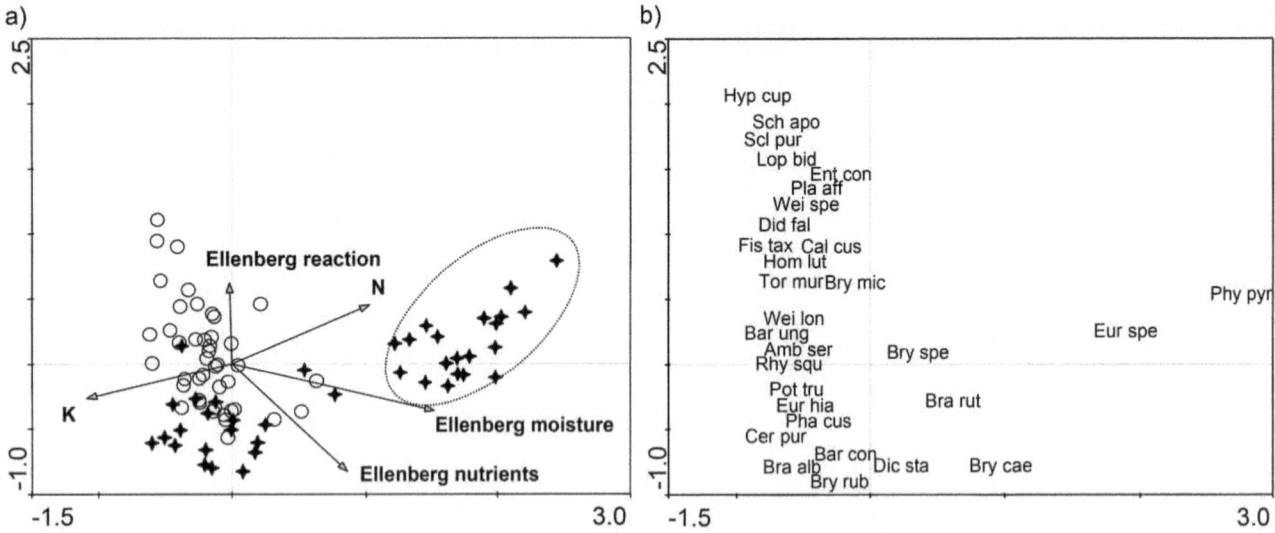

Figure 3. CCA-ordination of bryophyte species richness and environmental variables for stands (a) and species (b). CCA-ordination of bryophyte species and environmental factors in grasslands (n = 83). Stands and environmental factors (a) and species (b) of the same ordination are plotted separately to ease readability. Hainich-Dün: open circles; Schorfheide-Chorin: stars. Plots in the dotted circle are situated on either Gleyosols or Histisols. N and K are nutrient concentrations in aboveground vascular plant biomass and Ellenberg indicator values given are gained from vascular plant vegetation. Longer vectors indicate stronger correlations between variables and axes. See Table 1 for further details on single variables.

the mineral soils in Hainich-Dün. Against expectations, P concentration did not significantly affect bryophyte species richness, although P availability is an important driver of vascular plant species richness [38,50]. Taken together, our results show

Table 3. Spearman correlations of CCA-axes and environmental variables.

	Axis 1	Axis 2
Total species richness	−0.68**	–
Acrocarpous bryophytes	−0.51**	–
Pleurocarpous bryophytes	−0.71**	–
Cover of bryophytes	−0.52**	–
Fertilizer application	–	–
Mowing intensity	0.43**	–
Grazing intensity	−0.39**	−0.32**
Biomass	0.22*	−0.37**
Ca	–	0.57**
K	−0.58**	−0.27*
Mg	0.47**	0.23*
P	–	–
N	0.54**	0.47**
Ellenberg nutrient values	0.68**	−0.33**
Ellenberg moisture values	0.88**	–
Ellenberg light values	−0.41**	0.28*
Ellenberg reaction values	–	0.52**

Spearman correlations of CCA-axes and environmental variables (n = 83). Only significant correlations are shown. Levels of significance:
** = 0.0001<p<0.01,
* = 0.01<p<0.05.
Factors included in CCA factors are given in bold. Further details concerning single variables are given in Table 1.

that the Ellenberg indicator value for nutrients is the most useful and reliable proxy for environmental conditions which are detrimental to bryophyte vegetation in grasslands, irrespective of differences in soil conditions.

Growth Forms

Among different growth forms, pleurocarpous bryophytes were more strongly affected by productivity than acrocarpous species. Due to their creeping growth form most pleurocarpous species have higher demands for light and relatively slow growth rates compared with acrocarpous species. Only few species such as *Brachythecium rutabulum* and *Eurhynchium hians* tolerate such dark conditions under dense herb layers. However, many acrocarpous species are also disadvantaged by highly productive conditions. Thus, generally bryophyte species were replaced by a couple of nitrophilous acrocarpous species such as *Phascum cuspidatum, Pottia truncata* and several *Bryum* species which are typical for arable land. These fast colonizing species depend on small disturbance patches such as wheel tracks, which are common in meadow swards.

Conclusions

Bryophyte species richness in grasslands differed strongly between the regions especially due to differences in soil conditions and humidity. Moreover, in both regions our results demonstrate a strong negative impact of productivity and high nutrient levels on bryophyte vegetation in agricultural grasslands. Land-use intensity and in particular fertilizer application had negative effects on bryophytes, especially on pleurocarpous species. Nevertheless, moisture conditions of drained fen soils are assumed to have partially overruled relationships between land-use measures and species richness, at least in one region. Thus, site differences and indirect effects of land use such as drainage of fen soils were more important than direct measures of land-use intensity. However, both moisture and nutrient availability were strongly associated with each other. The mean Ellenberg indicator value for nutrients turned out to be the most powerful predictor to describe negative relationships between productivity and bryophyte species richness.

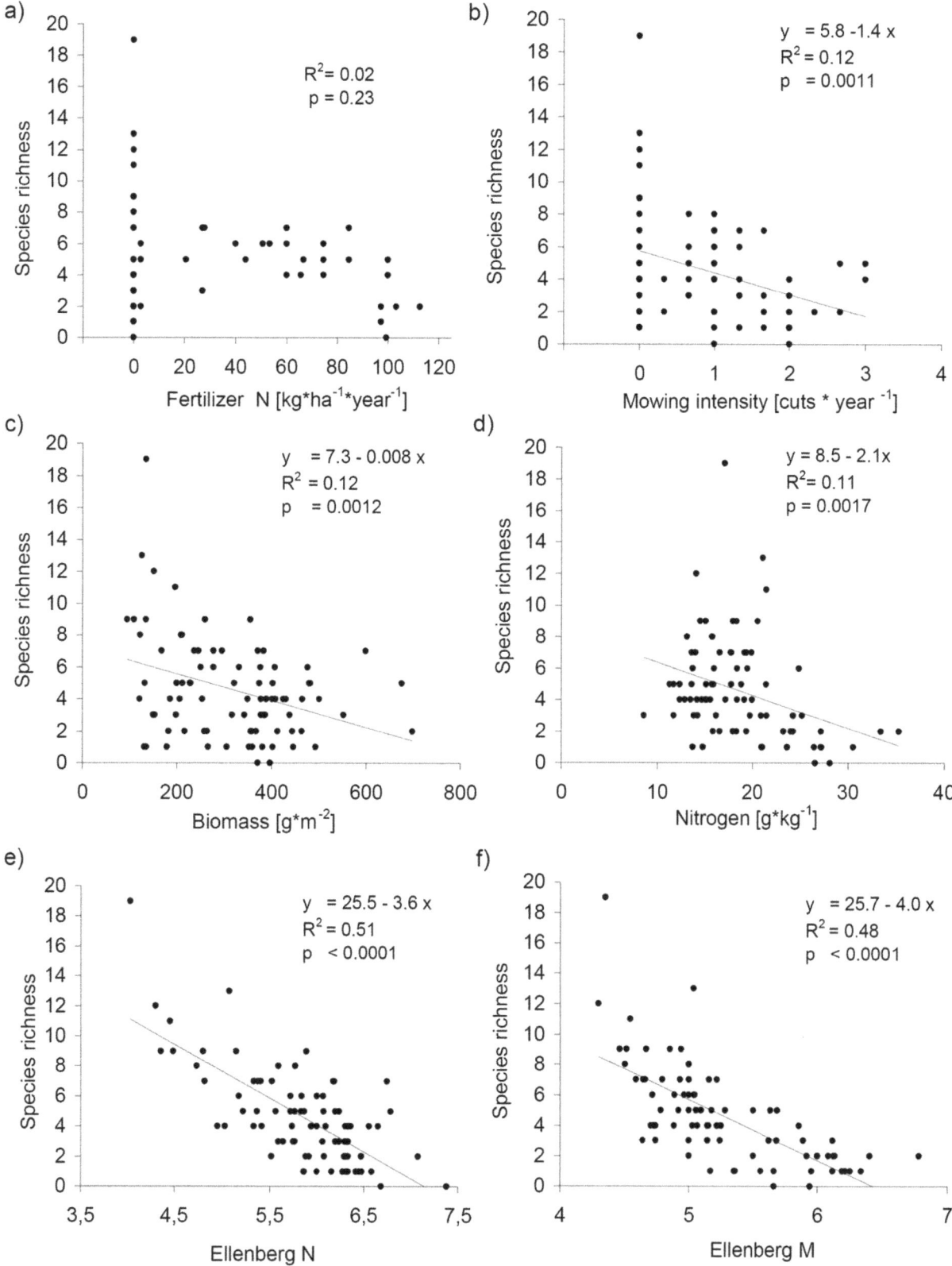

Figure 4. Relationships between bryophyte species richness and intensities of land-use and productivity. Relationships between total bryophyte species richness and intensities of land-use practices, mean annual amount of N fertilizer [kg*ha^{-1}*year^{-1}]; mean annual number of cuts [Cuts*year^{-1}] from 2006–2008, aboveground community biomass of vascular plants [g*m^{-2}], Nitrogen concentrations in biomass [[g/kg^{-1}]], mean Ellenberg indicator values for vascular plants for nutrients (N) and for moisture (M) on 85 grassland plots.

Finally, our results underlined that land-use intensification to increase grassland yield is responsible for low diversity of bryophytes in agricultural landscapes.

Supporting Information

Table S1 Abbreviations and full names of bryophyte species in NMDS ordination. Abbreviations and full names of bryophyte species in NMDS ordination displayed in Figure 2.

Acknowledgments

We thank all project members, field teams and managers of the Exploratories, Sonja Gockel, Andreas Hemp and Martin Gorke and Simone Pfeiffer for their work in maintaining the plot and project infrastructure. Furthermore, we thank François Buscot, Dominik Hessenmöller, the late Elisabeth Kalko, Eduard Linsenmair, Jens Nieschulze, Ingo Schöning, Ernst-Detlef Schulze and Wolfgang W. Weisser for their role in setting up the Biodiversity Exploratories. Further we thank Ariel Bergamini and Ewald Weber for review and critical remarks on an earlier draft of this manuscript. The work has been funded by the DFG Priority Program 1374 "Infrastructure-Biodiversity-Exploratories" (grants FI 1246/6-1, FI 1246/9-1, HO 3830/2-1). Field work permits were obtained from the responsible state environmental offices of Thuringia and Brandenburg (according to § 72 BbgNatSchG).

Author Contributions

Conceived and designed the experiments: MF NH DP. Performed the experiments: JM VHK. Analyzed the data: JM VHK DP. Contributed reagents/materials/analysis tools: DP TK. Wrote the paper: JM VHK TK NH DP MF.

References

1. Barnosky AD, Hadly EA, Bascompte J, Berlow EL, Brown JH, et al. (2012) Approaching a state shift in Earth's biosphere. Nature 486: 52–58.
2. Pärtel M, Zobel M, Zobel K, van der Maarel E (1996) The species pool and its relation to species richness: evidence from Estonian plant communities. Oikos 75: 111–117.
3. Knops JMH, Tilman D, Haddad NM, Naeem S, Mitchell CE, et al. (1999) Effects of plant species richness on invasion dynamics, disease outbreaks, insect abundances and diversity. Ecology Letters 2: 286–293.
4. Löbel S, Dengler J, Hobohm C (2006) Species richness of vascular plants, bryophytes and lichens in dry grasslands: the effects of environment, landscape structure and competition. Folia Geobotanica 41: 377–393.
5. Poschlod P, Bakker JP, Kahmen S (2005) Changing land use and its impact on biodiversity. Basic and Applied Ecology 6: 93–98.
6. Cornelissen JHC, Lang SI, Soudzilovskaia NA, During HJ (2007) Comparative cryptogam ecology: a review of bryophyte and lichen traits that drive biogeochemistry. Annals of Botany 99: 987–1001.
7. Bortoluzzi E, Epron D, Siegenthaler A, Gilbert D, Buttler A (2006) Carbon balance of a European mountain bog at contrasting stages of regeneration. New Phytologist 172: 708–718.
8. Turetsky M (2003) The Role of Bryophytes in Carbon and Nitrogen Cycling. – The Bryologist 106: 395–409.
9. Beringer J, Lynch AH, Chapin FS, Mack M, Bonan GB (2001) The representation of arctic soils in the land surface model: the importance of mosses. Journal of Climate 14: 3324–3335.
10. Keizer PJ, Van Tooren BF, During HJ (1985) Effects of bryophytes on seedling emergence and establishment of short-lived forbs in chalk grassland. Journal of Ecology 73: 493–504.
11. Van Tooren BF (1988) The fate of seeds after dispersal in chalk grassland: the role of the bryophyte layer. Oikos 53: 41–48.
12. Zamfir M (2000) Effects of bryophytes and lichens on seedling emergence of alvar plants: evidence from greenhouse experiments. Oikos 88: 603–611.
13. Ellenberg H, Weber HE, Düll R, Wirth V, Werner W, et al. (2001) Zeigerwerte von Pflanzen in Mitteleuropa. 3rd edn., Scripta Geobotanica 18: 1–248.
14. Zechmeister HG, Schmitzberger I, Steurer B, Peterseil J, Wrbka T (2003) The influence of land-use practices and economics on plant species richness in meadows. Biological Conservation 114: 165–177.
15. Hill MO, Preston CD, Bosanquet SDS, Roy DB (2007) BRYOATT: attributes of British and Irish mosses, liverworts and hornworts. Cambridge, Centre for Ecology and Hydrology.
16. Hejcman M, Száková J, Schellberg J, Šrek P, Tlustoš P, et al. (2010) The Rengen Grassland Experiment: bryophytes biomass and element concentrations after 65 years of fertilizer application. Environmental Monitoring and Assessment 166: 653–662.
17. Reich PB (2009) Elevated CO_2 reduces losses of plant diversity caused by nitrogen deposition. Science 326: 1399–1402.
18. Bergamini A, Pauli D (2001) Effects of increased nutrient supply on bryophytes in montane calcareous fens. Journal of Bryology 23: 331–339.
19. Bergamini A, Peintinger M, Schmid B, Urmi E (2001) Effects of management and altitude on bryophyte diversity and composition in montane calcareous fens. Flora 196: 180–193.
20. Peintinger M, Bergamini A (2006) Community structure and diversity of bryophytes and vascular plants in abandoned fen meadows. Plant Ecology 185: 1–17.
21. Losvik MH (2006) Thick moss layers and high cover of grasses: potential threats to herb diversity in hay meadows in Norway. Norwegian Journal of Geography 60: 312–316.
22. Austrheim G, Mysterud A, Hassel K, Evju M, Økland RH (2007) Interactions between sheep, rodents, graminoids and bryophytes in an oceanic alpine ecosystem of low productivity. Ecoscience 14: 178–187.
23. Watson EV (1960) A quantitative study of the bryophytes of chalk grassland. Journal of Ecology 48: 397–414.
24. During HJ, Van Tooren BF (1990) Bryophyte interactions with other plants. Botanical Journal of the Linnean Society 104: 79–98.
25. Van Tooren BF, Ode B, During HJ, Bobbink R (1990) Regeneration of species richness in the bryophyte layer of Dutch chalk grasslands. Lindbergia 16: 153–160.
26. Vanderpoorten A, Delescaille LM, Jacquemart AL (2003) The bryophyte layer in a calcareous grassland after a decade of contrasting mowing regimes. Biological Conservation 117: 11–18.
27. Haworth BJ, Ashmore MR, Headley AD (2007) Effects of nitrogen deposition on bryophyte species composition of calcareous grasslands. Water, Air and Soil Pollution 7: 111–117.
28. Aerts R, Chapin FS (2000) The mineral nutrition of wild plants revisited: A re-evaluation of processes and patterns. Advances in Ecological Research 30: 1–67.
29. Chytry M, Hejcman M, Hennekens SM, Schellberg J (2009) Changes in vegetation types and Ellenberg indicator values after 65 years of fertilizer application in the Rengen Grassland Experiment, Germany. Applied Vegetation Science 12: 167–176.
30. Fischer M, Bossdorf O, Gockel S, Hänsel F, Hemp A, et al. (2010) Implementing large-scale and long-term functional biodiversity research: the Biodiversity Exploratories. Basic and Applied Ecology 11: 473–485.
31. Koperski M, Sauer M, Braun W, Gradstein SR (2000) Referenzliste der Moose Deutschlands. Schriftenreihe Vegetationsk. 34: 1–519.
32. Ter Braak CJF (1986) Canonical correspondence analysis: a new eigenvector technique for multivariate direct gradient analysis. Ecology 67: 1167–1179.
33. Klaus VH, Hölzel N, Boch S, Müller J, Socher SA, et al. (2012) Direct and indirect effects of productivity estimates on plant species richness in grasslands: Regional differences preclude simple generalization of productivity-biodiversity relationships. Submitted.
34. Olde Venterink H, Karde I, Kotowski W, Peeters W, Wassen MJ (2009) Long-term effects of drainage and hay-removal on nutrient dynamics and limitation in the Biebrza mires, Poland. Biogeochemistry 93: 235–252.
35. Mälson K, Backéus I, Rydin H (2008) Long-term effects of drainage and initial effects of hydrological restoration on rich fen vegetation. Applied Vegetation Science 11: 99–106.
36. Olde Venterink H, Wassen MJ, Belgers JDM, Verhoeven JTA (2001) Control of environmental variables on species density in fens and meadows: importance of direct effects and effects through community biomass. Journal of Ecology 89: 1033–1040.
37. van der Wal R, Pearce ISK, Brooker RW (2005) Mosses and the struggle for light in a nitrogen-polluted world. Oecologia 142: 159–168.
38. Klaus VH, Kleinebecker T, Hölzel N, Boch S, Müller J, et al (2011) Nutrient concentrations and fibre contents of plant community biomass reflect species richness patterns along a broad range of land-use intensities among agricultural grasslands. Perspect. Plant Ecol. Evol. Syst. 13: 287–295.
39. Socher SA, Prati D, Boch S, Müller J, Klaus VH, et al (2012) Direct and productivity-mediated indirect effects of fertilization, mowing and grazing intensities on grassland plant species richness. Journal of Ecology doi: 10.1111/j.1365-2745.2012.02020.x.
40. Blüthgen N, Dormann CF, Prati D, Klaus VH, Kleinebecker T, et al. (2012) A quantitative index of land-use intensity in grasslands: integrating mowing, grazing and fertilization. Basic and Applied Ecology, Basic and Applied Ecology 13: 207–220.
41. Jeffrey DW, Pigott CD (1973) The response of grasslands on sugar-limestone in Teesdale to application of phosphorus and nitrogen. Journal of Ecology 61: 85–92.
42. Mickiewicz J (1976) Influence of mineral fertilization on the biomass of moss. Polish Ecological Studies 2: 57–62.
43. Virtanen R, Johnston AE, Crawley MJ, Edwards GR (2000) Bryophyte biomass and species richness on the Park Grass Experiment, Rothamsted, UK. Plant Ecology 151: 129–141.

44. Bergamini A, Pauli D, Peintinger M, Schmid B (2001) Relationships between productivity, number of shoots and number of species in bryophytes and vascular plants. Journal of Ecology 89: 920–929.

45. Aude E, Ejrnæs R (2005) Bryophyte colonisation in experimental microcosms: the role of nutrients, defoliation and vascular vegetation. Oikos 10: 323–330.

46. Diekmann M (2003) Species indicator values as an important tool in applied plant ecology – a review. Basic and Applied Ecology 4: 493–506.

47. Mundel G, Behrendt A (1997) Vergleichende Untersuchungen zum Nähr-stoffgehalt von Moorböden in der Niedermoorlandschaft "Havelländisches Luch". Archives of Agronomy and Soil Science 41: 277–293.

48. Broll G, Schreiber KF (1994) Magnesiumgehalte von Grünlandstandorten unter dem Einfluss von Extensivierung und Flächenstilllegung. VDLUFA-Schriften-reihe, 38, 915–918.

49. Olde Venterink H, Wassen MJ, Verkroost AWM, de Ruiter PC (2003) Species richness-productivity patterns differ between N-, P-, and K-limited wetlands. Ecology 84: 2191–2199.

50. Wassen MJ, Olde Venterink H, Lapshina ED, Tanneberger F (2005) Endangered plants persist under phosphorus limitation. Nature 437: 547–550.

17

Mosaic-Level Inference of the Impact of Land Cover Changes in Agricultural Landscapes on Biodiversity

Francisco Moreira[1]*, João P. Silva[1,2,3], Beatriz Estanque[2], Jorge M. Palmeirim[2], Miguel Lecoq[4], Márcia Pinto[3], Domingos Leitão[4], Ivan Alonso[3], Rui Pedroso[3], Eduardo Santos[3], Teresa Catry[3], Patricia Silva[4], Inês Henriques[1], Ana Delgado[1]

1 Centre for Applied Ecology "Prof. Baeta Neves", Institute of Agronomy, Technical University of Lisbon, Lisbon, Portugal, 2 Centre for Environmental Biology, Faculty of Sciences, University of Lisbon, Lisbon, Portugal, 3 Institute for Nature Conservation and Biodiversity, Lisbon, Portugal, 4 SPEA – Society for the Protection and Study of Birds, Lisbon, Portugal

Abstract

Changes in land use/land cover are a major driver of biodiversity change in the Mediterranean region. Understanding how animal populations respond to these landscape changes often requires using landscape mosaics as the unit of investigation, but few previous studies have measured both response and explanatory variables at the land mosaic level. Here, we used a "whole-landscape" approach to assess the influence of regional variation in the land cover composition of 81 farmland mosaics (mean area of 2900 ha) on the population density of a threatened bird, the little bustard (*Tetrax tetrax*), in southern Portugal. Results showed that ca. 50% of the regional variability in the density of little bustards could be explained by three variables summarising the land cover composition and diversity in the studied mosaics. Little bustard breeding males attained higher population density in land mosaics with a low land cover diversity, with less forests, and dominated by grasslands. Land mosaic composition gradients showed that agricultural intensification was not reflected in a loss of land cover diversity, as in many other regions of Europe. On the contrary, it led to the introduction of new land cover types in homogenous farmland, which increased land cover diversity but reduced overall landscape suitability for the species. Based on these results, the impact of recent land cover changes in Europe on the little bustard populations is evaluated.

Editor: Rohan H. Clarke, Monash University, Australia

Funding: Field work between 2004 and 2006 was financed by Project LIFE02NAT/P/8476: Conservation of the little bustard in Alentejo. JPS was partly supported by grant SFRH/BD/28805/2006 from Fundação para a Ciência e a Tecnologia. The funders had no role in study design, data collection and analysis, decision to publish, or preparation of the manuscript.

Competing Interests: The authors have declared that no competing interests exist.

* E-mail: fmoreira@isa.utl.pt

Introduction

Mediterranean ecosystems are amongst those ecosystem types predicted to undergo the greatest biodiversity changes in the long term [1]. The drivers for these changes include modifications in atmospheric carbon dioxide, climate, vegetation, and land use, but the latter is expected to play the main role [1]. In fact, the landscapes of the Mediterranean basin, particularly in Southern Europe, are changing at a fast pace (e.g., [2,3]), with potential consequences for biodiversity that represent a major research topic (e.g., [4,5]). In Mediterranean Europe, agricultural landscapes are particularly prone to change due to two major contrasting drivers: (i) abandonment of farming activities on marginal land, leading to loss of agricultural fields, shrub encroachment and afforestations of former agricultural land, and (ii) agricultural intensification in the most productive land, with consequences including the replacement of dry crops by irrigated crops, and loss of fallow land, pastures, and other non-crop habitats (e.g., [4,6–8]).

Within Mediterranean Europe, vast regions of the Iberian Peninsula are covered by agricultural landscapes known as pseudosteppes, characterised by a mosaic of land covers including cereal crops, dry legumes, ploughed fields, and grasslands (pastures and fallows) [9,10]. These land mosaics sustain populations of several bird species with unfavourable conservation status [6,9]. One such species is the little bustard *Tetrax tetrax*, a medium-sized ground-nesting bird that has undergone a major decline in most of its Palaearctic range [11]. More than half of the world's population now resides in the Iberian Peninsula [11,12], where grasslands of different types (pastures, natural steppe and fallow fields) are its prime breeding habitat (e.g., [13–17]).

Little bustard populations, like those of other steppe bird species, are negatively impacted by both agricultural intensification and abandonment [10,17,18]. Both processes have impacts on land use and land cover patterns which result in changes in habitat availability and quality for the species. Thus, different farmland mosaic compositions are expected to drive regional variation in the population density of this species within its range. Studies in Spain and France showed a positive influence of land cover diversity on the occurrence of little bustard males (e.g., [13,15,18]). However, the Iberian regions where male densities are highest do not correspond to diverse landscapes, and are in fact dominated by vast expanses of grassland pastures or fallow land [12,14,19].

These latter findings are apparently in contradiction with the positive correlation of land cover diversity and male density found elsewhere.

Most of these previous studies were conducted at local scales, using properties of individual sites/patches or the landscape context surrounding each site as explanatory variables. However, to evaluate the implications for biodiversity of changes in land cover and land use it is necessary to understand the influence of the properties of whole land mosaics on the status of species, assemblages and ecological processes [20–22]. Agricultural landscapes are mosaics of different land covers and land uses which offer a range of habitats for plant and animal species [20]. These mosaics have properties that may influence animal populations, related to: (i) the total extent of a specific habitat of the target study species, (ii) the composition of the mosaic, a known driver of the species composition of faunal assemblages, and (iii) the spatial configuration of elements in the mosaic [20], which becomes more relevant as fewer patches of adequate habitat remain in the landscape [22].

In the present study, we used a "whole land mosaic" approach [20,21] to explore the relationship between the regional variation in land mosaic composition and the density of little bustard. This type of evaluation, which can also be applied at the community level (e.g. biodiversity indices or species richness), requires landscape-level inference, which can be obtained only when both the response (population density) and explanatory variables characterise the landscape mosaic as a sample unit. In fact, landscape inference enables an assessment of the status of a species for the whole mosaic (not just a patch), and is more responsive to processes or patterns functioning at broad spatial scales [21]. Our aim here is to describe how regional variations in land mosaic composition influence little bustard densities, by characterizing land cover composition and little bustard density in 81 land mosaics in southern Portugal. More specifically, we hypothesise that little bustard male density should be higher in land mosaics dominated by the land cover providing adequate habitat for the species (grasslands) rather than to a higher land cover diversity.

Materials and Methods

Ethics Statement

No specific permits were required for the described field studies. Bird counts were carried out along public roads where no permission was required along private roads where, whenever possible, land owners were asked for permission. Little bustards are a protected species, but fieldwork was restricted to bird counts for which no specific permissions are required.

Study Areas

The Portuguese little bustard population is estimated at 10000–20000 individuals [11], mostly concentrated in the province of Alentejo, where we focused our sampling efforts. We defined 81 study sites (farmland mosaics) with areas ranging from 1657 to 9997 ha (mean = 2910 ha, median = 2502 ha, total area = 235740 ha) (Fig. 1), as follows.

Firstly, sites previously classified as Special Protection Areas (SPA) or Important Bird Areas (IBA) for steppe birds [23] (n = 14) were selected. These areas have a landscape mostly composed of open agricultural land, although a few also included forested and shrubland patches. Site limits corresponded to the total SPA/IBA area limits (or only the farmland portion, if the site also included forested areas) except if they were very large (over 10000 ha); in this case two to five mosaics of median size (1700 to 3600 ha) were

defined in each SPA/IBA, as it would have been impossible to cover the whole area (due to logistic and time constraints).

Secondly, additionally to the IBA/SPAs, a set of 2500 ha land mosaics (n = 67) was defined by first overlaying a grid of 10×10 km squares over the Corine Land Cover map 1990 of the province of Alentejo (1:100000) [24] and then selecting those squares with more than 40% of open agricultural and pastoral land area; land covers with potential little bustard habitat. In each of the selected 10×10 km units, we randomly selected two 5×5 km squares taking into account the level of sampling effort that could be made to cover the four sub-regions within Alentejo (Alto, Centro, Litoral and Baixo) (Fig. 1). Again, the total number of mosaics selected was dictated by the level of sampling effort that could be undertaken during a short time period (the male display period) every year.

The 81 sites were sampled for little bustards during 2003 (7 sites), 2004 (20 sites), 2005 (17 sites) and 2006 (37 sites). In each year, sampled sites were geographically stratified across the main sub-regions (Alto, Centro and Baixo), to avoid an association between year and sub-region (Fig. 1).

Little Bustard Counts

In each of the 81 sites, little bustard density was evaluated using a network of point count locations covering the whole area. Firstly, we used 1:25000 scale maps and field checks to identify the available road network that crossed each site. Following standard procedures to count little bustard males [12,18,26], points were then placed along the whole network of accessible non-paved roads crossing each sampling site, with the distance between survey points being set at a minimum of 600 m, to avoid double counts, and with the additional constraint that each point was at least 300 m from the site boundaries, villages and farmsteads, to minimise potential disturbance effects on little bustards. Because of differences in site area and road network density, the number of survey points per site ranged from 16 to 72 (mean = 29.1, total number of points = 2326). This corresponded to an average density of 1.05 points/km^2 (median = 1, range = 0.48–1.54).

Each site was surveyed one to three times (mean = 1.4, median = 1) during April and May, which corresponds to the time of the breeding season when males are most active and conspicuous [13]. The number of counts was dictated by logistic constraints. Roads were travelled by car, and the total number of males detected within 250 m of each survey point (an area of 19.6 ha) during 5 minutes was recorded. This radius was selected because it is the distance at which any calling male is most likely to be detected, and this survey method has been widely used in other studies [16,25,26]. GPS point coordinates and a map (or aerial photographs) overlaid with the boundary of the search circles facilitated the assessment of the survey area and the bird counts. Visual confirmation of males detected by their calls was made to ensure that they were within the 250 m radius of the survey point. Birds that flew from the sampled area as the observer approached the survey point were also counted, after checking their landing location to avoid double counts. All surveys were carried out within the first three hours after dawn or three hours before dusk, coinciding with the known peaks in male activity [13,27]. When more than one count was carried out per site, there was at least a one-week interval between successive counts. The sequence of road itineraries was changed from count to count to keep points from being sampled systematically at the same time of the day.

Land Cover Composition

The landscape composition in each site was determined by overlaying the CORINE Land Cover map for 2006, derived from

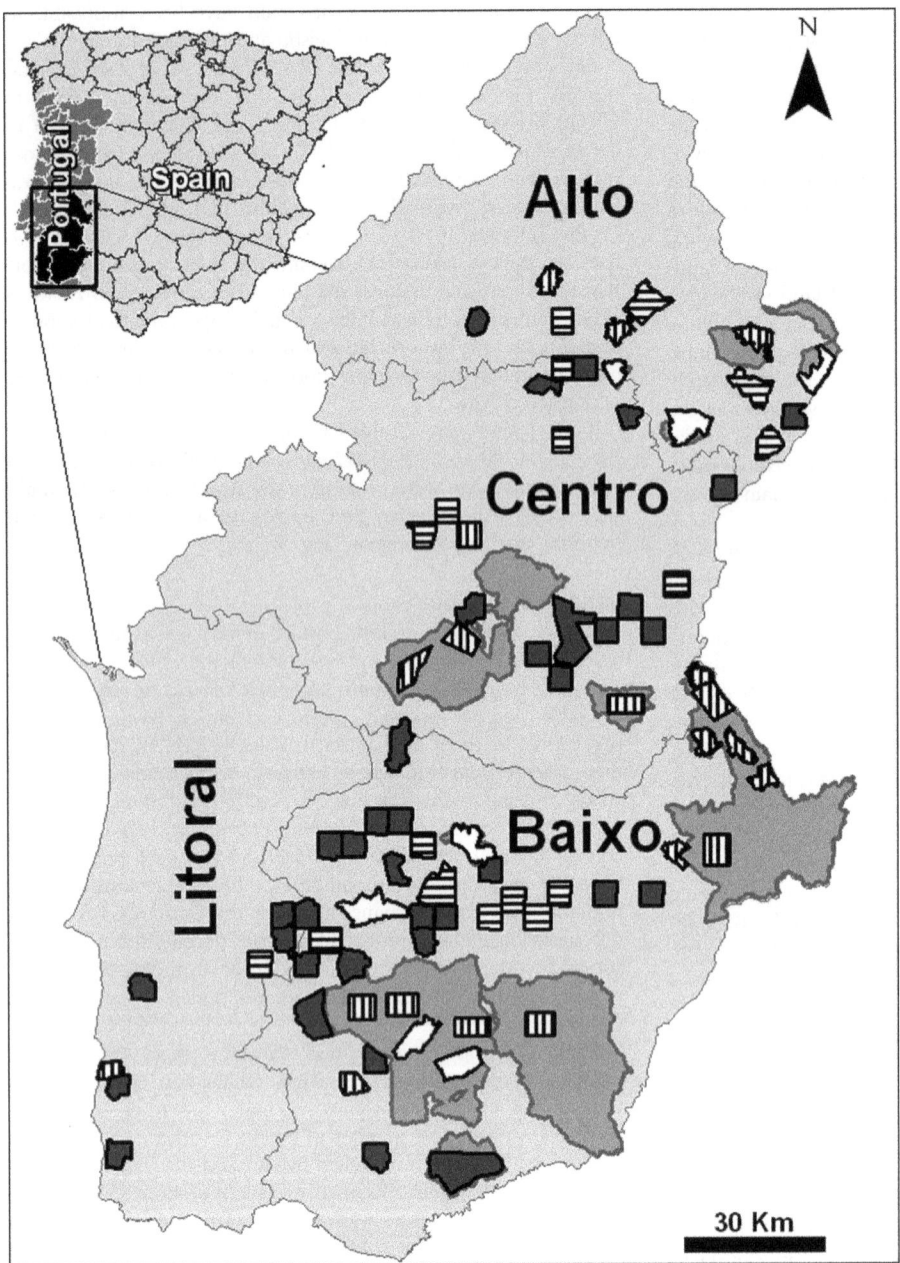

Figure 1. Location of the studied land mosaics for characterising little bustard densities in four regions of Alentejo (Alto, Centro, Baixo and Litoral), southern Portugal. Different stipple patterns correspond to different years of sampling: 2003 (white), 2004 (vertical pattern), 2005 (horizontal pattern) and 2006 (dark grey). Important Bird Areas (IBA) and Special Protection Areas (SPA) with importance for steppe birds are shown in light grey.

SPOT 4 satellite imagery [28] with site boundaries in a Geographic Information System (GIS). For the purposes of this study, the level-3 land cover nomenclature was simplified into seven classes (Table S1). Other land cover categories were much scarcer in the study areas and did not include potential habitats for little bustards (they mostly belonged to categories "artificial surfaces" and "wetlands"), and were discarded from the analyses.

One of the drawbacks of using the CORINE classification system is that it does not allow the distinction of the different uses within dry crops, the most common land cover type in the sampled sites (mean cover = 58.7%, median = 63.2%, range = 0.5–98.9%, n = 81). This information is highly relevant,

as this broad category includes both uses that are highly suitable for breeding little bustards (such as pastures and fallow land) and less suitable ones (e.g. ploughed fields, cereal crops, and sunflower fields) [13,29,30]. To overcome this problem, the dry crop information from CORINE mapping was complemented with information collected during the field surveys. For this purpose, in each sampled point the land cover composition in the surrounding buffer was visually estimated to the nearest 12.5% by dividing the 250-m radius circle in 8 "slice" sections and recording the dominant land cover (covering the largest proportion of the area) in each section, for the following categories: (1) *grassland*s (fallow fields, permanent grasslands, and set-aside fields), (2) *ploughed fields*, (3) *cereal and hay fields*, (4) *dry legumes* (including chick pea and alfalfa). In most cases, the dominant cover was easily identified due to the large size of the fields. In sites counted more than once, the cover could change from count to count due to agricultural activities (e.g. ploughing of fields), thus availability was averaged. The relative proportion of these four land uses was estimated for each site and these estimates were used to replace the estimated total area covered by dry crops (derived from CORINE) by its four components (see Table 1).

Data Analyses

Estimates of male density obtained for each point were averaged to yield a mosaic-level density, expressed as males/km^2, for each of the 81 sites. For sites sampled more than once, the mean male density was estimated for each sample point before estimating the mosaic-level average density. The existence of spatial dependence in the pattern of regional variation in male density was tested through a spatial correlogram based on the Moran's I autocorrelation coefficient [31], using function *correlog* of the *ncf* package [32] of the R language [33]. Its significance was tested using 1000 permutations and the progressive Bonferroni correction [34].

For each site, in addition to the 10 variables expressing land cover, we calculated two land cover diversity estimates: (i) land cover richness, expressed as the number of different land cover types in the site, and (ii) land cover equitability, estimated as the Shannon diversity index divided by its maximum possible value (natural log of the number of land cover types). Equitability expresses the relative proportion of all existing land covers in a given site, and varies between close to 0 (when one land cover is vastly dominant) and 1 (when all land covers occur in similar proportions).These two variables were estimated using the *vegan* package [35] for R. As some of the 12 variables used for describing land cover composition were intrinsically interdependent (multicollinearity), we used principal components analysis (PCA) based on a correlation matrix to describe the main land cover gradients [34]. The angular transformation was applied to variables expressing proportions prior to PCA. We retained only Principal Components (PCs) with an eigenvalue larger than 1, as factors with variances smaller than unity are no better than a single variable. These new PC variables (expressing site coordinates in the selected components) have the advantages of being uncorrelated with each other and of summarizing most of the information contained in the original variables. To obtain simpler and more interpretable components, the factors were rotated using the varimax criteria, thus minimizing the number of variables with high loadings on a given factor [34]. The site coordinates along the PCs were mapped in the GIS, to visualise the spatial patterns of land cover composition across the region.

To model the influence of landscape composition (PC) variables on little bustard density, we used mixed effects models [36]. Function *lme* of the *nlme* package [37] was used to fit the mixed models in R, with year as a random effect. We followed Zuur et al. [38] (chapter 5) and started with a model where the fixed component contained all explanatory variables and different variance structures for the random part were sought, owing to heterogeneity in residuals. Once the optimal random structure was found, model building then concentrated on finding the optimal fixed structure (using Maximum Likelihood estimation), by backward selection of variables, based on likelihood ratio tests. The final model was re-run using restricted maximum likelihood estimation (REML). Model assumptions and model fit of the final model were assessed using the proportion of variance explained (r^2), histograms and qqplots of residuals, and plots of residuals versus fitted values and explanatory variables. The existence of spatial dependence in residuals was tested through a spatial correlogram based on the Moran's I.

Results

Regional Variation in Little Bustard Densities

Little bustards were present in 68 of the 81 sampled land mosaics (Fig. 2A), and male regional density ranged from 0 to 9.73 males/km^2 (mean = 2.25±0.258, n = 81). Sites with higher density (>5 males/km^2) were concentrated mainly in Baixo and Alto Alentejo (see Fig. 1). The species was scarce or absent in Litoral Alentejo and in parts of Central Alentejo. A significant, although weak, positive spatial autocorrelation in bustard densities existed in nearby sites until a lag distance of ca. 10 km, and it declined progressively until a significant negative value was registered at lag distance of ca. 50 km (Fig. 2B).

Table 1. Explanatory variables and descriptive statistics (mean and range) across the 81 sampled sites.

Variable (short name)	Description	Mean (range)
Grasslands (Grass)	Proportion of grasslands, derived from CORINE corrected by field data	0.342 (0.001–0.776)
Cereal (Cereal)	Proportion of cereal, derived from CORINE corrected by field data	0.208 (0–0.620)
Irrigated crops (Irrigcrops)	Proportion of irrigated annual crops, derived from CORINE	0.123 (0–0.955)
Agro-forestry (Agrof)	Proportion of agro-forestry systems, derived from CORINE	0.104 (0–0.452)
Permanent crops (Permcrops)	Proportion of permanent crops, derived from CORINE	0.090 (0–0.400)
Shrublands (Shrub)	Proportion of shrublands, derived from CORINE	0.039 (0–0.266)
Ploughed (Plough)	Proportion of ploughed fields, derived from CORINE corrected by field data	0.031 (0–0.142)
Forests (For)	Proportion of forests, derived from CORINE	0.027 (0–0.214)
Mixed systems (Mixed)	Proportion of mixed systems, derived from CORINE	0.016 (0–0.191)
Dry legumes (Dryleg)	Proportion of dry legume crops, derived from CORINE corrected by field data	0.005 (0–0.074)
Richness (Rich)	Number of land cover types	6.8 (4–10)
Equitability (Equit)	Equitability of land cover types	0.73 (0.15–0.93)

Land cover variables are ordered by decreasing mean.

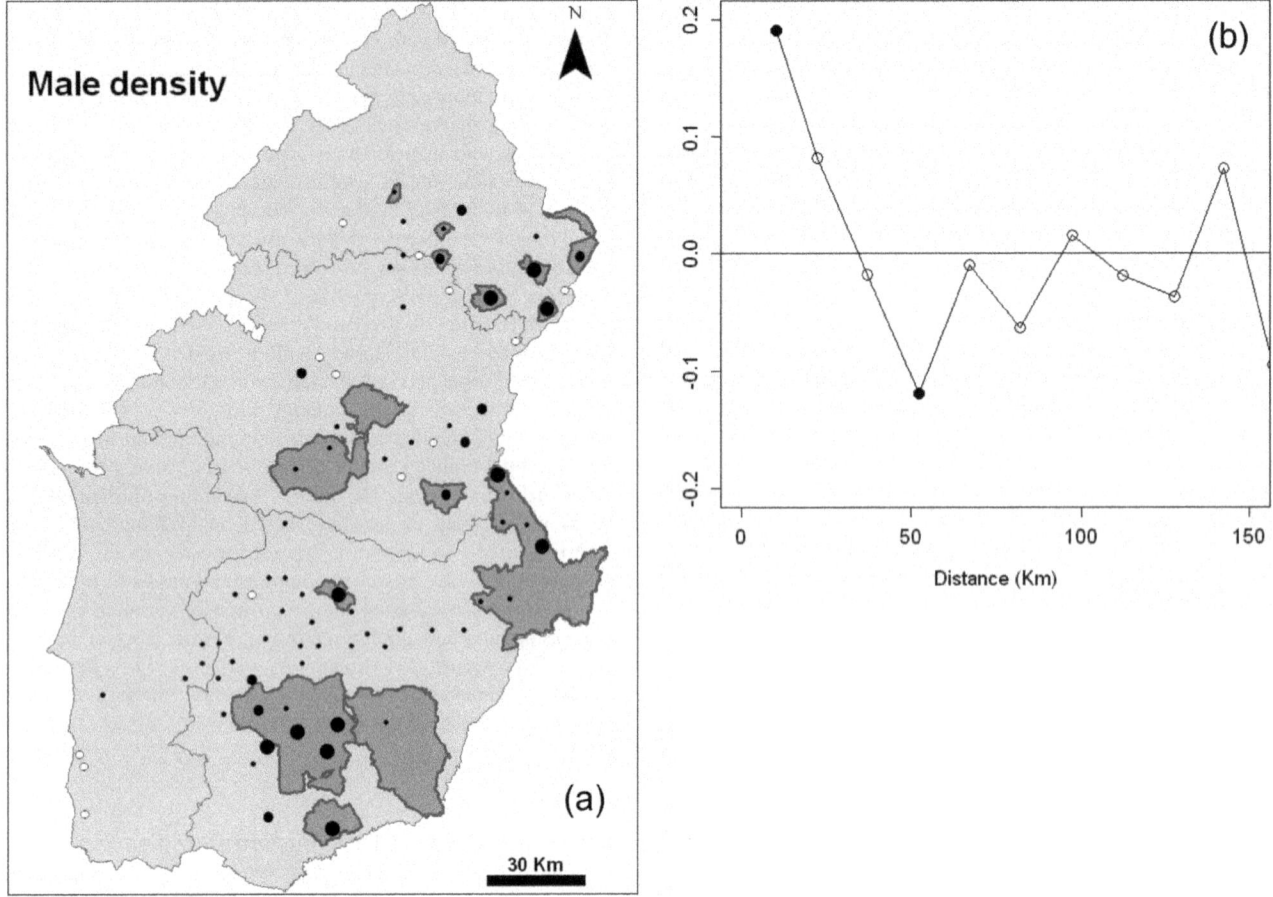

Figure 2. Spatial autocorrelation in little bustard density patterns. (a) Little bustard male density across the studied land mosaics in Southern Portugal. Important Bird Areas (IBA) and Special Protection Areas (SPA) with importance for steppe birds are shown in dark grey. Codes for male densities: small white dots (no males recorded), small black dots (0.01–2.99 males/km^2), medium-sized black dots (3.00–4.99 males/km^2), and large black dots (5.00–9.73 males/km^2). (b) Spatial correlogram of little bustard male densities. Dark symbols represent correlation statistics significant (p<0.05) after progressive Bonferroni correction.

Regional Variation in Land Cover Composition

The most common land cover type in the studied sites was grassland, which occupied in average ca. 34% of the total area (Table 1). Cereal, irrigated crops and agro-forestry systems all had a mean coverage higher than 10%. The mean number of land cover types per site was 6.8, and mean land cover equitability was 0.73 (Table 1).

The 12 original variables were summarised into four Principal Components (Table 2) with an eigenvalue larger than 1, and these accounted for 63.7% of total data variance. The first PC (PC1) represented a gradient of sites ranging from low to high land cover diversity (expressed as richness and equitability), where this increase was also associated to a higher cover by permanent crops and mixed systems. The spatial distribution of the PC 1 scores (Fig. 3A) showed a concentration of sites with low land cover diversity on southern sites and a few clusters in Centro and Baixo Alentejo. PC 2 represented a gradient ranging from sites with a high proportion of irrigated crops to sites with more grasslands. The spatial distribution of the PC 2 scores (Fig. 3B) showed that sites with more irrigated crops occurred along the coast of Litoral Alentejo, the western and central part of Baixo Alentejo and also in specific sites in Centro and Alto Alentejo. The third PC represented a gradient ranging from sites with higher proportions of cereal and ploughed fields to sites with

more forests and agro-forestry systems. The spatial distribution of the PC 3 scores (Fig. 3C) showed a large cluster of sites with more cereal and ploughed fields in Baixo Alentejo. The fourth PC was mostly a gradient of decreasing proportion of shrublands and increasing proportion of legume fields. Shrublands were more common in Litoral Alentejo, southern Baixo Alentejo and Centro Alentejo (Fig. 3D).

Land Cover Predictors of Little Bustard Density: Model Building

The initial model (AIC = 335.1, Table S2), including the random factor (year) and the 4 PC variables, showed heterogeneity in the residual patterns, mainly because of an increased residual spread along with PC 2 as well as different residual spread per year. Thus, residual heterogeneity was allowed by exploring different variance structures [38]. The comparison of several alternatives (Table S2) showed that the model including a combination of variance structures allowing a different spread per year and an exponential increase with PC 2 had the lowest AIC (324.0) and represented a significant improvement compared to the initial model (Likelihood ratio test = 19.1, p<0.001).

There was no significant spatial autocorrelation in both initial and varComb model residuals at any lag distance, showing that the existing spatial correlation in male densities was induced by

Table 2. Principal component loadings, eigenvalues and explained variance (% var.) for varimax rotated PC axes 1 to 4 describing patterns in land cover composition across the 81 study sites.

variable	PC 1	PC 2	PC 3	PC 4
Rich	**0.82**	-0.05	-0.09	-0.13
Permcrops	**0.71**	-0.26	0.05	0.31
Equit	**0.65**	0.40	0.13	0.23
Mixed	**0.51**	0.24	-0.06	-0.50
Irrigcrops	-0.15	**-0.89**	0.06	-0.04
Grass	-0.39	**0.69**	-0.07	-0.16
Plough	0.16	0.02	**-0.72**	0.04
Agrof	0.29	0.45	**0.68**	0.16
Cereal	0.24	0.18	**-0.68**	0.47
For	0.39	-0.21	**0.65**	-0.15
Shrub	-0.14	0.25	0.13	**-0.63**
Dryleg	-0.00	0.21	-0.06	**0.53**
Eigenvalue	2.371	1.949	1.929	1.392
% var.	19.7	16.2	16.0	11.6

Variables with correlation coefficients higher than 0.50 are highlighted in bold.

exogenous processes [31,34], namely the similarities in land cover composition in nearby sites. Thus, the explanatory variables in the model effectively accounted for the spatial dependence. Model building then continued with the fixed part, where the backward selection of the variables resulted in a model with the first three PCs. In this final model, the random component results showed that the random intercept had a variance of 0.50 and the correlation between sites sampled in a given year was quite low (intraclass correlation was 0.08). The estimates for the separate standard deviations per stratum (year) showed that residual variability was the highest in 2005 and the lowest in 2006. Finally, residual spread increased also as a function of $e^{(0.82* \ PC \ 2)}$. In the fixed component, the results (Table 3) showed that the more important variable explaining bustard density was PC 2, with densities positively correlated with this variable, meaning that the species was more abundant in land mosaics dominated by grasslands and with lower proportion of irrigated crops. Both PC 1 and PC 3 had negative coefficients, showing that higher densities were attained in mosaics with lower land cover diversity (and less permanent crops and mixed systems) and a lower proportion of forests and agro-forestry systems (and more cereal and ploughed fields). This model explained 48% of the regional variability in little bustard density, and there was no significant spatial autocorrelation in the residuals (Fig. 4).

Discussion

In the present study, we used a "whole land mosaic" approach [20] to explore the relationship between the regional variation in land cover composition and the population density of a threatened bird in 81 land mosaics spread across southern Portugal. This large scale approach provides the best evaluation of biodiversity or population responses to changing land cover composition, the main driver of biodiversity changes in Mediterranean landscapes,

and is recommended for conservation strategies for landscape mosaics [22]. However it is seldom used, at least in agricultural landscapes [20]. One assumption of this approach is that the mosaic-scale density of little bustards is a reliable indicator of landscape suitability, which may not be the case for all species [39].

Regional Variation in Male Densities

The widespread occurrence of the little bustard and the population densities measured in Alentejo suggest that, within an Iberian context, the region as a whole is suitable for the species. In fact, the regional density of ca. 2 males/km^2 estimated in the current study is similar to that observed in many areas in Spain (e.g., [21,25,29]), and is well above the densities observed in Western France [40]. Exceptional regional mean densities of over 5 males/km^2, rare in other regions of the Iberian Peninsula, occurred at 12 land mosaics, of which four were in the Castro Verde region, where a population of 3,400 to 5,000 males was estimated [41]. There was spatial autocorrelation in measured densities, with nearby sites (until ca. 10 km away) tending to share a high or low density of little bustard males. This spatial dependence could be caused by endogenous (e.g. behaviour, contagion, dispersal) or exogenous (environmental gradients) processes [34,38]. Although conspecific attraction has been described in this species at a local scale [42], we would not expect to find a biological basis for bustard average density in one land mosaic to be influenced by densities in the surrounding mosaics due to behavioural processes, because of the large grain size (thousands of hectares) used in this study. Thus, this spatial dependence was more likely induced by exogenous processes, namely the spatially structured patterns in land cover composition. This is corroborated by the fact that spatial dependence disappeared once the effect of land cover was taken into account in the models.

Landscape Patterns: Agricultural Intensification does not Decrease Land Cover Diversity

Both intensification and agricultural abandonment in farmed landscapes usually have significant impacts on landscape composition and configuration (e.g., [6,7]). Their consequence is almost always a trend towards simplification and increased homogeneity, through for example removal of field boundaries and non-crop elements, simplified crop rotations, loss of fallow fields, reduction of crop diversity, or increased field size [6,43]. This loss of landscape heterogeneity is usually seen as detrimental for biodiversity [43], but there are important exceptions. In Eastern Europe, high biodiversity value grasslands occur as very homogeneous land covers, and increasing agricultural intensification levels will lead to a higher land cover diversity [44–45]. This positive correlation between land cover diversity and agricultural intensification was also observed in the current study, where the main gradient of regional variation in landscape composition associated increasing land cover diversity (richness and equitability) with the increased cover by permanent crops and mixed systems. Many of these permanent crops consisted of irrigated olive groves and vineyards more prevalent in an agricultural intensification context. In addition, the obvious gradient of intensification reflected in the second axis of the PCA, expressing the replacement of grasslands by irrigated annual crops, was not related to land cover diversity confirming that, in this geographic context, increased agricultural intensification is not necessarily reflected in a decrease of land cover diversity.

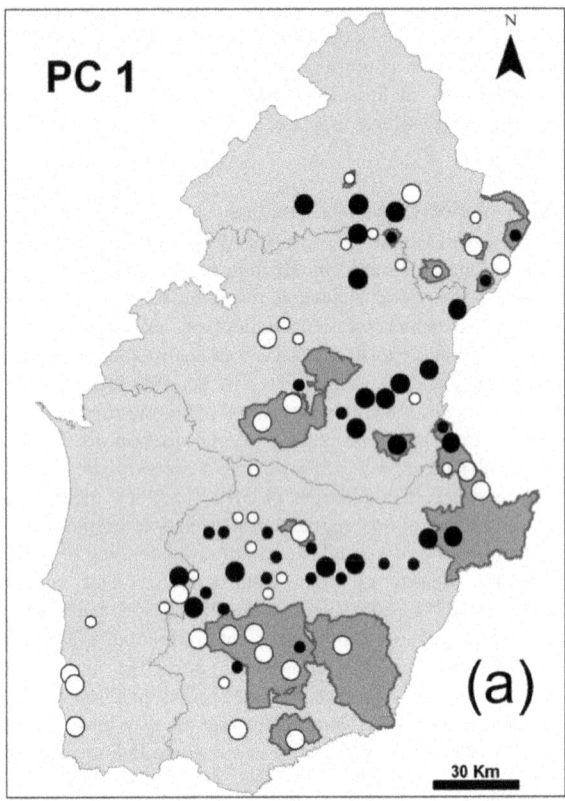

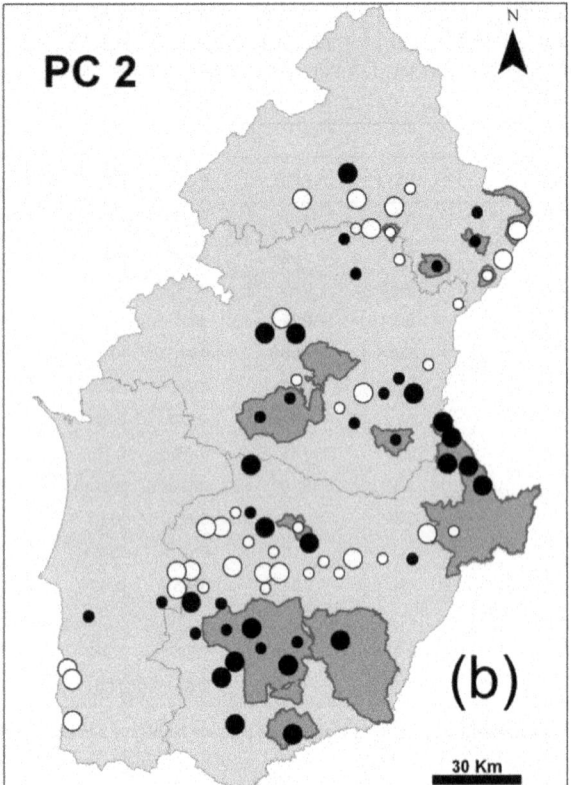

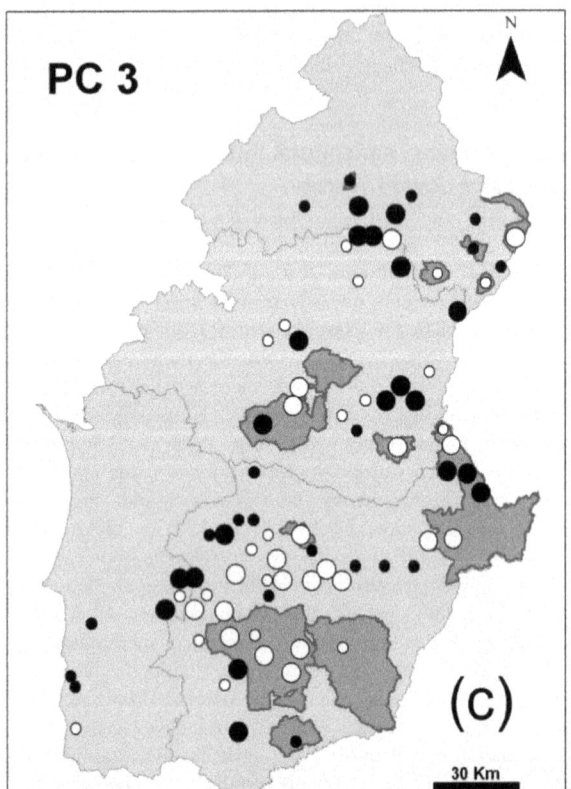

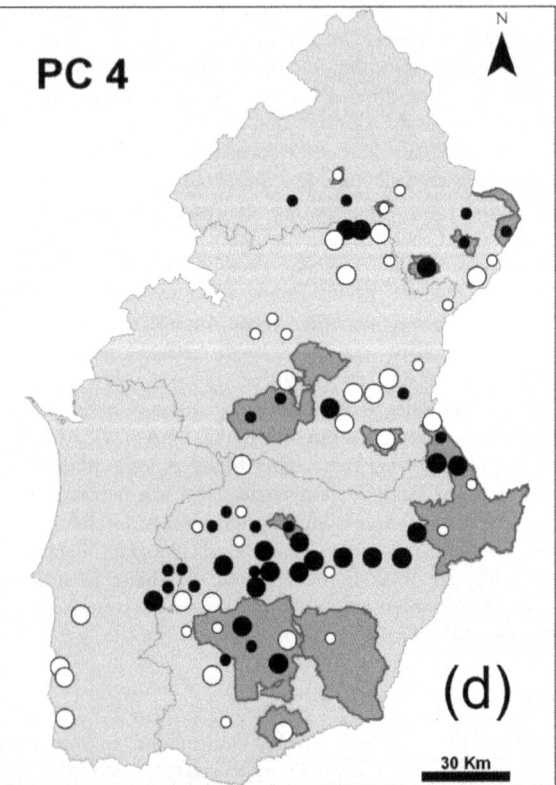

Figure 3. Site coordinates along the four first axes of a Principal Components Analysis to summarise land cover information in the 81 study sites. For each axis, each symbol denotes the four quartiles of site coordinates: large white dots (first quartile), small white dots (second quartile), small black dots (third quartile) and large black dots (forth quartile). (a) PC 1; (b) PC 2, (c) PC 3, (d) PC 4.

Table 3. Coefficients of explanatory variables (land cover PCs) (± standard errors) in the fixed part of the linear mixed model, and their significance.

Variable	Coefficient	P-value
PC 1	-0.48±0.142	0.0011
PC 2	0.73±0.111	<0.001
PC 3	-0.46±0.161	0.0054

Model AIC = 308.7 and r^2 = 48.1%.

Little Bustard Densities are Influenced by Both Landscape Diversity and Amount of Grasslands

Our study showed that land cover composition explained ca. 50% of the regional variability in little bustard densities across agricultural land mosaics in southern Portugal. The initial hypothesis that male little bustard density should be higher in landscape mosaics dominated by grasslands, rather than those with higher land cover diversity, was confirmed, with the main driver of population densities being the proportion of grasslands in the land mosaic (and, inversely, the proportion of irrigated crops). Several smaller scale studies have shown the importance of grasslands as the main habitat for displaying males, and where a higher male density can be found (e.g., [9,13,14,17,18,29]). In contrast, irrigated crops are usually unsuitable for displaying males [13,15,46].

Although grasslands were a key component in the land cover composition for promoting higher bustard densities, other land cover variables were found to influence population density. In land mosaics with higher land cover richness and diversity, which in our geographical context also had a higher cover by permanent crops and mixed systems, male density declined. This shows an avoidance of the species by diverse land mosaics and contrasts with the results of some studies made elsewhere (e.g., [15,47]). This apparent contradiction is likely explained by the fact that in other studies higher land cover diversity was usually associated to an increased prevalence of grasslands within a patchy landscape, which was not the case in our land mosaics in southern Portugal, and suggests that the context in southern Portugal is similar to the one of Eastern Europe, where increasing agricultural intensification levels lead to a higher land cover diversity harmful for specialist (often endangered) species in these low-intensity agricultural landscapes [44,45]. The conclusion that little bustards prefer homogeneous grassland landscapes is corroborated by recent study showing that they occurred in higher densities in larger grassland fields in a region in southern Portugal [19]. Finally, densities were also higher in land mosaics with a higher proportion of cereal and ploughed fields, and a lower proportion of unsuitable forest covers. Cereal fields may be suitable for other parts of the yearly cycle [48,49], or for nesting females [50], thus the existence of some cereal fields in a grassland landscape context might provide additional food and habitat resources to little bustards.

The unexplained regional variability in male density can be due to different unmeasured factors. Habitat quality could play a major role, and it can be expressed as variation in vegetation structure and food availability (e.g., [13,15,50]), grazing intensity, human disturbance (e.g., [48]) or a more suitable spatial configuration of the different land cover types. The variable size of our land mosaics may also explain some of this regional variability, if little bustards responded differently at different scales (e.g. 2500 ha c.f. ca. 10000 ha (the size of our largest land mosaic)). This potential

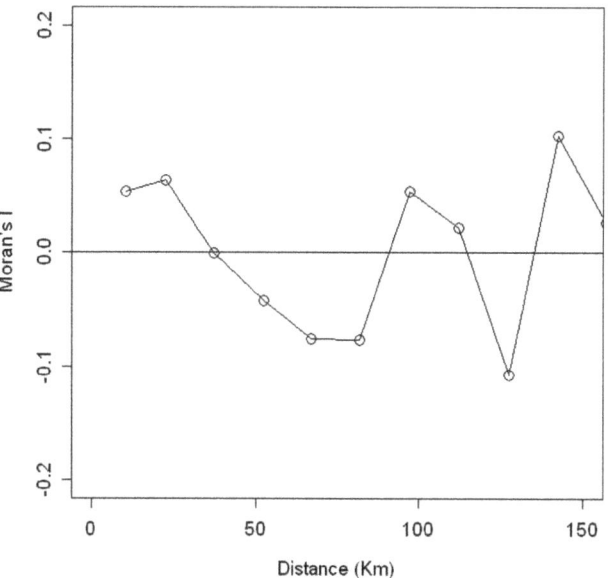

Figure 4. Spatial correlogram of the normalized residuals of the mixed effects model of the relationships between little bustard male densities and land cover variables.

scale effect is however unlikely in our dataset, as 80% of the mosaics had a similar size.

Implications of Land Cover Changes in the Mediterranean for Little Bustard Populations

Land cover changes have strong implications on biodiversity patterns, particularly in the Mediterranean region (e.g., [1,8]). Feranec et al. [3] described recent land cover changes (1990–2000) in European landscapes and identified the main landscape processes occurring during this period: urbanization, intensification of agriculture, extensification of agriculture, afforestations, deforestation and construction of water bodies. Of these processes, the ones more common in Portugal were afforestations (increase of forest cover due to natural regeneration and plantations), intensification of agriculture (mostly changes of arable land to vineyards, orchards, greenhouses and other irrigated crops.) and deforestation (loss of forest cover by clear-cutting, forest fires, etc.). The results of the current study suggest that the first two processes have caused habitat degradation and loss for little bustards in the last decades, whereas the impact of the latter depends on the type of land cover change that forests are experiencing (if there forests have been replaced by agricultural land, that may have been beneficial for the species). A more detailed study carried out for the period 1985–2000 in Portugal [51] revealed a 4% increase in permanent crops, a 28% decline in the area of pastures and a 2.8% increase in forests. As a whole, landscape fragmentation has increased (more polygons and less area per polygon). Land cover diversity had a large increase, more noticeable in the southern part of the country where this study was undertaken. This was accompanied by a large decline in the land cover dominance index in the region [51]. All these changes also point out to a likely degradation of overall suitability of the landscape mosaic for the little bustard populations, that is likely to continue in the near future and raises concern on the impact of these changes on population size and trends, particularly in Portugal and Spain, which hold more than half of the world's population of this species [52]. Thus, agri-environmental policies aimed to conserve little bustard populations should aim at maintaining or promoting vast expanses of grasslands in agricultural land mosaics.

Supporting Information

Table S1   Land cover level 3 CORINE categories pooled for creating each category used in this study.

Table S2   Comparison of different variance structures in the random part of the model. For each structure, the AIC of the model is given. The fixed part of the model included all explanatory variables.

Acknowledgments

We would like to thank João Correia and Sr. Santos for their field assistance. The Comissão de Coordenação e Desenvolvimento Regional – Alentejo provided important logistic support in 2004. Vincent Bretagnolle, Jeremy Wilson and two anonymous referees provided comments to earlier versions of this manuscript that significantly improved the work.

Author Contributions

Conceived and designed the experiments: FM JPS DL. Performed the experiments: FM JPS BE ML MP IA RP ES TC PS IH AD. Analyzed the data: FM JPS JMP. Wrote the paper: FM JPS JMP.

References

1. Sala O, Chapin F, Armesto J, Berlow E, Bloomfield J, et al. (2000) Global biodiversity scenarios for the year 2100. Science 287: 1770–1774.
2. Hill J, Stellmes M, Udelhoven T, Roder A, Sommer S (2008) Mediterranean desertification and land degradation mapping related land use change syndromes based on satellite observations. Global and Planetary Change 64: 146–157.
3. Feranec J, Jaffrain G, Soukup T, Hazeu G (2010) Determining changes and flows in European landscapes 1990–2000 using CORINE land cover data. Applied Geography 30: 19–35.
4. Moreira F, Russo D (2007) Modelling the impact of agricultural abandonment and wildfires on vertebrate diversity in Mediterranean Europe. Landscape Ecology 22: 1461–1476.
5. Vilà M, Burriel JA, Pino J, Chamizo J, Llach E, et al. (2003) Association between Opuntia species invasion and changes in land-cover in the Mediterranean region. Global Change Biology 9: 1234–1239.
6. Moreira F, Viedma O, Arianoutsou M, Curt T, Koutsias N, et al. (2011) Landscape-wildfire interactions in southern Europe: implications for landscape management. Journal of Environmental Management 92: 2389–2402.
7. Stoate C, Boatman ND, Borralho R, Rio Carvalho C, Snoo GR, et al. (2001) Ecological impacts of arable intensification in Europe. Journal of Environmental Management 63: 337–365.
8. Stoate C, Baldi A, Beja P, Boatman ND, Herzon I, et al. (2009) Ecological impacts of early 21st century agricultural change in Europe – A review. Journal of Environmental Management 91: 22–46.
9. Delgado A, Moreira F (2000) Bird assemblages of an Iberian cereal steppe. Agriculture, Ecosystems and Environment 78: 65–76.
10. Suárez F, Naveso MA, De Juana E (1997) Farming in the drylands of Spain: birds of the pseudosteppes. In: Pain DJ, Pienkowski MW, editors. Farming and birds in Europe. The Common Agricultural Policy and its implication for bird conservation. London: Academic Press. 297–330.
11. Goriup P (1994) Little Bustard Tetrax tetrax. In: Tucker GM, Heath MF, editors. Birds in Europe: their Conservation Status. Cambridge: BirdLife International. 236–237.
12. De Juana E, Martínez C (1996) Distribution and conservation status of Little bustard Tetrax tetrax in the Iberian Peninsula. Ardeola, 43, 157–167.
13. Martínez C (1994) Habitat selection by the Little Bustard Tetrax tetrax in cultivated areas of Central Spain. Biological Conservation, 67, 125–128.
14. Moreira F (1999) Relationships between vegetation structure and breeding bird densities in fallow cereal steppes in Castro Verde, Portugal. Bird Study, 46, 309–318.
15. Salamolard M, Moreau C (1999) Habitat selection by Little Bustard Tetrax tetrax in a cultivated area of France. Bird Study, 46, 25–33.
16. Wolff A, Paul JP, Martin JL, Bretagnolle V (2001) The benefits of extensive agriculture to birds: the case of the little bustard. Journal of Applied Ecology, 38, 963–975.
17. García J, Suárez-Seoane S, Miguélez D, Osborne P, Zumalacarregui C (2007) Spatial analysis of habitat quality in a fragmented population of little bustard (Tetrax tetrax): implications for conservation. Biological Conservation, 137, 45–56.
18. Morales MB, García JT, Arroyo B (2005) Can landscape composition changes predict spatial and annual variation of little bustard male abundance? Animal Conservation, 8, 167–174.
19. Silva JP, Palmeirim JM, Moreira F (2010) Higher breeding densities of the threatened little bustard Tetrax tetrax occur in larger grassland fields: Implications for conservation. Biological Conservation, 143, 2553–2558.
20. Bennett AF, Radford JQ, Haslem A (2006) Properties of land mosaics: implications for nature conservation in agricultural environments. Biological Conservation, 133, 250–264.
21. Radford JQ, Bennet AF (2007) The relative importance of landscape properties for woodland birds in agricultural environments. Journal of Applied Ecology, 44, 737–747.

22. Lindenmayer D, Hobbs R, Montague-Drake R, Alexandra J, Bennet A, et al. (2008) A checklist for ecological management of landscapes for conservation. Ecology Letters, 11, 78–91.
23. Costa LT, Nunes M, Geraldes P, Costa H (2003) Zonas Importantes para as Aves em Portugal. Sociedade Portuguesa para o Estudo das Aves, Lisboa, 160 pp.
24. Heymann Y, Steenmans C, Croissille G, Bossard M (1994) CORINE Land Cover. Technical Guide. Luxembourg, Office for Official Publications of the European Communities. 137 pp.
25. García de la Morena E, Bota B, Ponjoan A, Morales M (2006) El sisón común en España. I Censo Nacional 2005. SEO/Birdlife, Madrid. 155 pp.
26. Delgado A, Moreira F (2010) Between-year variations in little bustard Tetrax tetrax population densities are influenced by agricultural intensification and rainfall. Ibis 152: 633–642.
27. Jiguet F, Bretagnolle V (2001) Courtship behaviour in a lekking species: individual variations and settlement tactics in male little bustard. Behavioural Processes, 55, 107–118.
28. EEA (2007) CORINE Land Cover 2006. Technical Guidelines. European Environment Agency, Copenhagen, 66 pp.
29. Morales MB, Suárez F, García de la Morena E (2006) Réponses des oiseaux de steppe aux différents niveaux de mise en culture et d'intensification du paysage agricole: une analyse comparative de leurs effets sur la densité de population et la sélection de l'habitat chez l'outarde canepetiere Tetrax tetrax et l'outarde barbue Otis tarda. Revue d'Ecologie (Terre et Vie), 61, 261–270.
30. Santangeli A, Dolman P (2011) Density and habitat preferences of male little bustard across contrasting agro-pastoral landscapes in Sardinia (Italy). Journal of Wildlife Research, 57, 805–815.
31. Fortin MJ, Dale M (2005) Spatial analysis: a guide for ecologists. Cambridge University Press, Cambridge, 365 pp.
32. Bjornstad O (2009) ncf: spatial nonparametric covariance functions. R package version 1.1–3. Available: http://CRAN.R-project.org/package = ncf. Accessed 2011 Dec 26.
33. R Development Core Team (2009) R: A language and environment for statistical computing. R Foundation for Statistical Computing, Vienna, Austria. ISBN 3-900051-07-0. Available: http://www.R-project.org. Accessed 2011 Dec 26.
34. Legendre P, Legendre L (1998) Numerical Ecology. 2nd edn. Elsevier, Amsterdam. 853 pp.
35. Oksanen J, Blanchet FG, Kindt R, Legendre P, O'Hara RG, et al. (2010). vegan: Community Ecology Package. R package version 1.17–0. Available: http://CRAN.R-project.org/package = vegan. Accessed 2011 Dec 26.
36. Pinheiro JC, Bates DM (2000) Mixed-Effects Models in S and S-PLUS, Springer, 528 pp.
37. Pinheiro JC, Bates D, DebRoy S, Sarkar D, R Development Core Team (2009) nlme: Linear and Nonlinear Mixed Effects Models. R package version 3.1–92. Available: http://cran.r-project.org/web/packages/nlme/index.html. Accessed 2011 Dec 26.
38. Zuur A, Ieno E, Walker N, Saveliev A, Smith G (2009) Mixed effects models and extensions in ecology with R. Springer, New York, 574 pp.
39. Van Horne B (1983) Density as a misleading indicator of habitat quality. Journal of Wildlife Management, 47, 893–901.
40. Jolivet C, Bretagnolle V, Bizet D, Wolff A (2007) Statut de l'Outarde canepetière Tetrax tetrax en France en 2004 et mesures de conservation. Ornithos, 14, 80–94.
41. Moreira F, Leitão PJ, Morgado R, Alcazar R, Cardoso A, et al. (2007) Spatial distribuition patterns, habitat correlates and population estimates of steppe birds in Castro Verde. Airo, 17, 5–30.
42. Jiguet F, Bretagnolle V (2006) Manipulating lek size and composition using decoys: an experimental investigation of lek evolution models. The American Naturalist, 168, 758–768.
43. Benton TG, Vickery JA, Wilson JD (2003) Farmland biodiversity: is habitat heterogeneity the key? Trends in Ecology and Evolution, 18, 182–188.

44. Báldi A, Batári P (2011) Spatial heterogeneity and farmland birds: different perspectives in Western and Eastern Europe. Ibis, 153, 875–876.

45. Batári A, Fischer J, Báldi A, Crist TO, Tscharntke T (2011) Does habitat heterogeneity increases farmland biodiversity ? Frontiers in Ecology and Environment, 9, 182–188.

46. Martínez C, Tapia GG (2000) Density of the Little Bustard Tetrax tetrax in relation to agricultural intensification in Central Spain. Ardeola, 49, 301–304.

47. Campos B, López M (1996) Densidad y selección de hábitat del Sisón (Tetrax tetrax) en el Campo de Montiel (Castilla – La Mancha), España. In: Conservación de Las Aves Esteparias y su Hábitat (eds Fernández Gutiérrez J, Sanz-Zuasti J), 201–208. Junta de Castilla y León, Valladolid.

48. Silva JP, Pinto M, Palmeirim JM (2004) Managing landscapes for little bustard Tetrax tetrax: lessons from the study of winter habitat selection. Biological Conservation, 117, 521–528.

49. Silva JP, Faria N, Catry T (2007) Summer habitat selection of the threatened little bustard (Tetrax tetrax) in Iberian agricultural landscapes. Biological Conservation, 139, 186–194.

50. Morales MB, Traba J, Carriles E, Delgado MP, Garcia de la Morena E (2008) Sexual differences in microhabitat selection of breeding little bustards Tetrax tetrax: Ecological segregation based on vegetation structure. Acta Oecologica, 34, 345–353.

51. Freire S, Caetano M (2005) Assessment of land cover change in Portugal from 1985 to 2000 using landscape metrics and GIS. In: Proceedings of the GIS Planet 2005-II International Conference and Exhibition on Geographic Information, 30 May –2 June 2005, Estoril, Portugal, unpaginated CD-ROM.

52. Iñigo A, Borov C (2010) Little bustard (Tetrax tetrax), Action plan for the little bustard Tetrax tetrax in the European Union. SEO|BirdLife and BirdLife International.

18

Are Flying-Foxes Coming to Town? Urbanisation of the Spectacled Flying-Fox (*Pteropus conspicillatus*) in Australia

Jessica Tait[1,3], Humberto L. Perotto-Baldivieso[1], Adam McKeown[2], David A. Westcott[3]*

1 School of Energy, Environment and Agrifood, Cranfield University, Cranfield, Bedfordshire, United Kingdom, **2** CSIRO Sustainable Land and Water, Smithfield, QLD, Australia, **3** CSIRO Sustainable Land and Water, Atherton, QLD, Australia

Abstract

Urbanisation of wildlife populations is a process with significant conservation and management implications. While urban areas can provide habitat for wildlife, some urbanised species eventually come into conflict with humans. Understanding the process and drivers of wildlife urbanisation is fundamental to developing effective management responses to this phenomenon. In Australia, flying-foxes (Pteropodidae) are a common feature of urban environments, sometimes roosting in groups of tens of thousands of individuals. Flying-foxes appear to be becoming increasingly urbanised and are coming into increased contact and conflict with humans. Flying-fox management is now a highly contentious issue. In this study we used monitoring data collected over a 15 year period (1998–2012) to examine the spatial and temporal patterns of association of spectacled flying-fox (*Pteropus conspicillatus*) roost sites (camps) with urban areas. We asked whether spectacled flying-foxes are becoming more urbanised and test the hypothesis that such changes are associated with anthropogenic changes to landscape structure. Our results indicate that spectacled flying-foxes were more likely to roost near humans than might be expected by chance, that over the period of the study the proportion of the flying-foxes in urban-associated camps increased, as did the number of urban camps. Increased urbanisation of spectacled flying-foxes was not related to changes in landscape structure or to the encroachment of urban areas on camps. Overall, camps tended to be found in areas that were more fragmented, closer to human habitation and with more urban land cover than the surrounding landscape. This suggests that urbanisation is a behavioural response rather than driven by habitat loss.

Editor: Elissa Z. Cameron, University of Tasmania, Australia

Funding: This research was supported by a grant from Bat Conservation International to JT and funding to DAW from the Commonwealth of Australia's Department of the Environment, the National Environmental Research Program and the Rural Industries Research Development Corporation (through funding from the Commonwealth of Australia, the State of New South Wales and the State of Queensland under the National Hendra Virus Research Program). The funders had no role in study design, data collection and analysis, decision to publish, or preparation of the manuscript.

Competing Interests: The authors have declared that no competing interests exist.

* Email: David.Westcott@csiro.au

Introduction

By 2030 five billion humans are expected to live in urban areas while the global urban footprint is predicted to expand by 163% from c. 727,000 km^2 to c. 1,527,000 km^2 [1]. While already significant ecosystems in their own right, such urban expansion will place increasing pressure on other land uses, threatening native species and ecosystems, and becoming potent sources of invasive species and pathogens [2]. For example, much urban expansion is currently occurring in sensitive areas for biodiversity, e.g. in coastal lowlands or close to protected areas [1]. This means that consideration of urban systems is increasingly important in conservation planning and management. Urbanisation generally has a negative net effect on biodiversity, with many species becoming rare or locally extinct, in particular specialists, slow reproducers, and disturbance-sensitive species [2,3]. However, some species, often generalists, readily adapt to the urban landscape, some even reaching higher abundances in cities than in natural vegetation [4,5]. Although the presence of wildlife in urban areas can enhance human quality of life, some urban animal populations can prove problematic due to their impacts on

amenity, damage or their role as vectors of disease [6,7]. Understanding how particular species respond to urbanisation and identifying the processes leading to these responses is fundamental if we are to successfully manage the interaction between urbanisation, biodiversity and human welfare.

Urbanisation influences species distribution, abundance and movement. The urban mosaic can prove an attractive habitat for a wide range of taxa due to abundant food and shelter [2,3,8,9]. Urban areas can provide a refuge from hunting or predation pressure [10], from environmental disturbances such as drought or fire, as well as a more stable resource supply whilst natural vegetation is recovering [11]. Wildlife adapt to urban areas in a variety of ways, e.g. by adjusting their foraging, anti-predator behaviour, breeding behaviour and taking advantage of the climate associated with urban areas [12,13,14,15]. With increased urbanisation of wildlife populations comes increased contact with humans and attendant increases in opportunity for conflict, including amenity impacts such as noise, smell and vegetation damage [16], hazards such as vehicle collision or attacks on pets [17,18], and risk of disease transmission [6,7]. These 'human-wildlife conflicts' can lead to intense disagreement over the management of habitat and wildlife in and around urban areas [19].

Flying-foxes are large (up to c. 1 kg), colonially roosting bats which readily adapt to urban ecosystems. Roost sites (hereafter called camps) are found in and around many Australian towns and cities [20,21,22,23]. Hypotheses for why flying-foxes might use urban areas include loss of native habitat and urban expansion [20,24], changes in resource distribution due to plantings [25,26,27] and urban effects on local climate [21]. Evidence from other species suggests that urban areas may also provide refuge from predation [27], disturbance events such as droughts [28], bushfires [29] and post-cyclone effects [22,30,31]. It may also be possible that urban areas may be attractive because they offer a movement advantage, e.g. increase the ease of manoeuvring in flight due to the open nature of the habitat or ease of navigation due to landmarks and lighting, e.g. [32].

Roosting by flying-foxes in urban and peri-urban areas can result in contact and conflict with humans. The greatest concern is impact on amenity. It is not uncommon for flying-fox camps to contain 50,000 individuals [22,27], but even much smaller camps can be potent point sources of noise, odour and faeces, particularly when they occur within metres of residences. While smaller camps are often tolerated, larger camps in particular become a focus of community disquiet.

Alongside impact on amenity are concerns about the risks of disease transmission [33]. Bats host a high diversity of viruses and some of these are of agricultural and human health significance [33,34]. It would appear that in cases such as the paramyxoviruses (which include diseases such as measles, distemper, mumps, parainfluenza, Newcastle disease), these diseases have been present in bats for very long periods and have switched hosts to other mammals, including humans [34]. In Australia, there are two diseases of particular concern that are known to be carried by flying-foxes: Hendra virus and Australian bat lyssavirus. It has been suggested that the dynamics of diseases such as Hendra in flying-fox populations and the pattern of spillover events from flying-foxes to horses, may be influenced by urbanisation through the effects it has on habitat loss and through that, on resource distribution, connectivity between groups of flying-foxes, and increased interactions between flying-foxes and horses [35].

The negative impacts of flying-foxes, and the ever more strident calls for their 'control' that these impacts create, are at odds with the significance of their ecological role and conservation status. Flying-foxes are significant pollinators and seed dispersers in most vegetation types in their range [24,36]. Furthermore, two species, the spectacled flying-fox (Pteropus conspicillatus) and the grey-headed flying-fox (P. poliocephalus), are listed as vulnerable under Australia's Environment Protection and Biodiversity Conservation Act (1999). The resultant divergent and strongly held opinions on flying-foxes creates persistent tension between those who wish to see the animals conserved and those demanding they be controlled. Flying-fox management is a contentious and politicised issue in Australia.

To date, despite urbanised flying-foxes being a major management issue in northern and eastern Australia, there are few studies of urbanisation of flying-foxes. Identifying whether urbanisation of flying-fox populations is in fact occurring and the nature of its drivers is a fundamental step in developing effective management solutions. The aim of this study is to determine whether spectacled flying-foxes are urbanising and whether landscape features or change are associated with this. We use data from 15 years of monitoring of the spectacled flying-fox population to examine the spatial and temporal patterns of association of spectacled flying-fox camps with urban areas in the main part of their Australian range,

the Wet Tropics of north eastern Australia. Specifically we (1) ask whether spectacled flying-foxes are becoming more urbanised, (2) we test the hypotheses that any shift to urban areas is associated with anthropogenic changes to landscape structure or to an increase in the size or number of urban camps, and (3) whether the landscape characteristics of camp sites differ from those of the surrounding landscape and how this has changed over the study period.

Materials and Methods

Ethics Statement

This research was conducted under Animal Ethics Approvals from the CSIRO Ecosystem Sciences Animal Ethics Committee and complied with the Australian Code of Practice for the Care and Use of Animals for Scientific Purposes (2004; 2013). The work required no interaction with or handling of flying-foxes. The research was conducted under Scientific Purposes Permit #WTK03462308 and a 173P Authorisation from the Queensland Parks and Wildlife Service. The research was conducted on 50 separate land tenures. Details of the location of each camp have been lodged as a part of the National Flying-Fox Monitoring Program (NFFMP; http://www.environment.gov.au/node/ 16393) and camp locations, tenure, access information and camp sizes can be obtained from either from the NFFMP or from the authors.

Study Species

The spectacled flying fox is a phytophagous species, feeding primarily on floral resources and fruits in a wide range of vegetation communities, including closed forest, gallery forest, eucalypt open forest and woodland, coastal Melaleuca swamps, mangroves, vegetation in urban settings, and commercial fruit crops [37]. This highly mobile species forages at night and can disperse seeds and pollen over large distances [24]. By day the animals roost in camps, with an unknown proportion roosting solitarily or in small groups throughout the year [22]. In Australia, the spectacled flying-fox is found in the Wet Tropics of Queensland World Heritage Area between Townsville and Cooktown with small outlier populations north in the Iron and McIllwraith Ranges on Cape York [22,24] and to the south at Finch Hatton, near Mackay [38].

Study Site

The study included all known current and past spectacled flying-fox camps in the Wet Tropics region of Queensland, an area of approximately 9,000 km^2, on the north-east coast of Queensland (Figure 1). The region has a diverse, fragmented terrain of coastal plains and extensive uplands, typically 600–900 masl [39]. The vegetation is a complex mosaic of closed canopy rainforest and open eucalypt woodlands, with tropical savannas and grasslands on its drier margins and with clearing for agriculture on the wet fertile coastal floodplains, mid-montane tablelands and on the drier western slopes [39]. The region has a human population of approximately 252,000 [40]. Urbanisation is focused on several regional centres, Cairns, Mareeba, Atherton, Innisfail and Ingham, but there are also numerous small communities throughout the region.

Camp data

Data on the location and sizes of all flying-fox camps in the region were obtained during regular monitoring programs begun in 1998 and continuing today. Over the course of the study the number of camps surveyed increased from 30 to 50 as new camp

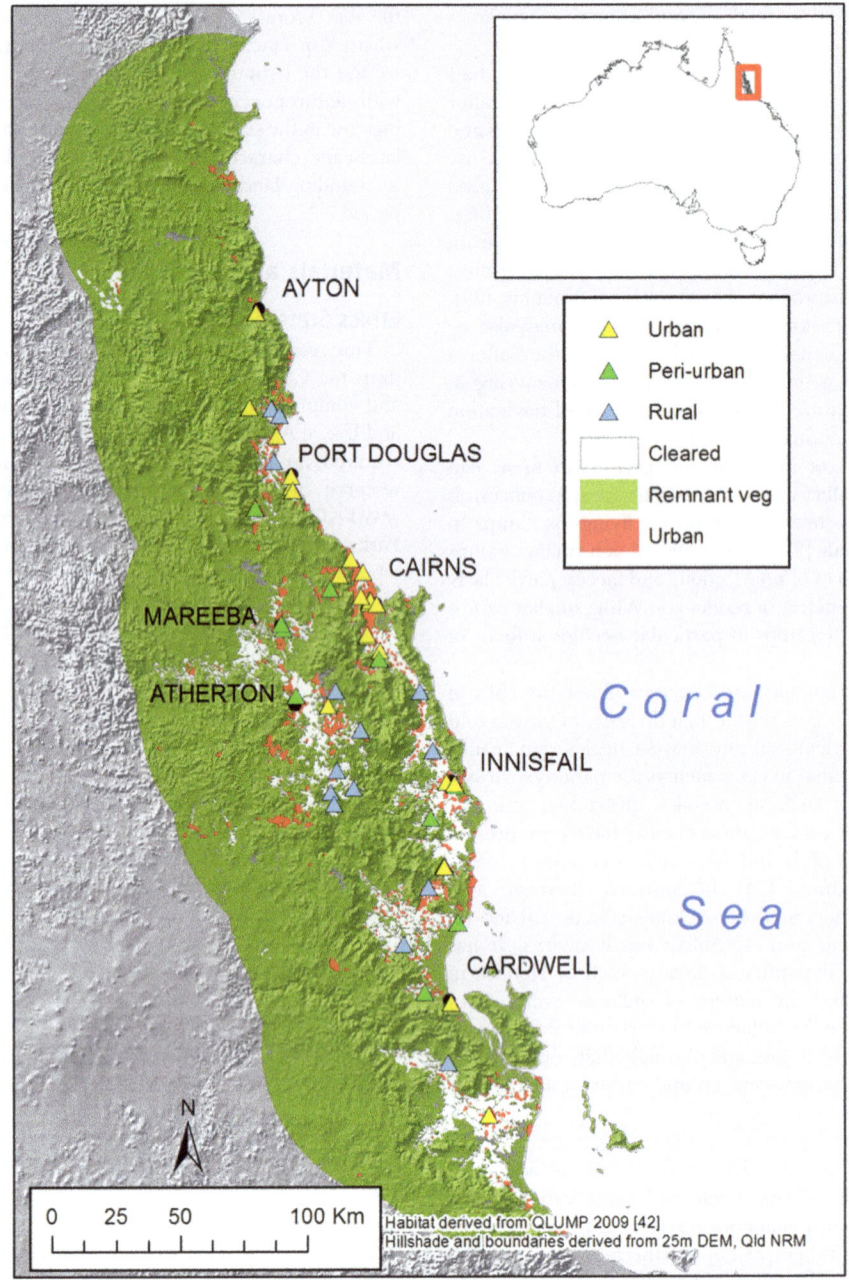

Figure 1. Location of the study area in the Wet Tropics Region of Northern Queensland, Australia. Spectacled flying-fox camps (triangles), towns (black dots) and urban areas (red shading) are also shown. Habitat mapping is derived from QLUMP 2009 [42] and the hillslopes and shading from the Qld. Dept. Natural Resources and Mines 25 m DEM (http://www.dnrm.qld.gov.au/mapping-data/data/topographic).

locations were identified. This increase resulted largely from the inclusion of historical camps. Once identified camps were not dropped from the surveys. These surveys were conducted under three programs. In 1998 and 1999 surveys were conducted in March and November while from 2000 to 2003 surveys were conducted in November only. These surveys were conducted by the Queensland Parks and Wildlife Service and involved positioning counters around the perimeter of camps to count the animals as they flew out of the camp at dusk, i.e. fly-out counts (see [41] for a more detailed description). Since May 2004 monthly, daytime, walk-through surveys of every camp in the study region have been conducted. In small camps (<1000 individuals),

surveyors attempted to count all flying-foxes in a camp. In larger camps the density of individuals was assessed by counting the number of roosting individuals in randomly-selected roost trees, the average of these was then extrapolated to give a camp size estimate by counting the number of roost trees. Each regional survey was typically completed within three consecutive days to minimise the chance of inter-camp movements and any resultant recounting of individuals [22]. The use of these different survey methods in the early and later phases of the data collection has the potential to introduce biases into the data. To avoid such issues our analyses do not rely on direct comparisons of abundance estimates derived from different methods. This is achieved either by restricting analyses to data derived from a single method or by calculating proportions based on samples collected using a single method, i.e. within years, and only comparing the proportions across years.

How urbanised is the spectacled flying-fox population in the Wet Tropics?

Count data from 1998–2012 was analysed to assess the proportion of the population associated with urban areas. Camps were assigned to three categories (urban, peri-urban and non-urban) depending on their distance from urban areas – defined as human habitation. Urban camps were defined as those surrounded by urban land use according to Queensland Government's land use classification (primary attributes), QLUMP 2009 [42] and on-ground assessment. Peri-urban camps were defined as those adjacent to urban land cover. Non-urban camps were those more than 250 m from any urban land use - a distance at which there is usually little concern about the presence of flying-foxes. When referring to urban and peri-urban camps together we use the term urban-associated.

To examine patterns of urbanisation across years we used data from all surveys, i.e. from 1998–2012. To examine seasonal patterns of camp use we restricted the analyses to the monthly data obtained between 2004 and 2012. To detect any seasonal trends in urbanisation, the monthly percentage of the population in urban-associated camps, consistency of occupation (proportion of surveys in which a camp was occupied), mean camp size and total population surveyed were analysed using monthly survey data from 2004 onwards. The proportion of the population using identified camps is known to vary through the year with a lower proportion in camps mid-year and a higher proportion during the warmer months when mating, birthing and raising of young occurs. We use the term population to refer to that part of the population using camps at any point in time.

Are spectacled flying-foxes being driven into urban areas by landscape change?

There were two QLUMP land use sampling periods during the time of the study, 1999 and 2009. These two sampling periods fell conveniently near the beginning and end of our study and so were used to document changes in land cover and use during the study period. QLUMP primary landuse attribute "intensive uses" was used as the first filter to determine urbanisation and the tertiary attribute was used to differentiate between non-urban uses, urban and rural residential uses. Percentage land cover assigned to particular land uses was extracted from QLUMP and overlaid with Regional Ecosystem data describing vegetation types [42] to produce the land cover categories: True urban, Rural Residential, Cleared, Rainforest, Sclerophyll and Other.

To determine whether areas surrounding camp sites had changed over time, landscape metrics which convey key information about landscape spatial structure were calculated in Fragstats 4.0 [43] for circular buffer zones with a diameter of 3.3 km around camps, this being half the average nearest neighbour distance between camps. These metrics were chosen to provide a description of the structure of the landscape in which camps occurred and were: forest mean patch area (MPA) a measure of the average size of forest patches in the buffer, forest patch density (PD) or the number of patches in the buffer, edge density (ED) or the length of patch edge as a function of area and a measure of the amount of interior patch habitat relative to edge habitat, and percentage urban cover (% urban) [44]. These measures for 1999 and 2009 were compared using Mann-Whitney U tests. In order

to test whether camps were found in more fragmented areas than would be expected by chance, buffers around camps and n = 33 random points in the landscape were compared with Mann-Whitney U tests.

Results

Surveys of spectacled flying-fox campsites across the Wet Tropics of Queensland from 1998–2012 indicate that the majority of the population were roosting in camps associated with urban areas (Figure 2). In each monthly survey between 5 and 16 camps were occupied (mean = 10, S.D. = 2.3). While a total of 30 camps were surveyed in the initial years and this number increased to 50 in the final years of the monitoring there was no significant increase over time in the number of camps found to be occupied each year ($r_p = 0.38$, p = 0.16, n = 15) (Table S1 in File S1). There was an increase over time in the number of urban camps occupied each year ($r_p = 0.64$, p<0.01, n = 15) but no significant trend in the numbers of peri-urban or non-urban camps (p>0.05 for both). The net result was an increase in urban-associated camps ($r_p = 0.5$, p<0.03, n = 15). The mean percentage of the population found in urban and peri-urban camps across the 111 surveys was 82% (±26 S.D.), though this varied across all surveys from 59% in the early years to 99% in the later years.

There was a distinct seasonal pattern in the proportion of the population that was encountered in urban-associated camps with this proportion consistently reaching high levels in May and June before declining slowly there after (Figure 3). The proportion of the population found in urban-associated camps was greatest during the period of the year when the population count was at its lowest, May–November (Figure 4), and was significantly greater than during the rest of the year (Mann-Whitney U test, Z = −4.95, p<0.0001). During this May–November period all camp types were less consistently occupied (Mann-Whitney U test, Non-urban: Z = −3.56, p<0.05; Urban-associated: Z = −3.58, p< 0.05), and had smaller mean camp sizes (Mann-Whitney U test, Non-urban: Z = 6.06, p<0.05; Urban-associated: Z = 4.64, p< 0.05), though the decrease in non-urban camp size is far more pronounced than that of the urban camps (Figure 5). Urban camps were occupied in 32% of month's surveyed (n = 16 camps), peri-urban camps 22% (n = 22) and non-urban camps 10% (n = 20).

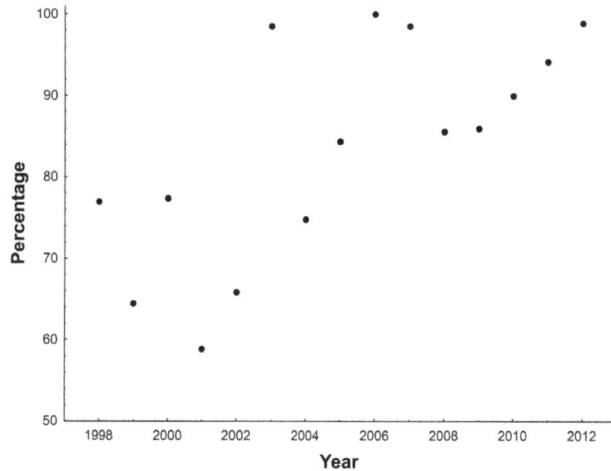

Figure 2. The percentage of the spectacled flying-fox population of the Wet Tropics found in urban associated camps during November surveys in each year of the study.

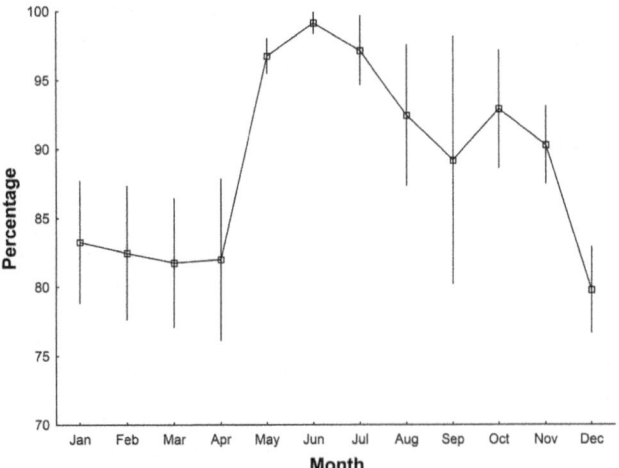

Figure 3. Changes through the year in the percentage (±S.E.) of the population occupying urban-associated camps. Monthly means calculated with survey data from 2004–2012.

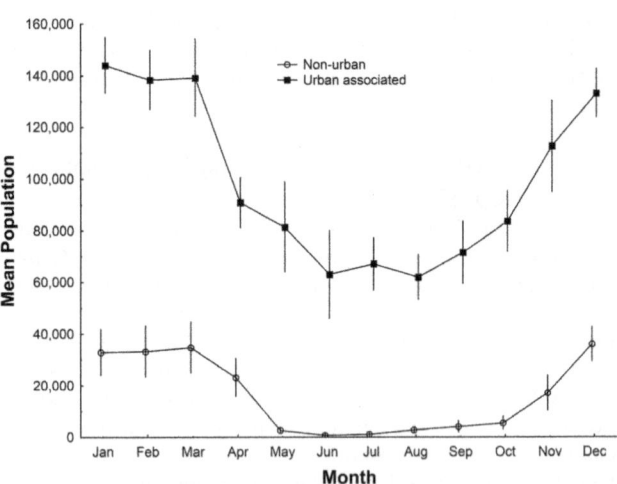

Figure 4. Changes through the year in the mean monthly population occupying urban associated and non-urban camps across the Wet Tropics calculated from 2004–2012 data. Monthly means calculated with survey data from 2004–2012.

There was an increase in the proportion of the population recorded using urban and peri-urban camps over the study period (Figure 2). This was the case irrespective of whether we considered the proportions recorded during November surveys only, the month that was surveyed in all years and in which the greatest proportion of the population are encountered in camps, ($r_p = 0.69$, p<0.005, n = 15), those for March and November surveys only, i.e. the months surveyed in all years except 2000–2003 ($r_p = 0.54$, p<0.01, n = 26), or for all months for which we have data ($r_p = 0.28$, p<0.005, n = 111). This trend appears to be due to a decrease in the number of non-urban camps used in any month (November only – r = −0.81, p<0.001, n = 15; March and November – r = −0.78, p<0.001, n = 26; all months – r = −0.57, <0.001, n = 111) and an increase in the number of urban camps used per month over time (November only - r = 0.7, p< 0.002, n = 15, [Figure 6], November and March - r = 0.75, p< 0.001, n = 26, all months – r = 0.56, p<0.001, n = 111).

A comparison of the landscape context of random points and the camps indicated that camps are significantly closer to urban areas than would be expected by chance (Wilcoxon signed rank test, Z = 33.26, p<0.0001). This effect did not arise during the study due to changes in where flying-foxes roosted. Individual camps did not move closer to urban areas over the period of the study. Nor did changes over time in the sub-set of camps occupied result in camps on average becoming closer to urban areas. Nor was the association of camps with urban areas due to urban expansion during the study. Urban land cover increased over the period of 1999 to 2009 from 1.3% to 1.4%, however this was not a statistically significant effect at the scale of the region (Table S2 in File S1). While there was a greater percentage of urban land cover around camps than around randomly chosen points in the landscape (Mann-Whitney U test, Z = 5.41, p<0.0001; Figure 7), comparison of land cover in 1999 and 2009 indicates no change in the extent of urban land use in the areas surrounding recorded camp locations (Mann-Whitney U test, Z = 0.31, p = 0.76). Similarly, areas surrounding occupied campsites did not become more fragmented over time than the randomly chosen locations, with no significant changes in MPA, PD or ED over time (p>0.05 in all cases). Despite this, camps do occur in locations that are more fragmented than random points in the landscape, with a smaller MPA, a significantly greater ED and PD (Mann-Whitney

U tests, Z = 4.11, 3.54 and 4.55, respectively, p<0.001 for all; Figure 7).

Discussion

We found that spectacled flying-foxes commonly roost near humans, as is the case for other *Pteropus* species in Australia, e.g. *P. poliocephalus*, *P. alecto*, *P. scapulatus* [20,21,23], and elsewhere, e.g. *P. giganteus* [45], and *P. dasymallus* [46]. Over the period of our study the majority of the spectacled flying-foxes found in camps were found in urban-associated camps, these camps were more consistently occupied than non-urban camps, and the proportion of the counted population encountered in urban camps increased. Furthermore, while the number of non-urban camps declined through the period of the study, the number of urban camps increased.

Of the hypotheses proposed for the association between flying-foxes and urban areas, our data only allow direct assessment of hypotheses related to the effect of changes in landscape structure.

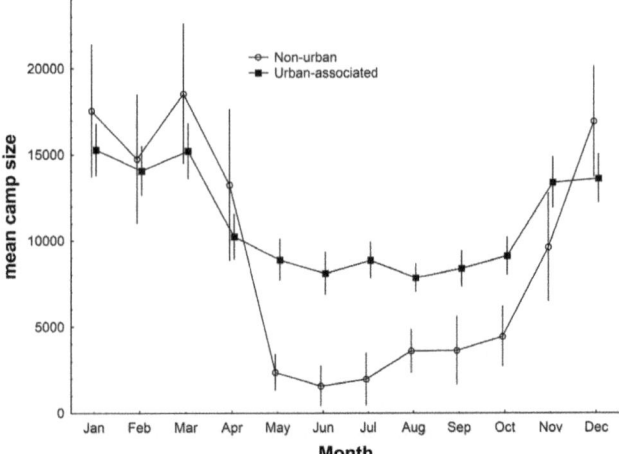

Figure 5. Changes through the year in the mean monthly camp size for urban-associated and non-urban camps. Calculated from 2004–2012 data.

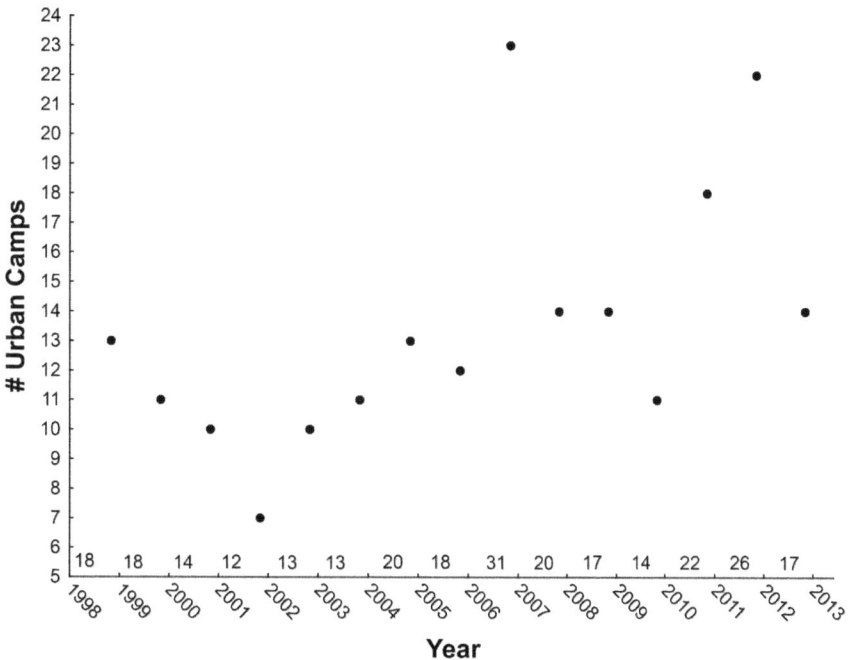

Figure 6. The number of urban associated camps recorded in surveys conducted in November of each year of the study. The number of occupied camps recorded in each year is indicated on the x axis.

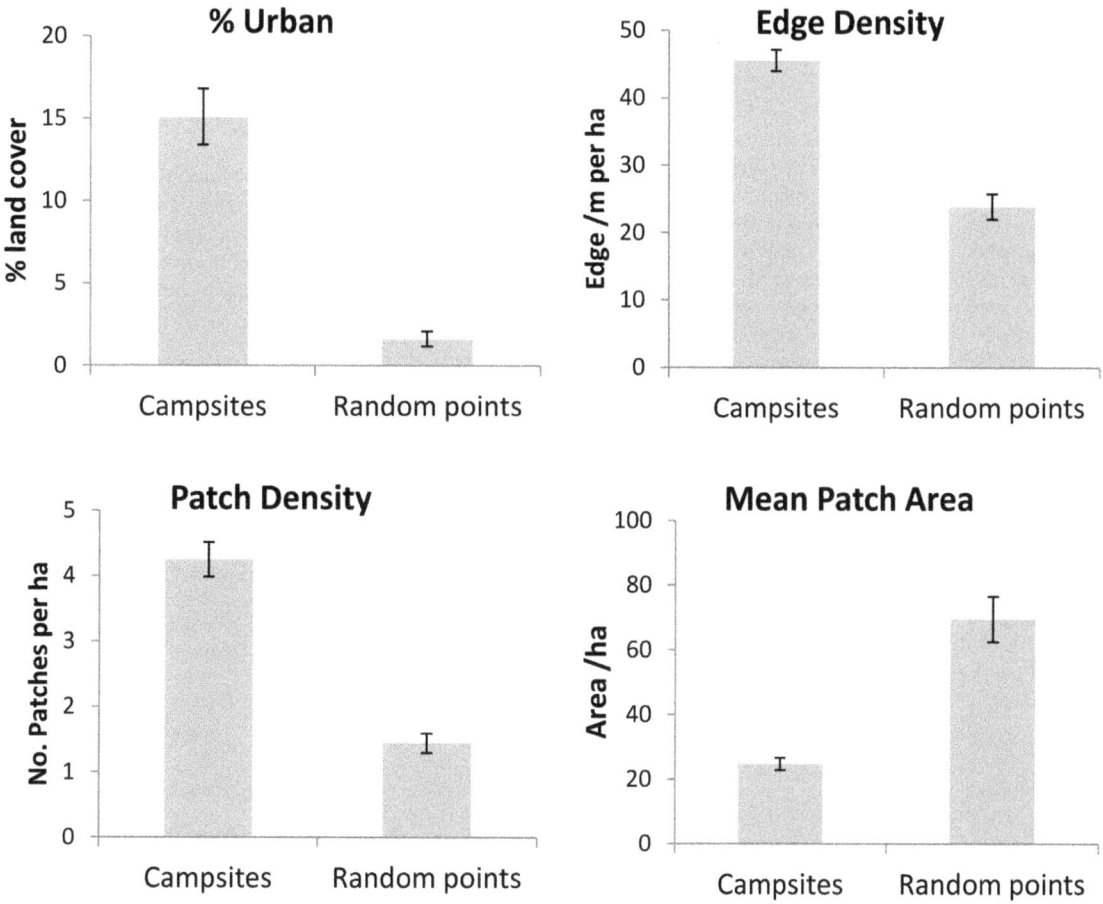

Figure 7. Landscape metrics (mean, ±SE) for buffer zones with a 3.3 km diameter surrounding all recorded campsites and random points in the landscape showing: percentage urban land cover, edge density, patch density and mean patch area.

Given that we found i) no significant changes in land cover at the scale of the landscape as a whole or, ii) in the immediate vicinity of camps, and, iii) no change in the proximity of individual camps to urban areas over the duration of the study, our findings for spectacled flying-foxes do not support the hypotheses that urbanisation is occurring due to urban expansion towards camps, habitat loss, roost site loss or habitat fragmentation [20,47,48,49]. Whether this also holds for other *Pteropus* species remains to be seen.

Our data does allow us to indirectly consider some alternative hypotheses for flying-fox urbanisation. Peaks in the urban-associated percentage of the population in 1998, 2000, 2003, 2006–7 and 2011–12 and are apparent in the November, November and March and all months' samples. The peaks suggest that there are periodic shifts in the distribution of the population towards urban areas. Although the drivers of these temporary shifts are unknown, explanations could include the attraction of fruiting or flowering events near urban areas [24], or disturbance events such as droughts [28], fires [29], cyclones [22,30,31], and human disturbance or culling at non-urban roosts or orchards [24]. While there were two major cyclonic disturbances during the study period, in 2006 and 2010 and preceding the peaks in 2007 and 2011 respectively, no disturbance events can be associated with the other peak years. Significantly, these cyclones also precede peaks in the number of occupied camps (Table S1 in File S1). Furthermore, changes in permitting for mitigation of flying-fox damage in orchards mean that there has been a decrease in disturbance and culling at non-urban associated camps over the period of the study [37]. Rather than being driven by increasing or episodic disturbance, the long-term trend of increasing urbanisation since 1998 suggests a longer-term population shift towards urban areas is occurring.

Another explanation for these results could be bias in the sampling of urban and non-urban camps; urban camps are easier to locate, access and survey and therefore potentially more likely to be monitored. Such bias is unlikely to explain our results for the following reasons. First, camps are never dropped from our monitoring, even when they haven't been occupied for long periods, so there is no shift of monitoring to an urban focus that might result from easier sampling. Second because non-urban camps are more likely to be overlooked than urban camps, we would expect an initial bias towards urban camps that would decrease as the population became better known through broad-scale searches, reports from the public [22] and telemetry studies [36,50]. Instead we have seen the shift towards urban camps increase. Consequently, we feel confident that our results are not due to a sampling bias favouring urban camps.

Our data shows a seasonal pattern of change in the size of the urban-associated proportion of the population encountered in camps, with a greater proportion found in these camps from May–Nov than Dec–Apr (Fig. 4). This period is also when the population count is at its lowest, and camps of all types are less reliably occupied and have a smaller mean size. It is possible that

without strong social reasons for aggregating (at this time mating has finished and females are pregnant), the animals disperse, possibly in association with reduced or more widely dispersed foraging resources. The decrease in non-urban camp size at this time of the year was far more pronounced than that of urban-associated camps, suggesting urban-associated camps may be core population centres (Fig. 5).

The urbanisation of spectacled flying-foxes documented here has significant implications for how the management of this, and other flying-fox species, is approached. Our results support the suggestion that flying-foxes are becoming increasingly urbanised and suggest that the conflict their presence in urban areas engenders is not going to go away. The lack of evidence for loss of habitat or roosting sites as a driver of this shift further suggests that spectacled flying-foxes are not being forced into urban areas, raising the possibility that their move is a behavioural response to the advantages offered by such locations. If this is the case then it is difficult to argue that moving problem urban camps on through the use of disturbance is likely to have any significant negative impacts on the population. Despite this, there is little evidence that past attempts to move urban camps have been successful or cost effective [23] and newspaper reports from Far North Queensland over the last century suggest that the common use of lethal methods was ineffective in deterring spectacled flying-foxes from urban areas (DAW, unpubl. data). We believe that this points to a need to explore new management options, particularly the identification of options that facilitate the co-existence of humans and flying-foxes. Management of the human side of the conflict is likely to prove more cost effective and successful. Identifying the actual drivers of urbanisation of flying-foxes will be significant for understanding and managing this process.

Supporting Information

File S1 Table S1. Patterns of occupancy of camps of different types over the duration of the study. **Table S2.** Proportion of the landscape in each landcover category across the period of the study, 1999 and 2009. There are no significant differences between the proportions in each category in 1999 and 2009 (p = 0.05).

Acknowledgments

Louise Shilton contributed to data collection. This paper was improved by comments from J. Parsons, C. Pavey and three anonymous reviewers.

Author Contributions

Conceived and designed the experiments: JT AM HLPB DAW. Performed the experiments: AM DAW. Analyzed the data: JT AM DAW. Contributed reagents/materials/analysis tools: JT AM HLPB DAW. Wrote the paper: JT AM HLPB DAW.

References

1. Seto KC, Fragkias M, Güneralp B, Reilly MK (2011) A Meta-Analysis of Global Urban Land Expansion. PLoS One 6: e23777.
2. Magle SB, Hunt VM, Vernon M, Crooks KR (2012) Urban wildlife research: Past, present, and future. Biol. Conserv. 155: 23–32.
3. Pickett STA, Cadenasso ML, Grove JM, Boone CG, Groffman PM, et al. (2011) Urban ecological systems: Scientific foundations and a decade of progress. Journal of Environmental Management 92: 331–362.
4. Moller AP, Diaz M, Flensted-Jensen E, Grim T, Ibanez-Alamo JD, et al. (2012) High urban population density of birds reflects their timing of urbanization. Oecologia 170: 867–875.
5. Davis A, Taylor CE, Major RE (2012) Seasonal abundance and habitat use of Australian parrots in an urbanised landscape. Landsc. Urban Plan. 106: 191–198.
6. Brearley G, Rhodes J, Bradley A, Baxter G, Seabrook L, et al. (2013) Wildlife disease prevalence in human-modified landscapes. Biol. Rev. Cambridge Philosophic. Soc. 88: 427–442.
7. Bradley CA, Altizer S (2007) Urbanization and the ecology of wildlife diseases. Trends Ecol. Evol. 22: 95–102.
8. Johnston R (2001) Synanthropic birds of North America. In: Marzluff J, Bowman R, Donnelly R, editors. Avian ecology and conservation in an urbanizing world. Massachusetts: Kluwer Academic Publishers. 49–69.

9. Puth LM, Burns CE (2009) New York's nature: a review of the status and trends in species richness across the metropolitan region. Divers. Distrib. 15: 12–21.

10. DeStefano S, DeGraaf RM (2003) Exploring the ecology of suburban wildlife. Frontiers in Ecology and the Environment 1: 95–101.

11. Davis A, Taylor CE, Major RE (2011) Do fire and rainfall drive spatial and temporal population shifts in parrots? A case study using urban parrot populations. Landsc. Urban Plan. 100: 295–301.

12. Slabbekoorn H, Peet M (2003) Birds sing at a higher pitch in urban noise: Great tits hit the high notes to ensure that their mating calls are heard above the city's din. Nature Clim. Change 424: 267–267.

13. McKinney RA, Raposa KB (2013) Factors influencing expanded use of urban marine habitats by foraging wading birds. Urban Ecosystems 16: 411–426.

14. Jokimäki J, Kaisanlahti-Jokimäki M-L, Suhonen J, Clergeau P, Pautasso M, et al. (2011) Merging wildlife community ecology with animal behavioral ecology for a better urban landscape planning. Landsc. Urban Plan. 100: 383–385.

15. Zuckerberg B, Bonter DN, Hochachka WM, Koenig WD, DeGaetano AT, et al. (2011) Climatic constraints on wintering bird distributions are modified by urbanization and weather. Journal of Animal Ecology 80: 403–413.

16. Urbanek RE, Allen KR, Nielsen CK (2011) Urban and suburban deer management by state wildlife-conservation agencies. Wildlife Society Bulletin 35: 310–315.

17. Forman R, Sperling D, Bissonette J, Clevenger A, Cutshall C (2003) Road ecology: science and solutions. Washington DC: Island Press. 504 p.

18. Curtis P, Hadidian J (2010) Responding to human–carnivore conflicts in urban areas. In: Gerth S, Riley S, Cypher B, editors. Urban carnivores: ecology, conflict, and conservation. Baltimore: The Johns Hopkins University Press. 201–212.

19. Redpath SM, Young J, Evely A, Adams WM, Sutherland WJ, et al. (2013) Understanding and managing conservation conflicts. Trends Ecol. Evol. 28: 100–109.

20. Markus N, Hall L (2004) Foraging behaviour of the black flying-fox (Pteropus alecto) in the urban landscape of Brisbane, Queensland. Wildl. Res. 31: 345–355.

21. Parris KM, Hazell DL (2005) Biotic effects of climate change in urban environments: The case of the grey-headed flying-fox (Pteropus poliocephalus) in Melbourne, Australia. Biol. Conserv. 124: 267–276.

22. Shilton LA, Latch PJ, McKeown A, Pert P, Westcott DA (2008) Landscape-scale redistribution of a highly mobile threatened species, Pteropus conspicillatus (Chiroptera, Pteropodidae), in response to Tropical Cyclone Larry. Austral Ecology 33: 549–561.

23. Roberts BJ, Eby P, Catterall CP, Kanowski J, Bennett G (2011) The outcomes and costs of relocating flying-fox camps: insights from the case of Maclean, Australia. In: Law B, Eby P, Lunney D, Lumsden L, editors. The Biology and Conservation of Australasian Bats. Mosman, NSW, Australia: Royal Zoological Society of NSW. 277–287.

24. Hall L, Richards G (2000) Flying foxes: Fruit and Blossom Bats of Australia. Sydney, NSW: University of NSW.

25. Eby P (1991) Seasonal movements of gray-headed flying-foxes, Pteropus poliocephalus (Chiroptera, Pteropodidae), from 2 maternity camps in northern New South Wales. Wildl. Res. 18: 547–559.

26. Eby P, Richards G, Collins L, Parry-Jones K (1999) The distribution, abundance and vulnerability to population reduction of a nomadic nectarivore, the grey-headed flying-fox Pteropus poliocephalus in New South Wales, during a resource concentration. Australian Zoologist 31: 240–253.

27. Parry-Jones KA, Augee ML (2001) Factors affecting the occupation of a colony site in Sydney, New South Wales by the Grey-headed Flying-fox Pteropus poliocephalus (Pteropodidae). Austral Ecology 26: 47–55.

28. Tidemann CR, Nelson JE (2004) Long-distance movements of the grey-headed flying fox (Pteropus poliocephalus). Journal of Zoology 263: 141–146.

29. Jenkins RKB, Andriafidison D, Razafimanahaka HJ, Rabearivelo A, Razafindrakoto N, et al. (2007) Not rare, but threatened: the endemic Madagascar flying fox Pteropus rufus in a fragmented landscape. Oryx 41: 263–271.

30. Craig P, Trail P, Morrell TE (1994) The decline of fruit bats in American Samoa due to hurricanes and overhunting. Biol. Conserv. 69: 261–266.

31. Esselstyn JA, Amar A, Janeke D (2006) Impact of Post-typhoon Hunting on Mariana Fruit Bats (Pteropus mariannus) 1. Pacific science 60: 531–539.

32. Meade J, Biro D, Guilford T (2005) Homing pigeons develop local route stereotypy. Proceedings of the Royal Society B: Biological Sciences 272: 17.

33. Halpin K, Hyatt AD, Plowright RK, Epstein JH, Daszak P, et al. (2007) Emerging viruses: Coming in on a wrinkled wing and a prayer. Clinical Infectious Diseases 44: 711–717.

34. Drexler JF, Corman VM, Muller MA, Maganga GD, Vallo P, et al. (2012) Bats host major mammalian paramyxoviruses. Nat Commun 3: 796.

35. Plowright RK, Foley P, Field HE, Dobson AP, Foley JE, et al. (2011) Urban habituation, ecological connectivity and epidemic dampening: the emergence of Hendra virus from flying foxes (Pteropus spp.). Proc. R. Soc. B-Biol. Sci. 278: 3703–3712.

36. Westcott DA, Dennis AJ, McKeown A, Bradford M, Margules C (2001) The Spectacled Flying fox, Pteropus conspicillatus, in the context of the world heritage values of the Wet Tropics World Heritage Area. Atherton, Queensland: CSIRO. 75 p.

37. Queensland Department of Environment and Resource Management (2010) National recovery plan for the spectacled flying fox Pteropus conspicillatus. Report to the Department of Sustainability, Environment, Water, Population and Communities, Canberra.

38. Parsons JG, Blair D, Luly J, Robson SKA (2009) Bat Strikes in the Australian Aviation Industry. J. Wildl. Manage. 73: 526–529.

39. Stork N, Turton S (2009) Living in a dynamic tropical forest Landscape. Oxford: Blackwell Publishing Ltd.

40. Queensland Treasury and Trade (2012) Population and Dwelling Profile, Far North Queensland. Office of Economic and Statistical Research.

41. Garnett S, Whybird O, Spencer HJ (1999) The Conservation status of the Spectacled Flying-fox Pteropus conspicillatus in Australia. Australian Zoologist 31: 38–54.

42. Queensland Government Information Service (2012).

43. McGarigal K, Cushman S, Ene E (2012) FRAGSTATS v4: Spatial Pattern Analysis Program for Categorical and Continuous Maps.

44. Perotto-Baldivieso HL, Melendez-Ackerman E, Garcia MA, Leimgruber P, Cooper SM, et al. (2009) Spatial distribution, connectivity, and the influence of scale: habitat availability for the endangered Mona Island rock iguana. Biodiversity and Conservation 18: 905–917.

45. Mahmood-ul-Hassan M, Gulraiz TL, Rana SA, Javid A (2010) The diet of Indian flying-foxes (Pteropus giganteus) in urban habitats of Pakistan. Acta Chiropt. 12: 341–347.

46. Nakamoto A, Kinjo K, Izawa M (2007) Food habits of Orii's flying-fox, Pteropus dasymallus inopinatus, in relation to food availability in an urban area of Okinawa-jima Island, the Ryukyu Archipelago, Japan. Acta Chiropt. 9: 237–249.

47. Williams NSG, McDonnell MJ, Phelan GK, Keim LD, Van der Ree R (2006) Range expansion due to urbanization: Increased food resources attract Grey-headed Flying-foxes (Pteropus poliocephalus) to Melbourne. Austral Ecology 31: 190–198.

48. Lunney D, Moon C (1997) Flying-foxes and their camps in the rainforest remnants of north-east NSW. In: Dargavel J, editor. Australia's everchanging forests III. Canberra: Centre for Resource and Environmental Studies, Australian National University. 247–277.

49. van der Ree R, McDonnell MJ, Temby I, Nelson J, Whittingham E (2006) The establishment and dynamics of a recently established urban camp of flying foxes (Pteropus poliocephalus) outside their geographic range. Journal of Zoology 268: 177–185.

50. McKeown A, Westcott DA (2012) Assessing the accuracy of small satellite transmitters on free-living flying-foxes. Austral Ecology 37: 295–301.

19

Global Agricultural Land Resources – A High Resolution Suitability Evaluation and Its Perspectives until 2100 under Climate Change Conditions

Florian Zabel*, Birgitta Putzenlechner, Wolfram Mauser

Department of Geography, Ludwig Maximilians University, Munich, Germany

Abstract

Changing natural conditions determine the land's suitability for agriculture. The growing demand for food, feed, fiber and bioenergy increases pressure on land and causes trade-offs between different uses of land and ecosystem services. Accordingly, an inventory is required on the changing potentially suitable areas for agriculture under changing climate conditions. We applied a fuzzy logic approach to compute global agricultural suitability to grow the 16 most important food and energy crops according to the climatic, soil and topographic conditions at a spatial resolution of 30 arc seconds. We present our results for current climate conditions (1981–2010), considering today's irrigated areas and separately investigate the suitability of densely forested as well as protected areas, in order to investigate their potentials for agriculture. The impact of climate change under SRES A1B conditions, as simulated by the global climate model ECHAM5, on agricultural suitability is shown by comparing the time-period 2071–2100 with 1981–2010. Our results show that climate change will expand suitable cropland by additionally 5.6 million km^2, particularly in the Northern high latitudes (mainly in Canada, China and Russia). Most sensitive regions with decreasing suitability are found in the Global South, mainly in tropical regions, where also the suitability for multiple cropping decreases.

Editor: Juergen P. Kropp, Potsdam Institute for Climate Impact Research, Germany

Funding: This research was carried out within the framework of the GLUES (Global Assessment of Land Use Dynamics, Greenhouse Gas Emissions and Ecosystem Services) Project, which has been supported by the German Ministry of Education and Research (BMBF) program on sustainable land management (FKZ 01LL0901E). (http://modul-a.nachhaltiges-landmanagement.de/en/). The funders had no role in study design, data collection and analysis, decision to publish, or preparation of the manuscript.

Competing Interests: The authors have declared that no competing interests exist.

* Email: f.zabel@lmu.de

Introduction

Natural constraints are limiting the land's suitability for agriculture and cultivation practices. They consist of prevailing local climatic, soil and topographic conditions determining the available energy, water and nutrient supply for agricultural crops. Besides natural conditions, complex interactions of social, economic, political, and cultural aspects determine whether and how land is used for agriculture. Agricultural land has become one of the largest terrestrial biomes on the planet, occupying approx. 40% of the land surface [1]. Thereby, a variety of different land use types and intensities determine heterogeneously distributed patterns, including e.g. the choice of crop varieties, irrigation practices, fertilization, terracing and the level of technological input [2]. Thus, natural constraints are to a limited extent suspended by human actions [3].

The demand for agricultural products is expected to increase by 70–110% by 2050, driven by a projected world population of 9 billion people, increasing meat consumption and a growing use for bio-based materials and biofuel [4–15].

An increase in agricultural production can be accomplished by agricultural intensification and expansion, while considering social and environmental externalities and changing climate conditions [5,16]. Bruinsma [16] concluded that additionally 1.2 million km^2 of converted land are projected to be necessary until 2030 and another 5% up to 2050 with most land expected to be transformed in South America and Sub Saharan Africa, while latest studies project an increase of cropland between 10-25% by 2050 compared to 2005 for different socio-economic and climate scenarios [17]. Nonetheless, the expansion of agricultural land into forested or protected areas must be viewed critically, in order to conserve valuable ecosystem services e.g. for regulating climate or conserving biodiversity [5–8].

Changing patterns of temperature and precipitation and man-made degradation affect the suitability of land for agricultural use. For example, 19-23 ha of suitable land are lost per minute due to soil erosion and desertification [18,19]. Additionally, the area of suitable land is decreasing due to urbanization, with an estimate of 1.5 million km^2 until 2030 [20,21].

When focusing on the natural potentials of land for agricultural use, suitability analyses give local evidence on todays and future availability and quality. Thus, they help answering questions for managing a transition towards a more environmentally efficient and sustainable land use and involve better information on the global scale impacts of land use decisions [1].

Table 1. List of investigated food, feed and energy crops.

Crop name
Barley (*hordeum vulgare*)
Cassava (*manihot esculenta*)
Groundnut (*arachis hypogaea*)
Maize (*zea mays*)
Millet (*pennisetum americanum*)
Oil palm (*elaeis guineensis*)
Potato (*solanum tuberosum*)
Rapeseed (*brassica napus*)
Paddy rice (*oryza sativa*)
Rye (*secale cereale*)
Sorghum (*sorghum bicolor*)
Soy (*glycine maximum*)
Sugarcane (*saccharum officinarum*)
Sunflower (*helianthus annus*)
Summer wheat (*triticum aestivum*)
Winter wheat (*triticum gestivum*)

The relationship between climate, soil, topography and agricultural suitability has long been recognized. As such, suitability analysis combine heterogeneous soil, terrain and climate information and determine whether specific crop requirements are fulfilled under the given local conditions and assumptions. A variety of regional suitability studies for specific crops exist [22–28], while only a few exist on a global scale and for a broad variety of crops [3,29–31].

In the meantime, global soil and topography data are available at high spatial resolution and global climate models have improved their capabilities and spatial resolution. Previous analysis showed that questions of scale play a major role in suitability analysis as coarse data affect the validity of results [32]. In this context, we present our results in modelling global crop-suitability using a fuzzy logic approach at a spatial resolution of 30 arc seconds. The results of this approach include the potentially suitable area for agriculture differentiated for 16 crops for rainfed and irrigated conditions, the start of the growing cycles and the number of crop cycles. We analyze global distribution of agricultural suitability and changes until 2100 considering the numbers of crop cycles.

Thereby, we identify changes, opportunities and challenges in global agriculture related to the expansion of agricultural land competing with protected and forested areas as ecosystem services.

Material and Methods

Local climate, soil and topography determine the natural suitability of land for agricultural use. Thereby, the climatic, soil and topographic requirements may vary over a wide range of different agricultural crops. This analysis investigates the suitability for the following 16 crops that are most important for the global economy, food security and biofuel issues (see Table 1).

We aggregated the world into 23 regions in order to regionally analyse the results (see Fig. 1). We applied a fuzzy-logic approach [33,34] in order to calculate the crops' suitability on the globe at a spatial resolution of 30 arc seconds (0.00833°, approx. 1 km^2 at the equator). The length of the growing cycle (*lgc*) and the '*membership functions*' that describe the crop-specific requirements for each of the crops during the growing period (Fig. 2) are derived from [35].

The membership functions representing climate constraints describe the degree of membership of each selected crop with regard to mean temperature and total precipitation during its

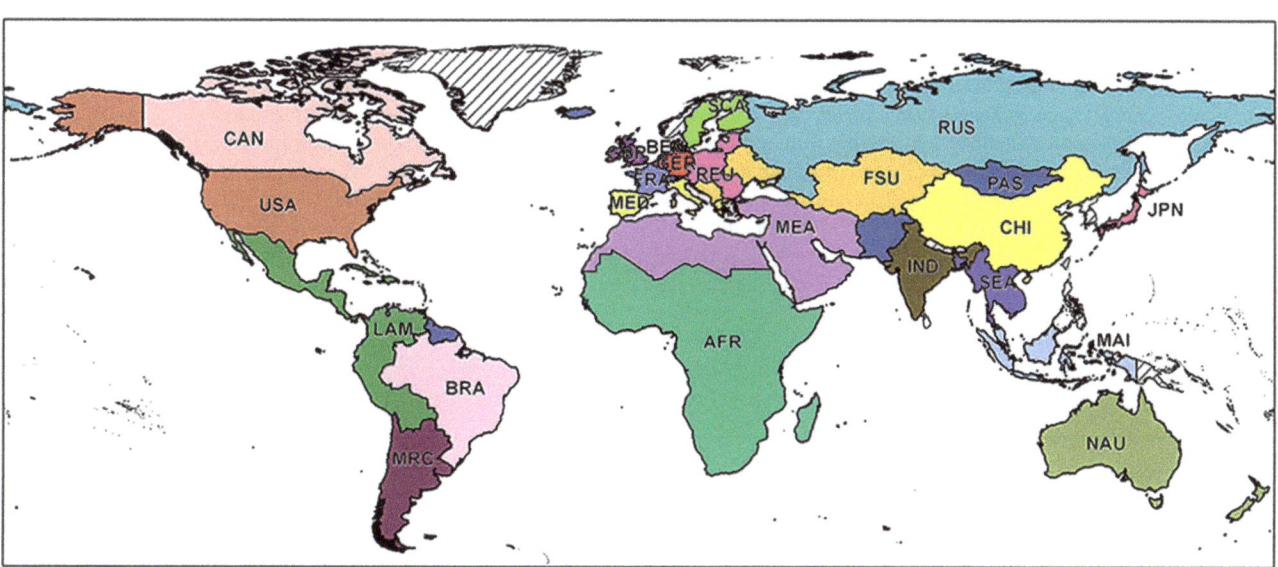

Figure 1. Map of the 23 world regions: AFR (Sub Saharan Africa), ANZ (Australia, New Zealand), BEN (Belgium, Netherlands, Luxemburg), BRA (Brazil), CAN (Canada), CHN (China), FRA (France), FSU (Rest of Former Soviet Union and Rest of Europe), GBR (Great Britain), GER (Germany), IND (India), JPN (Japan), LAM (Rest of Latin America), MAI (Malaysia, Indonesia), MEA (Middle East, North Africa), MED (Italy, Spain, Portugal, Greece, Malta, Cyprus), PAC (Paraguay, Argentina, Chile, Uruguay), ROW (Rest of the World), REU (Austria, Estonia, Latvia, Lithuania, Poland, Hungary, Slovakia, Slovenia, Czech Republic, Romania, Bulgaria), RUS (Russia), SCA (Finland, Denmark, Sweden), SEA (Cambodia, Laos, Thailand, Vietnam, Myanmar, Bangladesh), USA (United States of America).

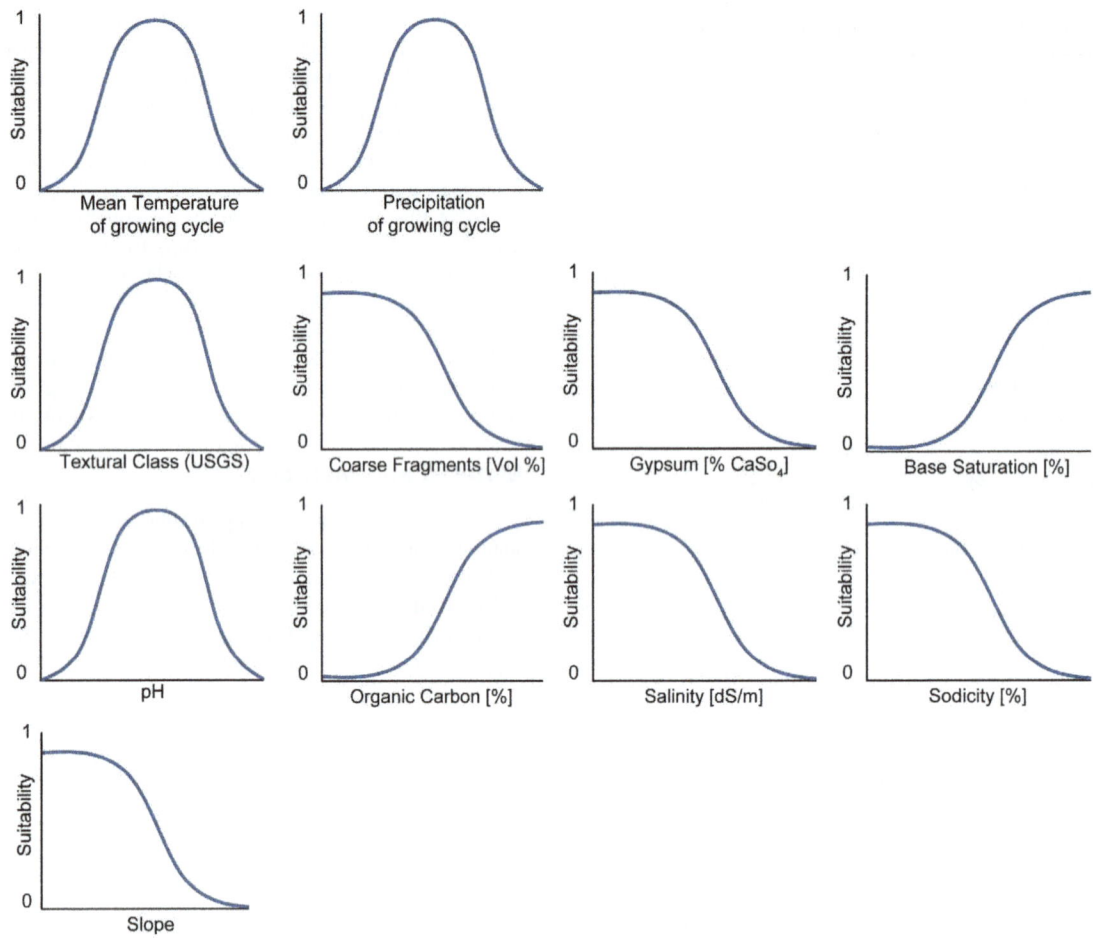

Figure 2. Membership functions for climatic, soil and topographic conditions.

respective growing cycle. Depending on the crop, membership functions have different curves according to [35]. Three shapes are in principle possible: 'more is better', 'less is better' and 'optimum'. For temperature e.g., the suitability is increasing from a minimum towards an optimal temperature and again decreasing until a maximum temperature is reached (Fig. 2). Eight soil parameters are considered: texture, proportion of coarse fragments and gypsum, base saturation, pH content, organic carbon content, salinity, and sodicity. Terrain is considered by the slope. The fuzzy-logic approach calculates *fuzzy values* based on the ecological rules (between 0 and 1), which determine the crops'

suitability on a specific location by the lowest membership value of all parameters.

An overview of the applied global datasets is given in Table 2. The climate data applied in this study are outputs from the global circulation model ECHAM5 of the Max-Planck Institute for Meteorology (MPI-M) [36,37]. It uses radiative forcing, sea surface temperature and sea ice concentrations from a 20th century/ SRES A1B scenario simulation. The 6-hourly dataset (temperature, precipitation) are converted to daily values for the climate period of 1981–2010 and 2071–2100. The daily data is spatially downscaled from its original resolution of 0.56° to 0.00833° (30

Table 2. Applied global datasets.

Parameter	Source	Detailed Description
Climate	ECHAM5	[37]
Soil	Harmonized World Soil Database (HWSD)	[41]
Topography	Space Shuttle Topography Mission (SRTM)	[39]
Crop-requirements	FAO Land Evaluation Part III: Crop Requirements	[35]
Irrigation	Global Map of Irrigation Areas (GMIA) v5.0	[44]
Protected Areas	International Union for Conservation of Nature (IUCN) Protected Areas	[45]
Forested Areas	GlobCover 2009	[46]

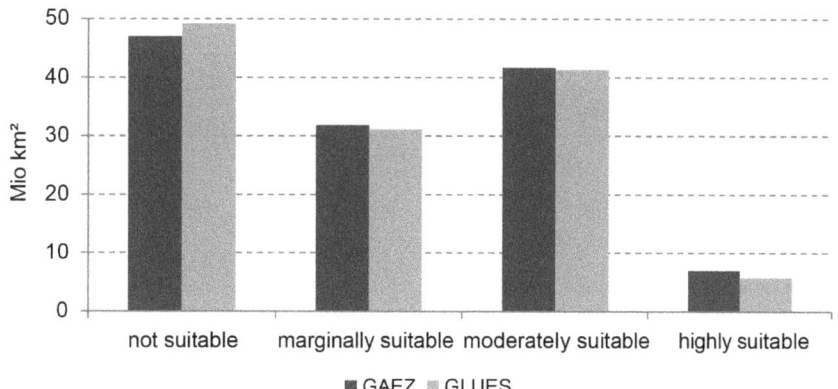

Figure 3. Global comparison of agriculturally suitable area between GAEZ (Baseline period 1961–1990) and GLUES (1961–1990).

arc seconds), based on an approach by [38], using sub-grid terrain information provided by the SRTM-dataset [39]. A bias-correction is executed during the downscaling procedure for temperature and precipitation based on monthly derived factors from the WorldClim dataset [40].

Mean temperature ($\bar{T}$) and total precipitation ($\bar{P}$) are calculated over the length of the growing cycle for each day of the year (doy) (see eq. 1 and eq. 2). Starting on the 1st of January ($doy = 1$), the growing cycle is shifted day by day until the 31st of December ($doy = 365$). The suitability value (S) is calculated for each doy as in eq. 3 for $\bar{T}$ and $\bar{P}$ according to the membership function (mf).

$$\bar{T}_{doy} = \overline{T_{doy}, \ldots, T_{doy+lgc}} \qquad (eq.1)$$

$$\bar{P}_{doy} = \sum_{doy}^{doy+lgc} P \qquad (eq.2)$$

$$S(\bar{T}_{doy}) = mf\left(\bar{T}_{doy}\right) \; ; \; S(\bar{P}_{doy}) = mf\left(\bar{P}_{doy}\right) \qquad (eq.3)$$

Since the natural suitability of crop growth is limited by the minimum value, the smaller value of the temperature and precipitation fuzzy value determines the climate suitability $S(C)$

which is calculated for each doy (eq. 4).

$$S(C_{doy}) = min\left\{S(\bar{T}_{doy}), S(\bar{P}_{doy})\right\} \qquad (eq.4)$$

Among all daily fuzzy values of $S(C)$ within the year, the maximum of $S(C)$ determines the climate suitability over the growing cycle and thus, the optimal start of the growing cycle (eq. 5) for cultivation of a single crop within the entire growing season.

$$S(C_{start\ of\ the\ growing\ cycle}) = max\left\{S(C_{doy\ 1}), \ldots, S(C_{doy\ 365})\right\} \qquad (eq.5)$$

In order to allow for the calculation of multiple cropping, the fuzzy values for each possible combination of days for the start of the growing cycle are tested as to how often they would fit within one year. The number of multiple cropping is selected that generates the highest accumulated value. Multiple cropping and the start of the growing cycle(s) are obtained for single, double and triple cropping. Hereby, the start of the growing cycle(s) in the context of this paper describes an optimal time for cultivation of a crop to reach the maximum suitability within a year. Crop mixing is not considered. Regarding temporal demands for technical field work, we assume a break of two weeks between crop cycles.

Moreover, the following assumptions are made: At least 20 mm of precipitation are required within the first two weeks of the growing season in order to provide enough soil moisture for germination. No day within the growing period must be below 5°C and below 1°C for winter crops. Vernalisation requirements are considered separately from the growing period for winter

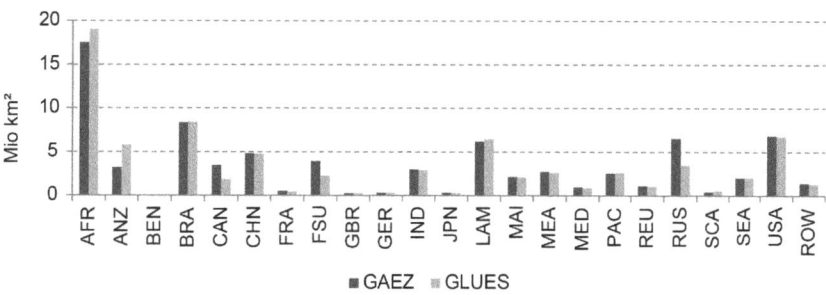

Figure 4. Comparison of total agriculturally suitable area of GAEZ (Baseline period 1961–1990) and GLUES (1961–1990) for different regions.

Table 3. Classified suitability considering rainfed conditions (1981–2010).

Not Suitable	Marginally Suitable	Moderately Suitable	Highly Suitable
49.8 million km^2	30.6 million km^2	41.3 million km^2	5.8 million km^2

crops: Vernalisation period starts 150 days before the start of the growing period. At least 20 days below 5°C must exist during the vernalisation period and there must not exist more than 3 days below −30°C. In order to consider permafrost conditions that exclude agricultural use, mean annual temperature must not be below 0°C. Mean daily incoming solar radiation must exceed 60 W/m^2 to provide enough energy for crop growth.

Thus, suitability values, number of crop cycles and the start of the growing cycle are calculated on each land surface pixel for both rainfed and irrigated conditions. For irrigated conditions, fuzzy values for precipitation are neglected during the calculation process. Due to a lack of global information on irrigation practices, we assume perennial irrigation on irrigated areas.

Besides climatic constraints, soil properties are limiting agricultural suitability. According to the membership functions (Fig. 2), the fuzzy values representing each of the soil properties are calculated. The minimum of the eight values represents the value of the soil suitability. Soil information was taken from the Harmonized World Soil Database (HWSD) [41], considering the topsoil (0–30 cm) of the dominant and all (up to 8) component soils at a spatial resolution of 30 arc seconds [42]. Within the calculation of soil suitability, fuzzy values of each of the component soils are calculated and weighted according to their share.

The suitability for crops to be cultivated is decreasing with increasing slope (see Fig. 2). The slope must not exceed 16% for the considered crops, except for oil palm and paddy rice. The slope was calculated and resampled to 30 arc seconds from Shuttle Radar Topography Mission (SRTM) data [39].

Across all climate, soil and topography fuzzy values, the lowest fuzzy value quantifies the crops' suitability at a certain location. The highest value across all crops determines the suitability for agriculture at a certain location.

This methodology does not allow for yield estimations, in which socio-economic and bio-physical aspects, which our approach does not consider, play an important role. However, this approach is well suited to draw conclusions about where areas are agriculturally suitable and how these areas may change with future climate conditions.

Results

The Earth surface consists of 510 million km^2 of which 149 million km^2 are land surface. Up to 60°S, excluding Antarctica, and considering a lack of input data, in total 127.5 million km^2 of land surface remain to be analyzed regarding their suitability for agriculture. We classified the results of the suitability analysis into

four categories: not suitable (0), marginally suitable (>0.0), moderately suitable (>0.33) and highly suitable (>0.75).

Comparison

Our results (further named GLUES in the Figures) highly correlate with existing studies, such as the GAEZ approach [29], when comparing the area of each of the four classified categories in each of the 23 World Regions (R^2 = 0.99).

The global aggregation of the classified areas and the regional distribution of not suitable and suitable areas show a high level of agreement (Fig. 3 and 4). Compared to the distribution of global cropland in the year 2000 [43], our approach identifies 95.5% of current cropland as suitable.

Rainfed

For the period 1981–2010, our suitability analysis shows that in total 77.7 million km^2 are potentially suitable for purely rainfed agricultural cultivation, while 49.8 million km^2 are not suitable for rainfed conditions (Table 3). Further, 30.6 million km^2 are marginally suitable, 41.3 million km^2 are moderately suitable and 5.8 million km^2 are highly suitable (Table 3).

Irrigation

Irrigated agriculture produces 40% of the world's food (FAO) on 3.1 million km^2 [44]. When considering irrigation, suitability is area weighted according to the fraction of rainfed and irrigated agricultural area (given by GMIA Version 5.0 [44]). Thereby, irrigation increases suitability on irrigated areas in global average by 0.13, adds 1.8 million km^2 of suitable land (Table 4) and allows for multiple cropping on 1.2 million km^2 (assuming sufficient water available for irrigation). Accordingly, huge areas e.g. in the Nile and Ganges delta are only becoming suitable due to irrigation. Overall, 79.6 million km^2 are suitable with spatially varying patterns (Fig. 5).

Figure 5 represents the global distribution of agricultural suitability as a result of local climate, soil and terrain conditions. In boreal regions, the growing season over all stages of phenology usually is too short for cultivation. The temperate zones seasonally have adequate temperatures and enough precipitation and often sufficient soil, while in subtropical regions, the annual distribution of precipitation strongly determines crop growth and soils often are alkaline. In inner tropics have adequate temperature and moisture throughout the year, but soil quality often restricts cultivation due to low organic content and acidity [3].

Table 4. Classified suitability considering rainfed and irrigated conditions (1981–2010).

Not Suitable	Marginally Suitable	Moderately Suitable	Highly Suitable
48.0 million km^2	31.8 million km^2	41.8 million km^2	5.9 million km^2

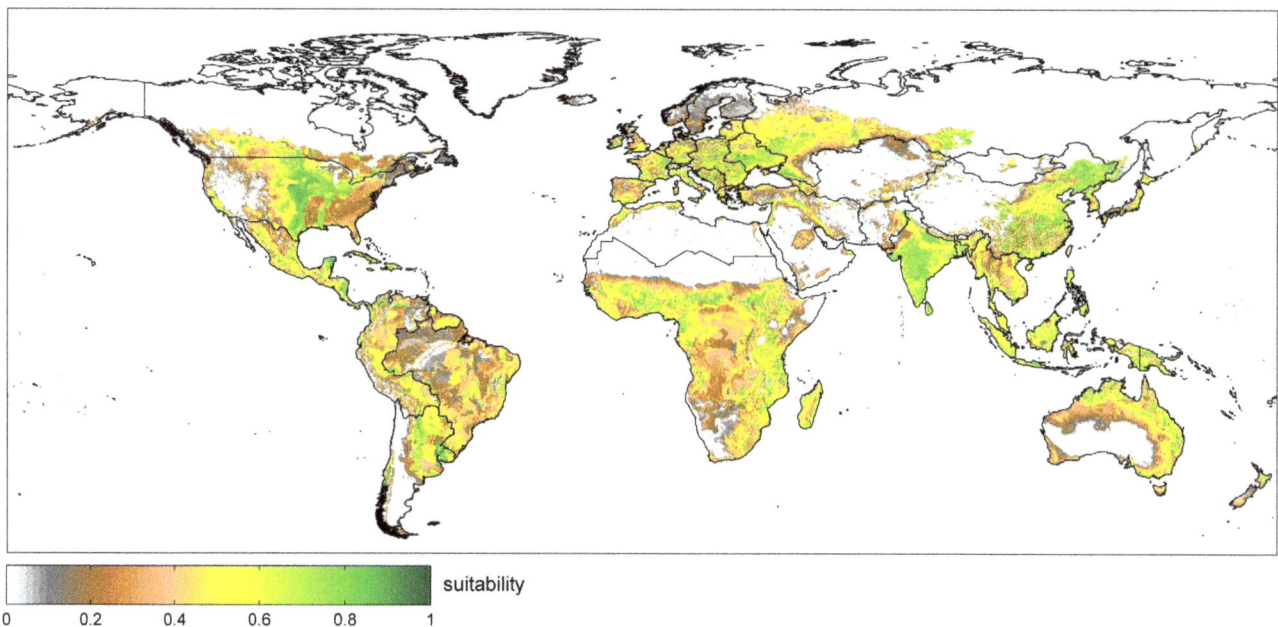

suitability

0   0.2   0.4   0.6   0.8   1

Figure 5. Agricultural suitability considering rainfed conditions and irrigated areas (1981–2010).

Protected Areas

Protected areas globally account for 8.3 million km^2. Information on actual protected areas is gathered from IUCN [45]. When excluding protected areas from the suitability calculation, 74.8 million km^2 remain suitable for cultivation. Thereby, protected areas are mainly situated in not suitable or marginally suitable areas (Table 5). Only 2% (0.2 million km^2) of the global protected area are located on land highly suitable for agriculture, 25% (2.1 million km^2) are on moderately suitable land, 30% (2.4 million km^2) on marginally suitable land while 43% (3.6 million km^2) are situated on unsuitable land. Overall, only 57% of global protected areas are suitable for agriculture.

Forested Areas

Dense forests are highly important to provide numerous ecosystem services. Densely forested areas account for 23.3 million km^2 according to GlobCover [46] and 23.5 million km^2 according to [47]. GlobCover defines forests as being dense when 75% of the pixel is forest [46]. Only 1.5 million km^2 or 6.2% of the global densely forested areas are currently protected.

4.9% (1.1 million km^2) of the densely forested areas (excluding forests within protected areas) are located in highly suitable land, 49.4% (11.1 million km^2) in moderately suitable land, 37.5% (8.4 million km^2) in marginally suitable land and only 8.2% (1.9 million km^2) are situated on unsuitable land. Overall, 92% of densely forested areas are potentially suitable for agriculture which indicates that global forests are subject to increasing societal stress.

Current Use of Suitable Land and Trade-Offs

When excluding both, protected areas and dense forests from the suitability calculation, 54.1 million km^2 remain suitable (Table 6). In comparison, currently used agricultural land (including pasture) today covers 49.1 million km^2, of which 15.5 million km^2 (status for 2011) are arable land (land under temporary and permanent crops; double-cropped areas are counted only once) [48]. Accordingly, 91% of all suitable land is already occupied by agriculture when today's protected and densely forested areas are preserved in the future. This illustrates that agricultural expansion is only possible by substituting other uses/covers of land which causes high social and ecological externalities. Figure 6 gives an overview of the current use/cover of suitable areas in the different regions of the world.

Figure 6 shows, that the current fraction of suitable area, which is not protected or dense forest is highly variable across regions. The most efficient use of current agriculturally suitable land is obvious in the USA, where only 2% of currently suitable land is not yet used or protected/dense forest.

In Africa, about 20% of the agriculturally suitable area is currently not used for agriculture or is statistically not recorded in the data of currently used agricultural land (Ramankutty *et al.*, 2008). This shows the extraordinary potentials of Africa for future expansion of agricultural land. However, agricultural expansion would always take place at ecological costs (e.g. conversion of tropical rainforest, grassland and savannah). In Latin America large suitable areas are protected or covered with dense forest and

Table 5. Classified suitability for 1981–2010 considering rainfed and irrigated conditions, excluding protected areas.

Not Suitable	Marginally Suitable	Moderately Suitable	Highly Suitable
44.4 million km^2	29.4 million km^2	39.7 million km^2	5.7 million km^2

Table 6. Classified suitability for 1981–2010 considering rainfed and irrigated conditions, excluding protected and densely forested areas.

Not Suitable	Marginally Suitable	Moderately Suitable	Highly Suitable
42.6 million km^2	21.0 million km^2	28.6 million km^2	4.6 million km^2

Table 7. Classified suitability for 2071–2100 considering rainfed and irrigated conditions, excluding protected and densely forested areas.

Not Suitable	Marginally Suitable	Moderately Suitable	Highly Suitable
37.8 million km^2	24.8 million km^2	30.2 million km^2	3.9 million km^2

the current fraction of remaining suitable area is smaller than in Africa, India is the prototype of a country, which is already using very large parts of its suitable agricultural land - and by for using the largest proportion (58%) of current cropland. Australia and larger parts of Asia still have reasonable land resources left for future expansion (Fig 6).

Future Change

For the investigation of future agricultural suitability for the time-period 2071–2100 as determined by the simulated climate effects of the SRES A1B emission scenario, we assume no changes in irrigated areas, soil properties, terrain or any adaptations, such as crop breeding. As result, when again excluding protected and densely forested areas, the global area being highly suitable for agriculture decreases from 4.6 to 3.9 million km^2, while marginally and moderately suitable areas increase (Table 7). In total, agriculturally suitable areas increase by 4.8 million km^2 due to the selected climate change scenario. However, most of the additional area is only marginally suitable for agricultural use.

Without excluding any areas, the impact of climate change increases the potentially suitable areas on the globe by 5.6 million km^2. Marginally suitable areas increase by 4.2 million km^2, moderately suitable areas increase by 2.3 million km^2, while highly suitable areas decrease by 0.8 million km^2 (Fig. 7).

A more regional analysis shows that the world is divided into regions that receive additional suitable land and regions where land that used to be suitable turns into not suitable land (Fig. 8). Regions in the northern hemisphere, such as Canada (+2.1 million km^2 of suitable land), Russia (+3.1 million km^2) and China (+0.9 million km^2), benefit most.

On the global scale, suitability improves on 18.7 million km^2 and worsens on 22.2 million km^2. In total, the area with decreasing suitability is 3.5 million km^2 more than the area with increasing suitability (Fig. 9). The highest absolute net loss of suitable areas is found in Sub-Saharan Africa.

Thereby, the globally averaged suitability value (averaged over all suitable areas), decreases from 0.41 to 0.39. The greatest losses of suitability are simulated for France and the Mediterranean (Fig. 10). The changing suitability is mapped in Fig. 11.

Growing Cycle and Multiple Cropping

The seasonal development of temperature and precipitation determines the length of the growing season, the start of the growing cycle and the potential number of annual cropping. Thus, the option of multiple cropping represents an important measure for farmers to increase production. Figure 12 shows the spatial distribution of the start of the growing cycle for the time period 1981–2010, exemplarily for maize.

Changing climate does not only affect the suitability of land, but also the start and length of the growing cycle. As an example, the start of the growing cycle for maize in Germany shifts in average 23 days earlier in time, when comparing the period of 2071–2100 with 1981–2010. The shift of growing cycles again influences the possibility for multiple cropping. Today's maximal achievable multiple cropping according to the course of temperature and precipitation is shown in Fig. 13.

Our results suggest that climate change has huge impacts on the areas suitable for multiple cropping under the assumed climate scenario. Until 2100, 6.0 million km^2 are globally lost for triple cropping until 2100, while the area which is suitable for double cropping increases by 2.3 million km^2. Multiplying the area with

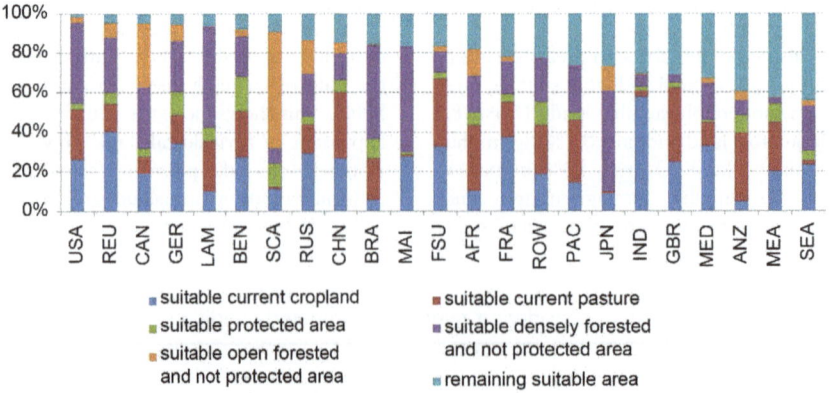

Figure 6. Current use of suitable areas (1981–2010), considering forest cover [46], protected areas [45] and current pasture and cropland (Ramankutty et al., 2008). If forested areas are agriculturally suitable and protected, they are attributed to 'suitable protected area'.

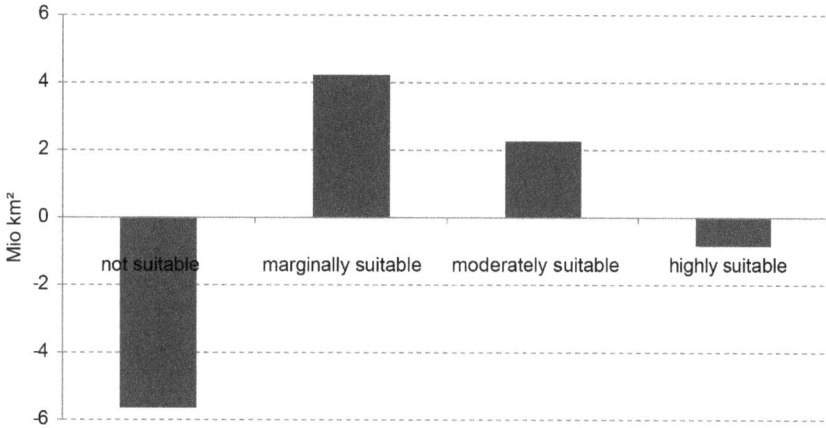

Figure 7. Global changes in agricultural suitability categories (million km²) between 1981–2010 and 2071–2100.

the number of cycles, this means a global decrease of 13.4 million km². Most of the increase in double cropping areas results from the transformation from triple to double cropping. Again, no change in irrigation is assumed in this calculation.

The largest decrease in multiple copping area can be found in Brazil (BRA) and in Sub-Saharan Africa (AFR), where areas suitable for triple cropping decrease by 1.7 (AFR) and 2.9 million km² (BRA) (Fig. 14), while the area for double cropping increases by 0.2 and 1.3 million km², respectively. In total, this means a decrease of multiple cropping area by 1.5 (AFR) and 1.6 million km² (BRA). This is equivalent to the amount of 4.7 and 6.1 million km² respectively, which are lost for agriculture, when multiplying the area with the number of possible crop cycles. This corresponds to 20.2 (AFR) and 28.8% (BRA) of today's potentially suitable area for multiple cropping. In the same manner, France (FRA) and the Mediterranean (MED) lose 24.1 (FRA) and 13.2% (MED) of their total equivalent area when considering the change of multiple cropping, which means a decrease by 93 (FRA) and 55% (MED) according to the multiple cropping area of 1981–2010. Regions where areas that potentially allow for more than one crop cycle increase due to climate change are CHI, IND, JPN, MEA, REU, RUS and USA, while the total area considerably increases mainly in the USA for both, double (0.35 million km²) and triple (0.12 million km²) cropping (Fig. 14).

Conclusions

The analyses of the present situation demonstrats that there is extraordinary potential e.g. for Sub Saharan Africa for future expansion of agricultural land without expanding into protected or forested areas. Further research is necessary to identify the environmental and social costs and consequences of agricultural expansion in these regions. Also further investigation is needed to give answers on how this land could be managed sustainable with benefit to local food systems and socio-economy.

Our results show at high spatial resolution how agricultural suitability may change until 2100 due to changing climate under the chosen scenario (SRES A1B), assuming no adaptation measurements by farmers. First, suitable areas increase especially in the northern regions such as Canada, China and Russia, where new land will be available for agricultural use. The increase in suitable areas mainly takes place in sparsely populated areas, which could imply a lack of labor for open up new agricultural land and prepare soils. Certainly, it will be related with high investment costs and it will take a long time to extend agriculture here. Secondly, global average suitability decreases under the chosen climate scenario. Especially the extend of highly suitable areas is reduced by the effect of climate change. Finally, suitable areas indirectly are reduced due to a substantial global reduction of the suitability for multiple cropping, especially in Sub Saharan Africa, and Brazil.

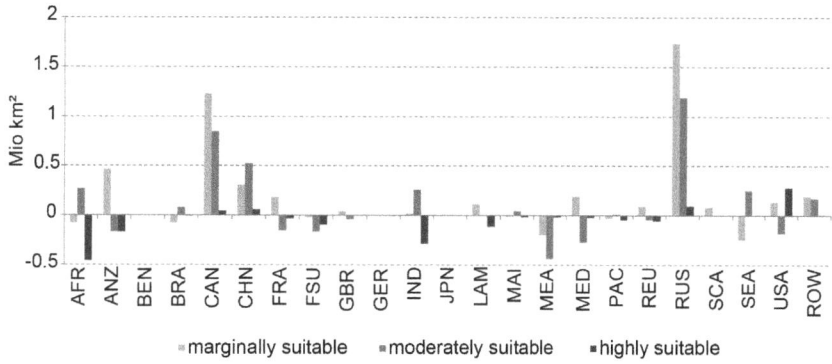

Figure 8. Regional change in agricultural suitability categories (million km²) between 1981–2010 and 2071–2100.

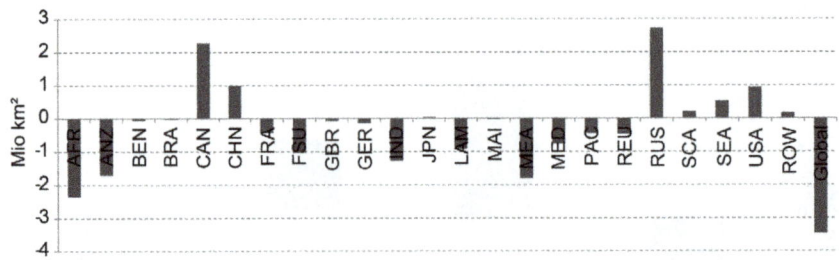

Figure 9. Regional change of agriculturally suitable area due to A1B climate change scenario between 1981–2010 and 2071–2100.

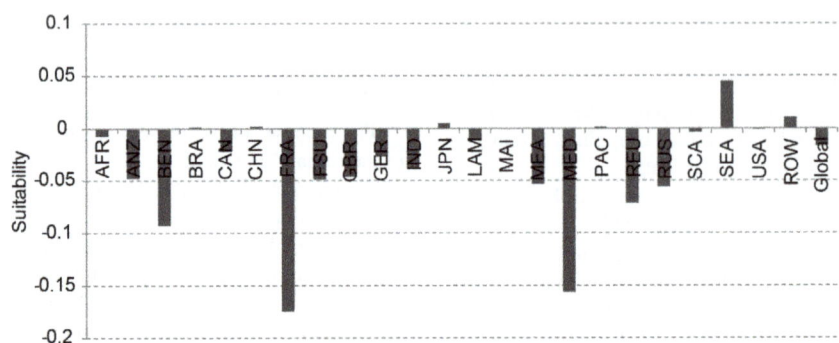

Figure 10. Regional changes in the average suitability between 1981–2010 and 2071–2100.

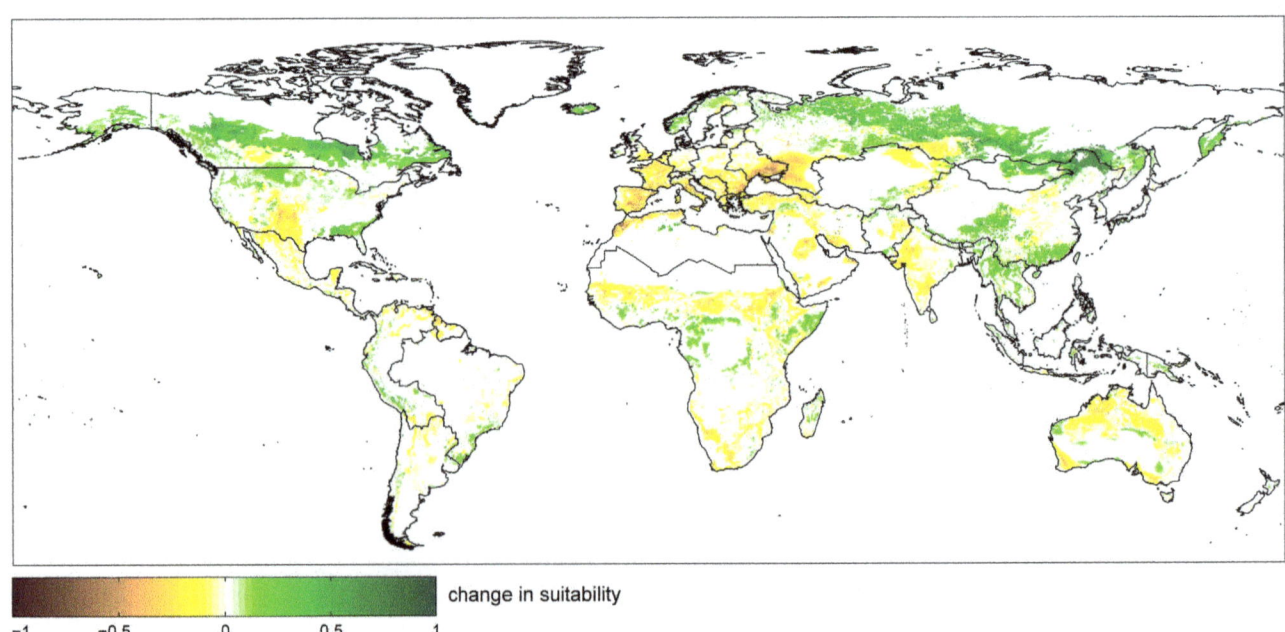

Figure 11. Change in agricultural suitability between 1981–2010 and 2071–2100. Green areas indicate an increase in suitability while brown areas show a decreasing suitability.

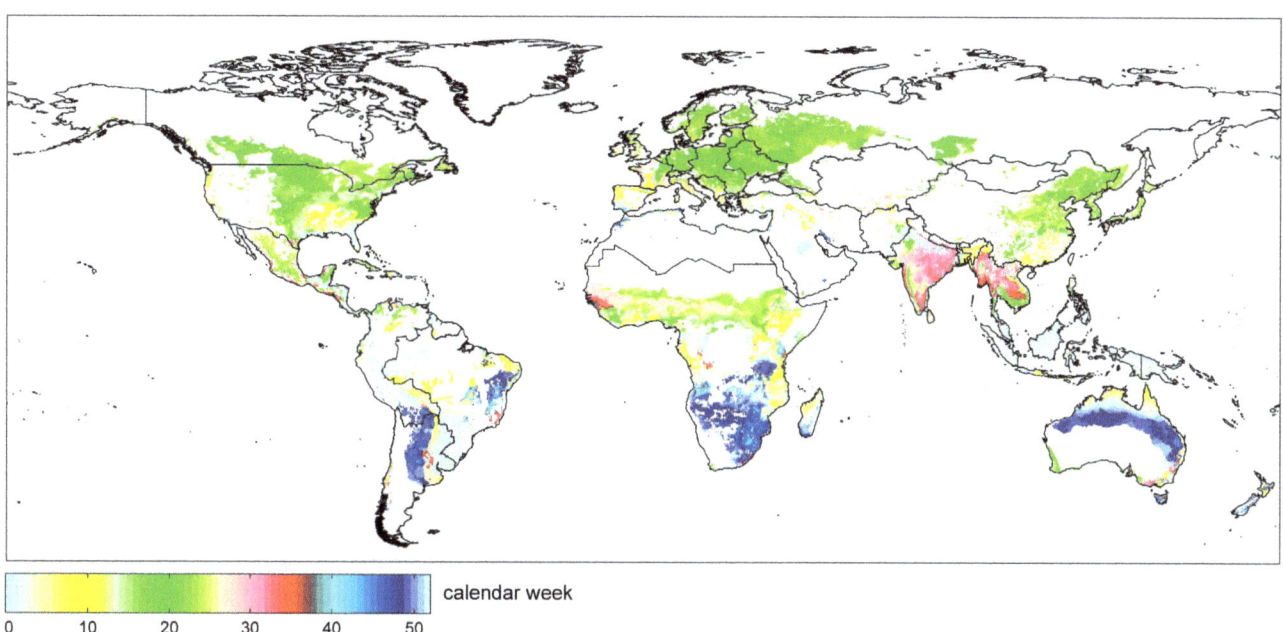

Figure 12. Start of the growing cycle for maize (1981–2010). The start of the growing cycle is illustrated for rainfed conditions and for irrigated conditions on predominantly irrigated areas (irrigated area > 50%). In case of multiple cropping, the map shows the start of the first growing cycle.

Overall, the Global North regionally increases suitability and the number of crop cycles, while the Global South and the Mediterranean area lose agriculturally suitable land without adaptations. This will decisively affect smallholder farmers as their options for adaptations through e.g. irrigation are limited.

Scientific knowledge on the geographical distribution has decisively being increased with the availability of global data sets, also based on remote sensing. The tensions between both limits of land expansion and intensification within the context of sustainable agricultural intensification stresses the ongoing debate on

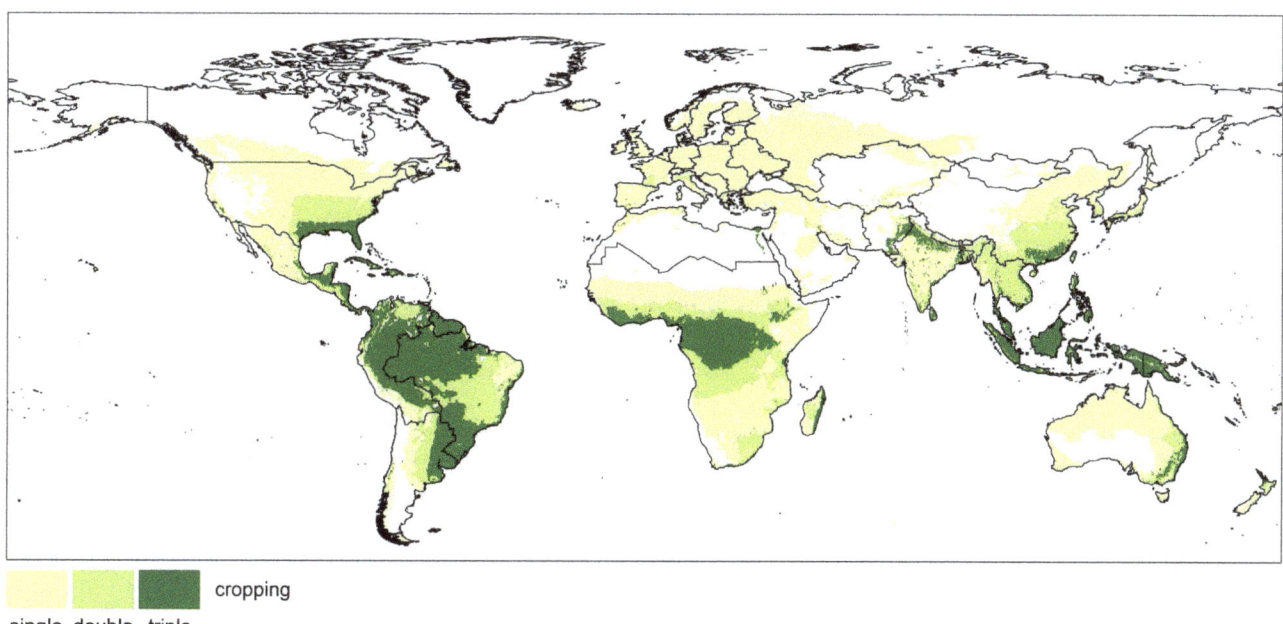

Figure 13. Suitable areas for single, double and triple cropping (1981–2010). Multiple cropping is illustrated for rainfed conditions and for irrigated conditions on predominantly irrigated areas (irrigated area > 50%).

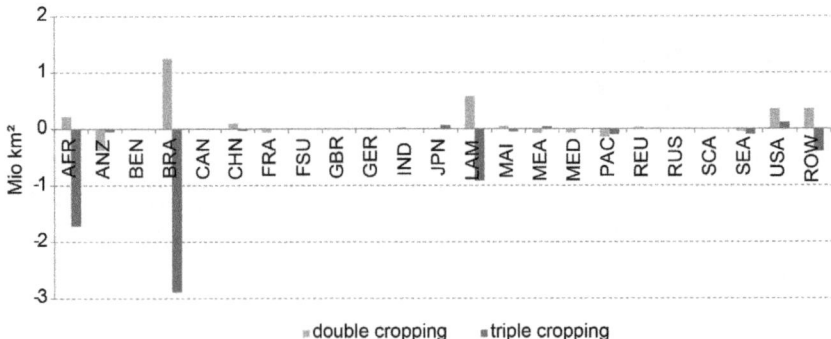

Figure 14. Change in area suitable for double and triple cropping (million km²) between 1981–2010 and 2071–2100.

global land management, considering the complex interplay and trade-offs between different uses of land and ecosystem services.

Acknowledgments

This research was carried out within the framework of the GLUES (Global Assessment of Land Use Dynamics, Greenhouse Gas Emissions and Ecosystem Services) Project. Thanks to all project members and to Jonas Maier who contributed to this study as a student assistant.

Author Contributions

Conceived and designed the experiments: FZ BP WM. Performed the experiments: FZ. Analyzed the data: FZ. Contributed reagents/materials/ analysis tools: FZ. Contributed to the writing of the manuscript: FZ.

References

1. Foley JA, DeFries R, Asner GP, Barford C, Bonan G, et al. (2005) Global Consequences of Land Use. Science 309: 570–574.

2. Václavík T, Lautenbach S, Kuemmerle T and Seppelt R (2013) Mapping global land system archetypes. Global Environmental Change 23: 1637–1647.

3. Ramankutty N, Foley JA, Norman J, McSweeney K (2002) The global distribution of cultivable lands: current patterns and sensitivity to possible climate change. Global Ecology and Biogeography 11: 377–392.

4. Alexandratos N, Bruinsma J (2012) World agriculture towards 2030/2050: the 2012 revision. ESA Working Paper No 12-03. Rome: FAO.

5. Tilman D, Balzer C, Hill J, Befort BL (2011) Global food demand and the sustainable intensification of agriculture. Proceedings of the National Academy of Sciences 108: 20260–20264.

6. Foley JA, Ramankutty N, Brauman KA, Cassidy ES, Gerber JS, et al. (2011) Solutions for a cultivated planet. Nature 478: 337–342.

7. Godfray HCJ, Beddington JR, Crute IR, Haddad L, Lawrence D, et al. (2010) Food Security: The Challenge of Feeding 9 Billion People. Science 327: 812–818.

8. Pretty J, Sutherland WJ, Ashby J, Auburn J, Baulcombe D, et al. (2010) The top 100 questions of importance to the future of global agriculture. International Journal of Agricultural Sustainability 8: 219–236.

9. Gregory PJ and George TS (2011) Feeding nine billion: the challenge to sustainable crop production. Journal of Experimental Botany.

10. Ray DK, Mueller ND, West PC, Foley JA (2013) Yield Trends Are Insufficient to Double Global Crop Production by 2050. PLoS ONE 8: e66428.

11. Vermeulen S, Zougmoré R, Wollenberg E, Thornton P, Nelson G, et al. (2012) Climate change, agriculture and food security: a global partnership to link research and action for low-income agricultural producers and consumers. Current Opinion in Environmental Sustainability 4: 128–133.

12. Kastner T, Rivas MJI, Koch W, Nonhebel S (2012) Global changes in diets and the consequences for land requirements for food. Proceedings of the National Academy of Sciences 109: 6868–6872.

13. Cassidy ES, West PC, Gerber JS, Foley JA (2013) Redefining agricultural yields: from tonnes to people nourished per hectare. Environmental Research Letters 8: 034015.

14. Erb K-H, Haberl H, Krausmann F, Lauk C, Plutzar C, et al. (2009) Eating the Planet: Feeding and fuelling the world sustainably, fairly and humanely – a scoping study. Social Ecology Working Paper. Alpen-Adria Universität Klagenfurt.

15. Spiertz JHJ, Ewert F (2009) Crop production and resource use to meet the growing demand for food, feed and fuel: opportunities and constraints. NJAS - Wageningen Journal of Life Sciences 56: 281–300.

16. Bruinsma J (2011) The resources outlook: by how much do land, water and crop yields need to increase by 2050? In: P. Conforti , editor. Looking ahead in world food and agriculture: Perspectives to 2050. Rome: FAO.

17. Schmitz C, van Meijl H, Kyle P, Nelson GC, Fujimori S, et al. (2014) Land-use change trajectories up to 2050: insights from a global agro-economic model comparison. Agricultural Economics 45: 69–84.

18. Pimentel D, Harvey C, Resosudarmo P, Sinclair K, Kurz D, et al. (1995) Environmental and Economic Costs of Soil Erosion and Conservation Benefits. Science 267: 1117–1123.

19. UNCCD (2014) Desertification - The Invisible Frontline. Bonn, Germany: United Nations Convention to Combat Desertification.

20. Seto KC, Fragkias M, Güneralp B, Reilly MK (2011) A Meta-Analysis of Global Urban Land Expansion. PLoS ONE 6: e23777.

21. Avellan T, Zabel F, Mauser W (2012) Are urban areas endangering the availability of rainfed crop suitable land? Remote Sensing Letters 3: 631–638.

22. Teka K, Haftu M (2012) Land Suitability Characterization for Crop and Fruit Production in Midlands of Tigray, Ethiopia. Momona Ethiopian Journal of Science 4: 12.

23. Kalogirou S (2002) Expert systems and GIS: an application of land suitability evaluation. Computers, Environment and Urban Systems 26: 89–112.

24. Baja S, Chapman DM, Dragovich D (2002) A Conceptual Model for Defining and Assessing Land Management Units Using a Fuzzy Modeling Approach in GIS Environment. Environmental Management 29: 647–661.

25. Braimoh AK, Vlek PLG, Stein A (2004) Land Evaluation for Maize Based on Fuzzy Set and Interpolation. Environmental Management 33: 226–238.

26. Kurtener D, Torbert HA, Krueger E (2008) Evaluation of Agricultural Land Suitability: Application of Fuzzy Indicators. In: O . Gervasi, B . Murgante, A . Laganûâ, D . Taniar, Y . Mun and M . Gavrilova, editors. Computational Science and Its Applications – ICCSA 2008. Springer Berlin Heidelberg. pp. 475–490.

27. Nisar Ahamed TR, Gopal Rao K, Murthy JSR (2000) GIS-based fuzzy membership model for crop-land suitability analysis. Agricultural Systems 63: 75–95.

28. Van Ranst E, Tang H, Groenemam R, Sinthurahat S (1996) Application of fuzzy logic to land suitability for rubber production in peninsular Thailand. Geoderma 70: 1–19.

29. IIASA/FAO (2012) Global Agro-ecological Zones (GAEZ v3.0) - Model Documentation. IIASA, Laxenburg, Austria and FAO, Rome, Italy: IIASA, FAO.

30. Fischer G, Hizsnyik E, Prieler S, Wiberg D (2011) Scarcity and abundance of land resources: competing uses and the shrinking land resource base. SOLAW Background Thematic Report. FAO.

31. Lane A, Jarvis A (2007) Changes in Climate will modify the Geography of Crop Suitability: Agricultural Biodiversity can help with Adaptation. Journal of SAT Agricultural Research 4: 12.

32. Avellan T, Zabel F, Mauser W (2012) The influence of input data quality in determining areas suitable for crop growth at the global scale – a comparative analysis of two soil and climate datasets. Soil Use and Management 28: 249–265.

33. Burrough PA (1989) Fuzzy mathematical methods for soil survey and land evaluation. Journal of Soil Science 40: 477–492.

34. Burrough PA, Macmillan RA, van Deursen W (1992) Fuzzy classification methods for determining land suitability from soil profile observations and topography. Journal of Soil Science 43: 193–210.

35. Sys CO, van Ranst E, Debaveye J, Beernaert F (1993) Land evaluation: Part III Crop requirements. Agricultural Publications. Brussels: G.A.D.C.

36. Bengtsson L, Hodges KI, Keenlyside N (2009) Will Extratropical Storms Intensify in a Warmer Climate? Journal of Climate 22: 2276–2301.

37. Jungclaus JH, Keenlyside N, Botzet M, Haak H, Luo JJ, et al. (2006) Ocean Circulation and Tropical Variability in the Coupled Model ECHAM5/MPI-OM. Journal of Climate 19: 3952–3972.

38. Marke T, Mauser W, Pfeiffer A, Zängl G, Jacob D, et al. (2013) Application of a hydrometeorological model chain to investigate the effect of global boundaries and downscaling on simulated river discharge. Environ Earth Sci: 1–20.

39. Farr TG, Rosen PA, Caro E, Crippen R, Duren R, et al. (2007) The Shuttle Radar Topography Mission. Reviews of Geophysics 45: RG2004.

40. Hijmans RJ, Cameron SE, Parra JL, Jones PG, Jarvis A (2005) Very high resolution interpolated climate surfaces for global land areas. International Journal of Climatology 25: 1965–1978.

41. FAO/IIASA/ISRIC/ISSCAS/JRC (2012) Harmonized World Soil Database (version 1.2). FAO, Rome, Italy and IIASA, Laxenburg, Austria.

42. Avellan T, Zabel F, Putzenlechner B, Mauser W (2013) A Comparison of Using Dominant Soil and Weighted Average of the Component Soils in Determining Global Crop Growth Suitability. Environment and Pollution 2: 11.

43. Ramankutty N, Evan AT, Monfreda C, Foley JA (2008) Farming the planet: 1. Geographic distribution of global agricultural lands in the year 2000. Global Biogeochemical Cycles 22: GB1003.

44. Siebert S, Henrich V, Frenken K, Burke J (2013) Global Map of Irrigation Areas version 5. Rheinische Friedrich-Wilhelms-University, Bonn, Germany/Food and Agriculture Organization of the United Nations, Rome, Italy.

45. IUCN (2008) Guidelines for Applying Protected Area Management Categories In: N. Dudley, editor. Gland, Switzerland: IUCN.

46. Bontemps S, Defourny P, Bogaert Ev, Arino O, Kalogirou V (2009) GLOBCOVER 2009, Products Description and Validation Report. ESA, Univ. catholique de Louvain.

47. Hansen MC, Potapov PV, Moore R, Hancher M, Turubanova SA, et al. (2013) High-Resolution Global Maps of 21st-Century Forest Cover Change. Science 342: 850–853.

48. FAOSTAT (2014) FAOSTAT Land USE module. Retrieved 24 February 2014. Available at: http://faostat.fao.org/site/377/DesktopDefault.aspx?PageID=377#ancor.

20

The Relative Impacts of Climate and Land-Use Change on Conterminous United States Bird Species from 2001 to 2075

Terry L. Sohl*

Earth Resources Observation and Science (EROS) Center, U.S. Geological Survey, Sioux Falls, South Dakota, United States of America

Abstract

Species distribution models often use climate data to assess contemporary and/or future ranges for animal or plant species. Land use and land cover (LULC) data are important predictor variables for determining species range, yet are rarely used when modeling future distributions. In this study, maximum entropy modeling was used to construct species distribution maps for 50 North American bird species to determine relative contributions of climate and LULC for contemporary (2001) and future (2075) time periods. Species presence data were used as a dependent variable, while climate, LULC, and topographic data were used as predictor variables. Results varied by species, but in general, measures of model fit for 2001 indicated significantly poorer fit when either climate or LULC data were excluded from model simulations. Climate covariates provided a higher contribution to 2001 model results than did LULC variables, although both categories of variables strongly contributed. The area deemed to be "suitable" for 2001 species presence was strongly affected by the choice of model covariates, with significantly larger ranges predicted when LULC was excluded as a covariate. Changes in species ranges for 2075 indicate much larger overall range changes due to projected climate change than due to projected LULC change. However, the choice of study area impacted results for both current and projected model applications, with truncation of actual species ranges resulting in lower model fit scores and increased difficulty in interpreting covariate impacts on species range. Results indicate species-specific response to climate and LULC variables; however, both climate and LULC variables clearly are important for modeling both contemporary and potential future species ranges.

Editor: Stephanie S. Romanach, U.S. Geological Survey, United States of America

Funding: TLS was supported by the U.S. Geological Survey's Climate and Land Use (CLU) Mission Area. The funders had no role in study design, data collection and analysis, decision to publish, or preparation of the manuscript.

Competing Interests: The author has declared that no competing interests exist.

* Email: sohl@ugs.gov

Introduction

Species distribution models (SDMs) are based on the assumption that presence at a given location is based on suitable environmental conditions to support the species' ability to find shelter, feed, and/or reproduce [1,2]. Such models have been widely used to model current species distributions, either to establish extant distributions or to understand the specific environmental variables that drive species distributions [3,4,5]. A central premise of many SDMs is that climate is a primary driving force of the distribution of species [6]. Projected climate data are frequently used with SDMs to explore potential future impacts of climate change on species distributions [7,8,9], based on the assumption that the basic physiological tolerances of species to environmental conditions are constant through time [10]. Jimenez-Valverde et al. [11] modeled typical climate conditions for 94 bird species in North America and noted the dominant signal of climate in shaping North American bird distributions. Thuiller et al. [12] modeled distributions of plants, birds, mammals, and reptiles in Europe and found that models using climate alone performed nearly as well as models that included both climate and landscape variables. Bucklin et al. [13] found that climate variables were strong predictors for contemporary species distribution modeling and that additional predictors (including land cover) were not essential.

Climate is obviously a primary driver of many SDMs. While land use and land cover (LULC) change is often used for modeling contemporary species distributions, it is not often used when examining future time frames [7]. Despite the results from the studies listed above, other studies have found that including LULC in bioclimatic models of species distribution can improve the explanatory power of SDMs [7,12]. Lee and Jetz [14] found that LULC projections were vital for future modeling, noting that loss of habitat is a high predictor of extinction for bird species. Barbet-Massin et al. [7] found that SDMs perform best if both climate and LULC are included. Sinclair et al. [15] were critical of SDMs for rarely including anthropogenic impacts on biological systems, suggesting that changing landscape patterns are likely to have at least as great an impact on species distributions as climate change. Many other studies have found that projected land use is a vital component of SDMs, with loss of predictive power when LULC is not included in the assessment [1,8,9,16].

While many state the need to use LULC data when projecting future change in species distributions, such data are often not available [17]. Riordan et al. [16] noted the disconnect between relatively high-resolution climate projections used in global assessment models and the generally coarse treatment of LULC, leading to high-resolution, projected LULC rarely being used (or available) for SDMs. Those studies that have used both projected climate and LULC data to model future species distributions have often relied on very coarse spatial-scale LULC data [1,7]. LULC is much more heterogeneous at local scales than climate. Those studies that do use LULC but only at very coarse scales miss the inherent spatial variability in LULC that is typically not found with climate data.

The goals of this paper are to examine the relative effects of climate and LULC change on bird species distributions in the conterminous United States, using maximum entropy modeling and projected climate and LULC data from 2001 to 2075. Specific research questions include:

- What are the relative influences of LULC and climate in modeling contemporary (2001) breeding bird distributions for the conterminous United States?

- What are the relative impacts of projected climate and projected LULC change on United States breeding bird distributions in the future (2075)?

- What are the specific impacts of climate and LULC on one focus species (Hooded Warbler (*Wilsonia citrina*) used to demonstrate individual species results)?

- What are the implications for the use of climate and LULC data in SDMs?

Materials and Methods

A maximum entropy model (Maxent) was used in conjunction with species presence data, current and projected climate and LULC data, and topographic data to model distributions for 50 diverse bird species in the conterminous United States. Twelve distinct modeling simulations were conducted for each species to disentangle the effects of climate and LULC on species distributions for both the "present" (2001 for this assessment) and for multiple scenarios in the future (2075). Although many supporting data were available through the year 2100, 2075 was selected as the assessed "future" year to accommodate the use of 30-year averaged climate data (as described below). The following provides a summary of data sources, the model structure, model parameterization, and the assessment framework.

Materials

The modeling approach described below required spatially explicit data on both bird species "presence", as well as environmental variables (covariates) that could be used to model species distributions. Note the goal was to assess long-term trends in changes to species distributions. Longer-term aggregate or average data values were thus used to temper the effects of seasonal or yearly variation, both for the species presence data (where distributions of single-season presence records could be impacted by unusual seasonal conditions such as drought or large disturbance events) and for the covariates (where single-season climate data, for example, may be unrepresentative of longer-term climate trends).

Species Presence Data – eBird. The source of bird species presence data was eBird, a "citizen-science" database [18,19]. eBird allows public entry of bird sightings, with recorded

information on the time and date of the sighting, location, observation protocol, quantity of each species, and observer information. As a citizen science database, there are potential issues (discussed below) related to the lack of a formal sampling protocol [20], but eBird offers several potential advantages for species distribution modeling including 1) large number of sample points (millions for some species), 2) global data (although observations are currently heavily biased towards North America and Europe), and 3) observations for all seasons. eBird data have been successfully used for a number of SDM assessments [20,21,22,23]. Hochachka and Fink [20] found strong linkages between individual species and land cover using eBird data, and that the data were valuable for examining distribution patterns at multiple scales.

Fifty bird species with breeding ranges partially or completely within the conterminous United States were selected for the assessment (**Table 1**). Species were selected to ensure variability in size of breeding range, geographic region, and preferred breeding habitat. The goal was to ensure that a variety of "real-world" model applications were represented. To minimize potential effects of annual variation in species presence, data from eBird entries from 1992 to 2012 were used to establish "current" breeding records. With current and projected land cover data available for every year from 1992 to 2100 (see below), a nominal date of 2001 (middle of the 1992 to 2010 period) was used to represent contemporary species distributions and tie the 1992 to 2010 species occurrences to one specific date of land-cover conditions. Data were also filtered by season to ensure records corresponded to breeding populations; for all species a consistent June 1 to July 15 observation period was used to represent likely "breeding" presence, a reasonable assumption for the species that were assessed. Some migratory species initially included in the assessment were removed from consideration based on dispersed patterns of eBird sightings for the June 1 to July 15 period, indicating post-breeding movement had already occurred by July 15 (e.g., Long-billed Curlew (*Numenius americanus*)). One species included, the American Goldfinch (*Spinus tristis*) generally begins breeding after this period, but is considered non-migratory and still is within breeding range. For the 50 species assessed, eBird sightings for the June 1 to July 15 period corresponded well to published breeding range maps from NatureServe [23].

eBird allows users to enter one of several potential observation protocols, including "stationary count", "traveling count", or "exhaustive area count". However, regardless of observation protocol, users only enter one geographic coordinate. A single coordinate for a "traveling count" where the travel distance was substantial could result in a data point that was many kilometers from the actual observation. For exhaustive area counts with a large search area, a single coordinate may similarly be some distance from the actual observation. To eliminate potential issues with unrepresentative locations of eBird sightings, all "traveling count" sightings with a travel distance of more than 2 km were eliminated (similar to Fink et al. [23]), as were all "exhaustive area count" sightings with a search area of more than 100 hectares.

Additional potential issues with eBird data include spatial bias in presence samples [25]. eBird observations, like other citizen science data, tend to be clustered around highly populated and/or easily accessible areas [15,18,19]. Sampling bias has a much stronger effect on presence-only models (used here) than on presence-absence models, as model results end up representing both presence, as well as the density of the sampling effort [26]. Spatial filtering is an effective means to reduce bias in sample data prior to use in species distribution modeling [27,28]. For this assessment, the seasonal 1992 to 2012 observations were spatially

Table 1. Species modeled and number of eBird sample points for each.

	Species	Scientific Name	Original	Final
1	**American Goldfinch**	*Spinus tristis*	236,217	2,663
2	**Anna's Hummingbird**	*Calypte anna*	32,047	427
3	**Baird's Sparrow**	*Ammodramus bairdii*	513	48
4	**Band-tailed Pigeon**	*Patagioenas fasciata*	17,415	407
5	**Black-capped Chickadee**	*Poecile atricapillus*	131,634	1,877
6	**Blue-winged Teal**	*Anas discors*	15,288	1,243
7	**Bobolink**	*Dolichonyx oryzivorus*	28,658	1,105
8	**Brown-headed Cowbird**	*Molothrus ater*	178,324	3,996
9	**Brown Thrasher**	*Toxostoma rufum*	61,661	2,254
10	**Cactus Wren**	*Campylorhynchus brunneicapillus*	4,714	215
11	**Carolina Wren**	*Thryothorus ludovicianus*	107,244	1,893
12	**Chestnut-collared Longspur**	*Calcarius ornatus*	1,426	105
13	**Dickcissel**	*Spiza americana*	29,479	1,411
14	**Downy Woodpecker**	*Picoides pubescens*	150,261	2,925
15	**Eastern Kingbird**	*Tyrannus tyrannus*	111,057	2,956
16	**Ferruginous Hawk**	*Buteo regalis*	1,587	238
17	**Gambel's Quail**	*Callipepla gambelii*	6,307	198
18	**Grasshopper Sparrow**	*Ammodramus savannarum*	23,254	1,323
19	**Gray Partridge**	*Perdix perdix*	616	129
20	**Gray Vireo**	*Vireo vicinior*	265	43
21	**Great Blue Heron**	*Ardea herodias*	141,552	3,449
22	**Great Horned Owl**	*Bubo virginianus*	11,130	1,487
23	**Green-winged Teal**	*Anas carolinensis*	6,726	530
24	**Hooded Warbler**	*Wilsonia citrina*	15,482	773
25	**Lark Bunting**	*Calamospiza melanocorys*	3,268	355
26	**Lark Sparrow**	*Chondestes grammacus*	20,978	1,467
27	**Northern Harrier**	*Circus cyaneus*	14,795	1,231
28	**Northern Pintail**	*Anas acuta*	4,269	466
29	**Orchard Oriole**	*Icterus spurius*	41,136	1,876
30	**Painting Bunting**	*Passerina ciris*	15,294	569
31	**Pied-billed Grebe**	*Podilymbus podiceps*	23,272	1,287
32	**Pileated Woodpecker**	*Dryocopus pileatus*	48,118	1,982
33	**Pygmy Nuthatch**	*Sitta pygmaea*	11,848	322
34	**Red-eyed Vireo**	*Vireo olivaceus*	138,887	2,389
35	**Red-headed Woodpecker**	*Melanerpes erythrocephalus*	22,809	1,593
36	**Red-tailed Hawk**	*Buteo jamaicensis*	101,388	3,715
37	**Ruby-throated Hummingbird**	*Archilochus colubris*	81,241	2,090
38	**Savannah Sparrow**	*Passerculus sandwichensis*	41,214	1,435
39	**Scissor-tailed Flycatcher**	*Tyrannus forficatus*	16,571	718
40	**Sedge Wren**	*Cistothorus platensis*	7,827	478
41	**Sharp-tailed Grouse**	*Tympanuchus phasianellus*	801	121
42	**Short-eared Owl**	*Asio flammeus*	760	139
43	**Sora**	*Porzana carolina*	6,687	649
44	**Tufted Titmouse**	*Baeolophus bicolor*	129,472	2,058
45	**Vesper Sparrow**	*Pooecetes gramineus*	16,387	1,164
46	**Western Kingbird**	*Tyrannus verticalis*	45,319	2,028
47	**Western Meadowlark**	*Sturnella neglecta*	36,755	1,825
48	**Western Tanager**	*Piranga ludoviciana*	33,108	1,127

Table 1. Cont.

	Species	Scientific Name	Original	Final
49	White-headed Woodpecker	*Picoides albolarvatus*	4,389	137
50	Yellow-headed Blackbird	*Xanthocephalus xanthocephalus*	16,794	1,031

"Original" represents conterminous United States observations from 1992 to 2012, from June 1 to July 15. "Final" represents points that have had 1) spatial filtering applied to reduce points in heavily sampled areas, and 2) removal of points with long travel distances (traveling count) or large search areas (search area count).

filtered to eliminate sample points within 20 km of any other sample point. The threshold of 20 km was chosen because it more aggressively reduced sampling density in the very dense eBird database than past studies [27,28], while still maintaining adequate numbers of points for modeling. The elimination of sample points based on observation protocol or sampling density greatly reduced the number of sample points used in the assessment, often by a factor of 20 or more (**Table 1**). However, the filtering successfully eliminated the high concentration of points in heavily populated areas while maintaining a relatively large number of observations for most species (minimum of 43 points, maximum of 3,996, mean of 1,313). Only two species had fewer than 100 sample points (Gray Vireo (*Vireo vicinior*) and Baird's Sparrow (*Ammodramus bairdii*)), at 48 and 43 points, respectively. The number of sample points were considered adequate, as Wisz et al. [29] and Hernandez et al. [30] examined the effect of sample size on species distribution models and found that Maxent outperformed other modeling techniques when sample sizes were small, with "reasonable" models possible with sample sizes as small as 10.

Land-Use and Land-Cover Data. A newly available suite of LULC projections for the conterminous United States was used [31,32]. The LULC projections were produced for the conterminous United States, with annual LULC maps from 1992 to 2100 for four Intergovernmental Panel on Climate Change (IPCC) Special Report on Emissions Scenarios (SRES) [33]. The spatial resolution of the data was 250 m, with 16 LULC classes. The four modeled SRES were the A1B, A2, B1, and B2 scenarios; however, complimentary climate data were not available for the B2 scenario, so only A1B, A2, and B1 were used in this assessment (see **table 2** for characteristics of the three IPCC SRES scenarios used in this assessment). To simplify the modeling and interpretation of model results, the original sixteen LULC classes were aggregated to eight basic LULC classes (**table 3**). Aggregated 2001 LULC data served as one of the covariates when constructing the initial models. Projected 2075 LULC data provided information on LULC change for the 2075 model simulations.

Error in LULC data obtained from remote sensing sources is a concern for SDMs [4]. The LULC projections described above used the 1992 National Land Cover Database (NLCD) [34] as the mapping starting point. The projections were thus subject to not only the inherent uncertainty in projecting future LULC conditions, but also carried the legacy of any mapping error in the original 1992 NLCD. Given the lack of a rigid sampling protocol in the citizen-science eBird data, locational inaccuracies may also be a factor for the species' presence data. To reduce the effects of potential locational or mapping error in the LULC and presence data, LULC covariates used in the model were "neighborhood" measures of abundance for a given LULC class, rather than per-pixel measures. Use of a neighborhood LULC measure provided not only site-level habitat information, but also provided information on habitat in the surrounding area.

Individual species have unique, scale-dependent responses to landscape structure [35,36,37]. In modeling one individual species, it would be preferable to identify the appropriate scale of analysis that captures that species' habitat preferences. However, the objective here was to identify relative influences of climate and LULC across 50 different species. Optimizing (varying) the scale of analysis for each individual species introduces another (unwanted) variable into the assessment. One set scale was thus selected to minimize the scale-dependent impacts on modeling results. In tests of multiple landscape scales for SDMs, Cunningham and Johnson [35] found that scales between 800 m and 1600 m were the most suitable a majority of 19 bird species tested. For this study, a 5×5 pixel (1,250 m × 1,250 m) window around each point was chosen within which counts were tallied for each LULC class. The neighborhood counts for each LULC variable served as the LULC covariates within the modeling framework. A "LULC diversity" measure was also calculated, tallying the number of different LULC classes within each 5×5 window. The LULC diversity measure was also used as a covariate, as a measure of local landscape heterogeneity. **Table 3** summarizes the LULC covariates (as well as climate and topographic covariates described below).

Climate Data. The goal of this work was to examine long-term trends in bird species distributions in response to climate and LULC change. Global circulation models often produce climate data with monthly and yearly summaries, with year-to-year variability inherent in the output data. However, the use of average conditions was preferable to modeling with a single year of future climate projections, to minimize annual variability and focus on long-term trends [39]. A suite of global circulation models (GCMs) was used to obtain 30-year averages of climate consistent with IPCC SRES characteristics. A downscaling methodology similar to Hay et al. [39] was used to downscale coarse-scale climate data to a 4-km resolution for the conterminous United States [40], with data ultimately resampled to 250 m to match other covariates. Downscaled output was produced for six GCMs (BCCR-BCM2, CCSM3, CSIRO3.0-Mk, CSIRO-Mk3.5, INM-CM3.0, and MIROC 3.2) that provided climate data consistent with the IPCC SRES storylines (see http://www.ipcc-data.org/gcm/montly/SRES_AR4/index.html). Monthly data on average temperature, minimum temperature, maximum temperature, and precipitation were produced from each of the models. Variable averages across the six GCMs were calculated to reduce the bias present in any one individual model.

Covariates available for use in this assessment included not only yearly averages for temperature and precipitation, but also monthly averages, and monthly and annual minimum and maximum temperatures. However, not all variables were used, in order to reduce potential effects of multicollinearity. Correlation between potential climate variables was very high, particularly between the various temperature variables (Pearson correlation coefficient $r > 0.90$ between nearly all paired temperature variables, such as monthly temperature and annual temperature).

Table 2. Relative socioeconomic characteristics of the three IPCC SRES scenarios used in this assessment.

	A1B	A2	B1
Primary focus	Economic growth	Economic growth	Environmental sustainability
Globalization or Regionalization	Global Convergence	Regional Development	Global Convergence
Global Population	Increase to 8.7 billion by 2050, then slow decline	Continuous increase to 15.1 billion by 2100	Increase to 8.7 billion by 2050, then slow decline
Gross Domestic Product Growth	Very High	Medium	High
Energy Use	Very High	High	Low
Energy Strategy	Balanced, fossil fuel and alternative fuels	Regionally variable, based on local resources	Push to alternative and post-fossil fuel energy
Pace of technology change	Rapid	Slow	Medium
Technology diffusion	Rapid	Slow, regional variability	Rapid
Economic equity	Homogenization, higher incomes	Fragmented, uneven, continued income gaps	Homogenization, but lower incomes than A1B
Environmental Protection	Focus on "management" of resources rather than "conservation"	Uneven environmental management, protection higher in affluent areas	Broad support for environmental conservation, efficiency gains for resource use

See Nakicenovic et al. (2000) for additional information on SRES characteristics and Sohl et al. (2014) for how these characteristics were interpreted to create the LULC projections used in this assessment.

To minimize multicollinearity effects and to simplify data analysis, only the 30-year climate averages of annual temperature and annual precipitation, averaged across the six GCMs, were used as climate covariates in this assessment.

Bradley et al. [41] noted that the use of LULC data in conjunction with climate variables often does little to improve SDM results, due to collinearity of LULC and climate data at regional scales. There was little evidence of highly correlated LULC and climate variables in this assessment. Pearson correlation coefficients were computed for all LULC and climate covariate pairs. The highest correlation was between precipitation and the shrubland count ($|r| = 0.39$), while no other LULC and climate variable pair had $|r|$ values higher than 0.29.

Topographic Data. The few studies that have used projected LULC data in conjunction with projected climate data to look at future species distributions have often restricted themselves to those two categories of data [7,8,16]. However, when developing SDMs for current conditions, modelers tend to use a wider array of input variables, with topography often playing a key role [1,42,43,44]. Because the objective of this study was to assess the relative impacts of climate and LULC in "real-world" modeling applications, topography variables were included as covariates in

Table 3. Covariates used as predictor variables within Maxent.

Variable Category	Variable Name	Description
Land Cover	Cropland Count	5×5 neighborhood count of "cropland" pixels
Land Cover	Forest Count	5×5 neighborhood count of "forest" pixels (all forest)
Land Cover	Grass Count	5×5 neighborhood count of "grassland" pixels
Land Cover	Hay Count	5×5 neighborhood count of "hay/pasture" pixels
Land Cover	Shrub Count	5×5 neighborhood count of "shrubland" pixels
Land Cover	Urban Count	5×5 neighborhood count of "urban" pixels
Land Cover	Water Count	5×5 neighborhood count of "water" pixels
Land Cover	Wetland Count	5×5 neighborhood count of "wetland" pixels (all wetland)
Land Cover	LULC Diversity	5×5 neighborhood count of the number of different LULC classes
Climate	Average Temp	Average annual temperature
Climate	Average Precip	Average annual (total) precipitation
Topography	Elevation	Elevation data from National Elevation Database
Topography	Slope	Slope data derived from National Elevation Database
Topography	Compound Topographic Index	Compound Topographic Index data derived from National Elevation Database

All data were mapped to a common geographic extent at 250-m resolution.

this assessment in recognition that SDMs often do not focus solely on LULC and climate. Three topographic variables were used, based on the USGS National Elevation Dataset for the conterminous United States [45]; 1) elevation, 2) slope, and 3) compound topographic index (a measure of "wetness" and high flow accumulation). Each variable was resampled to match the geographic extent and 250 m spatial resolution of the LULC and climate covariates.

Methods

Maximum Entropy Modeling Framework. MaxEnt model [46] (Version 3.3.1) running on a Windows desktop was used to model bird species distributions. Maxent was designed to model species distributions based on presence-only species data [26]. Maxent statistically minimizes entropy between the probability density of "presence" data, and probability density from "background" data, as defined in covariate space [26]. Maxent has been shown to be one of the most effective methodologies for modeling species distributions when presence-only data are used [2,26].

Maxent estimates suitability for a given species by fitting feature classes based on environmental covariates. The filtered eBird data for each of the 50 species served as presence points. Environmental covariates were the LULC, climate, and topographic variables described above and shown in **Table 3**. Modeled feature classes in Maxent potentially included linear, quadratic, product, hinge, threshold, and categorical [46]. Linear features model linear response to a covariate, while quadratic features model response to the variable squared. Product features model interactions between paired variables. Hinge features model piecewise constant responses, while threshold features model abrupt boundary relationships between covariates and response. Category features are binary indicators used to indicate positive or null response to each class within a categorical covariate (e.g., thematic land cover map). All variables in this assessment were presented as continuous variables, including nominally thematic LULC data that were represented as counts within a 5×5 neighborhood around each point. Categorical features were thus not used in this assessment, but the other five Maxent features were used in modeling species response to the covariates.

The most widespread method for testing model results is a random hold-out of sample data [47]; 75% of the filtered eBird samples were used for training the model while 25% were reserved for testing. Maxent uses "background" points as locations where presence was not recorded, with background points either selected at random from the geographic extent of the study area, or specifically provided by the model user. The relationship between presence and background points in Maxent can strongly influence model results. Spatial bias in the presence points can result in a selection of background points with a fundamentally different spatial distribution [27], resulting in a model that represents the sampling effort as much as species presence [48]. A number of options were available to correct for spatial bias issues [25,46,49]. Several studies have discussed the use of spatially filtering or discarding records in over-sampled efforts [27,48,50], the approach used here and described above, with Kramer-Schadt et al. [26] finding it better reduced both errors of commission and of omission compared to other methodologies. Because the eBird data were spatially filtered, no attempts were made to account for bias through other measures.

Choice of the study area extent also can influence Maxent results [51,52]. VanDerWal et al. [53] found that model performance suffered when background points were selected from either too restricted or too broad a geographic extent, in relationship to the presence points. Specifically, if background points are selected from too broad a geographic area, predictive models were dominated by coarse-scale determinants of distribution (such as climate) [53], while those that use too limited a geographic area underestimate the importance of these variables [52] To reduce the influence of a mismatch between background area and "presence" points, a consistent buffer was applied around 2001 (contemporary) input presence points to construct a unique geographic extent for each species. The buffer zone was used to definitively set the study area for each species, both for defining where background points could be selected by Maxent, and to set the complete geographic range for modeling both current and future distributions. Ideally a unique geographic region would be optimized for each species according to characteristics of the observation data [51], but to facilitate comparison across the 50 species, a consistent buffered extent was used for all species. VanDerWal et al. [53] used a 200-km buffer, but initial experimentation for this assessment found that to be too restrictive for changes in conterminous United States bird species range from 2001 to 2075, with some species' ranges shifting by more than 200 km. A 500-km buffer around input eBird points was used, restricting both the range from which background points could be selected, and restricting the prediction space for each species' range.

Remaining parameterization of Maxent largely followed model defaults. Anderson and Gonzalez [54] and Warren and Seifert [55] recommended species-specific tuning of Maxent settings, noting that the regularization value (used to restrict model "over-fitting" to input data) had a large effect on results. However, Phillips and Dudik [46] tested regularization values and found that "regularization parameters which are the defaults in MaxEnt software...are well suited for a wide range of presence-only datasets." The six feature types are also selectable, yet Syfert et al. [56] found little influence on model results by varying the feature types that are used. Phillips and Dudik [46] found that using the default 10,000 background points achieved similar model results as if all possible background sites were used; the default setting was thus used. Default settings were also used that enabled "clamping" of covariate and feature values for the 2075 model simulations. With the model trained on the 2001 covariate data, the potential existed for "novel" covariate values when the model was applied in 2075, using projected climate and LULC data. For the 2075 model simulations, the enabled clamping resulted in a rescaling of both covariate and feature values if their values were higher or lower than those found in the training data. Values higher than those encountered in the training data were rescaled to the training data maximum, while values lower than those encountered in the training data were rescaled to the training data minimum. The implications of the use of clamping are provided in the discussion section.

Parameterizing Maxent as described above, initial model simulations for each species were conducted using the filtered eBird data for presence points, and the 2001 LULC, 2001 climate, and topographic variables as covariates. Twelve model simulations were made in total for each species (**Table 4**). A base model simulation was done for 2001 using all variables (simulation 1), while additional simulations were done for 2001 with climate and topography (excluding LULC) (simulation 2) or land cover and topography (excluding climate) (simulation 3). The model developed for simulation 1 was applied for 2075 to examine potential future impacts of climate change, LULC change, or both (topographic variables are static in all simulations). For each of the three IPCC scenarios, simulations were done with all 2075 variables, with projected climate but static LULC, and with

Table 4. Twelve model simulations were conducted for each species, three for 2001 and nine for 2075.

Simulation	Description	Climate (Scenario)	LULC (Scenario)	Topo Data	Scenario
1	2001 All	2001	2001	Yes	-
2	2001 Climate	2001	-	Yes	-
3	2001 LULC	-	2001	Yes	-
4*	2075 A1B All	2075 A1B	2075 A1B	Yes	A1B
5*	2075 A1B Climate Change	2075 A1B	2001	Yes	A1B
6*	2075 A1B LULC Change	2001	2075 A1B	Yes	A1B
7*	2075 A2 All	2075 A2	2075 A2	Yes	A2
8*	2075 A2 Climate Change	2075 A2	2001	Yes	A2
9*	2075 A2 LULC Change	2001	2075 A2	Yes	A2
10*	2075 B1 All	2075 B1	2075 B1	Yes	B1
11*	2075 B1 Climate Change	2075 B1	2001	Yes	B1
12*	2075 B1 LULC Change	2001	2075 B1	Yes	B1

Model simulations variously include or exclude climate and LULC covariates in order to assess the individual effects of each.
*Simulations for 2075 used the model developed for run 1 (2001 "All"), applying 2075 climate and/or LULC data from the appropriate scenario.

projected LULC but static climate. Keeping either climate or LULC static from 2001 to 2075 allowed for the examination of the relative effects of projected climate versus projected land use change on future bird species distributions. Three 2001 simulations and nine 2075 simulations were thus conducted for each of the 50 species, resulting in 600 individual model simulations.

Assessing Model Results. Several different metrics were used to assess the relative impacts of climate and LULC change on bird species distributions. The three 2001 model simulations were assessed for model fit through a comparison of Area Under the Curve (AUC) of the Receiver Operating Characteristic (ROC). AUC values represent the probability that a randomly selected "presence" site will have a higher AUC value than a randomly chosen "background" site. Comparison of AUC scores was used to examine relative impacts on model fit when LULC or climate data were excluded from the analysis. A second criterion was the relative contributions of the covariates to model results, measured by relative changes in regularized training gain between variables. This information was provided as a "percent contribution" from Maxent. A third criterion was a comparison of modeled "suitable" range for each species. Elith et al. [26] cautions against cross-species comparisons using logistic output from Maxent, as probability of presence is relative to the sampling effort for a given species. However, changes in relative range for each individual species can be identified by applying a threshold value to Maxent's logistic output, to differentiate between likely presence and absence locations. The "maximum sensitivity plus specificity" threshold was used [15,38], a thresholding technique that limits both errors of commission and errors of omission and has been found to outperform other techniques [57,58].

The 2075 simulations were evaluated by assessing changes in "suitable" breeding range as compared to 2001. Net change in range area was determined for each species by first applying the "maximum sensitivity plus specificity" threshold to modeled output and then differencing the threshold results, with comparisons of net effects of climate change alone, LULC change alone, and both climate and LULC change from 2001 to 2075 (for each scenario).

Finally, results were examined in terms of species range and relationship to the conterminous United States study area. While data sources and analyses often stop at political boundaries, species

ranges obviously do not, and the use of conterminous United States borders for this work resulted in the modeling of truncated ranges for many species. Both 2001 and 2075 model results could be impacted dependent upon whether the entire range was modeled or if one or more maximum extent boundaries were artificially truncated [51,52,53]. Many SDM applications model truncated species distributions (see discussion below); assessing results on species range characteristics allowed for an examination of LULC and climate impacts across a variety of "real-world" modeling situations. For each of the assessment criteria discussed above, mean values were provided (**Table 5**) for species within the following "range classes": 1) "Single Truncated" (species with either the northern *or* southern extent artificially truncated by United States borders, 2) "Double Truncated" (species with ranges truncated at *both* the northern and southern United States border, and 3) "Whole Ranges" (species with >95% of current breeding ranges found within the conterminous United States, measured with NatureServe species distributions [24].

Results

2001 Models ("current" species' distributions)

The 2001 models were assessed for model fit using AUC scores. Values of 0.5 indicate model fit was no better than random, while increasing values above 0.5 indicated an improved model fit. **Figure 1** provides AUC scores for the three 2001 model simulations for each of the 50 species. AUC scores ranged from a low of 0.716 to a high of 0.987, with considerable variation among species, as well as among the three model simulations for a given species. Model simulations with all variables included (simulation 1) had the highest mean AUC score, at 0.891, and the highest AUC score for each of the 50 species. AUC scores were significantly lower (p<0.001; paired t-test) for both simulation 2 (climate, no LULC) and simulation 3 (LULC, no climate), with mean AUC scores of 0.863 and 0.874, respectively. Results indicate significantly poorer model fit when LULC data were excluded than if climate data were excluded (p<0.01; paired t-test). By range class AUC scores were significantly lower when either LULC or climate data was omitted, for all range classes (p< 0.01, paired t-test) (**Table 5A**). AUC scores overall were similar for the Single Truncated and Whole Range classes, but were much

Table 5. Impacts on assessment variables by range class.

	Single Truncated	Double Truncated	Whole Range
5(A) –2001 MODEL FIT (AUC Score – Mean Value)			
All Variables	0.916	0.839	0.906
LULC, No Climate	0.891	0.834	0.892
Climate, No LULC	0.891	0.799	0.888
5(B) –2001 VARIABLE CONTRIBUTION (in percent)			
Climate Variables	49.5%	52.8%	52.7%
Topography Variables	12.8%	6.0%	13.9%
Land Cover Variables	37.7%	41.2%	33.4%
5(C) –2001 Range (Mean Values – Percent of conterminous United States area)			
All Variables	23.0%	42.2%	26.6%
LULC, No Climate	28.4%	43.7%	30.1%
Climate, No LULC	31.1%	57.0%	32.8%
5(D) –2075 Breeding Range (Mean Values – Percent change from 2001)			
All Variables	−9.9%	+2.6%	+12.0%
LULC, No Climate	+3.8%	+3.5%	+1.5%
Climate, No LULC	−13.0%	+1.2%	+10.2%

Values represent mean values across all species in a class. "Single Truncated" (27 species) represents species with ranges artificially truncated at either the north *or* south by the United States border. "Double Truncated" (15 species) represents species with truncated ranges that extend to or past the United States/Canada border in the north and the United States/Mexico border in the south. "Whole Range" (8 species) represents species where >95% of the current range is found within the conterminous United States.

lower on average for the Double-Truncated class. Omission of LULC resulted in the lowest overall AUC score for every species in this class. The relative impact of LULC or climate data omission was more balanced for the other two range classes and varied by species.

Figure 2 depicts Maxent-provided proportional contributions of each covariate to the regularized training gain, aggregated across all 50 species for simulation 1 (all variables modeled). The climate covariates played an important role in shaping 2001 simulations, with annual temperature and precipitation providing 51.0% of the contribution to model results. Temperature was one

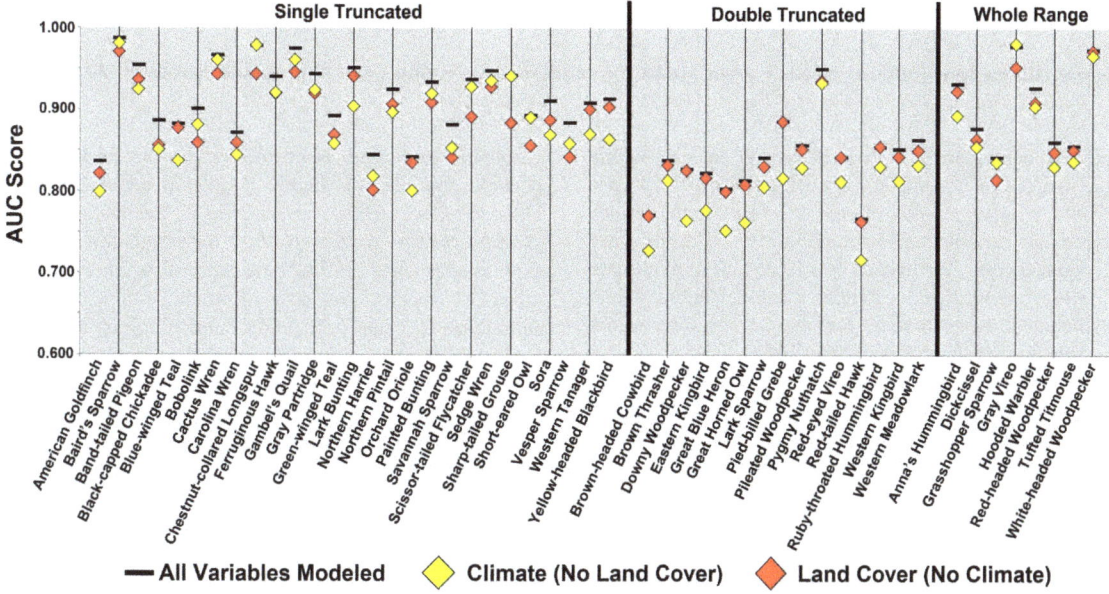

Figure 1. AUC scores for each species, for run 1 (all variables modeled), run 2 (Climate, no Land Cover), and run 3 (Land Cover no Climate). AUC scores are also parsed by range class.

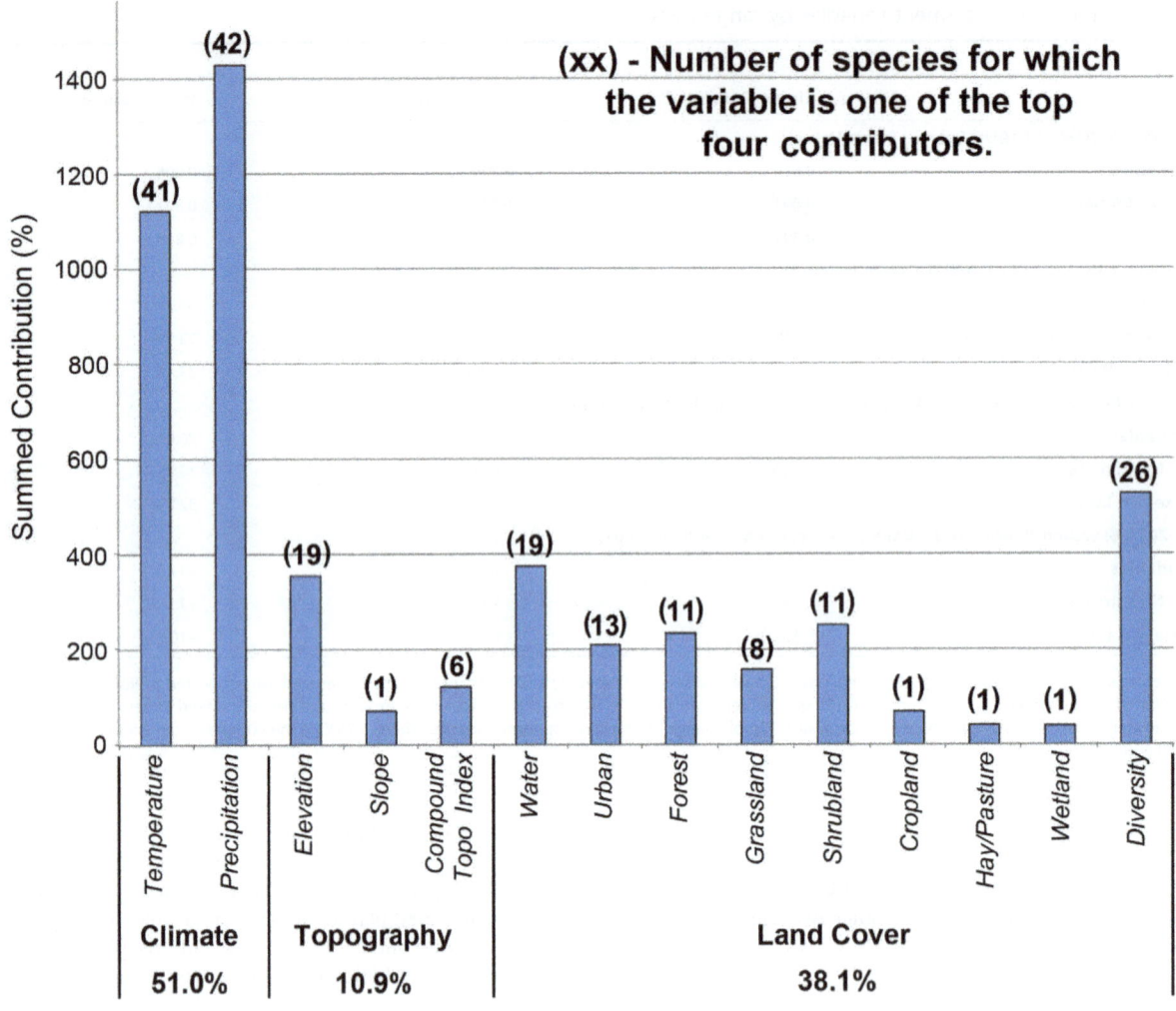

(xx) - Number of species for which the variable is one of the top four contributors.

Figure 2. Proportional contributions of each covariate to the regularized training gain, aggregated across all 50 species.

of the top four contributing covariates for 41 species, while precipitation was one of the top three covariates for 42 species. LULC variables in aggregate contributed 38.1% to model results, while topographic variables contributed 10.9%. Results vary among individual species, but overall, it is clear that both climate and LULC were important contributors to model output when both were included as covariates. Results were similar when categorizing species by range class (**Table 5B**).

Figure 3 provides a comparison of modeled "suitable" area for each species, among model simulations 1, 2, and 3, using the unique maximum sensitivity plus specificity threshold criteria for each species and simulation. Values are presented as a percentage of the total land surface for the conterminous United States with Maxent logistic output values above the threshold criterion. While the predicted suitable range for a given species was sometimes similar across each of the three 2001 model simulations for a species, in many cases, the area deemed to be suitable varied dramatically depending upon what variables were used as covariates. For 36 of the 50 species, the area deemed suitable was highest in simulation 2, when only climate and topographic

variables were used as covariates (LULC excluded). The area deemed suitable was nearly double in some cases (e.g., Great Horned Owl (*Bubo virginianus*), Yellow-headed Blackbird (*Xanthocephalus xanthocephalus*)) for simulation 2, as opposed to simulation1 when LULC data were also incorporated. For the other 14 species, the area deemed suitable was highest for simulation 3, when only LULC and topographic variables were used as covariates (climate excluded). Adding covariates to the model, be they LULC or climate, clearly acted to further define (and restrict) the area deemed to be suitable for species' habitation. Using climate data alone resulted in broad, overly generalized suitability ranges if LULC data were not used to help further define suitable landscapes. Results were similar when evaluating the three different range classes (**Table 5C**), with the smallest range consistently modeled when all variables were used as covariates. However, for the Double Truncated range class, the omission of LULC data from the model resulted in a much larger increase in range as compared to the other two range classes, while omitting climate data had little impact.

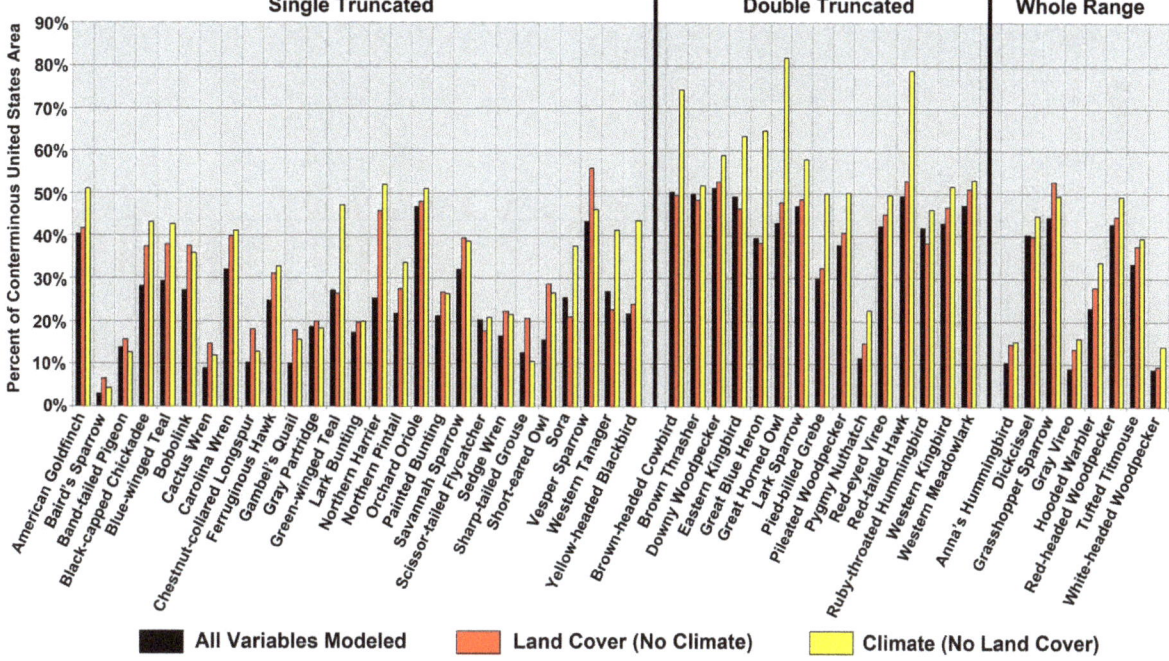

2001 - Area of "Suitable" Range
Presence threshold - Maximum sensitivty plus specificity

Figure 3. Area (percent of conterminous U.S. land mass) classified as suitable to support a given species, for model run 1 (all variables modeled), run 2 (Climate, no Land Cover), and run 3 (Land Cover no Climate). Suitability was determined by applying the maximum sensitivity plus specificity threshold to Maxent logistic output. Results are also parsed by range class.

2075 Models ("projected" species' distributions)

Figure 4 depicts projected changes in range for each of the 50 species, measured as change relative to the range modeled in 2001 (simulation 1), using the maximum sensitivity plus specificity threshold to differentiate between presence and absence. Range differences are provided for each of the 3 model simulations, for each of the 3 scenarios, with bar height providing the mean change in range across all three scenarios, and deviation bars providing the variation between scenarios. Depending upon species, modeled changes in range varied according to which covariates were used and between different IPCC scenarios. Changes in range varied from a near complete loss of all conterminous United States suitable range (Baird's Sparrow(*Ammodramus bairdii*)) to range expansions that nearly double the current range (Cactus Wren (*Campylorhynchus brunneicapillus*), Gambel's Quail (*Callipepla gambelii*), Gray Vireo (*Vireo vicinior*)). **Figure 4** indicates that the magnitude of projected changes in range was much more strongly impacted by projected climate change than by projected LULC change, when using a threshold to define suitability. When only LULC changed (climate static) from 2001 to 2075, changes in projected ranges from 2001 were highly significant ($p<0.001$; paired t-test) but were never more than 20% (either positive or negative). When only climate changed (LULC static) from 2001 to 2075, range changes were often quite dramatic, with 20 species showing range changes of 25% or more for a given scenario. Climate and LULC could either both influence species' distributions in the same direction, or a positive species response to one category of covariates could be offset by a negative species response to the other category.

Table 5(D) and Figure 4 show substantial differences in the relative effects of LULC and climate on 2075 model results,

depending upon range class. The most dramatic overall changes in range were in the Single Truncated class, where climate change obviously had a strong effect on model results. Climate change had a much more muted impact on the Double Truncated class, with low overall changes in range. Climate had moderate to strong impacts for the Whole Range class. The impacts of LULC change were much more consistent across range classes than were the impacts of climate change.

Species Focus - Hooded Warbler (*Wilsonia citrina*)

While it is impractical to individually discuss each of the 50 modeled species, the relative impacts of climate and LULC change on one species, the Hooded Warbler (*Wilsonia citrina*), are highlighted here to demonstrate specific impacts of climate and LULC. The Hooded Warbler is a forest-dependent species that primarily breeds in the eastern United States. **Figure 5** provides 1) a map of Maxent logistic output for 2001, using simulation 1 (all covariates modeled), and 2) changes in output for each 2075 scenario, and for each 2075 model simulation. The AUC score for simulation 1 (2001) indicated a high-level of model fit (AUC = 0.927), with precipitation, temperature, and forest count (in relative order) measured as the three covariates contributing the most to model results. For simulation 1, 23.1% of the conterminous United States was deemed "suitable" (threshold) range for the Hooded Warbler. The predicted range sharply increased to 27.9% in simulation 3 (climate excluded) and 33.9% for simulation 2 (LULC excluded), a pattern seen for many species (**Figure 3**).

Changes in predicted range by 2075 indicate a strong influence of both climate and LULC change (**Figure 5**). The economically focused A1B and A2 scenarios are similar, as a changing climate

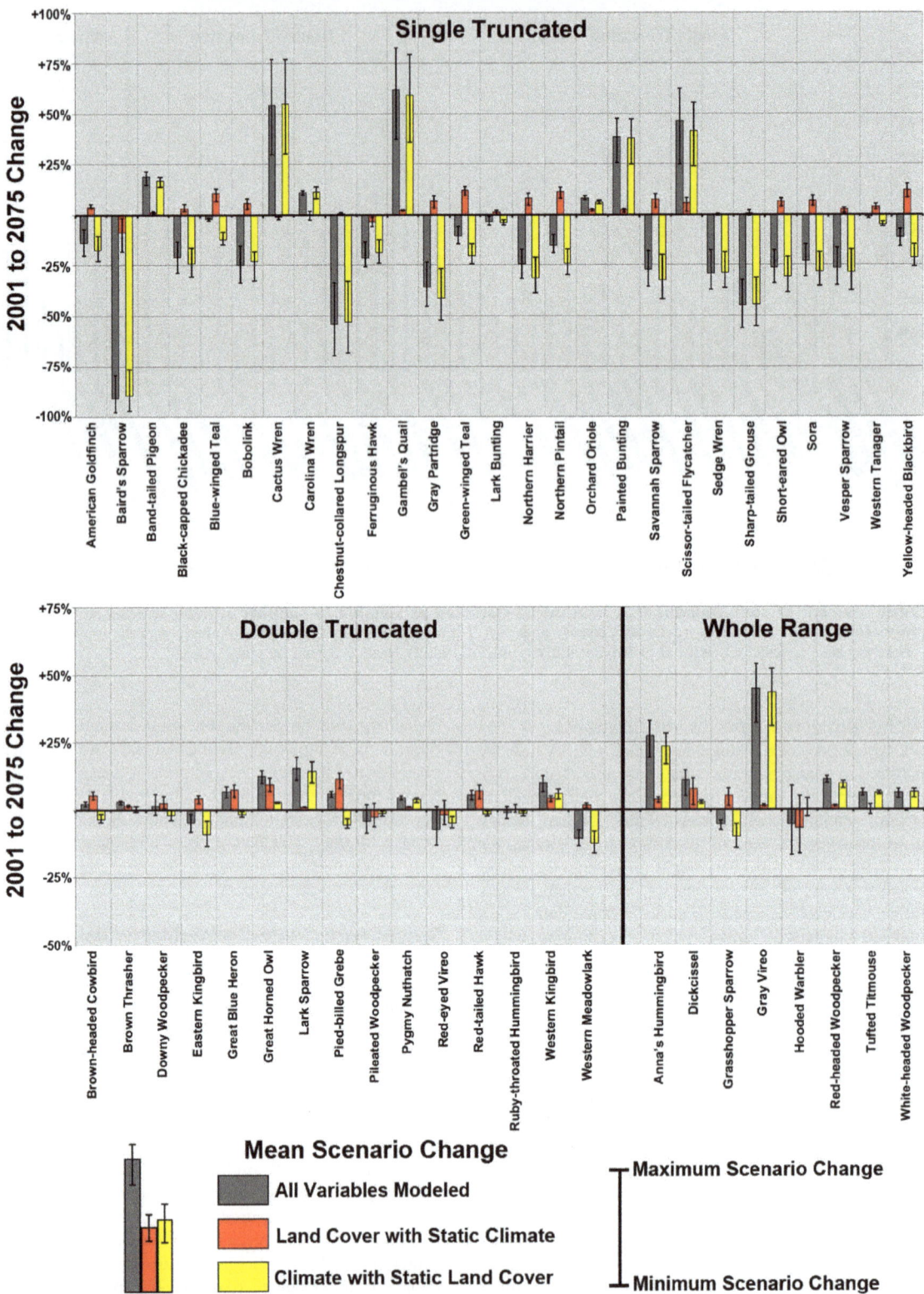

Figure 4. Changes (2001 to 2075) in area classified as suitable to support a given species. Change is presented as area change, relative to the contemporary (2001) modeled range. Bar height represents mean change across the 3 IPCC scenarios, while error bars represent scenario variability. Suitability was determined by applying the maximum sensitivity plus specificity threshold to Maxent logistic output. Results are also parsed by range class.

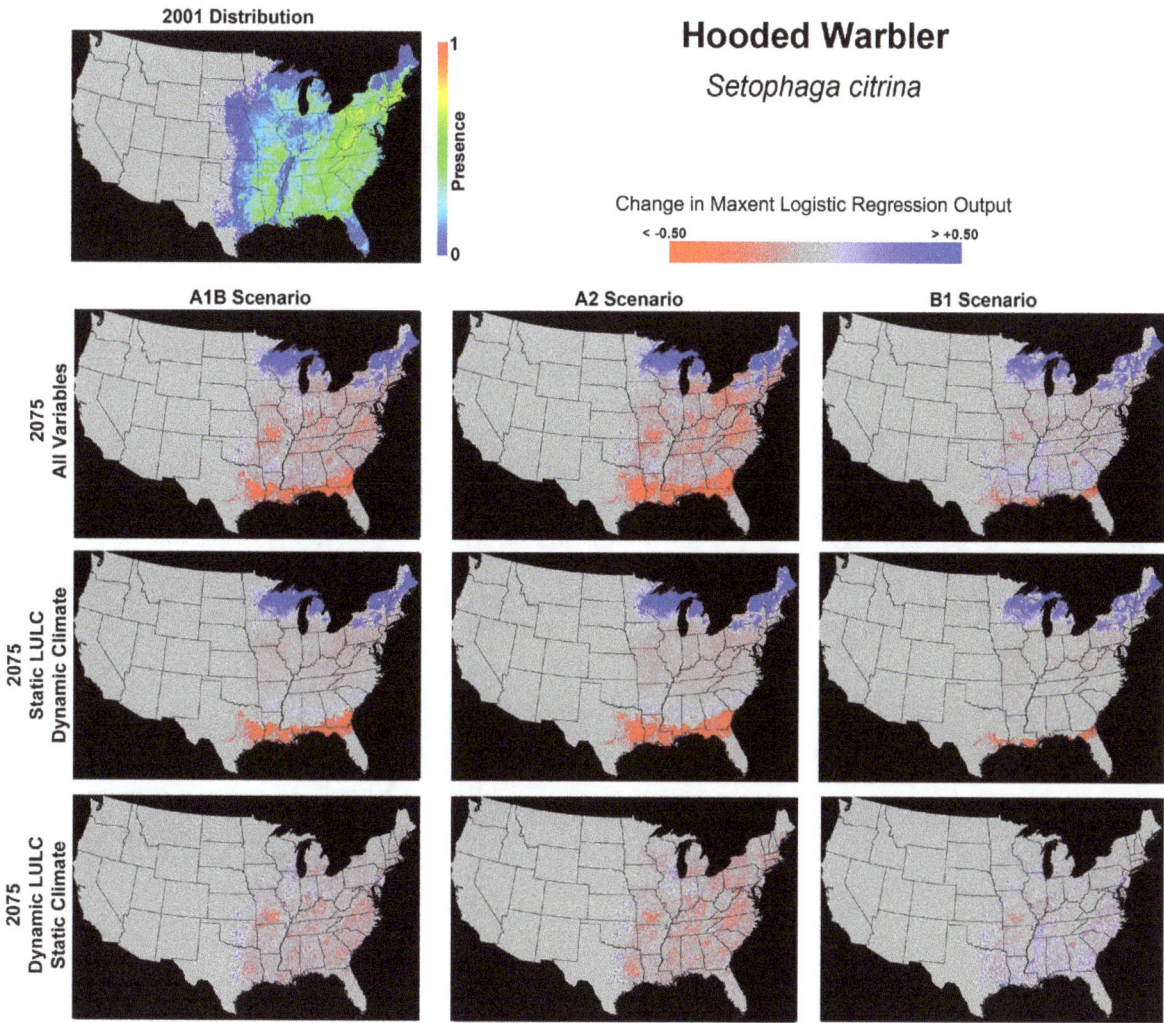

Figure 5. Maxent 2001 logistic output for the Hooded Warbler, and projected changes under each 2075 scenario and model run. Climate change results in broad northward shifts in species range across all scenarios. LULC change alters the local pattern of habitat suitability, with losses under the A1B and A2 scenarios, and general increases in the B1 scenario.

resulted in strong shifts in overall species range, with large contiguous bands of losses of range in the south and gains in the north. The effects of LULC change are more fragmented, but substantial forest loss results in local areas of decline throughout much of the eastern United States. The effects of climate change are more muted for the environmentally focused B1 scenario, with less severe shifts to the north. While local areas of forest loss do result in range declines in the B1 scenario, afforestation and forest regeneration result in higher presence scores in many locations.

Figure 6 displays modeling results for the Hooded Warbler for both 2001 and 2075 (A2 scenario) for a smaller area within their current breeding range. At this scale the relative impacts of both LULC and climate are evident for both current (2001) modeling, and for future projections. **Figure 6(a)** and **6(e)** show LULC change from 2001 to 2075, characterized by substantial expansion of urban and agricultural lands, at the expense of forest land. **Figures 6(b)**, **6(c)**, and **6(d)** show model results with LULC and topography as covariates, all variables as covariates (LULC, climate, and topography), and climate and topography as covariates, respectively. Without the use of climate data, suitability was highly heterogeneous, but higher elevation areas that currently do not support Hooded Warbler populations (e.g., parts

of the upper-left quadrant) often had high suitability values even with the use of topographic information (**Figure 6(b)**). Without the use of LULC data, suitability was less heterogeneous and the cooler high-elevation areas were characterized by lower values, yet areas of dense anthropogenic land-use that are unsuitable for Hooded Warbler breeding were often characterized as highly suitable (**Figure 6(d)**). The use of both LULC and climate data, in conjunction with topographic data, resulted in a heterogeneous distribution of suitability values, capturing both the influence of cooler high-elevation areas as well as areas of dense anthropogenic land-use (**Figure 6(c)**).

For 2075 model simulations, **Figure 6(g)** shows the impacts on Hooded Warbler range when both projected climate and projected LULC are used in the model. Range expansion occurred towards higher elevations as a warming climate results in more suitable breeding conditions. LULC change, primarily urbanization and agricultural expansion, resulted in large but heterogeneous losses of breeding range, counter-balancing range gains due to climate change. **Figure 6(f)** shows a model simulation with projected 2075 LULC but a static 2001 climate. Without the use of projected climate data, the range expansion due to warming was not captured (**Figure 6(i)**). **Figure 6(h)**

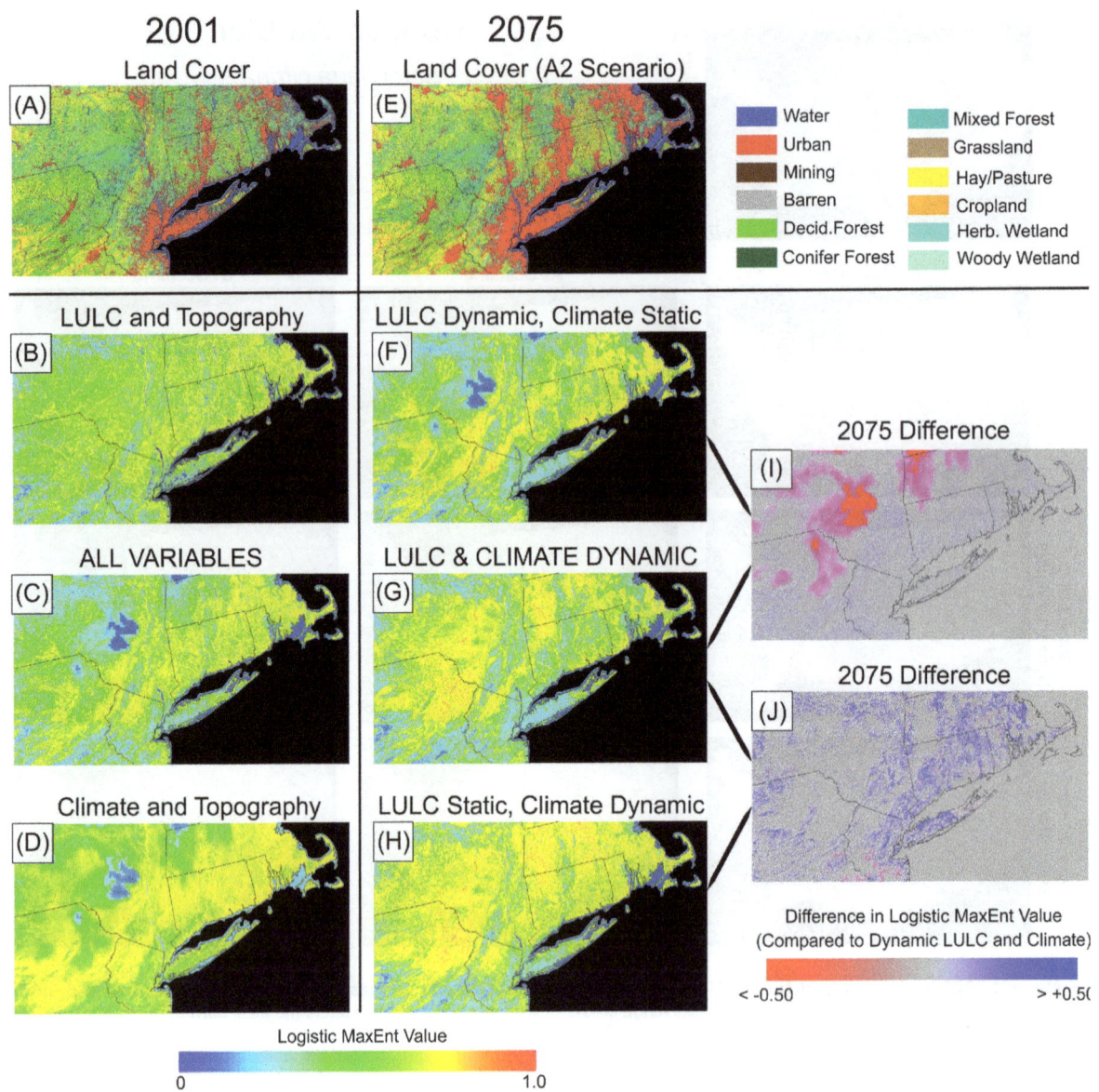

Figure 6. LULC and model results for a portion of the Hooded Warbler's range. Panels on the left (6 a–d) depict 2001 LULC, and the three model runs for 2001. Panels in the middle (6 e–h) depict 2075 LULC for the A2 scenario, and the three model runs for 2075 for that scenario. The two panels on the right depict differences in model results (compared to 6 g, when all variables were modeled) if 1) climate was held static for 2075 (6i), and 2) LULC was held static for 2075 (6j).

shows a model simulation with projected 2075 climate but a static 2001 LULC, where many climatically suitable areas were still noted as suitable for breeding despite the substantial loss of forest habitat (**Figure 6(j)**).

Specific results for all 50 species, including range maps as provided in Figure 6, are accessible through a companion website (http://landcover-modeling.cr.usgs.gov/sdm.php).

Discussion

Research Questions

What are the relative influences of LULC and climate in modeling contemporary (2001) breeding bird distributions? Clearly both climate and LULC change impact current bird species distributions, with relative impacts that are species specific. 2001 model fit was generally better with LULC

simulations (climate excluded) than for climate simulations (LULC excluded), yet climate data covariates contributed more to model results than LULC data. One story that arises from these seemingly conflicting results is one of scale. Results suggest that climate data alone, without constraints afforded by the use of habitat (LULC) data, provide a "broad-brush" picture of suitability for a given species. LULC data alone excel at providing local-level insight to site-level habitat suitability. Given the inherent heterogeneity of the moderate-scale LULC data used here compared to variations in climate across geographic space, it is not surprising that climate alone offers only a general characterization of species range. For the 2001 species models, the modeled area deemed to be suitable to support a species was generally much higher when climate data were used without LULC data (**Figure 3**). The addition of LULC data to climate-based model simulations greatly restricted modeled species ranges in most cases. Prince et al. [59] similarly described climate as determining overall potential carrying capacity for a species, but noted the impact of climate change itself may be overestimated, as other factors that determine local suitability must be assessed. These results suggest that the use of climate data, without supporting LULC data, likely results in errors of commission, where climatically suitable regions are labeled as appropriate for supporting a given species, despite underlying LULC conditions that make actual presence unlikely. Araujo and Peterson [9] discussed such commission errors in bioclimatic envelope modeling, attributing overzealous predictions of range to an incomplete model; in this case, climate-only models for 2001 are "incomplete" without supporting LULC data.

Actual species range in relationship to the modeled study area influenced 2001 model results. Model fit was negatively impacted if the study area was largely contained within the actual species range. This was the case for the Double Truncated species, where ranges spanned all latitudes in the conterminous United States and were truncated at both the northern and southern borders. Climate data, in particular temperature data, thus did little to improve model fit, as species occurrences already spanned most potential climate regimes within the conterminous United States. With the resultant small impact of climate data, overall model fit suffered (**Figure 1; Table 5a**) and the addition of climate data did little to improve model results over the LULC and topography model (**Figure 1**). Modeled species range was also influenced by the relationship between actual range and study area. As noted above, the use of climate data without LULC data often resulted in errors of commission, but these errors were magnified for the Double Truncated species, with an over-prediction in suitable range as compared to the complete model with both climate and LULC data (**Figure 3**).

What are the relative impacts of projected climate and projected LULC change on breeding bird distributions? For modeled species ranges, projected changes in climate provided more dramatic shifts in future species' ranges than did projected LULC change. LULC change alone altered suitable range by no more than 20% for any species, yet climate change resulted in shifts of 50% or more for several species. Differences between the three different scenarios were often substantial, with some scenarios projecting double the range shift compared to other scenarios. However, the overall storyline was climate change impacting net changes in species range more than projected LULC change.

The relationship of the actual species range to the study area obviously affected 2075 results, with climate impacts often over- or under-estimated in relationship to LULC impacts, depending upon species. The Single Truncated class contained species where either the northern or southern extent of their actual range was artificially truncated by the borders of the study area. With a warming climate, for species with ranges truncated along their southern extent but not the north, the models thus predicted overall range expansion to the north, without capturing the (presumed) range contraction due to climate change at the species' southern range extent. Conversely, for species with ranges truncated along their northern extent but not the south, the models predicted overall range contraction, capturing contraction in the south but failing to capture (presumed) expansion in the north. By only capturing "half of the story" (i.e., either capturing range expansion in the north or range contraction in the south), these results provided an unrealistically high impacts of climate on *net* range, either positive or negative. While the impacts of climate on *net* change were thus likely overestimated for these species, *gross* change was likely underestimated, as half of the story was

"missing". For the Double-Truncated class, the relative impacts of climate change to LULC change were likely underestimated. For these species, the impact of climate on species range was artificially dampened by the truncation of northern and southern range boundaries, areas where range could potentially expand or contract, respectively, due to a warming climate. For the Whole Range species, the relative impacts of climate change versus land use change vary, with net change values only providing part of the story as evidenced when assessing results for the focus species, the Hooded Warbler.

What are the specific impacts of climate and LULC on one focus species, the Hooded Warbler? The presented results for the Hooded Warbler mirrored those for many of the 50 modeled species. Net change in breeding range area showed relatively little change, while geographic patterns change dramatically. Climate change resulted in a broad overall shift in range to the north and to higher elevations, while LULC change resulted in heterogeneous, local-scale changes in habitat suitability. The breeding distributions of the Hooded Warbler have been found to be highly correlated with climate variables [60]. The species was unknown as a breeder in Canada until 1949, but with a warming climate they have started to breed in increasing numbers in extreme southern Ontario [60]. Melles et al. [60] modeled the relationship between climate and habitat covariates and the Hooded Warbler range, and found strong relationships between range expansion to the north and changes in climate over the last few decades, with habitat availability acting as a constraint on expansion. Naujokaitis-Lewis et al. [61] examined the potential impacts of climate change out to 2080 on Hooded Warblers and projected breeding range shifts to the north with characteristics dependent upon which GCM was assessed. However, they also found that land-use pressures around the Great Lakes were limiting factors to range expansion, and recommended future work that focused on "the development of more realistic (habitat) loss scenarios". The newly available LULC projections used in this work allowed for such an analysis.

As discussed, most future projections use projected climate data but ignore future LULC change. **Figure 6** clearly indicates that for a species such as the Hooded Warbler where climate change drives a broad overall shift in range to the north and to higher elevations, the modeled extent of suitable range at a local level can potentially be misrepresented without the use of projected LULC data. In this case, habitat loss due to urbanization and agricultural expansion would be missed without the use of LULC data, resulting in an over-prediction of suitable range (**Figure 6j**). Alternatively, without the use of projected LULC data, suitable range may be under-predicted if beneficial LULC change occurs (e.g., Grasshopper Sparrow (*Ammodramus savannarum*) results within the Eastern United States, where projected clearing of forest land in most scenarios resulted in more suitable habitat conditions by 2075). Exclusion of climate data can also result in a misrepresentation of modeled range. For the Hooded Warbler, range expansion to higher elevations was missed for the 2075 model excluding climate data (**Figure 6i**).

What are the implications for the use of climate and LULC data in SDMs? For contemporary species modeling or for projected changes in species range, both climate and LULC data should ideally be used. In general, model fit consistently increases with the use of both climate and LULC data, while predicted suitable range decreases. The implication is that information is missing from SDMs if both climate and LULC are not used as covariates. For modeling of current species range, areal summaries of modeled range (**Figure 3**) as well as spatially explicit maps of modeled range (**Figure 6b, c, and d**) show that

SDMs relying on climate data without LULC data provide only a broad-brush, generalized species range, while LULC data alone provide site-level information on habitat suitability while omitting climatic thresholds of unsuitability. Dependent on application, a broad-brush generalization of a species range may be adequate. However, it should be recognized that the results likely over-represent the area of suitable range and fine-scale detail is unlikely to be obtained.

Exclusion of LULC data is primarily an issue for projections of future species' range. As shown here, bioclimatic modeling where LULC information is not included or is considered static likely results in a misrepresentation of future species' range. For example, Hooded Warbler results provided here and in past studies indicate that while climate drives broad-scale shifts in range, SDMs likely misrepresent the extent of future range shifts if LULC change is not taken into account. Given how little projected LULC data is used in modeling future species distributions, quantitative estimates of range shifts are likely overestimated if habitat loss dominates projections of LULC change, or underestimated if habitat gain dominates projections of LULC change. Bioclimatic models that do not use any form of LULC information, even static LULC information for the future, likely overestimate suitable ranges (**Figure 3**).

The relationship between the study area and the actual species range also needs to be strongly considered, both in the project design and assessment phases. The methodology used here mimics that of many modeling applications. Of all the citations included in this paper where species distribution and/or probability-of-occurrence modeling was done, over two-thirds (21 of 31) of model applications assessed only partial/truncated species ranges. Many of the recommendations referenced in this paper with regard to model parameterization and handling scale issues [35,38,46,53], spatial bias and other issues with presence data [23,48,50], and relative influences of climate and LULC [1,8,12,16,59,60] were derived from studies where only partial ranges were assessed. Despite the prevalence of modeling of partial ranges, the results here indicate that caution is needed in project design, both for accurate modeling of species range, and for the interpretation of modeling results. Modeling of an entire species' range may improve model fit and enable a more direct interpretation of results, yet is often not practical due to data or processing limitations. While modeling a partial range is thus unavoidable in many cases, model results should be interpreted within the context of the overall project design and the relationship between species range and the study area. Modeling results may still be "valid" when using truncated ranges, but if the intent is to study the impacts of climate change on species distribution, for example, then the use of a "double-truncated" study area boundary would obviously be a poor choice, as the effects of climate would likely be artificially muted. If the intent is to quantify specific impacts of LULC and climate, disentangling the relative effects of LULC, climate, and other covariates would be complicated by the modeling of truncated ranges.

Comparison to Existing Research

The conceptual approach behind Barbet-Massin et al. [7] modeling of bird species in Europe and Matthews et al. [1] modeling of eastern U.S. bird species are likely the most similar work to this assessment. Each assessed a large number of species across broad geographic regions, and both incorporated projected climate and projected LULC data. Similar to Barbet-Massin et al. [7], this assessment found that LULC-based models alone predicted smaller overall shifts in future range size than did climate-based models. However, these results differ from multiple studies that discussed the relative influence of climate versus LULC, including components of Barbet-Massin et al. [7]. Barbet-Massin et al. [7] found that modeling accuracy was higher with climate-only variables than with habitat-only variables; in this study, the opposite was true in the majority of species that were assessed. Thuiller et al. [12] found that the inclusion of LULC covariates improved explanatory power of bioclimatic models, but that the "addition of land cover variables to pure bioclimatic models does not improve their predictive accuracy". In this assessment, on average, AUC scores declined more in the absence of LULC data than in the absence of climate data, while for all 50 species, 2001 model fit was improved when LULC data were included as a covariate as opposed to models with only climate and topographic data.

The differences in results may potentially be explained by 1) the difference in scale between the different assessments, 2) variations in the number of climate covariates, and 3) the use of topographic data within this assessment. Barbet-Massin et al. [7] used much coarser, 0.5-degree resolution LULC data, and noted that "such a resolution was probably too rough to precisely account for habitat factors." Thuiller et al. [12] also used a very coarse spatial resolution (50-km grid cells) and noted results may differ at finer resolutions. Bucklin et al. [13] similarly found that LULC variables provided little benefit in SDMs, but noted that both thematic and spatial resolution improvements over their LULC data source may have provided different results. Barbet-Massin et al. [7] and Thuiller et al. [12] also noted the lack of a measure of fragmentation or landscape heterogeneity in their assessments; in this study a LULC diversity measure was used to represent heterogeneity. This assessment also used only two climate covariates, while Barbet-Massin et al. [7], Thuiller et al. [12], and Bucklin et al. [13] each used eight climate covariates. Additional research is needed to assess the optimum combination of covariates and how covariate choice impacts results, particularly for heavily correlated climate covariates. The use of topographic data in this assessment also may have impacted the relative impacts of LULC and climate data. The information content provided by topographic data alone, or topographic data in combination with LULC data (e.g., "product features" within Maxent that assess 2-way interactions between covariate pairs) may partially mimic or replace the information content that is provided by climate variables [42].

Matthews et al. [1] also modeled U.S. bird species, and also used projected climate data, projected LULC data, and topographic data to assess future changes in distributions. They modeled the eastern portion of the United States at a very coarse spatial resolution (20-km grid cells), with changes in tree species representing the only modeled form of LULC change. Similar to their results, the predictive power (indicated by goodness-of-fit measures) of the models described here decreased when only climate and elevation data were used as predictor variables (i.e., LULC excluded). Even with the differences in spatial resolution, both studies found that modeling with only climate and topographic variables leads to generalized species distribution maps that lack fine-scale detail. Matthews et al. [1] noted that modeling with climate and topographic data alone makes the resultant models much more susceptible to over-prediction of future impacts of climate change on species range. It should also be noted that their use of the eastern United States as the study area resulted in artificial truncation of nearly all modeled species ranges.

Caveats and Future Research

In assessing potential future changes in species ranges, caution has been recommended when attempting to apply a contemporary model to future climate conditions [26,46]. Transferability of model results is confounded when novel conditions (i.e., specific combinations of covariates not found in the original model's training data) are found for future dates or for other geographic regions. In this assessment novel conditions were most likely to occur with higher temperatures due to climate change. However, the use of the 500-km buffer to establish the study area for each species, as well as the use of the clamping feature in Maxent, resulted in a muted influence of novel conditions on model results. Selecting background points within a 500-km buffer of a species' current range enabled the collection of background points with higher (points selected south of the breeding range) or lower (points selected north of the breeding range) average temperatures than those found in the breeding range. Thus temperature was often used as a threshold feature in species' models, with conditions modeled as unsuitable if temperature in a given location was above or below a modeled tolerance level for the species. For example, the breeding range for the Bobolink (*Dolichonyx oryzivorus*) covers much of the northern United States, but they are absent in southern areas. An examination of the 2001 model developed for the Bobolink shows average temperature used as a threshold feature. For a species such as this, with the southern end of its breeding range currently within the conterminous United States, novel conditions potentially introduced by a warming climate were unimportant, as the model already ensured exclusion of the species as a breeder in areas with temperatures above threshold values found in the training data.

Novel conditions could potentially be a problem for species with current breeding ranges extending to the United States and Mexico border. No background data south of the border were used to train the model. With a warming climate, for some species, it is likely that local temperatures in the projected climate data exceeded any temperatures found in the training data. The clamping feature in Maxent was used to control novel conditions in situations such as this, effectively rescaling novel covariate values to maximum values found in the training data. Clamping thus eliminated statistical issues with applying models into a novel prediction space, but by rescaling extreme values in the projection space, the model may effectively be dampening the impact of future change on future species distributions for 2075. Clamping of novel temperatures, for example, could result in the model incorrectly representing far southern portions of a species range as "suitable" for breeding, when in fact a temperature tolerance limit has been reached and has pushed the southern limit of the breeding range north of the United States and Mexico border.

There are additional potential caveats in interpreting results of this assessment. These results are based on one modeling methodology (Maxent), with one defined method for parameterization. Many papers have focused on the effects of different parameterization settings when using the Maxent model [27,48,56], and it was not the intention of this paper to revisit how different parameterizations affect Maxent results. The results presented here were also conducted at one specific spatial scale, with one specific suite of covariates and bird presence data. It was impractical to perform comprehensive analyses across all possible permutations of modeling frameworks, parameterization settings, spatial scales, thematic scales, temporal resolution, and data sources; results may differ for assessments where these components are altered. There was no attempt to rigorously address all potential sources of modeling uncertainty in this assessment. Conlisk et al. [38] attempted to disentangle all sources of uncertainty in SDMs, concluding that the modeling framework itself is the most important source of uncertainty. Ideally multiple models would be used to also disentangle effects of the modeling frameworks themselves, but resources were unavailable for a multi-model assessment given the large number of species, and multiple combinations of dates, covariates, and scenarios. Other potential drawbacks to the approach used here is an oversimplified representation of the driving forces behind species distributions. One final area that needs further exploration is the correction of bias for eBird data. Spatial bias was mitigated by spatially filtering the data. However, given the number and diversity of species modeled, a consistent filtering threshold of 20 km was used for all species; no attempts were made to tailor the filtering protocol to the spatial data characteristics for each species, nor were attempts made to quantify the reduction in spatial bias in this assessment. Additional potential sources of error and bias in eBird data that were not accounted for include accuracy of geographic data entry and highly variable observation and identification skills among eBird participants [62,63].

Conclusion

This work represents the first assessment of the effects of climate and LULC for bird species in the conterminous United States using both 1) newly available LULC projections of high-spatial and thematic resolution and 2) climate and LULC projections that are both consistent with IPCC SRES scenario frameworks. While modeling results clearly indicate a species-dependent determination of the relative impacts of climate and LULC change on both current and future range, it is clear that SDMs benefit by including both climate and LULC covariates. The use of climate data alone likely results in errors of commission and an over-prediction of current range. For future modeling of species range, the use of climate change information without corresponding LULC change may result in the misrepresentation of future range either positively or negatively, dependent upon whether projected LULC change was harmful or beneficial to a species. The inclusion of LULC data in SDMs 1) significantly increased measures of model fit, and 2) "tempered" predicted ranges from climate-only modeling frameworks by providing fine-scale information on local habitat suitability. When modeling future shifts in range, climate had the dominant impact on range shifts, yet LULC change was dominant for many species. Relationship of the species' range to the geographic bounds of the study area also clearly impacts whether climate or LULC has the dominant effect on modeled species range, and needs to be considered at both the design and assessment stages of a study.

All LULC projections used for this assessment are available at http://landcover-modeling.cr.usgs.gov. The computed predictor variables (covariates) used in this assessment, all range maps for 2001 and 2075 for each of the fifty modeled species, and a spreadsheet of all quantitative data reported in this paper are accessible at http://landcover-modeling.cr.usgs.gov/sdm.php. While this paper has focused on generalized results across the 50 modeled species, detailed model results for each of the 50 modeled species also are included herin. eBird data used as presence locations for this work may be obtained by through http://ebird.org.

Acknowledgments

Funding for this research was provided by the U.S. Geological Survey's Climate and Land Use Program. I thank Alisa Gallant for her work on downscaling the GCM data that were used in this assessment.

Author Contributions

Conceived and designed the experiments: TLS. Performed the experiments: TLS. Analyzed the data: TLS. Contributed reagents/materials/analysis tools: TLS. Contributed to the writing of the manuscript: TLS.

References

1. Matthews SN, Iverson LR, Prasad AM, Peters MP (2011) Changes in potential habitat of 147 North American breeding bird species in response to redistribution of trees and climate following predicted climate change. Ecography 34: 933–945.

2. Brambilla M, Ficetola GF (2012) Species distribution models as a tool to estimate reproductive parameters: a case study with a passerine bird species. Journal of Animal Ecology 81: 781–787.

3. Root T (1988) Energy constraints on avian distributions and abundances. Ecology 69(2): 330–339.

4. Thogmartin WE, Knutson MG, Sauer JR (2006) Predicting regional abundance of rare grassland birds with a hierarchical spatial count model. The Condor 108: 25–46.

5. Rahbek C, Gotelli NJ, Colwell RK, Entsminger GL, Rangel TF, et al. (2007) Predicting continental-scale patterns of bird species richenss with spatially explicit models. Proceedings of the Royal Society B (274): 165–274.

6. Pearson RG, Dawson TP (2003) Predicting the impacts of climate change on the distribution of species: are bioclimate envelope models useful? Global ecology and Biogeography 12: 361–371.

7. Barbet-Massin M, Thuiller W, Jiguet F (2012) The fate of European breeding birds under climate, land-use and dispersal scenarios. Global Change Biology 18: 881–890.

8. Jongsomjit D, Stralberg D, Gardali T, Salas L, Wiens J (2013) Between a rock and a hard place the impacts of climate change and housing development on breeding birds in California. Landscape Ecology 28: 187–200.

9. Watling JI, Bucklin DN, Speroterra C, Brandt LA, Mazzotti FJ, et al. (2013) Validating predictions from climate envelope models. PLoS ONE 8(5): e63600. doi:10.1371/journal.pone.0063600.

10. Araujo MB, Peterson AT (2012) Uses and misuses of bioclimatic envelope modeling. Ecology 93(7): 1527–1539.

11. Jimenez-Valverde A, Barve N, Lira-Noriega A, Maher SP, Nakazawa Y, et al. (2011) Dominant climate influences on North American bird distributions. Global Ecology and Biogeography 20: 114–118.

12. Thuiller W, Araujo MB, Lavorel S (2004) Do we need land-cover data to model species distributions in Europe? Journal of Biogeography 31: 353–361.

13. Bucklin DN, Basille M, Benscoter AM, Brandt LA, Mazzotti FJ, et al. (2014) Comparing species distribution models constructed with different subsets of environmental predictors. Diversity and Distributions. doi:10.1111/ddi.12247.

14. Lee TM, Jetz W (2011) Unravelling the structure of species extinction risk for predictive conservation science. Proceeding of the Royal Society B(278): 1329–1338.

15. Sinclair SJ, White MD, Newell GR (2010) How useful are species distribution models for managing biodiversity under future climates? Ecology and Society 15(1): 8. [online]. Available: http://www.ecologandsociety.org/vol15/iss1/art8/.

16. Riordan EC, Rundel PW (2014) Land use compounds habitat losses under projected climate change in a threatened California ecosystem. PLoS ONE 9(1): e86487. doi:10.1371/journal.pone.0086487.

17. Pearson RG, Dawson TP, Liu C (2004) Modelling species distributions in Britain: A hierarchical integration of climate and land-cover data. Ecography 27: 285–298.

18. Sullivan BL, Wood CL, Iliff MJ, Bonney RE, Fink D, et al. (2009) eBird: A citizen-based observation network in the biological sciences. Biological Conservation 142: 2282–2292.

19. Sullivan BL, Aycrigg JL, Barry JH, Bonney RE, Bruns N, et al. (2014) The eBird enterprise: An integrated approach to development and application of citizen science. Biological Conservation 169: 31–40.

20. Hochachka W, Fink D (2012) Broad-scale citizen science data from checklists: prospects and challenges for macroecology. Frontiers in Biogeography 4(4): 150–154.

21. Lerman SB, Nislow KH, Nowak DJ, DeStefano S, King DI, et al. (2014) Using urban forest assessment tools to model bird habitat potential. Landscape and Urban Planning 122: 29–40.

22. Hurlbert AH, Liang Z (2012) Spatiotemporal variation in avian migration phenology: Citizen science reveals effects of climate change. PLoS ONE 7(2): e31662. doi:10.1371/journal.pone.0031662.

23. Fink D, Hochachka WM, Zuckerberg B, Winkler DW, Shaby B, et al. (2010) Spatiotemporal exploratory models for broad-scale survey data. Ecological Applications 20(8): 2131–2147.

24. Ridgeby RS, Allnutt TF, Brooks T, McNicol DK, Mehlman DW, et al. (2003) Digital distribution maps of the birds of the Western Hemisphere, version 1.0 NatureServe, Arlington, Virginia USA.

25. Yackulic CB, Chandler R, Zipkin EF, Royle JA, Nichols JD, et al. (2013) Presence-only modeling using MaxEnt: when can we trust the inferences? Methods in Ecology and Evolution 4: 236–243.

26. Elith J, Phillips SJ, Hastie T, Dudik M, Chee YE, et al. (2011) A statistical explanation of MaxEnt for ecologists. Diversity and Distributions 17: 43–57.

27. Kramer-Schadt S, Niedballa J, Pilgrim JD, Schroder B, Lindenborn J (2013) The importance of correcting for sampling bias in MaxEnt species distribution models. Diversity and Distributions 19: 1366–1379.

28. Boria RA, Olson LE, Goodman SM, Anderson RP (2014) Spatial filtering to reduce sampling bias can improve the performance of ecological niche models. Ecological Modelling 275: 73–77.

29. Wisz MS, Hijmans RJ, Li J, Peterson AT, Graham CH, et al. (2008) Effects of sample size on the performance of species distribution models. Diversity and Distributions 14: 763–773.

30. Hernandez PA, Graham CH, Master LL, Albert DL (2006) The effect of sample size and species characteristics on performance of different species distribution modeling methods. Ecography 29: 773–785.

31. Sohl TL, Sleeter BM, Zhu Z, Sayler KL, Bennett S, et al. (2012) A land-use and land-cover modeling strategy to support a national assessment of carbon stocks and fluxes: Applied Geography 34: 111–124.

32. Sohl TL, Sayler KL, Bouchard MA, Reker RR, Friesz AM, et al. (2014) Spatially explicit modeling of 1992 to 2100 land cover and forest stand age for the conterminous United States. Ecological Applications 24(5): 1015–1036. Available: http://dx.doi.org/10.1890/13-1245.1.

33. Nakicenovic N, Alcamo J, Davis G, de Vries HJM, Fenhann J, et al. (2000) Special Report on Emissions Scenarios (SRES). Intergovernmental Panel on Climate Change (IPCC). Cambridge University Press, Cambridge, UK. 570 p.

34. Vogelmann JE, Howard SM, Yang L, Larson CR, Wylie BK, et al. (2001) Completion of the 1990s National Land Cover Data Set for the conterminous United States. Photogrammetric Engineering and Remote Sensing 67: 650–652.

35. Cunningham MA, Johnson DH (2006) Proximate and landscape factors influence grassland bird distributions. Ecological Applications 16(3): 1062–1075.

36. Bakker KK, Naugle DE, Higgins DF (2002) Incorporating landscape attributes into models for migratory grassland bird conservation. Conservation Biology 16(6): 1638–1646.

37. Fearer TM, Prisley SP, Stauffer DF, Keyser PD (2007) A method for integrating the Breeding Bird Survey and Forest Inventory and Analysis databases to evaluate forest bird-habitat relationships at multiple spatial scales. Forest Ecology and Management 243: 128–143.

38. Conlisk E, Syphard AD, Franklin J, Flint L, Flint A, et al. (2013) Uncertainty in assessing the impacts of global change with coupled dynamic species distribution and population models. Global Change Biology 19: 858–869.

39. Hay LE, Markstrom SL, Ward-Garrison C (2011) Watershed-scale response to climate change through the twenty-first century for selected basins across the United States. Earth Interactions 15: 1–37.

40. Wu Y, Liu S, Gallant AL (2012) Predicting impacts of increased CO^2 and climate change on the water cycle and water quality in the semiarid James River Basin of the Midwestern USA. Science of the Total Environment 430: 150–160.

41. Bradley BA, Olsson AD, Wang O, Dickson BG, Pelech L, et al. (2012) Species detection vs. habitat suitability: Are we biasing habitat suitability models with remotely sensed data? Ecological Modelling 244: 57–64.

42. Kery M, Gardner B, Monnerat C (2010) Predicting species distributions from checklist data using site-occupancy models. Journal of Biogeography 37: 1851–1862.

43. Moreno R, Zamora R, Molina JR, Vasquez A, Herrera MA (2011) Predictive modeling of microhabitats for endemic birds in South Chilean temperate forests using Maximum entropy (Maxent). Ecological Informatics 6: 364–370.

44. Johnston KM, Freund KA, Schmitz OJ (2012) Projected range shifting by montane mammals under climate change: implications for Cascadia's National Parks. Ecosphere 3(11): 97. Available: http://dx.doi.org/10.1890/ES12-00077.1.

45. U.S. Geological Survey (1999) USGS 30-Meter Resolution, One-Sixtieth Degree National Elevation Dataset for CONUS, Alaska, Hawaii, Puerto Rico, and the U.S. Virgin Islands. U.S. Geological Survey (USGS) Earth Resources Observation and Science (EROS) Center, Sioux Falls, SD.

46. Phillips SJ, Dudik M (2008) Modeling of species distributions with Maxent: new extensions and a comprehensive evaluation. Ecography 31: 161–175.

47. Bahn V, McGill BJ (2012) Testing the predictive performance of distribution models. Oikos 000: 001–011, doi:10.1111/j.1600-0706.2012.00299.x.

48. Phillips SJ, Dudik M, Elith J, Graham CH, Lehmann A, et al. (2009) Sample selection bias and presence-only distribution models: implications for background and pseudo-absence data. Ecological Applications 19(1): 181–197.

49. Dudik M, Schapire RE, Phillips SJ (2005) Correcting sample selection bias in maximum entropy density estimation. Advances in Neural Information Processing Systems 18. MIT Press, Cambridge, Massachusetts, USA. 320–330.

50. Veloz SD (2009) Spatially autocorrelated sampling falsely inflates measures of accuracy for presence-only niche models. Journal of Biogeography 36: 2290–2299.

51. Elith J, Kearney M, Phillips SJ (2010) The art of modelling range-shifting species. Methods in Ecology and Evolution 1: 330–342.

52. Barve N, Barve V, Jimenez-Valverde A, Lira-Noriega A, Maher AP, et al. (2011) The crucial role of the accessible area in ecological niche modeling and species distribution modeling. Ecological Modelling 222: 1810–1819.

53. VanDerWal J, Shoo LP, Graham C, Williams SE (2009) Selecting pseudo-absence data for presence-only distribution modeling: How far should you stray from what you know? Ecological Modelling 220: 589–594.

54. Anderson RP, Gonzalez Jr. I (2011) Species-specific tuning increases robustness to sampling bias in models of species distributions: An implementation with MaxEnt. Ecological Modelling 222: 2796–2811.

55. Warren DL, Seifert SN (2011) Ecological niche modeling in Maxent: the importance of model complexity and the performance of model selection criteria. Ecological Applications 21(2): 335–342.

56. Syfert MM, Smith MJ, Coomes DA (2013) The effects of sampling bias and model complexity on the predictive performance of MaxEnt species distribution models. PLoS One 8(2): e55158. doi:10.1371/journal.pone.0055158.

57. Liu C, Berry PM, Dawon TP, Pearson RG (2005) Selecting thresholds of occurrence in the prediction of species distributions. Ecography 28: 385–393.

58. Liu C, White M, Newell G (2013) Selecting thresholds for the prediction of species occurrence with presence-only data. Journal of Biogeography (40): 778–789.

59. Prince K, Lorrilliere R, Barbet-Massin M, Jiguet F (2013) Predicting the fate of French bird communities under agriculture and climate change scenarios. Environmental Science & Policy 33: 120–132.

60. Melles SJ, Fortin MJ, Lindsay K, Badzinski D (2011) Expanding northward: influence of climate change, forest connectivity, and population processes on a threatened species' range shift. Global Change Biology 17: 17–31.

61. Naujokaitis-Lewis IR, Curtis JMR, Tischendorf L, Badzinski D, Lindsay K, et al. (2013) Uncertainties in coupled species distribution-metapopulation dynamics models for risk assessments under climate change. Diversity and Distributions 19: 541–554.

62. Yu J, Wong WK, Hutchinson R (2010) Modeling Experts and Novices in Citizen Science Data for Species Distribution Modeling. Proceedings of the 2010 IEEE International Conference on Data Mining, (pp. 1157–1162), Washington, DC: IEEE Computer Society.

63. Dickinson JL, Zuckerberg B, Bonter DN (2010) Citizen Science as an Ecological Research Tool: Challenges and Benefits. Annual Review of Ecology, Evolution, and Systematics 41: 149–172.

21

Temporal Changes in Randomness of Bird Communities across Central Europe

Swen C. Renner[1,2*¤], Martin M. Gossner[3], Tiemo Kahl[4], Elisabeth K. V. Kalko[1†], Wolfgang W. Weisser[3], Markus Fischer[5], Eric Allan[6]

1 Institute of Experimental Ecology, University of Ulm, Ulm, Germany, **2** Smithsonian Conservation Biology Institute, National Zoological Park, Front Royal, Virginia, United States of America, **3** Terrestrial Ecology Research Group, Department of Ecology and Ecosystem Management, Centre for Food and Life Sciences Weihenstephan, Technische Universität München, Freising, Germany, **4** Chair of Silviculture, University of Freiburg, Freiburg, Germany, **5** Institute of Plant Sciences and Botanical Garden, University of Bern, Bern, Switzerland, **6** Institute of Plant Sciences, University of Bern, Bern, Switzerland

Abstract

Many studies have examined whether communities are structured by random or deterministic processes, and both are likely to play a role, but relatively few studies have attempted to quantify the degree of randomness in species composition. We quantified, for the first time, the degree of randomness in forest bird communities based on an analysis of spatial autocorrelation in three regions of Germany. The compositional dissimilarity between pairs of forest patches was regressed against the distance between them. We then calculated the y-intercept of the curve, i.e. the 'nugget', which represents the compositional dissimilarity at zero spatial distance. We therefore assume, following similar work on plant communities, that this represents the degree of randomness in species composition. We then analysed how the degree of randomness in community composition varied over time and with forest management intensity, which we expected to reduce the importance of random processes by increasing the strength of environmental drivers. We found that a high portion of the bird community composition could be explained by chance (overall mean of 0.63), implying that most of the variation in local bird community composition is driven by stochastic processes. Forest management intensity did not consistently affect the mean degree of randomness in community composition, perhaps because the bird communities were relatively insensitive to management intensity. We found a high temporal variation in the degree of randomness, which may indicate temporal variation in assembly processes and in the importance of key environmental drivers. We conclude that the degree of randomness in community composition should be considered in bird community studies, and the high values we find may indicate that bird community composition is relatively hard to predict at the regional scale.

Editor: Francisco Moreira, Institute of Agronomy, University of Lisbon, Portugal

Funding: The work has been funded by the German Science Foundation (www.dfg.de) Priority Program 1374 Infrastructure Biodiversity-Exploratories (Ka 1241/19-1, Re 1733/6-1). The funders had no role in study design, data collection and analysis, decision to publish, or preparation of the manuscript.

Competing Interests: The authors have declared that no competing interests exist.

* Email: swen.renner@boku.ac.at

¤ Current address: Institute of Zoology, University of Natural Resources and Life Sciences, Vienna, Austria

† Deceased

Introduction

Understanding the processes determining the species richness, diversity, and abundance of organisms remains a key challenge. Deterministic processes such as those driven by habitat structure and heterogeneity [1–3], species-specific ecological traits [4,5], seasonality [6], or resource availability such as food [7], have been shown to be important in many ecosystems. Stochastic processes such as neutral dynamics [8] may, however, also explain a proportion of the species richness and diversity of communities [9]. As both types of processes are likely to be important in driving community assembly it is crucial to determine their relative importance and to understand to what extent stochastic processes shape the community structure of organisms. If random processes play a significant role, predicting changes in community composition may be challenging.

Among animals, birds are ecologically well-known and easy to identify and thus form part of many ecological studies. Many studies have explored the drivers of bird community composition, including land management intensification, climate change [10], alterations in habitat structural parameters [1,2], and changes in resource availability [7], or nest site availability [11]. However the role of stochastic processes in affecting bird communities has seldom been explored or quantified [12].

Land use intensification has resulted in substantial declines in bird diversity [5,10,13]. In particular, forest conversion from mainly hardwood to softwood species has resulted in declines of many bird species in areas with high human population pressure [6]. Land use intensification is also likely to alter assembly processes and therefore change the relative importance of deterministic and stochastic processes in driving community composition. An increase in management intensity might be expected to increase the importance of deterministic processes because only species adapted to high land use intensity can persist. This would imply that environmental filters play a larger role, and community composition would be expected to become more homogenous, in these landscapes. Moreover, high land use intensity is likely to reduce redundancy [14] which might also decrease the importance of random processes in driving community composition. We would therefore expect a higher degree of randomness in community composition at low management intensity than at high management intensity. Some studies have suggested that randomness is reduced in more disturbed sites [15–17], but the relative importance of deterministic and random processes in shaping communities under different land use intensities has not been studied. Moreover it is unclear whether these effects are consistent across differently managed forest habitats.

Community composition may turn over substantially between years and the importance of different assembly processes may also change over time [18]. Long-term studies on birds have shown that relative abundance, species composition and species diversity can vary considerably between years [12]. Factors driving species turnover between years include extreme weather events, habitat fragmentation [19], or population processes such as immigration, dispersal, or mortality [20]. If the strength of these processes, in determining species composition, varies over time, then the importance of stochastic processes might also vary over time. However, how the importance of random processes varies over time has not been quantified.

Calculating the proportion of community composition that is determined by random processes is challenging. The traditional approach is to use a null model to determine the deviation of observed composition from that expected by chance. Creating a null model is, however, technically not trivial and in some cases may even be impossible [9,17]. An alternative approach was proposed by Brownstein et al. [17], who suggested using the y-intercept (the 'nugget') from a regression of community similarity against spatial distance as a measure of the proportion of the community composition determined by random processes. Differences in species composition between sites will be due to spatially autocorrelated environmental differences, dispersal limitation and chance. The effects of chance are expected to be the same across a spatial gradient, so that at zero spatial distance only the effects of chance remain to drive differences in species composition. The nugget is conceptually the dissimilarity in species composition at zero geographic distance, which would be expected to be zero if geographic distance explained the variation in species composition, i.e. if there was no influence of stochastic processes. Values of the nugget greater than zero can therefore be interpreted as indicating the influence of random or chance processes, with larger nuggets indicating a larger role for random processes in determining community composition. We use the terms 'chance' and 'randomness' synonymously to refer to those processes which result in community compositions that are unpredictable from the (spatially autocorrelated), biotic and abiotic environment.

We analyse the relative strength of random processes in driving bird community assembly across local and regional scales and across time. We study bird communities in three regions of Germany, in forests varying in management intensity, across five consecutive years (2008 to 2012). We hypothesise that random processes will be important in driving bird community composition and will vary over time but their importance will be reduced under intensive forest management. We also hypothesise that the degree of randomness found in bird communities is not "stable" and in fact varies substantially over time.

Materials and Methods

Study regions and sites

Our study is part of the large-scale and long-term research platform 'Biodiversity Exploratories' (a detailed description of the study area, selection of study regions and sites and classification procedures is given in [21]).

In total, we studied 150 forest plots in three regions of Germany: the south-west region (Schwäbische Alb; approximate centre coordinates: 48.4° North, 9.5°East, altitude 500 to 800 m a.s.l.), the central region (Hainich-Dün; 51.1° North, 10.4°East, 285 to 550 m), and the north-east region (Schorfheide-Chorin; 53.0° North, 13.9°East, 3 to 140 m). Each plot was covered by forest which was homogenous in terms of canopy tree species composition, soil, and mean slope ≤20% [21]. Forest plots were between 250 m and 45 km apart within each of the three regions (Figure 1), which means we estimate effects at small spatial scales. A large number of small spatial distances are necessary to provide a reliable estimate of the nugget [17]. The lack of larger spatial distances (>100 km) is problematic if the dissimilarity values do not asymptote or do not reach 1, and if there is substantial dispersal limitation. Therefore we did not analyse the nuggets between the three regions. Plots within each region span a large gradient in management intensity and cover a large area (Figure 2).

Land use intensification

To understand whether the degree of randomness in community composition is linked to forest management we split all forest plots per region into whether they were managed at high or low intensity. We used a compound land use intensity index (details outlined in Appendix S1; [22]), which is based on the harvested tree volume, abundance of coniferous trees (as an indicator of "naturalness", conifers are not part of the natural forest in these study areas) and volume of dead wood with saw cuts (as an indicator of disturbance). Using this index, we divided all plots per region into the 25 with management intensity higher than the median and the 25 with management intensity lower than the median.

Bird surveys

At each of the 150 sites we surveyed birds by standardized audio-visual point-counts and recorded all birds exhibiting territorial displays (singing and calling activity) for five minutes per point count locality and time period. We used 50-m-fixed-radius point counts and noted all males of each bird species during the five-minute interval. Each site was visited five times between 15 March and 15 June (first surveying period 15–30 March; 2nd 15–30 April; 3rd 1–15 May; 4th 16–31 May; 5th 1–15 June) in 2008 to 2012.

A minimum of five and a maximum of 15 sites were surveyed per day by one observer from sunrise to 11:00 h; occasionally the evening chorus was surveyed after 17:00 h to sunset (<20 times out of 750 events per year). The sequence in which sites were visited was randomized. Each song or call heard on a site was interpreted as one male territorial display behaviour. The maximum number of birds displaying site^{-1} year^{-1} (i.e. the maximum number of individuals per species observed in any of the five surveys) was used as a measure of the relative abundance of each bird species. We considered a species as present in any given site if it was recorded at least once during a survey round within any given year. Aerial species (swifts and swallows) were excluded from analysis, since they had been surveyed irregularly and are

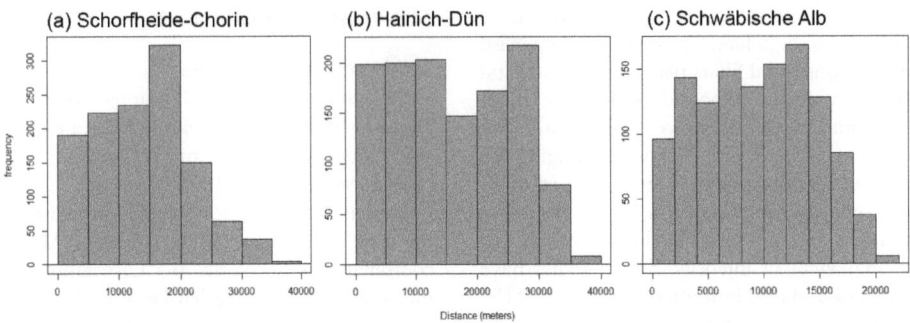

Figure 1. Histograms of Euclidian distance in meters between all 1225 possible distance of forest sites for each of the three regions. Note the different scales in each histogram.

biased towards beech forests (where detectability for aerial species is higher than in spruce forests).

The data is accessible through the Biodiversity Exploratories database http://www.biodiversity-exploratories.de/intranet/ (follow the link "BExIS"; registration required).

Data analyses

First we used generalized linear mixed effect models (GLMM) to test whether bird species richness or relative abundance was affected by time, region, site, or management intensity and the interaction between these factors. The model was specified as: species richness ~year * region * management intensity + (year | site id). The model was fit with Poisson errors. In a second model, we analysed the response of relative abundance instead of species richness.

We calculated the relative degree of randomness in bird community composition following Brownstein et al. [17]. This approach calculates the nugget from the relationship between dissimilarity in species composition and geographic distance and uses this as a measure of the proportion of the species composition explained by chance. This is conceptually the dissimilarity in species composition at zero geographic distance, which would be expected to be zero if geographic distance explained the variation in species composition, i.e. if there was no influence of stochastic processes. We calculated dissimilarity in bird species composition between sites using the Jaccard index D', with EstimateS 8.2 [23], and did this separately for each of the three regions (north-east, central, and south-west).

We calculated the nuggets for each year separately and across years. For the analysis across years we used all of the species observed per site across the five years and used this cumulative species list to calculate dissimilarity. We then determined the distance in meters between all sites within each of the three regions (Euclidian distance from each site centroid to site centroid).

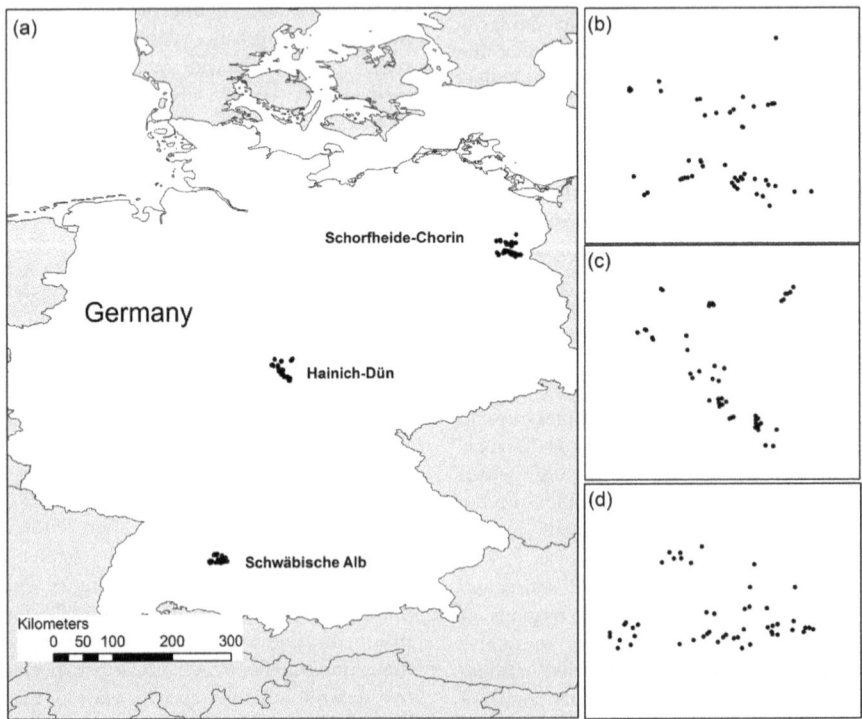

Figure 2. Spatial distribution of all 150 sites across the three regions. (a) Locations of the three regions within Germany and distribution of the sites within Schorfheide-Chorin (b), Hainich-Dün (c), and Schwäbische Alb (d).

Dissimilarities in species composition between all possible 1,225 pairs of experimental sites per region were related to the geographic distance between them. A non-linear least squares equation was used to model the relationship between the Jaccard D' dissimilarities ($1-D$'), as the y-variable, and spatial distance. We therefore fitted a spatial autocorrelation dissimogram (Figure 3) [24]: $D'_{spatial} = a \cdot e ^\wedge (-b \cdot e ^\wedge(-c \cdot d))$, where a, b and c are fitted parameters estimated with the Gauss-Newton algorithm, and d is distance between plots. We used package nls2 [25] in R [26] to fit all of the models.

We used different algorithms to explore the suitability of different equations for calculating the nugget [17]. Only two formulae resulted in models which converged (Gompertz and Negative exponential), all others resulted in many fewer models converging (Appendix S2). From the fitted curves (examples in Figure 3) we determined the nugget ($a \cdot e ^\wedge -b$) as the y-intercept and the Asymptote (a), which represents the dissimilarity at infinite distance (i.e., the fitted maximum dissimilarity).

Nuggets >1.0 were excluded from the analysis, as they indicate poorly fitting models. We also calculated the amount of variance explained by each of the models, using a pseudo R^2. We calculated this as the square of the Pearson correlation coefficient for the correlation between model fitted values and the original data. If the pseudo R^2 is low then a small amount of the variation in community composition is explained by geographic distance. In general, pseudo R^2 approaches may not be entirely appropriate for non-linear models but they do convey an idea of the goodness-of-fit.

Further statistical analysis

To understand whether species richness and stochasticity are related, we assessed associations of species richness and the nuggets using a linear model in R (R command lm).

Detectability and occupancy of sites by bird species might affect our analysis: low detectability of species might bias our results by increasing variation in species composition between sites and therefore increasing our estimate of the degree of randomness in species composition. Species cannot always be detected even when they are present at a site but repeated surveys (typically ≥3 repetitions) at a given site reduce this detection bias [27,28]. To further assess whether our data is biased through detection probability, we calculated the detectability (estimate of ψ; [27]) of each bird species in each plot to determine if low detectability could have an influence on our estimate of the degree of randomness in community composition. We applied the "multi-season" model in PRESENCE 6.1 [28] and calculated overall

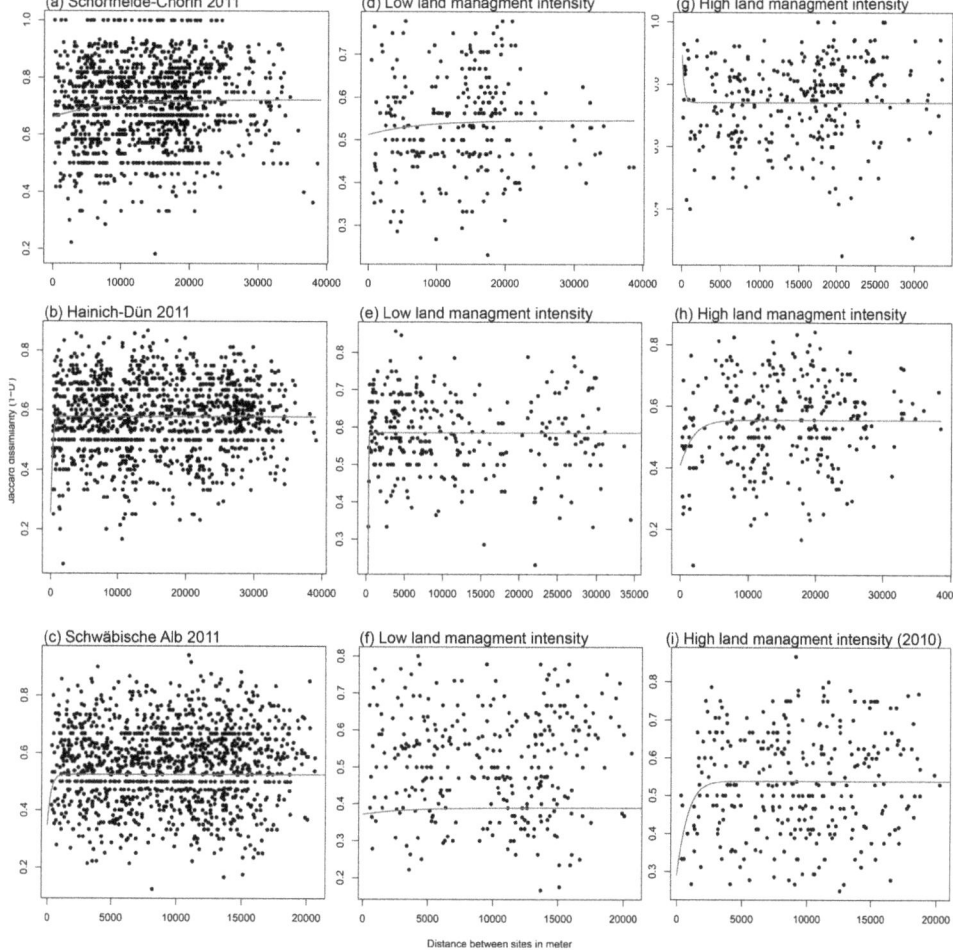

Figure 3. Dissimogram for the bird communities showing the Jaccard dissimilarity between each possible site pair _versus_ Euclidian distance between the forests sites. (a–c) all sites in 2011, **(d–f)** sites with low land use intensity, and **(g–i)** sites with high land use intensity. The x-axis is the distance between each pair of sites. The _grey line_ is the fitted line estimated based on non-linear least squares. Note the different scales in each x-axis.

detectability for each species over five years and five repetitions within each year.

In addition to calculating the detectability of individual species, we calculated the inter-annual turnover in species composition for each plot. We might expect that if the nuggets are driven by measurement error, i.e. high nuggets are due to the fact that we have failed to completely sample the local bird community, then excluding plots with high turnover values will reduce the size of the nuggets. We first determined for each plot all the species seen in any of the five years (i.e. the cumulative species richness) and then the species which were seen across the whole five-year period (i.e. those observed in four or five years in the plot). We then calculated the turnover between the species seen in ≥4 years and those seen in <4 years for each plot. Finally, we repeated our calculations of the nuggets, excluding those plots with turnover values of 60%, 70%, or 80%.

Results

In total, we observed 82 bird species in the three regions over the five consecutive years. The species richness of birds varied considerably between the three regions and across time (Figure 4a). Species richness was significantly lower in the south-west region compared to the other two regions (GLMM: $p \leq 0.01$; Figure 4a; detailed information in Appendix S3), and management intensity decreased species richness. The relative abundance of birds in the three regions showed a similar pattern to species richness with significant differences between years and the regions and also lower abundance in the south-west ($p \leq 0.02$). Management intensity reduced abundance ($p \leq 0.02$). In general, inter-annual variation was higher than the between region variation for both species richness and relative abundance.

Degree of randomness in bird community composition

A large proportion of bird community composition was explained by random processes (Table 1, example in Figure 3). Using the cumulative species richness per site, the nuggets from the dissimogram ranged between 0.25 and 0.86 (Gompertz equation). This suggests that random factors alone cause high turnover between communities.

In contrast to our hypothesis we found substantial variation in the degree of randomness in bird communities within the same site across years. Over three years (2010 to 2012), during which the observers and effort were constant, the nuggets calculated varied between 0.393 and 0.763 in the central region (Hainich-Dün), from 0.859 to 0.926 in the south-west (Schwäbische Alb) and from

0.689 to 0.808 in the north-east (Schorfheide-Chorin) (Table 1). The 95% confidence intervals and the lower and upper limits of the mean nuggets per forest sites over five years indicate significant temporal variation (Table 2).

Management intensity did not affect the degree of randomness in community composition (Table 1). However, the variation in the nugget over time was somewhat higher at low management intensity (0 to 0.93), than at high management intensity (0.24 to 0.75) (Table 1). Bird species richness and abundance were not related to the nugget (linear model: species richness Adjusted $R^2 = -0.083$, F = 0.001, P = 0.974, abundance Adjusted $R^2 = -0.004$, F = 0.942, P = 0.351). We therefore did not find any consistent changes in the nugget based on diversity or land use intensification.

Non-linear least square model-fit was relatively low and the highest pseudo R^2 value observed was 13.4% (Table 3). Therefore, space explained only a small portion of the variation in species richness. However, the nuggets and pseudo R^2 values were not correlated (linear model; F = 0.440, P = 0.528), indicating that low pseudo R^2 are not the only reason for the high nuggets that we found. Differences between the converging models with Gompertz or Negative Exponential functions were negligible and the nuggets calculated by the two methods diverged by less than 1% from each other (except in three cases where they diverged by 11%, 6%, and 3%). This indicates that our results were not sensitive to the particular function used to model the relationship between distance and dissimilarity.

Detectability and plot-based species turnover

The mean detectability of the 82 bird species in the 150 forest sites was $\psi = 0.57$ (i.e. on average, 57% of individuals per species were detected). This level of detectability is relatively high, indicating that our sampling was fairly complete. To further assess the influence of sampling on the nuggets we repeated our analysis excluding those plots which experienced high temporal turnover in species composition. This made little difference to the nuggets, which remained high even when plots with high species turnover were excluded (Table 1): the confidence intervals of the nuggets with high turnover plots included those calculated by all other analysis (compare Table 2).

Discussion

Our analyses suggested that random processes were important in structuring our forest bird communities, with around half of the variation in species composition between communities explained

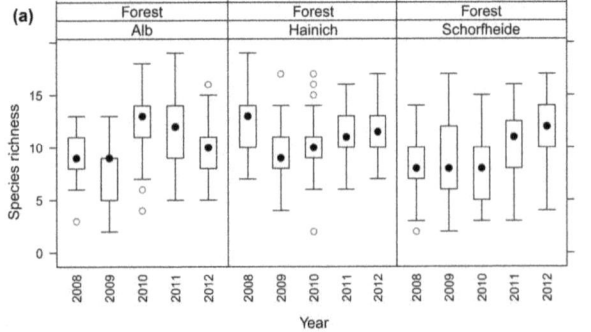
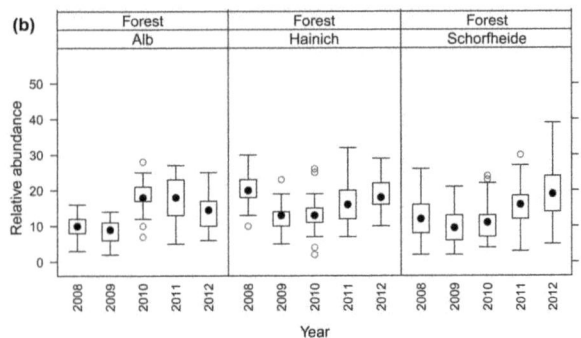

Figure 4. Temporal variation in the bird communities across five consecutive years (2008–2012) in the three study regions. (a) Temporal variation in species richness (counted bird species), and (**b**) temporal variation in relative abundance. The *Black dot* represents the median, the *Box* 1ˢᵗ and 3ʳᵈ quartile, *whiskers* 95% confidence intervals and *grey circles* outliers.

Table 1. Summary of estimated nuggets for the bird communities from 2008 to 2012.

Year(s)	Gompertz	Negative Exponential	High land use intensity	Low land use intensity	60% cut-off[b]	70% cut-off[b]	80% cut-off[b]
Cumulative species over years Schorfheide-Chorin (north-east) 2008–2012[a]	0.6660	0.6689	n/a	0.5323	0.6073	n/a	n/a
2008	0.8603	n/a	n/a	0.9336			
2009	0.6606	0.6603	n/a	0.3285			
2010	0.6604	0.6604	0.7511	0.5310			
2011	0.6604	0.6604	n/a	0.5120			
2012	n/a	n/a	n/a	0.0860			
Mean (2008–2012)	0.7104	0.6604	0.7511	0.4782			
± s. d. (2008–2012)	0.0999	0.0001	n/a	0.3113			
Minimum (2008–2012)	0.6604	0.6603	0.7511	0.0860			
Maximum (2008–2012)	0.8603	0.6604	0.7511	0.9336			
Cumulative species over years Hainich-Dün (centre) 2008–2012[a]	0.5077	0.5072	0.4910	0.5182	0.3983	0.4859	0.5093
2008	0.4430	0.4424	0.4511	0.4093			
2009	0.5390	0.5376	n/a	0.5554			
2010	0.5543	0.5547	n/a	n/a			
2011	0.2573	0.1950	0.4093	0.0000			
2012	0.4474	0.4463	0.2481	n/a			
Mean (2008–2012)	0.4482	0.4352	0.3695	0.3216			
± s. d. (2008–2012)	0.1183	0.1437	0.1072	0.2879			
Minimum (2008–2012)	0.2573	0.1950	0.2481	0.0000			
Maximum (2008–2012)	0.5543	0.5547	0.4511	0.5554			
Cumulative species over years Schwäbische Alb (south-west) 2008–20	n/a	0.5384	0.4416	n/a	0.4945	0.2243	n/a
2008	0.4369	0.4365	0.3487	0.3715			
2009	n/a	n/a	n/a	n/a			
2010	0.3486	0.3351	0.2894	n/a			
2011	n/a	0.1300	n/a	n/a			
2012	0.6261	0.6261	0.5329	0.5535			
Mean (2008–2012)	0.4705	0.3819	0.3903	0.4625			
± s. d. (2008–2012)	0.1418	0.2068	0.1270	0.1287			

Table 1. Cont.

Year(s)	Gompertz	Negative Exponential	High land use intensity	Low land use intensity	60% cut-off[b]	70% cut-off[b]	80% cut-off[b]
Minimum (2008–2012)	0.3486	0.1300	0.2894	0.3715			
Maximum (2008–2012)	0.6261	0.6261	0.5329	0.5535			

"n/a" indicates that the model did not converge or that the values for nuggets were >1 or <0.
[a]Calculated from Jaccard Dissimilarity (1-D') calculated with the cumulative species per plot over five years, i.e. all of the species observed at least once on the plot during this time period.
[b]The analysis was restricted to those plots with low inter-annual species turnover to determine if this influenced the high nuggets. Plots with turnover values higher than 60%, 70%, or 80% were excluded, see methods for details on the calculation of turnover.

by chance. This might suggest an important role for stochastic processes in affecting bird community structure [9,17,29–31]. Our analysis assumes that the major environmental factors driving bird community composition are spatially autocorrelated. This is likely to be true for factors such as climate and weather [6,10] but might not be the case for land use [1,2,6], forest structure [2] or food and nest site availability [7]. In addition, the approach by Brownstein et al. [17] has limitations: on the one hand the approach will work only with small spatial distances (<100 km), otherwise the distance decay and dispersal limitations will overrule the observable effects at these small distances. On the other hand, at very small spatial extent ("initial similarity" at <1 km) other ecological factors such as extent, latitude, or body size might add up towards random factors [32]. Therefore distances between 1 and 100 km are best suited for calculating the nugget. However, with these caveats in mind, our analysis does suggest that a large fraction of bird community composition is not predictable from the environment.

Effect of spatial extent and dispersal ability

The values for the degree of randomness in community composition which we found were higher than those found by the only other study to quantify the degree of randomness in species composition, using this particular method. Brownstein et al. [17] found less randomness in community composition in sessile plant communities. This difference in the degree of randomness found in the two studies might relate to the different dispersal abilities of the two groups of organism. In plants, dispersal is much more limited than for birds [33]. Even sedentary bird species can have flexible home ranges and do not typically remain in a fixed area [34,35 36]. Birds can therefore easily disperse between forest sites and the foraging range of even medium-sized passerines is typically <2 ha and in rare cases it can exceed 10 ha [37]. Bird species may therefore have been widely distributed within the three regions, meaning that even geographically distant plots could still have a similar bird community composition. The low pseudo R^2 values we found indicate that spatial distance might not explain much of the variation in species composition and this might have led to less reliable estimates of the nugget. Including more distant plots in the analysis might have led to a better relationship between space and compositional turnover. However, Brownstein et al. [17] also found that distance explained little variance in dissimilarity in their plant communities (1.2% to 23%), so this does not seem to be the major cause of the difference from our results. Future studies analysing bird communities at larger spatial extents are necessary to see if the high degree of randomness we find is a result of the spatial extent of our study. Our analysis nevertheless suggests that, at least at this spatial scale, a substantial degree of variation in bird community composition is not predictable from the environment.

Effects of species turnover and detectability

As well as being affected by the degree of randomness in community composition, the large nuggets we found could also be driven by observer bias and low detectability of species. Even well trained and experienced ornithologists can miss up to 10% of bird species during surveys [38,39]. While such errors might play a role in our results they cannot explain the high temporal variation because the same observers carried out all surveys in the central region and four out of five years in the north-eastern region. The spatial grain of our sampling might also play a role and larger plots might have resulted in more predictable species composition because rare species would be more likely to be missed in small plots. However the opposite is also possible if large plots contain more microhabitats and therefore have more variable species

Table 2. Confidence Intervals (CI, 95%) with lower and upper limit of CI for mean nuggets over the five consecutive years.

Region	Mean	CI	lower	upper
Schorfheide-Chorin (north-east)	0.710	0.138	0.573	0.848
Hainich-Dün (centre)	0.448	0.131	0.317	0.580
Schwäbische Alb (south-west)	0.471	0.288	0.183	0.758

compositions [17]. We were not able to test for the effect of spatial grain here and further studies are needed to determine the effect of spatial grain on the predictability of bird community composition. To better assess the issue of detection probabilities, we calculated detectability ψ for all of the bird species and found that our mean detection probabilities ($\psi = 0.57$) were comparatively high compared to some other studies on bird communities, with mean detectability of $0.15 \leq \psi \leq 0.43$ [40,41]. Issues in detectability are typically reduced by increasing number of repetitions [27,28], where each repetition increases the chance that the bird community is sampled completely (or at least more completely). Because we have a comparatively high number of repetitions per site, we assume that an even higher effort to repeat surveys would not decrease the degree of randomness (cf. temporal variation below). In addition, excluding plots with high levels of species turnover across time from our analysis did not significantly reduce the nuggets, which would be expected if incomplete sampling of species drove the high nuggets. These results indicate that sampling error is not the only factor driving our estimates for the large degree of randomness in species composition. Our results

do therefore suggest that a substantial fraction of the variation in local bird community composition is driven by random processes.

Temporal and spatial variation

The degree of randomness in bird community composition varied substantially between years. We also found temporal variation in the species richness and abundance of the bird communities, which suggests that there was high turnover in bird community composition between years. Other studies have shown spatial variation in the degree of randomness [17], but temporal variation has not been quantified so far. Temporal variation in the degree of randomness might arise because the deterministic drivers of bird community composition, such as weather or seasonality in resource availability can vary considerably between years. The degree of randomness in community composition would vary between years if the strength of these deterministic drivers also varies across time, i.e. so that climate strongly determines bird community composition in some years but has a relatively weak effect on community composition in other years. In the years in which it has a weak effect, composition might vary more

Table 3. Pseudo R^2 between original and fitted values in non-linear least square analysis.

Year(s)	All forest	High land-use intensity	Low land-use intensity	60% Cut-off[b]	70% Cut-off[b]	80% Cut-off[b]
Schorfheide-Chorin (north-east) 2008–2012[a]	0.016	0.000	0.019	0.091	n/a	0.006
2008	0.000	0.002	0.000			
2009	0.011	0.004	0.010			
2010	0.010	0.000	0.048			
2011	0.010	0.000	0.005			
2012	0.008	0.000	0.005			
Hainich-Dün (centre) 2008–2012[a]	0.130	0.069	0.071	0.121	0.093	0.129
2008	0.072	0.027	0.050			
2009	0.054	n/a	0.063			
2010	0.056	0.003	0.006			
2011	0.007	0.030	0.007			
2012	0.043	0.085	0.099			
Schwäbische Alb (south-west) 2008–2012[a]	n/a	0.022	n/a	0.018	0.002	n/a
2008	0.006	0.016	0.000			
2009	n/a	0.003	n/a			
2010	0.001	0.019	n/a			
2011	n/a	n/a	0.000			
2012	0.005	0.020	0.004			

[a]Calculated from Jaccard Dissimilarity (1-D') calculated with the cumulative species per plot over five years, i.e. all of the species observed at least once on the plot during this time period.
[b]The analysis was restricted to those plots with low inter-annual species turnover to determine if this influenced the high nuggets. Plots with turnover values higher than 60%, 70%, or 80% were excluded, see methods for details on the calculation of turnover.

stochastically between communities. Increasing detectability (by increasing the number of repetitions [27,28]), could in theory reduce this temporal variation in the degree of randomness. Each additional repetition increases the chance that the bird community is more completely sampled. We have a comparatively high number of repetitions per site and per year and a therefore comparatively high ψ. Therefore a more exhaustive sampling protocol, with even more frequent sampling per year, would be unlikely to reduce the temporal variation we found. The high temporal variation in the degree of randomness indicates that future studies need to more often consider temporal variation in the drivers of community composition.

We also found spatial variation and large differences in the degree of randomness between our three regions. This agrees with other studies, which have observed spatial variation in randomness [17,38]. The three regions varied in both species numbers and relative abundance of bird species and this may have caused the variation in degree of randomness between the regions. These results show that conclusions about the importance of random processes in driving community composition should be based on wide spatial and temporal sampling.

Effects of land use intensity

Management intensity did not have an effect on the nugget, i.e. our results suggest that increased land use intensification does affect the degree of randomness in community composition. We expected that high management intensity would reduce the influence of random processes in structuring bird communities, however we did not find evidence for this. Management intensity did, however, affect both species richness and abundance of birds, which indicates that it is an important driver of bird communities. However it does not seem to alter the degree of randomness, which remains high even in more intensively managed forests.

Conclusion

The Brownstein model provides a simple method for calculating the degree of randomness in community composition from spatially explicit data. Using this method we find that a large proportion of the variation in bird community composition between sites is driven by random processes. This means that bird community composition may not be predictable from the environment alone, at least at this spatial scale, and that predicting

shifts in local bird community composition in response to global change may be difficult. If stochastic processes do play a large role in determining bird community composition, it also means that birds may not be ideally suited as indicators of diversity for other groups of organisms. We conclude that determining the degree of randomness is important for analyses of community structure, particularly as we suggest that random processes may play a surprisingly large role in driving community composition.

Supporting Information

Figure S1 Bird species numbers (*dots*) and abundance (*circles*) in relation to land use intensity (W_i) in forest sites over all three regions.

Appendix S1 Explanations of management intensity index.

Appendix S2 Formula used to calculate the "nugget".

Appendix S3 Summary statistics of GLMM results.

Appendix S4 Relevant data as used for nls() modelling.

Acknowledgments

We thank the past and current managers of the three Exploratories, S Gockel, A Hemp, K Wells, K Wiesner, and M Gorke, as well as S Pfeiffer and I Mai for their work in maintaining the project infrastructure and the local management teams for the incredible support in the field. We thank KE Linsenmair and F Buscot, for their role in setting up the Biodiversity Exploratories project. Field work permits were issued by the responsible state environmental offices of Baden-Württemberg, Brandenburg, and Thüringen.

Author Contributions

Conceived and designed the experiments: SCR. Performed the experiments: SCR EA. Analyzed the data: SCR EA. Contributed reagents/materials/analysis tools: SCR EKVK. Contributed to the writing of the manuscript: SCR MMG TK WWW MF EA. Conceptual discussion: SCR MMG TK WWW MF EA.

References

1. Bradbury RB, Hill RA, Mason DC, Hinsley SA, Wilson JD, et al. (2005) Modelling relationships between birds and vegetation structure using airborne LiDAR data: a review with case studies from agricultural and woodland environments. Ibis 147: 443–452.

2. Goetz SJ, Steinberg D, Betts MG, Holmes RT, Doran PJ, et al. (2010) LiDAR remote sensing variables predict breeding habitat of a Neotropical migrant bird. Ecology 91: 1569–1576.

3. Gossner MM, Getzin S, Lange M, Pašalić E, Türke M, et al. (2013) The importance of heterogeneity revisited from a multi-scale and multi-taxa approach. Biol Cons 166: 212–220.

4. Elith J, Leathwick JR (2009) Species distribution models: ecological explanation and prediction across space and time. Ann Rev Ecol Evol System 40: 677–697.

5. Flynn DFB, Gogol-Prokurat M, Nogeire T, Molinari N, Trautman Richers B, et al. (2009) Loss of functional diversity under land management intensification across multiple taxa. Ecol Let 12: 22–33.

6. Jetz W, Wilcove DS, Dobson AP (2007) Projected Impacts of micro-climate and land management change on the global diversity of birds. PLoS Biology 5: e157.

7. Renner SC, Baur S, Possler A, Winkler J, Kalko EKV, et al. (2012) Food preferences of winter bird communities in different forest types. PLoS ONE 7: e53121.

8. Hubbell SP (2005) Neutral theory in community ecology and the hypothesis of functional equivalence. Func Ecol 19: 166–172.

9. Gotelli NJ, Ulrich W (2012) Statistical challenges in null model analysis. Oikos 121: 171–180.

10. Eglington SM, Pearce-Higgins JW (2012) Disentangling the relative importance of changes in climate and land-use intensity in driving recent bird population trends PLOS ONE 7: e30407.

11. Kennedy CM, Marra PP, Fagan WF, Neel MC (2010) Landscape matrix and species traits mediate responses of Neotropical resident birds to forest fragmentation in Jamaica. Ecol Monogr 80: 651–669.

12. Sauer JR, Hines JE, Fallon JE, Pardieck KL, Ziolkowski DJ, et al. (2012) The North American breeding bird survey, results and analysis 1966–2011. Version 1.213.2011. USGS Patuxent Wildlife Research Centre, Laurel, MD.

13. Fischer J, Lindenmayer DB (2007) Landscape modification and habitat fragmentation: a synthesis. Global Ecol Biogeogr 16: 265–280.

14. Laliberté E, Wells JA, DeClerck F, Metcalfe DJ, Catterall CP, et al. (2010) Land management intensification reduces functional redundancy and response diversity in plant communities. Ecol Let 13: 76–86.

15. Drake JA, Zimmerman CR, Purucker T, Rojo C (1999) On the nature of assembly trajectory Ecological assembly rules: perspectives, advances, retreats (eds.: E Weiher, PA Keddy), 233–250. Cambridge University Press, Cambridge.

16. Honnay O, Verhaeghen W, Hermy M (2001) Plant community assembly along dendritic networks of small forest streams. Ecology 82: 1691–1702.

17. Brownstein G, Steel JB, Porter S, Gray A, Wilson C, et al. (2012) Chance in plant communities: a new approach to its measurements using the nugget from spatial autocorrelation. J Ecol 100: 987–996.

18. Chalcraft DR, Williams JW, Smith MD, Willig MR (2004) Scale dependence in the species-richness-productivity relationship: the role of species turnover. Ecology 85: 2701–2708.

19. Stouffer PC, Johnson EI, Bierregaard RO, Lovejoy TE (2011) Understory bird communities in Amazonian rainforest fragments: species turnover through 25 years post-isolation in recovering landscapes. PLOS ONE 6: e20543.
20. Hanski I (1999) Metapopulation ecology. Oxford University Press.
21. Fischer M, Bossdorf O, Gockel S, Hänsel F, Hemp A, et al. (2010) Implementing large-scale and long-term functional biodiversity research: The Biodiversity Exploratories. Basic Appl Ecol 11: 473–485.
22. Kahl T, Bauhus J (2014). An index of forest management intensity based on assessment of harvested tree volume, tree species composition and dead wood origin. Nat Cons 7: 15–27.
23. Colwell RK (2006) EstimateS: Statistical estimation of species richness and shared species from samples. Version 8. Available: http://viceroy.eeb.uconn.edu/estimates/ Accessed 2014 Apr 24.
24. Mistral M, Buck O, Meier-Behrmann DC, Burnett DA, Barnfield TE, et al. (2000) Direct measurement of spatial autocorrelation at the community level in four plant communities. J Veg Sci 11: 911–916.
25. Grothendieck G (2013) nls2: Non-linear regression with brute force R package version 02 Available: http://CRAN.R-project.org/package=nls2 Accessed 2014 Apr 24.
26. R Development Core Team (2011) R: A language and environment for statistical computing R Foundation for Statistical Computing, Vienna Available: http://www.R-project.org/ Accessed 2014 Apr 24.
27. MacKenzie DI, Nichols JD, Lachman GB, Droege S, Royle JA, et al. (2002) Estimating site occupancy rates when detection probabilities are less than one. Ecology 83: 2248–2255.
28. Hines J (2013) Program PRESENCE 61 USGS Available: http://www.mbr-pwrc.usgs.gov/software/bin/setup_presence.zip Accessed 2014 Apr 24.
29. Hutchinson GE (1959) Homage to Santa-Rosalia or why are there so many kinds of animals. Am Nat 93: 145–159.
30. Hutchinson GE (1961) The paradox of the plankton. Am Nat 95: 137–145.
31. Hubbell SP, Foster RB (1986) Biology, chance, and history and the structure of tropical rain forest tree communities Community Ecology (eds.: J Diamond, TJ Case), pp 314–329. Harper Row, New York.
32. Soininen J, McDonald R, Hillebrand H (2007) The distance decay of similarity in ecological communities. Ecography 30: 3–12.
33. Gillies CS, St Clair CC (2008) Riparian corridors enhance movement of a forest specialist bird in fragmented tropical forest. Proc Nat Acad Sciences 105: 19774–19779.
34. Haskell JP, Ritchie ME, Olff H (2002) Fractal geometry predicts varying body size scaling relationships for mammal and bird home ranges. Nature 418: 527–530.
35. McLoughlin PD, Ferguson SH, Messier F (2000) Intraspecific variation in home-range overlap with habitat quality: a comparison among brown bear populations. Evolution Ecol 14: 39–60.
36. Kubiczek K, Renner SC, Böhm SM, Kalko EKV, Wells K (2014), Movement and ranging patterns of the Common Chaffinch in heterogeneous forest landscapes. PeerJ 2: e36837. Glutz von Blotzheim UN (1988) Handbuch der Vögel Mitteleuropas. Aula Verlag, Wiesbaden.
37. Farmer RG, Leonard ML, Horn AG (2012) Observer effects and avian call count survey quality: rare-species biases and overconfidence. Auk 129: 76–86.
38. Alldredge MW, Pacifici K, Simons TR, Pollock KH (2008) A novel field evaluation of the effectiveness of distance and independent observer sampling to estimate aural avian detection probabilities. J Applied Ecol 45: 1349–1356.
39. Suarez-Rubio M, Leimgruber P, Renner SC (2011) Does exurban development influence bird species richness and diversity? J Ornithol 152: 461–471.
40. Boulinier T, Nichols JD, Hines JE, Sauer JR, Flather CH, et al. (2001) Forest fragmentation and bird community dynamics: inference at regional scales. Ecology 82: 1159–1169.

22

Does Fire Influence the Landscape-Scale Distribution of an Invasive Mesopredator?

Catherine J. Payne[1], Euan G. Ritchie[1], Luke T. Kelly[2], Dale G. Nimmo[1]*

1 Centre for Integrative Ecology, School of Life and Environmental Sciences, Deakin University, Melbourne, Victoria, Australia, **2** Australian Research Council Centre of Excellence for Environmental Decisions, School of Botany, University of Melbourne, Melbourne, Victoria, Australia

Abstract

Predation and fire shape the structure and function of ecosystems globally. However, studies exploring interactions between these two processes are rare, especially at large spatial scales. This knowledge gap is significant not only for ecological theory, but also in an applied context, because it limits the ability of landscape managers to predict the outcomes of manipulating fire and predators. We examined the influence of fire on the occurrence of an introduced and widespread mesopredator, the red fox (*Vulpes vulpes*), in semi-arid Australia. We used two extensive and complimentary datasets collected at two spatial scales. At the landscape-scale, we surveyed red foxes using sand-plots within 28 study landscapes – which incorporated variation in the diversity and proportional extent of fire-age classes – located across a 104 000 km^2 study area. At the site-scale, we surveyed red foxes using camera traps at 108 sites stratified along a century-long post-fire chronosequence (0–105 years) within a 6630 km^2 study area. Red foxes were widespread both at the landscape and site-scale. Fire did not influence fox distribution at either spatial scale, nor did other environmental variables that we measured. Our results show that red foxes exploit a broad range of environmental conditions within semi-arid Australia. The presence of red foxes throughout much of the landscape is likely to have significant implications for native fauna, particularly in recently burnt habitats where reduced cover may increase prey species' predation risk.

Editor: R. Mark Brigham, University of Regina, Canada

Funding: Funding for the site-scale study was provided by the Victorian government's Department of Environment and Primary Industries, under the Mallee Hawkeye project (Contract Number 313764). Funding and logistical support for the landscape-scale study was provided by Land and Water Australia, the Mallee Catchment Management Authority, Parks Victoria, Department Sustainability and Environment Victoria, Department Environment and Heritage SA, Lower Murray-Darling Catchment Management Authority, Department Environment and Climate Change NSW, Australian Wildlife Conservancy and Birds Australia. The funders had no role in study design, data collection and analysis, decision to publish, or preparation of the manuscript.

Competing Interests: The authors have declared that no competing interests exist.

* Email: dale@deakin.edu.au

Introduction

Predators shape ecosystems worldwide [1]. They can exert top-down regulation of lower trophic levels [2] and induce trophic cascades which flow through entire ecosystems [3]. Predators introduced to areas outside of their native range can have a particularly strong effect on native species [4], and have caused population declines and extinctions in a range of ecosystems [5]. Many invasive predators are 'mesopredators': smaller predator species that increase in abundance or activity following the removal of apex predators [6]. For example, in Australia, persecution of the native apex predator, the dingo (*Canis dingo*), has led to increases in the density or activity of invasive mesopredators (e.g. the red fox [*Vulpes vulpes*]) throughout large portions of the continent [3].

Fire is another globally significant process that affects environments worldwide [7]. Fire influences ecosystems via bottom-up control by altering the availability of key resources for biota. Fire incinerates plant matter, altering vegetation structure [8,9], which in turn affects the distribution and abundance of animals [10].

Invasive mesopredators and fire share an important characteristic from a conservation perspective: both can be manipulated through management interventions. Invasive mesopredators are managed using lethal control and exclusion fencing, and fire using suppression or prescribed burning. However, management of mesopredators and fire usually occurs in isolation, without consideration of the potential effects of fire *on* mesopredators [11]. It is important to rapidly address this significant knowledge gap because some fire regimes may exacerbate the effects of invasive mesopredators by simplifying vegetation and amplifying predation risk [12,13]. For example, interactions between fire regimes and invasive mesopredators have been hypothesised as a cause of lower survival of reptile species in recently-burned areas [14], and a contributor to the collapse of small mammal communities in northern Australia [15].

The red fox is one of the world's most widely distributed mesopredators. It is common in both the northern and southern hemispheres. Foxes, and a second introduced mesopredator, the feral cat (*Felis catus*), are widely regarded as the primary cause of extinctions and declines of Australia's marsupial fauna [5]. Evidence for the negative impact of foxes has been demonstrated through predator-control experiments that have shown that prey species increase in both range and activity when foxes are removed [16,17]. Further evidence comes from dietary studies showing foxes eat a wide range of native mammal, reptile, bird, and invertebrate prey [18–20].

Despite indications that foxes may inhibit the recovery of native species following fire [12,21], whether foxes are themselves influenced by fire remains poorly known. This knowledge gap limits the ability of land managers to consider the effects of fire management on red foxes, which could have negative ramifications for native biodiversity. While foxes are widely considered as habitat generalists, they do display local variability in occurrence related to habitat or landscape structure [22]. For example, in some regions, foxes prefer heterogeneous landscapes [22], as they are able to use multiple landscape elements on a daily or seasonal basis [23,24]. Fire management in many regions seeks to maximise landscape heterogeneity by creating mosaics of fire ages (i.e. 'patch mosaic burning'; [25]). Does such management inadvertently favour invasive mesopredators?

The few studies that have explored the topic have focused on relatively short temporal scales (<30 years and often <10 years post fire) or small spatial scales (but see [26]). However, in some ecosystems, post-fire vegetation recovery continues for a century or more after fire [27]. Consequently, animal species respond to fire over similarly long time-frames [28]. The effects of fire can also occur across multiple spatial scales [29]; while time since fire may affect a species' occurrence at any *point* in the landscape, the area and composition of fire-ages within a 'whole' landscape can play a critical role in affecting species' landscape-level distributions [30]. This is likely to be especially true for large, mobile species, such as the red fox.

In addition to the effects that fire may have on species' occurrence, other environmental factors may be locally important. With regard to foxes, this includes climate [26], the distribution of vegetation types [31], and the distance to roads [24] and agricultural land [22]. Foxes rely on free standing water for drinking, particularly when temperatures are high (>30°C), as is common in many semi-arid environments. Hence, as annual rainfall decreases (aridity intensifies) permanent water may be reduced in its availability and limit fox occurrence. Foxes are often thought of as edge specialists [22]. They often prefer to hunt in open areas such as resource-rich agricultural fields or structurally simple vegetation types adjacent to more complex vegetation which provides cover during the day [22,32]. Their ability to hunt may be further enhanced where roads create easy access and increased visibility in otherwise structurally complex habitats [24,33].

Here, we examine what drives the occurrence (reporting rate) of red foxes in semi-arid Australia at multiple spatial scales, with a particular emphasis on the role of fire. We conducted two large-scale natural experiments. First, we explored landscape-scale patterns of fox occurrence in relation to the properties of fire mosaics; namely, the amount and diversity of fire age-classes within each of 28 study landscapes (each 12.6 km^2). Second, we explored site-scale patterns of fox occurrence in relation to fire history at 108 sites stratified along a century-long post-fire chronosequence. In both cases, we also quantified the influence of other environmental variables such as vegetation type and distance to agricultural land. Our aims were: 1) to determine the drivers of fox distribution in semi-arid Australia; and 2) to understand the specific role of fire in influencing fox occurrence at large scales relevant to fire and mesopredator management.

Materials and Methods

Study region

This study was undertaken in the Murray Mallee region of south-eastern Australia (Fig. 1). The climate in the region is semi-arid, with mean annual rainfall of 200–350 mm and average daily maximum temperatures are 30–33°C in summer and 15–18°C in winter (Australian Bureau of Meteorology; http://www.bom.gov.au). The vegetation is predominantly 'tree mallee' characterised by an overstorey of *Eucalyptus* species (<5–8 m) with a multi-stemmed growth form [34]. Two vegetation types are common throughout region [35]. 'Triodia Mallee' has a canopy of *Eucalyptus dumosa* and *E. socialis* with an understorey of *Triodia scariosa* and mixed shrubs, and occurs mainly on sandier soils typical of dunes. 'Chenopod Mallee' has a canopy of *E. oleosa* and *E. gracilis* with an open understorey of chenopod species, and occurs on heavier soils typical of swales.

Mallee vegetation is fire-prone with large fires (i.e. > 100,000 ha) occurring somewhere in the region on a bidecadal basis [36], although individual sites can go long periods without fire (i.e. >100 years; [37]). Fire is actively managed in the region through prescribed burning and suppression for both asset protection and conservation objectives [36]. Most wildfires are ignited by lightning strikes and are stand-replacing, essentially resetting vegetation succession to 'year-zero' (Fig. 2; [8]).

Site selection

We refer to two datasets in this study derived from two different natural experiments that differed in both their spatial grain and extent. We refer to these as 'landscape-scale' and 'site-scale' datasets throughout, in reference to the spatial scale of the response and predictor variables (i.e. the spatial grain) of the respective datasets.

Landscape-scale data. The landscape-scale dataset consists of 28 study landscapes, each with a 4 km diameter circle (12.6 km^2; Fig. 1), distributed throughout a 104, 000 km^2 study area. These landscapes were selected as part of a broad-scale natural experiment: the Mallee Fire and Biodiversity Project. Study landscapes were selected to allow a comparison of the effects of different approaches to patch mosaic burning on biodiversity, with a particular emphasis on the role of the area and diversity of fire-ages ('pyrodiversity', see [25]). Thus, landscapes were stratified according to number and spatial extent of fire-age classes within the landscape [29]. The fire history of the region was mapped using the ENVI package [38] and then converted to shape files for use in ArcMap version 9.2 [39]. Only fires that occurred post-1971 were mapped due to limited availability of Landsat imagery prior to this time (see [36]).

Site-scale data. We collected site-scale data within a subset of the 28 study landscapes located within the region's largest national park; Murray Sunset National Park (6630 km^2; Fig. 1). Ten sites were established within each of 10 of the original study landscapes. Sites were distributed to incorporate a range of fire-age classes (range = 7–105 years), as well as capturing geographic and topographic variation. We established an additional landscape, containing 12 sites, following an experimental burn during the study (fire age = 0 years), resulting in 11 landscapes containing 112 sites. We omitted four sites to comply with ethics permits due to their close proximity to active nesting sites of the endangered malleefowl (*Leipoa ocellata*). This resulted in a total of 108 sites being surveyed. All sites were a minimum of 200 m apart and typically >100 m from the edge of a fire-age class.

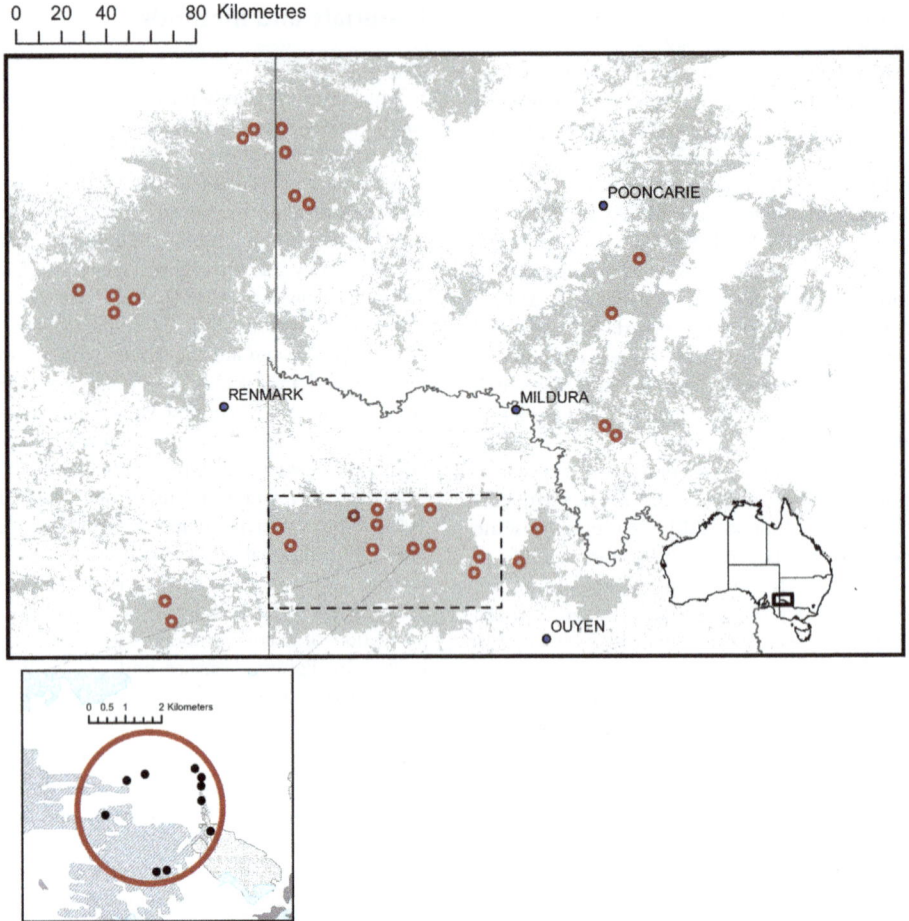

Figure 1. Map of study area showing all landscapes (circles) considered in this study (grey shading indicates mallee vegetation; majority of white areas indicates agricultural land used for grazing and cereal crops). The dashed box shows the spatial extent of the site-scale study. An inset shows an example of a study landscape including the position of 10 sites within where site-scale data were collected. Within the inset, different hatching represents different fire ages.

Predator surveys

Landscape-scale data. We surveyed large mammalian predators using track surveys from three sand-plots within each study landscape (n = 84 in total). Each sand-plot was a 100 m×2 m area smoothed out by dragging a weight along an unsealed vehicle track. The locations of sand-plots within landscapes were chosen to incorporate variation in the topography (dunes and swales) within each landscape. Sand-plots were typically >500 m apart. We checked each sand-plot for tracks once per day by walking along the transect and identifying tracks to species level for three consecutive days in spring (October–November 2007), and again in summer (January–March 2008), resulting in six survey nights for each sand-plot and thus 18 survey nights per landscape. Following checking, sand-plots were smoothed over in preparation for the following day. If the sand-plot was heavily disrupted on one day (due to weather or vehicle disturbance), it was surveyed for an additional day.

Site-scale data. We used camera traps (Passive ScoutGuard 550; ScoutGuard IR Cameras, Australia) to survey for mammalian predators at the site-scale during April–July 2012. We installed one camera per site and deployed each for a minimum of 15 nights. We attached cameras to a post at a height of 0.5 m and positioned them facing southward. Vegetation was removed within the immediate area of the camera to reduce false triggering. A

15 second video was taken each time the camera motion sensor was triggered. To attract predators to the front of the camera from the local vicinity, we placed a scent lure of tuna oil soaked into chemical wadding inside a bait holder made from PVC piping with steel mesh at one end. We positioned the lure 3 m from the base of the camera post, and secured it to the ground with a peg.

Predictor variables

Landscape-scale data. Six predictor variables were chosen to represent the properties of the study landscapes (Table 1). Three of these variables represent the fire history of the landscape: (1) the extent of recently burnt vegetation in the landscape (<10 years since fire; 'recently burned'); (2) the extent of long unburnt vegetation (unburned since 1972; 'long unburned'); and (3) the diversity of fire-ages within a landscape ('fire diversity'). Fire diversity was calculated as the Shannon-Wiener diversity index of the proportional cover of fire age-classes within each landscape.

Three predictor variables were chosen to describe properties of the study landscapes other than fire history. We used a measure of mean solar radiation ('solar radiation') as a surrogate for aridity across the region. The solar radiation variable represents the total amount of solar energy falling on a horizontal space per day (MJ/m^2). We derived these values from a gridded data set (5 km resolution) extending over 18 years (1990–2008; Australian

(a)

(b)

Figure 2. Examples of mallee vegetation with differing fire histories. (a) A recently burned site; (b) A long unburned site.

Bureau of Meteorology http://www.bom.gov.au, 2009). Solar radiation was the mean of the 18 yearly averages of the grids that overlaid each landscape. Solar radiation is negatively correlated with annual rainfall and positively correlated with temperature. We used the proportional extent of *Triodia* mallee vegetation ('Triodia Mallee') within the landscape to capture differences in vegetation types. The extent of mallee vegetation in the study area was mapped in previous work (see [35]). Finally, we used the distance from the centre point of each landscape to the closest area of contiguous non-mallee vegetation ('distance to agricultural land'), to capture the context of landscapes with respect to landscape modification. The area surrounding each reserve is comprised almost entirely of grazing land and grain crops. We calculated distance to agricultural land using ArcGIS [39].

Site-scale data. Eight predictor variables were chosen at the site level (Table 1). The fire history of sites was represented by the time since the last fire ('time since fire'; range: 0–105 years). This was determined using two methods. Recent fire history (since 1972) was calculated using the fire history maps (see [36]). Fire-ages for sites burnt prior to the availability of satellite imagery (i.e. before 1972) were estimated using regression models of the relationship between stem diameter and tree age, and then using stem diameter to estimate the age of trees in areas where fire history was unknown (see [37] for detailed methods). This extended the time since fire axis from 0–32 years to 0–105 years.

Vegetation type was considered as a categorical variable with two levels: Triodia Mallee or Chenopod Mallee ('vegetation type'). We again considered the effects of landscape modification by including the distance of sites to both the border of the National Park (1.72–21.28 km; 'distance to edge') and dirt roads (range: 28–1044 m; 'distance to road'). We used park boundary as a proxy for an edge habitat because the park forms abrupt boundaries with cleared agricultural land and other non-mallee vegetation. We calculated distance variables using ArcGIS [39]. Aridity (solar radiation) was not considered at this scale as the data were collected from a single reserve.

Four additional predictor variables were included to describe vegetation structure at the sites. We established vegetation transects in representative areas 15 m from each camera location.

Table 1. Predictor variables included in models using the landscape-scale and the site-scale datasets.

Dataset	Predictor variable	Description
Landscape-scale data	Recently burned	Extent of landscape burned within 10 years of surveys
	Long unburned	Extent of landscape not burned since 1972 (>35 years since fire)
	Fire diversity	Shannon-Wiener diversity index of the extent of three fire age classes (0–10 years, 11–35 years and >35 years)
	Solar radiation	Long-term average monthly gridded solar exposure (MJ/m²) from 1990–2008 for each landscape
	Triodia Mallee	Extent of landscape comprised of vegetation type in which *Triodia scariosa* typically occurs
	Distance to agricultural land	Distance from the centre of each landscape to contiguous non-mallee vegetation (m)
Site-scale data	*Time since fire*	Amount of time since a site last experienced fire (years)
	Bare ground cover	Cover of bare ground present
	Triodia cover	Cover of *Triodia scariosa* <1 m
	Eucalypt cover	Cover of eucalypt shrubs <1 m
	Shrub cover	Cover of non-eucalypt shrubs <1 m
	Vegetation type	Broad vegetation classification (Triodia Mallee or Chenopod Mallee)
	Distance to edge	Distance from each site to the nearest park boundary (m)
	Distance to road	Distance from each site to the nearest road (m)

We recorded substrate type and vegetation structure at 1 m intervals along a 50 m transect using a 2 m structure pole (2 cm diameter) held vertically above the ground. The four variables considered in the analysis represent the cover of open, bare ground ('bare ground cover'), spinifex ('*Triodia* cover'), eucalypt shrubs (defined as Eucalypt trees <3 m in height 'eucalypt cover') and non-eucalypt shrubs ('shrub cover'). Bare ground was included because it gives an approximation of the 'openness' of the vegetation at the ground level. The cover of spinifex, eucalypt shrubs, and non-eucalypt shrubs were included as they form the majority of the ground and understorey structural complexity, and are known to drive fauna in the region [34,40].

Response variables

For both datasets, the response variable was the 'reporting rate' of foxes. At the landscape-scale, we defined reporting rate as the number of nights that a fox was recorded as 'present' and 'absent', respectively, at a sand pad over the 18 nights of sampling per landscape (i.e. three sand-plots surveyed for six nights in each landscape). Likewise, at the site-scale, reporting rate is the number of nights that foxes were and were not detected at the site, respectively, over the course of sampling (i.e. 15 nights).

Statistical analysis

We used generalised linear mixed models (GLMMs) with the Laplace approximation [41] to examine the relationship between response and predictor variables at both landscape and site-scales. In landscape-scale models, we included 'reserve' as a random effect to account for spatial clustering of landscapes in conservation reserves (Fig. 1). Similarly, in the site-scale models, we included 'landscape' as a random effect to account for potential spatial correlation due to the clustering of sites into landscapes. Because we were studying the reporting rate of red foxes, a proportion, we modelled the response variable (at both scales) using a binomial distribution of errors and a logit link function.

For the landscape-scale dataset, we developed a set of candidate models that included all combinations of the six landscape-scale predictor variables. At the site-level, we developed two separate sets of models. As fire affects the variables used to describe vegetation structure (e.g. *Triodia* cover, bare ground cover; [8]), including both fire and vegetation structure variables in the same model could result in unreliable parameter estimates due to colinearity between predictor variables [42]. Thus, one model set (model set 1) included time since fire, vegetation type, distance to edge and distance to road, and a second model set (model set 2) included the vegetation structure variables (bare ground cover, *Triodia* cover, eucalypt cover and shrub cover). All combinations of predictors within the two sets of models were considered, meaning all eight site-level variables were in the same number of models overall. All variables included within a model set had low levels of colinearity (i.e. r <0.5). We tested both datasets for overdisperson using Pearson's residuals [43], and found no evidence of overdispersion.

We compared each set of candidate models using Akaike's Information Criterion corrected for small sample sizes (AICc; [44]). To compare the level of support for each model relative to the most parsimonious model, we calculated the difference (Δ_i) between the AIC$_c$ value of the best model (lowest AIC$_c$ value) and the AIC$_c$ value of each candidate model [44]. We considered models with Δ_i<2 to have substantial support [44]. We also calculated the Akaike weight (w_i) for each model. By summing these weights to calculate predictor weights ($\sum w_i$) for each variable, we were able to explore the influence of individual predictor variables at both the landscape and site level.

When there was no clear 'best model' (i.e. the most parsimonious model was not strongly weighted [w_i<0.9]), we used model averaging to determine the direction and magnitude of the effect of each predictor variable [44]. We considered a variable as important when the associated 95% confidence interval of the averaged estimate did not overlap with zero. We performed all statistical analyses in R version 2.15.1 [45] using the lme4 package [41] and the MuMIn package [46].

Ethics statement

The landscape-scale data were collected with approval from animal ethics committees at La Trobe University (approval number AEC06/07[L]V2) and Deakin University (approval number A41/2006), and permits from the Department of Sustainability and Environment, Victoria (permit 10003791), the Department of Environment and Heritage, South Australia (permit 13/2006), and the National Parks and Wildlife Service, NSW (license number S12030). The site-scale data were collected in accordance with the regulations of the Deakin University Animal Ethics Committee (approval number B10-2012) and in accordance with Department of Sustainability and Environment, Victoria (approval number 10006279).

Results

At the landscape-scale, we recorded fox tracks in 24 of 28 (86%) study landscapes. We detected foxes on 3.32±0.49 (mean ± standard error) of 18 nights per landscape over the total sampling period. Other large-bodied, mammalian predators were uncommon: we detected cats at only 7 of 28 landscapes (25%). At the site-scale, we observed foxes at 62 of 102 (61%) sites (six cameras failed to reach the full 15 day survey period due to fault and were excluded from further analysis i.e. n = 102) and found the species to be widely distributed across the study area. We did not detect any cats at the site-scale over the 15 night sampling period.

At the landscape-scale, all models were a poor fit for the data and explained <6.5% of the variation in the data (% deviance explained). At the site-scale, all models explained <3.5% of the variation in the data. For both datasets, model selection indicated there was a similar level of support for several models (Δ_i<2; Table 2), including the intercept-only model (i.e. only an intercept terms, no predictor variables), which received substantial support at both scales. As no single model was supported as being clearly best (i.e. w_i>0.9; Table 2), we employed multi-model inference using model averaging to estimate the size, direction and uncertainty of parameter effects for fox explaining reporting rate in both datasets.

The model-averaged coefficients for each predictor variable, in both datasets, were small and uncertain. The 95% confidence intervals of all predictor variables overlapped with zero (Fig. 3). The $\sum w_i$ for all predictor variables was low: <0.5 and <0.6 for the landscape- and site-scale datasets respectively.

Graphical exploration of the data further highlights that fox activity was not strongly linked to key predictor variables (Fig. 4). In summary, the data shows that neither fire, nor any other predictor variable measured, affected the reporting rate of foxes at either the landscape- or site-scale.

Discussion

Introduced mesopredators and fire are two processes that shape ecosystems around the world [4,7]. Here, we have shown that a widespread and ecologically devastating mesopredator, the red fox [5], is largely unaffected by fire and is an extreme habitat

Table 2. Model selection results for red fox reporting rate for landscape-scale and sits-scale datasets.

Candidate model	df	LogLik	AIC$_c$	Δ_i	w_i	%Dev
Landscape-scale dataset						
Null model (intercept only)	2	−32.37	69.2	0.00	0.14	0.00
Distance to agricultural land	3	−31.21	69.4	0.21	0.12	3.57
Distance to agricultural land + Triodia Mallee	4	−30.53	70.8	1.58	0.06	5.68
Triodia Mallee	3	−31.99	71.0	1.76	0.06	1.17
Fire diversity	3	−32.10	71.2	1.99	0.05	0.82
Site-scale dataset						
Bare ground cover	3	−65.90	138.0	0.00	0.13	2.16
Bare ground cover + *Triodia* cover	4	−65.07	138.6	0.51	0.10	3.39
Triodia cover	3	−66.26	138.8	0.73	0.09	1.62
Null model (intercept only)	2	−67.36	138.8	0.79	0.09	0.00
Bare ground cover + eucalypt cover	4	−65.67	139.8	1.71	0.06	2.50

Models are shown for which $\Delta_i < 2.0$.

generalist in semi-arid Australia. This result was confirmed using two large, complementary datasets, collected at different times and characterised by differing spatial scales and sampling strategies.

Fire and the red fox

Our findings show that fire does not exert a strong influence on the distribution of the red fox in semi-arid mallee ecosystems. Despite conducting two intensive natural experiments across a broad geographic region, we did not detect a relationship between the reporting rate of foxes and fire history at either the landscape- or site-scale. At the landscape-scale, the red fox was recorded equally often in landscapes dominated by recently burned or long unburned vegetation, and in landscapes with a single fire age-class as those with a diversity of fire ages. At the site-scale, the red fox has a similar reporting rate in recently burned sites as in sites unburned for over a century. The post-fire preferences of the red fox are thus extremely broad, both spatially and temporally (also see [12,47]).

Fire causes significant changes to vegetation structure over century-long time frames in mallee ecosystems [8]. In doing so, fire affects the distribution of a large range of fauna species [34]. Indeed, work conducted within the same study landscapes has shown the large and long-term effects fire has on birds, reptiles, and small mammals [28–30]. The lack of a response to fire by foxes is therefore not typical of native fauna in the region. It also suggests that foxes are not restricted to areas with particular soil or vegetation attributes for denning. This is consistent with foxes not being affected by any of the vegetation attributes measured (e.g. *Triodia* cover, shrub cover etc.).

A related way that fire could influence foxes is by altering the distribution of prey resources. As mentioned above, the distribution of many prey species are significantly affected by fire in the study region (e.g. birds, mammals, reptiles). Thus, foxes occupy a range of post-fire ages despite the strong influence of fire on the type and abundance of prey available. Red foxes have a broad and generalist diet [48], being able to consume a wide range of prey including both vertebrates and invertebrates, and even vegetation [19,20]. Furthermore, foxes are capable of prey switching to capitalize on the most abundant prey source available [18,49], thereby reducing their reliance on any particular prey item. This flexibility in their diet is likely to be a key component of their life

history that allows them to occur within such a broad range of post-fire conditions.

One objection to our findings at the site-scale may be that the local site is not a relevant spatial scale to characterize the effects of fire, as foxes are a relatively large and mobile species. Given the large estimated home ranges of foxes in other parts of arid Australia (e.g. 8–33 km^2; [31]), foxes may select broader areas (i.e. kms^2) that capture their resource requirements across entire landscapes, and this might include a large area of a particular fire-age, or multiple fire ages. Such use of multiple habitat types by foxes has been demonstrated in other systems [23,24]. Our landscape-scale study characterized land mosaics at a large scale relevant to the home range of foxes (12.6 km^2), and still failed to detect any relationship between fox activity and fire history. Therefore, our results suggest that the lack of relationships between fox reporting rate and fire history does not stem from spatial scaling issues. Instead, foxes are resilient towards the effects of fire at multiple temporal and spatial scales.

Climate and distance to modified land

In addition to fire, we examined other variables that could influence the distribution of the red fox. Here, we again found red foxes to be flexible to a broad range of ecological conditions. Foxes displayed no response to an aridity gradient across the study region. This lack of response to aridity is unsurprising, as the geographic range of the red fox spans the northern hemisphere and much of Australia, suggesting the species is capable of coping with a range of climatic conditions.

Despite foxes occupying a broad climatic niche in space, fluctuations in populations do occur in response to extreme weather events. For example, fox populations in arid areas rise rapidly following high rainfall events, in response to increased prey availability [50]. Our site-scale study was carried out during a year of record high rainfall (Australian Bureau of Meteorology, Ouyen Station). Considered in isolation, this may suggest that the wide distribution of the fox was partly due to a productivity-related increase in food resources (predominantly populations of native and introduced rodents; [40]). However, the landscape-scale data were collected near the end of a severe, decade-long drought. Foxes were widely distributed across the region despite the drought. This indicates that, in semi-arid Australia, foxes can be

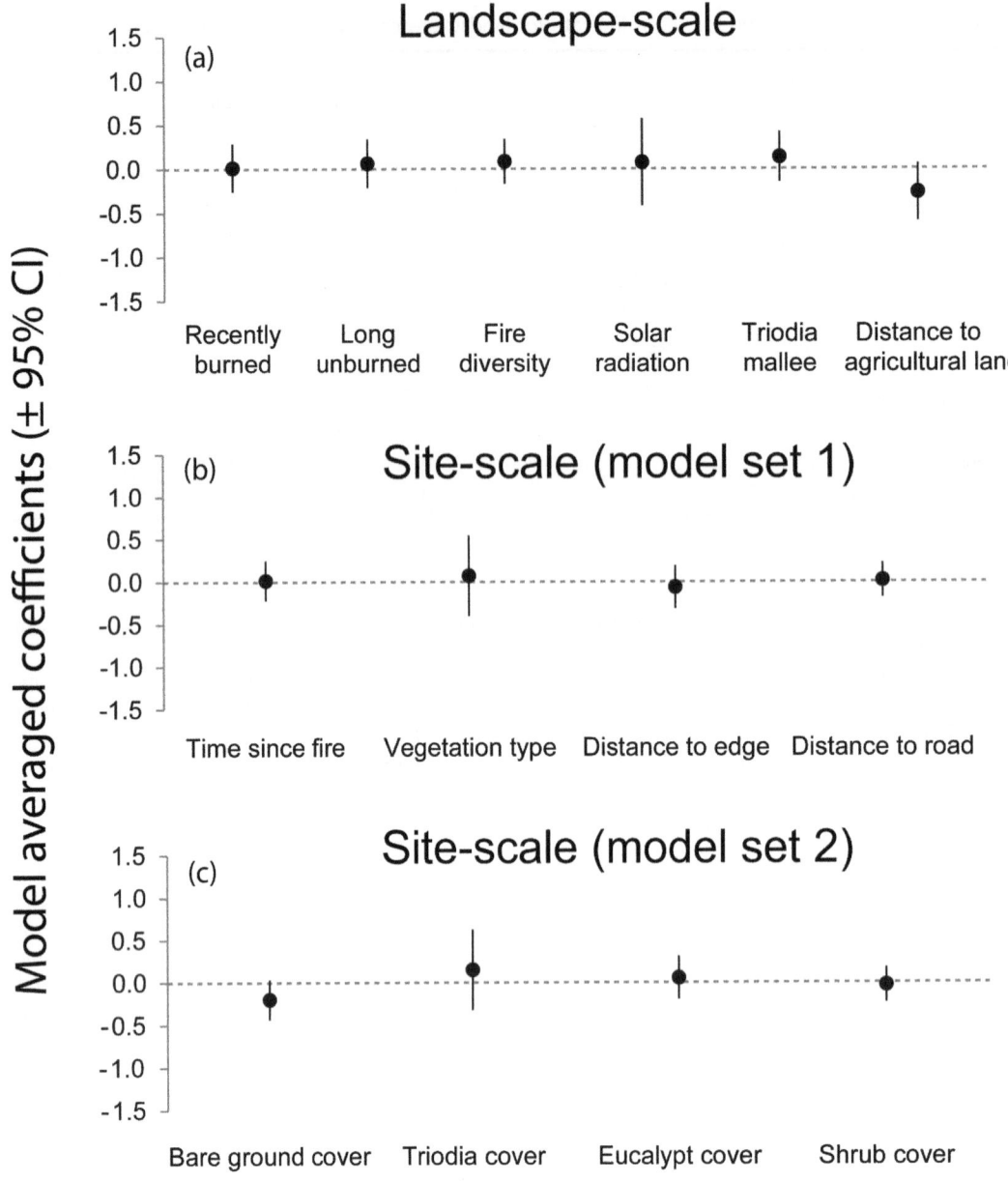

Figure 3. Model-averaged regression coefficients and 95% confidence intervals of models describing the reporting rate of foxes at both the landscape-scale (a) and site-scale (b and c).

widespread during a broad range of climatic conditions and despite fluctuations in their prey populations which accompany climactic extremes [40].

Some studies have found that foxes are positively associated with edges between fragments and modified land (e.g. agricultural land) [22,32]. Our results indicate that foxes do not show a preference for edge habitats in mallee ecosystems, despite our sites and landscapes capturing a broad gradient of distances to agricultural land, from <2 km to >30 km. Edge habitats may be more important for foxes in highly fragmented landscapes, where they occur with small remnant patches of wooded vegetation which provide the only available cover [22]. While the mallee region has been subject to large amounts of land clearing, there are still relatively large intact areas of native vegetation. Edges may be less important in this region because the interior mallee vegetation provides sufficient shelter and prey.

Nevertheless, it is also possible that edge effects occur closer to the agricultural boundary than we sampled (i.e. <2 km).

The use of roads and tracks by foxes is also well documented [51,52]. Foxes have been found to be more abundant along roadsides [33]. In the mallee system, however, we found similar reporting rates at varying distances (28–1044 m) from roads, indicating foxes use areas well away from roads equally as often as sites close to roads. One hypothesis for the use of roads by foxes is that they provide 'runways' which facilitate movement and allow access to foraging areas that would be otherwise difficult to reach [51,52]. In contrast to environments with a dense understory, mallee vegetation is relatively open, and is unlikely to limit the movement of foxes to roadsides. This may explain the lack of preference for sites near roads in the current study.

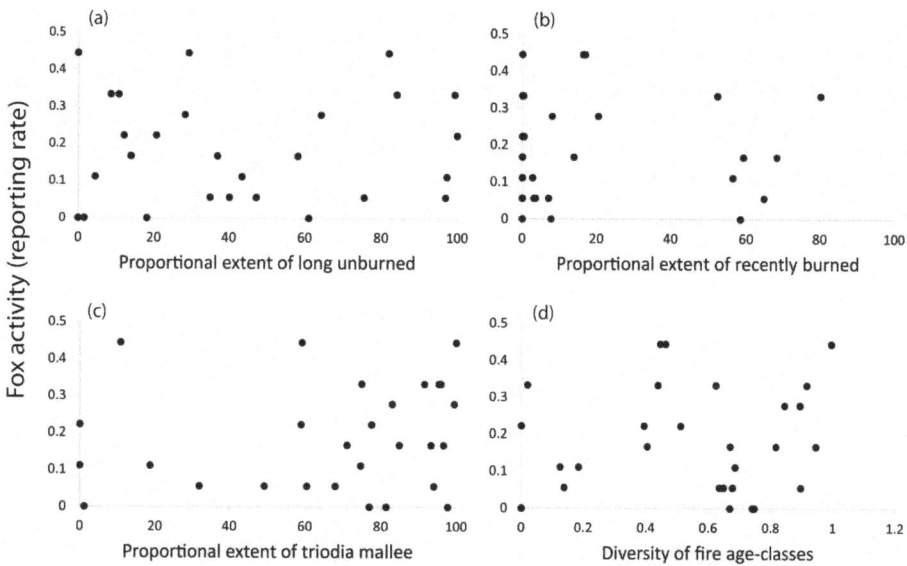

Figure 4. Relationships between the reporting rate of the red fox and the properties of fire mosaics. Circles are raw data points.

Implications

Fire is used as a conservation tool in Australia and around the world [25]. This study suggests it is unlikely that any particular approach to fire management will alter the reporting rate of the red fox in the semi-arid mallee systems of Australia. However, the presence of foxes in recently burned sites and landscapes is a concern. Predation by invasive mesopredators has been hypothesized as a cause of low post-fire survival in reptiles [14] and mammals [12,13], due to the reduced cover available in burnt habitats. Although we found no effect of fire history on red fox occurrence, it is possible that predation pressure differs across fire ages due to increased predation risk in recently burned areas. Thus, assessing predation pressure directly across a range of post-fire ages is an important area for further research.

The loss of apex predators can cause smaller predators to increase in abundance, expand their range, and change their temporal activity; this is known as 'mesopredator release' [2,6]. Red foxes have been shown to select particular habitats which may allow them to avoid dominant predators (e.g. coyotes; [53]). As such, one further explanation for the lack of obvious habitat selection by foxes in this system may be the lack of regulating predators. In other Australian systems, the presence of the dingo, Australia's largest terrestrial apex predator, has been shown to affect fox distributions [3]. Dingoes are largely extinct from the study area but were once common, and as such there is no direct

regulation of the abundance or distribution of foxes via biotic interactions. Thus, one potential way to control red foxes in mallee communities is by reinstating dingoes as the apex predator. As this is likely to be a controversial idea owing to the proximity of mallee vegetation to agricultural land and livestock, trialing reintroductions in a controlled and experimental way would be an important first step towards a proof of concept, and a potential solution to this complex conservation issue.

Acknowledgments

Thanks to all members of the Mallee Fire and Biodiversity Team, particularly Andrew Bennett, Mike Clarke, Lisa Farnsworth and Lauren Brown, and to the many volunteers, agency staff, and land owners who assisted with logistics and field work.

Author Contributions

Conceived and designed the experiments: DGN CJP EGR LTK. Performed the experiments: DGN CJP LTK. Analyzed the data: DGN CJP. Contributed to the writing of the manuscript: CJP DGN LTK EGR.

References

1. Estes JA, Terborgh J, Brashares JS, Power ME, Berger J, et al. (2011) Trophic downgrading of planet Earth. Science 333: 301–306.
2. Ritchie EG, Johnson CN (2009) Predator interactions, mesopredator release and biodiversity conservation. Ecology Letters 12: 982–998.
3. Letnic M, Ritchie EG, Dickman CR (2012) Top predators as biodiversity regulators: the dingo *Canis lupus dingo* as a case study. Biological Reviews 87: 390–413.
4. Salo P, Korpimäki E, Banks PB, Nordström M, Dickman CR (2007) Alien predators are more dangerous than native predators to prey populations. Proceedings of the Royal Society B: Biological Sciences 274: 1237–1243.
5. Johnson C (2006) Australia's Mammal Extinctions: a 50000 year history. New York: Cambridge University Press.
6. Crooks KR, Soule ME (1999) Mesopredator release and avifaunal extinctions in a fragmented system. Nature 400: 563–566.
7. Bowman DMJS, Balch JK, Artaxo P, Bond WJ, Carlson JM, et al. (2009) Fire in the Earth system. Science 324: 481–484.
8. Haslem A, Kelly LT, Nimmo DG, Watson SJ, Kenny SA, et al. (2011) Habitat or fuel? Implications of long-term, post-fire dynamics for the development of key resources for fauna and fire. Journal of Applied Ecology 48: 247–256.
9. Smit IPJ, Asner GP, Govender N, Kennedy-Bowdoin T, Knapp DE, et al. (2010) Effects of fire on woody vegetation structure in African savanna. Ecological Applications 20: 1865–1875.
10. Letnic M, Tamayo B, Dickman CR (2005) The responses of mammals to La Nina (El Nino Southern Oscillation) - associated rainfall, predation and wildfire in central Australia. Journal of Mammalogy 86: 689–703.
11. Driscoll DA, Lindenmayer DB, Bennett AF, Bode M, Bradstock RA, et al. (2010) Fire management for biodiversity conservation: Key research questions and our capacity to answer them. Biological Conservation 143: 1928–1939.
12. Arthur AD, Catling PC, Reid A (2012) Relative influence of habitat structure, species interactions and rainfall on the post-fire population dynamics of ground-dwelling vertebrates. Austral Ecology 37: 958–970.

13. Sutherland EF, Dickman CR (1999) Mechanisms of recovery after fire by rodents in the Australian environment: a review. Wildlife Research 26: 405–419.

14. Smith AL, Bull CM, Driscoll DA (2012) Post-fire succession affects abundance and survival but not detectability in a knob-tailed gecko. Biological Conservation 145: 139–147.

15. Woinarski JCZ, Legge S, Fitzsimons JA, Traill BJ, Burbidge AA, et al. (2011) The disappearing mammal fauna of northern Australia: context, cause, and response. Conservation Letters 4: 192–201.

16. Kinnear JE, Onus ML, Bromilow RN (1988) Fox control and rock-wallaby population dynamics. Wildlife Research 15: 435–450.

17. Risbey DA, Calver MC, Short J, Bradley JS, Wright IW (2000) The impact of cats and foxes on the small vertebrate fauna of Heirisson Prong, Western Australia. II. A field experiment. Wildlife Research 27: 223–235.

18. Catling PC (1988) Similarities and contrasts in the diets of foxes, *Vulpes vulpes*, and cats, *Felis catus*, relative to fluctuating prey populations and drought. Wildlife Research 15: 307.

19. Glen AS, Fay AR, Dickman CR (2006) Diets of sympatric red foxes *Vulpes vulpes* and wild dogs *Canis lupus* in the Northern Rivers Region, New South Wales. Australian Mammalogy 28: 101–104.

20. Risbey DA, Calver MC, Short J (1999) The impact of cats and foxes on the small vertebrate fauna of Heirisson Prong, Western Australia. I. Exploring potential impact using diet analysis. Wildlife Research 26: 621–630.

21. Letnic M and Dickman CR (2005) The responses of small mammals to patches regenerating after fire and rainfall in the Simpson Desert, central Australia. Austral Ecology 30: 24–39.

22. Graham CA, Maron M, McAlpine CA (2012) Influence of landscape structure on invasive predators: feral cats and red foxes in the brigalow landscapes, Queensland, Australia. Wildlife Research 39: 661–676.

23. Lucherini M, Lovari S, Crema G (1995) Habitat use and ranging behaviour of the red fox (Vulpes vulpes) in a Mediterranean rural area: is shelter availability a key factor? Journal of Zoology 237: 577–591.

24. Meek PD, Saunders G (2000) Home range and movement of foxes (*Vulpes vulpes*) in coastal New South Wales, Australia. Wildlife Research 27: 663–668.

25. Parr CL, Andersen AN (2006) Patch mosaic burning for biodiversity conservation: a critique of the pyrodiversity paradigm. Conservation Biology 20: 1610–1619.

26. Southgate R, Paltridge R, Masters P, Ostendorf B (2007) Modelling introduced predator and herbivore distribution in the Tanami Desert, Australia. Journal of Arid Environments 68: 438–464.

27. Gosper CR, Yates CJ, Prober SM (2013) Floristic diversity in fire-sensitive eucalypt woodlands shows a 'U'-shaped relationship with time since fire. Journal of Applied Ecology 50: 1187–1196.

28. Watson SJ, Taylor RS, Nimmo DG, Kelly LT, Haslem A, et al. (2012) Effects of time since fire on birds: How informative are generalized fire response curves for conservation management? Ecological Applications 22: 685–696.

29. Nimmo DG, Kelly LT, Spence-Bailey LM, Watson SJ, Taylor RS, et al. (2013) Fire mosaics and reptile conservation in a fire-prone region. Conservation Biology 27: 345–353.

30. Kelly LT, Nimmo DG, Spence-Bailey LM, Taylor RS, Watson SJ, et al. (2012) Managing fire mosaics for small mammal conservation: a landscape perspective. Journal of Applied Ecology 49: 412–421.

31. Moseby KE, Stott J, Crisp H (2009) Movement patterns of feral predators in an arid environment – implications for control through poison baiting. Wildlife Research 36: 422–435.

32. Catling P, Burt R (1995) Why are red foxes absent from some eucalypt forests in eastern New South Wales? Wildlife Research 22: 535–545.

33. Towerton AL, Penman TD, Kavanagh RP, Dickman CR (2011) Detecting pest and prey responses to fox control across the landscape using remote cameras. Wildlife Research 38: 208–220.

34. Bradstock RA, Cohn JS (2002) Fire regimes and biodiversity in semi-arid mallee ecosystems. In: R. A Bradstock, J. E Williams and M. A Gill, editors. Flammable Australia: The fire regimes and biodiversity of a continent. Cambridge: Cambridge University Press. 238–258.

35. Haslem A, Callister KE, Avitabile SC, Griffioen PA, Kelly LT, et al. (2010) A framework for mapping vegetation over broad spatial extents: A technique to aid land management across jurisdictional boundaries. Landscape and Urban Planning 97: 296–305.

36. Avitabile SC, Callister KE, Kelly LT, Haslem A, Fraser L, et al. (2013) Systematic fire mapping is critical for fire ecology, planning and management: A case study in the semi-arid Murray Mallee, south-eastern Australia. Landscape and Urban Planning 117: 81–91.

37. Clarke MF, Avitabile SC, Brown L, Callister KE, Haslem A, et al. (2010) Ageing mallee eucalypt vegetation after fire: insights for successional trajectories in semi-arid mallee ecosystems. Australian Journal of Botany 58: 363–372.

38. ITT (2005) ENVI. Version 4.2. Boulder, Colorado: ITT Industries.

39. Environmental Systems Research Institute (2007) Arc View. Version 9.2. Redlands, California: ESRI.

40. Kelly LT, Dayman R, Nimmo DG, Clarke MF, Bennett AF (2013) Spatial and temporal drivers of small mammal distributions in a semi-arid environment: The role of rainfall, vegetation and life-history. Austral Ecology 38: 786–797.

41. Bates D, Maechler M, Bolker B (2012) lme4: Linear mixed-effects models using S4 classes. R package (Version 0.999999-0). Available: http://CRAN.R-project.org/package=lme4.

42. Quinn GP, Keough MJ (2002) Experimental design and data analysis for biologists. New York: Cambridge University Press.

43. Zuur AF, Ieno EN, Walker N, Saveliev AA, Smith GM (2009) Mixed effects models and extensions in ecology with R. New York: Springer.

44. Burnham KP, Anderson DR (2002) Model selection and multi-model inference. New York: Springer.

45. R Development Core Team (2012) R: A language and environment for statistical computing. Vienna, Austria. Available: http://www.R-project.org/: R Foundation for Statistical Computing.

46. Bartoń K (2012) MuMIn: Multi-model inference. R package (Version 1.7.11). Available: http://CRAN.Rproject.org/package=MuMIn.

47. Catling PC, Coops N, Burt RJ (2001) The distribution and abundance of ground-dwelling mammals in relation to time since wildfire and vegetation structure in south-eastern Australia. Wildlife Research 28: 555–565.

48. White JG, Gubiani R, Smallman N, Snell K, Morton A (2006) Home range, habitat selection and diet of foxes (*Vulpes vulpes*) in a semi-urban riparian environment. Wildlife Research 33: 175–180.

49. Leckie FM, Thirgood SJ, May R, Redpath SM (1998) Variation in the diet of red foxes on Scottish moorland in relation to prey abundance. Ecography 21: 599–604.

50. Pavey CR, Eldridge SR, Heywood M (2008) Population dynamics and prey selection of native and introduced predators during a rodent outbreak in arid Australia. Journal of Mammalogy 89: 674–683.

51. Carter A, Luck GW, McDonald SP (2012) Ecology of the red fox (Vulpes vulpes) in an agricultural landscape. 2. Home range and movements. Australian Mammalogy 34: 175–187.

52. Frey SN, Conover MR (2006) Habitat use by meso-predators in a corridor environment. The Journal of Wildlife Management 70: 1111–1118.

53. Gosselink TE, Deelen TRV, Warner RE, Joselyn MG (2003) Temporal habitat partitioning and spatial use of coyotes and red foxes in east-central Illinois. The Journal of Wildlife Management 67: 90–103.

PERMISSIONS

All chapters in this book were first published in PLOS ONE, by The Public Library of Science; hereby published with permission under the Creative Commons Attribution License or equivalent. Every chapter published in this book has been scrutinized by our experts. Their significance has been extensively debated. The topics covered herein carry significant findings which will fuel the growth of the discipline. They may even be implemented as practical applications or may be referred to as a beginning point for another development.

The contributors of this book come from diverse backgrounds, making this book a truly international effort. This book will bring forth new frontiers with its revolutionizing research information and detailed analysis of the nascent developments around the world.

We would like to thank all the contributing authors for lending their expertise to make the book truly unique. They have played a crucial role in the development of this book. Without their invaluable contributions this book wouldn't have been possible. They have made vital efforts to compile up to date information on the varied aspects of this subject to make this book a valuable addition to the collection of many professionals and students.

This book was conceptualized with the vision of imparting up-to-date information and advanced data in this field. To ensure the same, a matchless editorial board was set up. Every individual on the board went through rigorous rounds of assessment to prove their worth. After which they invested a large part of their time researching and compiling the most relevant data for our readers.

The editorial board has been involved in producing this book since its inception. They have spent rigorous hours researching and exploring the diverse topics which have resulted in the successful publishing of this book. They have passed on their knowledge of decades through this book. To expedite this challenging task, the publisher supported the team at every step. A small team of assistant editors was also appointed to further simplify the editing procedure and attain best results for the readers.

Apart from the editorial board, the designing team has also invested a significant amount of their time in understanding the subject and creating the most relevant covers. They scrutinized every image to scout for the most suitable representation of the subject and create an appropriate cover for the book.

The publishing team has been an ardent support to the editorial, designing and production team. Their endless efforts to recruit the best for this project, has resulted in the accomplishment of this book. They are a veteran in the field of academics and their pool of knowledge is as vast as their experience in printing. Their expertise and guidance has proved useful at every step. Their uncompromising quality standards have made this book an exceptional effort. Their encouragement from time to time has been an inspiration for everyone.

The publisher and the editorial board hope that this book will prove to be a valuable piece of knowledge for researchers, students, practitioners and scholars across the globe.

LIST OF CONTRIBUTORS

Rocío Tarjuelo, Manuel B. Morales, Juan Traba and M. Paula Delgado
Terrestrial Ecology Group (TEG), Department of Ecology, Universidad Auto´noma de Madrid, Madrid, Spain

Mohsen Ahmadi
Department of Environmental Sciences, Faculty of Natural Resources, University of Tehran, Karaj, Iran

José Vicente López-Bao
Research Unit of Biodiversity (UO/CSIC/PA), Oviedo University, Mieres, Spain
Grimsö Wildlife Research Station, Dep. of Ecology, Swedish University of Agricultural Sciences (SLU), Riddarhyttan, Sweden

Mohammad Kaboli
Department of Environmental Sciences, Faculty of Natural Resources, University of Tehran, Karaj, Iran

Jonathan Storkey
AgroEcology Department, Rothamsted Research, Harpenden, Hertfordshire, United Kingdom

Pierre Stratonovitch
Computational and Systems Biology Department, Rothamsted Research, Harpenden, Hertfordshire, United Kingdom

Daniel S. Chapman
Centre for Ecology and Hydrology, Edinburgh, United Kingdom

Francesco Vidotto
University of Turin, Grugliasco, Italy

Mikhail A. Semenov
Computational and Systems Biology Department, Rothamsted Research, Harpenden, Hertfordshire, United Kingdom

Morten Lauge Pedersen
Department of Civil Engineering, Aalborg University, Aalborg, Denmark

Klaus Kevin Kristensen
Ringkjøbing-Skjern Municipality, Ringkøbing, Denmark, 3 Section of Freshwater Biology

Nikolai Friberg
Norwegian Institute for Water Research, Oslo, Norway

Arika Ligmann-Zielinska
Department of Geography, Michigan State University, East Lansing, Michigan, United States of America

Daniel B. Kramer
James Madison College, Michigan State University, East Lansing, Michigan, United States of America
Department of Fisheries and Wildlife, Michigan State University, East Lansing, Michigan, United States of America

Kendra Spence Cheruvelil
Department of Fisheries and Wildlife, Michigan State University, East Lansing, Michigan, United States of America
Lyman Briggs College, Michigan State University, East Lansing, Michigan, United States of America

Patricia A. Soranno
Department of Fisheries and Wildlife, Michigan State University, East Lansing, Michigan, United States of America

Zhang Zhaoyong
State Key Laboratory of Desert and Oasis Ecology, Xinjiang Institute of Ecology and Geography, Chinese Academy of Sciences, Urumqi, China
University of the Chinese Academy of Sciences, Beijing, China

Jilili Abuduwaili
State Key Laboratory of Desert and Oasis Ecology, Xinjiang Institute of Ecology and Geography, Chinese Academy of Sciences, Urumqi, China

Hamid Yimit
Key Laboratory of Xingjiang Arid Land Lake Environment and Resource, Xinjiang Normal University, Urumqi, China

Zongqiang Wei
School of Geographic and Oceanographic Science, Nanjing University, Nanjing, China
School of Environmental and Land Resource Management, Jiangxi Agricultural University, Nanchang, China

Shaohua Wu
School of Geographic and Oceanographic Science, Nanjing University, Nanjing, China

Xiao Yan
School of Environmental and Land Resource Management, Jiangxi Agricultural University, Nanchang, China

Shenglu Zhou
School of Geographic and Oceanographic Science, Nanjing University, Nanjing, China

Anne-Sophie Pellier
Center for International Forestry Research (CIFOR), Bogor, Indonesia, Borneo Futures initiative, People and Nature Consulting International, Jakarta, Indonesia

Jessie A. Wells
ARC Centre of Excellence for Environmental Decisions, Centre for Biodiversity & ConservationScience, University of Queensland, Brisbane, Australia
Borneo Futures initiative, People and Nature Consulting International, Jakarta, Indonesia

Nicola K. Abram
Durrell Institute for Conservation and Ecology, School of Anthropology and Conservation, Marlowe Building, University of Kent, Canterbury, Kent, United Kingdom
Borneo Futures initiative, People and Nature Consulting International, Jakarta, Indonesia

David Gaveau
Center for International Forestry Research (CIFOR), Bogor, Indonesia
Borneo Futures initiative, People and Nature Consulting International, Jakarta, Indonesia

Erik Meijaard
Center for International Forestry Research (CIFOR), Bogor, Indonesia
ARC Centre of Excellence for Environmental Decisions, Centre for Biodiversity & Conservation Science, University of Queensland, Brisbane, Australia
Borneo Futures initiative, People and Nature Consulting International, Jakarta, Indonesia

Yan Wang
College of Water Science, Beijing Normal University, Beijing, China
Nanjing Institute of Environmental Science, Ministry of Environmental Protection, Nanjing, China

Jixi Gao
College of Water Science, Beijing Normal University, Beijing, China
Nanjing Institute of Environmental Science, Ministry of Environmental Protection, Nanjing, China

Jinsheng Wang
College of Water Science, Beijing Normal University, Beijing, China

Jie Qiu
Nanjing Institute of Environmental Science, Ministry of Environmental Protection, Nanjing, China

Sami Khanal
School of Environment and Natural Resources, Ohio State University, Wooster,
OH, United States of America

Robert P. Anex
Dept. of Biological Systems Engineering, University of Wisconsin-Madison, Madison, Wisconsin, United States of America

Christopher J. Anderson
Dept. of Agronomy, Iowa State University, Ames, Iowa, United States of America

Daryl E. Herzmann
Dept. of Agronomy, Iowa State University, Ames, Iowa, United States of America

Jan O. Engler
Zoological Research Museum Alexander Koenig, Bonn, Germany,
Department of Wildlife Sciences, University of Gö ttingen, Gö ttingen, Germany

Niko Balkenhol
Department of Wildlife Sciences, University of Gö ttingen, Gö ttingen, Germany

Katharina J. Filz
Department of Biogeography, Trier University, Trier, Germany
Museum of Natural History Dortmund, Dortmund, Germany

Jan C. Habel
Department of Ecology and Ecosystemmanagement, Technical University Munich, Freising-Weihenstephan, Germany

Dennis Rödder
Zoological Research Museum Alexander Koenig, Bonn, Germany
Department of Wildlife Sciences, University of Gö ttingen, Gö ttingen, Germany

Aritz Ruiz-González
Department of Zoology and Animal Cell Biology, University of the Basque Country, UPV/EHU, Vitoria-Gasteiz, Spain
Systematics, Biogeography and Population Dynamics Research Group, Lascaray Research Center, University of the Basque Country, UPV/EHU, Vitoria-Gasteiz, Spain

Conservation Genetics Laboratory, National Institute for Environmental Protection and Research, ISPRA, Ozzano dell'Emilia, Bologna, Italy

Mikel Gurrutxaga
Systematics, Biogeography and Population Dynamics Research Group, Lascaray Research Center, University of the Basque Country, UPV/EHU, Vitoria-Gasteiz, Spain,
Department of Geography, University of the Basque Country, UPV/EHU,
Vitoria-Gasteiz, Spain

Samuel A. Cushman
U.S. Forest Service, Rocky Mountain Research Station, Flagstaff, AZ, United States of America

María José Madeira
Department of Zoology and Animal Cell Biology, University of the Basque Country, UPV/EHU, Vitoria-Gasteiz, Spain
Systematics, Biogeography and Population Dynamics Research Group, Lascaray Research Center, University of the Basque Country, UPV/EHU, Vitoria-Gasteiz, Spain

Ettore Randi
Conservation Genetics Laboratory, National Institute for Environmental Protection and Research, ISPRA, Ozzano dell'Emilia, Bologna, Italy
Department 18/Section of Environmental Engineering, Aalborg University, Aalborg, Denmark

Benjamin J. Gómez-Moliner
Department of Zoology and Animal Cell Biology, University of the Basque Country, UPV/EHU, Vitoria-Gasteiz, Spain
Systematics, Biogeography and Population Dynamics Research Group, Lascaray Research Center, University of the Basque Country, UPV/EHU, Vitoria-Gasteiz, Spain

Dorsaf Kerfahis
Department of Biological Sciences, Seoul National University, Seoul, Republic of Korea, School of Chemical and Biological Engineering, Interdisciplinary Program of Bioengineering, Seoul National University, Seoul, Republic of Korea

Binu M. Tripathi
Department of Animal and Plant Sciences, University of Sheffield, Sheffield, United Kingdom

Junghoon Lee
School of Chemical and Biological Engineering, Interdisciplinary Program of Bioengineering, Seoul National University, Seoul, Republic of Korea

David P. Edwards
Department of Animal and Plant Sciences, University of Sheffield, Sheffield, United Kingdom

Jonathan M. Adams
Department of Biological Sciences, Seoul National University, Seoul, Republic of Korea

Nadja K. Simons
Terrestrial Ecology Research Group, Department of Ecology and Ecosystem Management, School of Life Sciences Weihenstephan, Technische Universität München, Freising, Germany

Martin M. Gossner
Terrestrial Ecology Research Group, Department of Ecology and Ecosystem Management, School of Life Sciences Weihenstephan, Technische Universität München, Freising, Germany

Thomas M. Lewinsohn
Department of Animal Biology, Institute of Biology, University of Campinas, Campinas, Sao Paulo, Brazil

Steffen Boch
Institute of Plant Sciences, University of Bern, Bern, Switzerland

Markus Lange
Max-Planck-Institute for Biogeochemistry, Jena, Germany

Jörg Müller
Institute of Biochemistry and Biology, University of Potsdam, Potsdam, Germany

Esther Pašalić
Terrestrial Ecology Research Group, Department of Ecology and Ecosystem Management, School of Life Sciences Weihenstephan, Technische Universität München, Freising, Germany

Stephanie A. Socher
Institute of Plant Sciences, University of Bern, Bern, Switzerland,

Manfred Türke
Terrestrial Ecology Research Group, Department of Ecology and Ecosystem Management, School of Life Sciences Weihenstephan, Technische Universität München, Freising, Germany

Markus Fischer
Institute of Plant Sciences, University of Bern, Bern, Switzerland
Biodiversity and Climate Research Centre, Senckenberg Gesellschaft für Naturforschung, Frankfurt/Main, Germany

Wolfgang W. Weisser
Terrestrial Ecology Research Group, Department of Ecology and Ecosystem Management, School of Life Sciences Weihenstephan, Technische Universität München, Freising, Germany

Xuan Fang
Key Laboratory of Virtual Geographic Environment, Ministry of Education, School of Geography Science, Nanjing Normal University, Nanjing, China

Guoan Tang
Key Laboratory of Virtual Geographic Environment, Ministry of Education, School of Geography Science, Nanjing Normal University, Nanjing, China

Bicheng Li
Research Center of Soil and Water Conservation and Ecological Environment, Chinese Academy of Sciences, Yangling, Shaanxi, China

Ruiming Han
School of Geography Science, Nanjing Normal University, Nanjing, China

Hai-Ning Liu
College of Resources and Environment Science, Hunan Normal University, Changsha, China

Li-Dong Gao
Hunan Provincial Center for Disease Control and Prevention, Changsha, China,

Gerardo Chowell
Division of International Epidemiology and Population Studies, Fogarty International Center, National Institutes of Health, Bethesda, Maryland, United States of America
Simon A. Levin Mathematical, Computational & Modeling Sciences Center, School of Human Evolution and Social Change, Arizona State University, Tempe, Arizona, United States of America

Shi-Xiong Hu
Hunan Provincial Center for Disease Control and Prevention, Changsha, China

Xiao-Ling Lin
College of Resources and Environment Science, Hunan Normal University, Changsha, China

Xiu-Jun Li
School of Public Health, Shandong University, Jinan, China

Gui-Hua Ma
College of Resources and Environment Science, Hunan Normal University, Changsha, China

Ru Huang
College of Resources and Environment Science, Hunan Normal University, Changsha, China

Hui-Suo Yang
Center for Disease Control and Prevention of Beijing Military Region, Beijing, China

Huaiyu Tian
College of Resources and Environment Science, Hunan Normal University, Changsha, China

Hong Xiao
College of Resources and Environment Science, Hunan Normal University, Changsha, China

Chuan-Hung Chiu
Department of Bioenvironmental systems engineering, National Taiwan University, Taipei, Taiwan

Tzai-Hung Wen
Department of Geography, National Taiwan University, Taipei, Taiwan

Lung-Chang Chien
Division of Biostatistics, University of Texas School of Public Health at San Antonio Regional Campus, San Antonio, Texas, United States of America; Research to Advance Community Health Center, University of Texas Health Science Center at San Antonio Regional Campus, San Antonio, Texas, United States of America

Hwa-Lung Yu
Department of Bioenvironmental systems engineering, National Taiwan University, Taipei, Taiwan

Jessica Tait
School of Energy, Environment and Agrifood, Cranfield University, Cranfield, Bedfordshire, United Kingdom

Humberto L. Perotto-Baldivieso
School of Energy, Environment and Agrifood, Cranfield University, Cranfield, Bedfordshire, United Kingdom

Adam McKeown
CSIRO Sustainable Land and Water, Smithfield, QLD, Australia

David A. Westcott
CSIRO Sustainable Land and Water, Atherton, QLD, Australia

Florian Zabel, Birgitta Putzenlechner and Wolfram Mauser
Department of Geography, Ludwig Maximilians University, Munich, Germany

Terry L. Sohl
Earth Resources Observation and Science (EROS) Center, U.S. Geological Survey, Sioux Falls, South Dakota, United States of America

Swen C. Renner
Institute of Experimental Ecology, University of Ulm, Ulm, Germany
Smithsonian Conservation Biology Institute, National Zoological Park, Front Royal, Virginia, United States of America

Martin M. Gossner
Terrestrial Ecology Research Group, Department of Ecology and Ecosystem Management, Centre for Food and Life Sciences Weihenstephan, Technische Universität München, Freising, Germany

Tiemo Kahl,
Chair of Silviculture, University of Freiburg, Freiburg, Germany

Elisabeth K. V. Kalko
Institute of Experimental Ecology, University of Ulm, Ulm, Germany

Wolfgang W. Weisser
Terrestrial Ecology Research Group, Department of Ecology and Ecosystem Management, Centre for Food and Life Sciences Weihenstephan, Technische Universität München, Freising, Germany,

Markus Fischer
Institute of Plant Sciences and Botanical Garden, University of Bern, Bern, Switzerland

Eric Allan
Institute of Plant Sciences, University of Bern, Bern, Switzerland

Catherine J. Payne
Centre for Integrative Ecology, School of Life and Environmental Sciences, Deakin University, Melbourne, Victoria, Australia

Euan G. Ritchie
Centre for Integrative Ecology, School of Life and Environmental Sciences, Deakin University, Melbourne, Victoria, Australia

Luke T. Kelly
Australian Research Council Centre of Excellence for Environmental Decisions, School of Botany, University of Melbourne, Melbourne, Victoria, Australia

Dale G. Nimmo
Centre for Integrative Ecology, School of Life and Environmental Sciences, Deakin University, Melbourne, Victoria, Australia

Index

9 781682 865071